Lecture Notes in Computer Science   2825

Edited by G. Goos, J. Hartmanis, and J. van Leeuwen

Springer
Berlin
Heidelberg
New York
Hong Kong
London
Milan
Paris
Tokyo

Werner Kuhn   Michael Worboys
Sabine Timpf (Eds.)

Spatial
Information Theory

Foundations of Geographic Information Science

International Conference, COSIT 2003
Kartause Ittingen, Switzerland, September 24-28, 2003
Proceedings

 Springer

Series Editors

Gerhard Goos, Karlsruhe University, Germany
Juris Hartmanis, Cornell University, NY, USA
Jan van Leeuwen, Utrecht University, The Netherlands

Volume Editors

Werner Kuhn
University of Münster, Institute for Geoinformatics
Robert-Koch-Straße 26-28, 48149 Münster, Germany
E-mail: kuhn@ifgi.uni-muenster.de

Michael Worboys
University of Maine
NCGIA, Department of Spatial Information Engineering
5711 Boardman Hall, Orono, ME 04469-5711, USA
E-mail: worboys@spatial.maine.edu

Sabine Timpf
University of Zurich-Irchel, Department of Geography
Winterthurerstraße 190, 8057 Zurich, Switzerland
E-mail: timpf@geo.unizh.ch

Cataloging-in-Publication Data applied for

Bibliographic information published by Die Deutsche Bibliothek
Die Deutsche Bibliothek lists this publication in the Deutsche Nationalbibliografie;
detailed bibliographic data is available in the Internet at <http://dnb.ddb.de>.

CR Subject Classification (1998): E.1, I.2, F.1, H.2.8, H.1, J.2

ISBN 978-3-540-20148-9 Springer-Verlag Berlin Heidelberg New York

Springer-Verlag Berlin Heidelberg New York
a member of BertelsmannSpringer Science+Business Media GmbH

http://www.springer.de

© Springer-Verlag Berlin Heidelberg 2003

Typesetting: Camera-ready by author, data conversion by Olgun Computergrafik
Printed on acid-free paper   SPIN: 10953631   06/3142   5 4 3 2 1 0

Preface

COSIT, the series of Conferences on Spatial Information Theory, has been around for more than ten years. Its hallmarks are a fruitful interdisciplinary dialogue between computational and human perspectives on spatio-temporal information and a thorough review process that selects the best papers while giving all authors detailed feedback on how to develop their work. A clear profile of the COSIT community has emerged from the series of conference proceedings, all published as Springer Lecture Notes in Computer Science, and from the permanent web site at http://www.cosit.info, containing links to the conference web sites and proceedings, a history and program of the series, an impact study, interviews with participants, and pictures.

The proceedings of this sixth conference provide ample evidence that COSIT is healthy and maturing, while retaining its youth. Out of the 61 submissions, the program committee selected 26 papers for presentation, in discussions based on at least three double-blind reviews and one or more meta-review from PC members for each paper. Classical COSIT themes, such as spatial reasoning (about distances and directions, regions and shapes) or vagueness are being further refined; topics like wayfinding and landmarks are boosted by new synergies between cognitive and computational approaches; and the study of ontologies for space and time, a subject since the first COSIT, is gaining more depth. COSIT is adding new teams and talents every time: out of the 54 authors of accepted papers this year, 32 (or 59%) had never before published a COSIT paper!

Around the paper sessions, two keynote addresses, tutorials, a pre-conference workshop, an interactive poster session, a doctoral colloquium, and social events turned this year's conference into yet another highlight of the series. Following the cherished COSIT tradition of cloistered meeting places, the conference was held at Kartause Ittingen in Switzerland, a wonderful site that no doubt inspired participants to embark on work for the next meeting in 2005.

We thank the many people who made COSIT 2003 such a success: all those who submitted work and participated at the meeting, the reviewers, the program committee, the local organizing committee, and the staff of Kartause Ittingen. Thanks

September 2003   Werner Kuhn
                                                            Mike Worboys
                                                            Sabine Timpf

Chairing Committee

Sabine Timpf, University of Zurich, Switzerland (General Chair)
Werner Kuhn, University of Münster, Germany (Program Chair)
Michael F. Worboys, University of Maine, USA (Program Chair)

Program Committee

Anthony G. Cohn, University of Leeds, UK
Michel Denis, Université Paris-Sud, France
Max J. Egenhofer, University of Maine, USA
Andrew U. Frank, Technical University Vienna, Austria
Christian Freksa, University of Bremen, Germany
Stephen Hirtle, University of Pittsburgh, USA
Benjamin Kuipers, University of Texas at Austin, USA
David M. Mark, University at Buffalo, USA
Daniel R. Montello, University of California at Santa Barbara, USA
Barbara G. Tversky, Stanford University, USA

Reviewers

Gary Allen
John Bateman
Brandon Bennett
Mark Blades
Gilberto Camara
Eliseo Clementini
Helen Couclelis
Matteo Cristani
Matt Duckham
Martin Erwig
Carola Eschenbach
Sara Fabrikant
Leila De Floriani
Scott Freundschuh
Mark Gahegan
Antony Galton
Christopher Gold
Christopher Habel
Kathleen Hornsby
Marinos Kavouras
Roberta Klatzky

Markus Knauff
Lars Kulik
Gerard Ligozat
Paola Magillo
Harvey Miller
Reinhard Moratz
Bernhard Nebel
Dimitris Papadias
Eric Pederson
Jonathan Raper
Martin Raubal
Anthony Richardson
John Rieser
Thomas Röfer
Christoph Schlieder
Michel Scholl
Barry Smith
John Stell
Holly Taylor
Frank Tendick
Andrew Turk

David Uttal   Wai-Kiang Yeap
Laure Vieu   Benjamin Zhan
Rob Weibel
Karl Wender

Organizing Committee

Sabine Timpf, University of Zurich, Switzerland
Urs-Jakob Rüetschi, University of Zurich, Switzerland
Elisabeth Cottier (conference secretary), University of Zurich, Switzerland
Martina Forster (conference secretary), University of Zurich, Switzerland
Matt Duckham (doctoral colloquium), University of Maine, USA
Kurt Brassel (excursion), University of Zurich, Switzerland
Doris Simon (logo), Fachhochschule Karlsruhe, Germany
Martin Steinmann (poster design), University of Zurich, Switzerland

Table of Contents

IV Computational Approaches

V Reasoning about Regions

VI Vagueness

VII Visualization

VIII Landmarks and Wayfinding

Desiderata for a Spatio-temporal Geo-ontology

Antony Galton

School of Engineering and Computer Science, University of Exeter, UK
A.P.Galton@exeter.ac.uk

Abstract. We survey the manifold variety of kinds of phenomena which come within the purview of geography and GI Science, and identify three key desiderata for a fully spatio-temporal geo-ontology which can do justice to those phenomena. Such a geo-ontology must (a) provide suitable forms of representation and manipulation to do justice to the rich network of interconnections between field-based and object-based views of the world; (b) extend the field-based and object-based views, and the forms of representation developed to handle them, into the temporal domain; and (c) provide a means to develop different views of spatio-temporal extents and the phenomena that inhabit them, especially with reference to those phenomena such as storms, floods, and wildfires which seem to present dual aspects as both object-like and process-like.

In recent years, many researchers in geographical information (GI) science have recognised the importance of ontological concerns in their field of study, of giving a correct account of the types of entity or phenomenon that come under their purview, and the relationships and interactions between them. In parallel with this there has been an ever-increasing awareness of the importance of time in GI science, not just as an optional adjunct but as an integral part of what GI science is about. Putting these two concerns together, we see the need for an ontology for GI science that takes full cognizance of the temporal dimension, in other words for a spatio-temporal geo-ontology.

It has increasingly been suggested that such an ontology must embrace a fully spatio-temporal, four-dimensional view of the world [1]. In order to achieve such a view, it may be necessary to reconceptualise many of the existing categories in our geographical ontology, and in this paper I explore some of the issues that must be taken into consideration if we are to do this. The approach I shall take is to survey in very general terms the manifold variety of kinds of phenomena which come within the scope of geography and GI science, and which must therefore be encompassed by a comprehensive geo-ontology. From this survey will emerge three key desiderata for a geo-ontology that is to do justice to those phenomena.

1 Spatial Objects and Fields

Our survey of geographical phenomena begins with *various kinds of objects*: Mount Fuji, the Atlantic Ocean, the Galapagos Islands, New York, the Sydney Opera House, Switzerland, the Niagara Falls, the 8.35 a.m. train from London King's Cross to Leeds on 21st February 2003, the Gobi desert, the inhabitants of Quebec, the Gulf Stream, my house, . . . — the list goes on and on, comprising a wide range of degrees of materiality, of

W. Kuhn, M.F. Worboys, and S. Timpf (Eds.): COSIT 2003, LNCS 2825, pp. 1–12, 2003.

naturalness, of definiteness, and of stability. An object is defined by some abstract notion of *identity* which (a) defines (to an appropriate degree of precision) its spatial extent at any one time, and (b) enables it to persist through changes in spatial location and other attributes. Thus considered, an object has spatial parts but not temporal parts, rather existing as an entirety at each moment of its history—in philosophical terminology, spatial objects are *continuants*.

Our notion of an object seems to be all-embracing: the world is often taken to be the totality of *things*. This is a consequence of the fact that both our everyday thinking and our armoury of technical conceptual tools are pervasively *object-oriented*. However, this does not do justice to the nature of what the world contains. As one writer has put it, 'Remove the animals, artifacts, and perhaps the trees, and just try to individuate what remains: infinitely variegated but not very cleanly divided rock outcroppings, muskeg, bramble patches, cloud formations, lichen, and the rest.' [2] To describe all these *non-object* features of reality (and notice how language traps me into saying 'features' when really I want to say something like 'featuredness') we require mass nouns, not count nouns, and, in a more technical setting, we need *fields* rather than objects. Thus we need to take the field view seriously—it is not just a relict of the days when the most sophisticated computer output device was the line-printer so that it was natural to handle geographical data in raster format [3].

Thus the next item on our survey is *various kinds of fields*. A (spatial) field is a mapping from spatial locations to values from some range. The variousness of fields is attested by such examples as temperature, pressure, humidity, elevation, albedo, wind-speed, water-speed, soil type, vegetation, land use, population density, and sovereignty, as well as more recondite things such as 'distance to the nearest airport', all of which can be mapped, in principle, as field values. Like objects, fields change over time, but unlike with objects there do not seem to arise any issues of identity.

In GI Science the object-based view and the field-based view have generally been presented as the poles of a sharp dichotomy, between which it has been a matter of debate as to which should be preferred. As Schuurman put it [4], 'despite a lack of agreement on the level of ontological certainty possible in GIS, ontology talk in GIS always seems to lead back to field versus object views of the world'. It is generally recognised nowadays that both views are necessary, and following Plewe (as reported in [5]), we can interpolate intermediate forms of representation between the pure object view and the pure field view. This is indicated only programmatically by Plewe, and the question is how to handle such intermediate forms technically. One possibility is suggested by the idea of an 'object-field', i.e., a field whose values are objects [6], introduced as a way of handling in a field-like way the 'planar enforced' structures variously referred to as adjacent polygon models or partitions. A possible overall classification might be as illustrated in Figure 1. In addition, we must recognise the many ways in which objects can give rise to fields and vice versa. In general we need to codify ways of deriving new information from existing data.

There are many instances in which, from a mass of field data, we wish to pick out certain features as especially salient from some point of view, and to confer upon them the status of 'objects'. One example is the maxima and minima of continuous fields (e.g., mountain peaks as maxima of the elevation field, highs and lows in the atmospheric

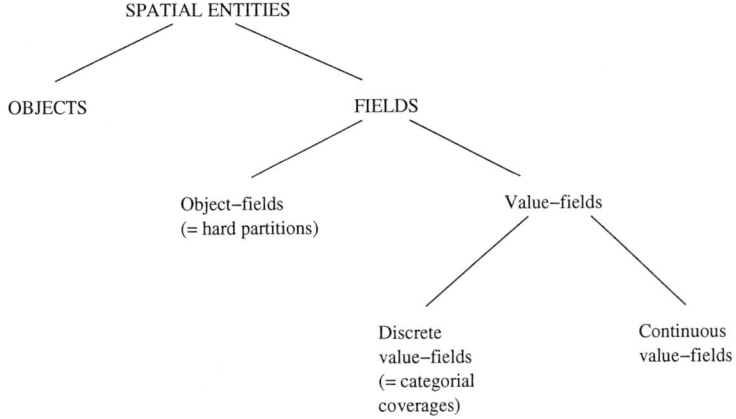

Fig. 1. Classification of objects and fields

pressure field). Another is the derivation of an object field from a discrete value field: the field carves up space into regions that are homogeneous with respect to the field attribute. Any or all of these regions may be singled out as objects (depending, perhaps, on how coherent they are). If the field values describe vegetation cover, then a maximal connected region within which the vegetation is *woodland*, say, might be singled out as *a wood*. A discrete value field can be derived from a continuous value field by banding [6]. This is what happens when the areas with elevations within a particular range are colour-coded on a map: the continuous range of elevations is divided into bands (e.g., Sea level to 100m, 100m–200m, and so on).

Fields can be created from objects, too. A classic example is population density: individual people are objects, each of which is assigned to a location (ideally, and ignoring movement). The mean population density for a region is the number of people assigned to that region divided by its area. The population density at a point is not uniquely defined: one has to agree on some standard area around that point within which to measure the mean population density, and depending on the size and shape of that area, different values will be obtained. Thus there are many ways of deriving a population density field from a given distribution of objects—which does not mean it is not a useful thing to do, only that one must be careful in interpreting the results.

Finally, objects can be created from other objects, either by collecting them together or by subdividing them. An aggregation, or agglomeration [7], consists of a number of individual objects brought together in thought to be considered as a single unit. Such bringing together may be justified in various ways: spatial proximity, causal interaction, coherent motion, shared history, administrative fiat. Some of these involve time and therefore would not be available in a purely static model. Aggregations raise interesting questions concerning identity and will be discussed more fully below. Another way of deriving new objects from old is by considering *parts*—connected components, or parts defined by some attribute, or parts only weakly connected to the rest.

All these considerations lead to the conclusion that the field-based and object-based views of the world are not separate monolithic systems but intimately related by an intricate network of interconnections. Some of these are well understood, but others have barely begun to be explored, even in the purely static case. A prime desideratum for geo-ontology is to sort out this network of interconnections and to devise suitable forms of representation and manipulation to do justice to them.

2   Temporal Objects and Fields

The object/field dichotomy arises for time as well as space, and this extension of the dichotomy into the temporal domain constitutes our second key desideratum. The analogy between the spatial and temporal cases was pointed out by Worboys [8]; for further discussion, see [9]. In principle the dichotomy can apply in any situation where there is a *locational framework* populated with data. For the present survey, it adds two further components to the 'furniture' of the world and its history.

First, the temporal analogue of objects: these are *events* and *processes* of various kinds. As a category, processes are extremely problematic and have given rise to much discussion; I have more to say on this below. The clearest case of 'temporal objects' are events with a relatively sharply delineated location in space and time: the sinking of the Titanic, the World Cup final in 1966, the opening of the Channel Tunnel. But there are also numerous happenings which are in various ways more ill-defined: the floods which affected central Europe in the summer of 2002, last night's thunderstorm, the coastal erosion in north Norfolk, the collapse of the communist regimes in Eastern Europe, the dismantling of the Berlin Wall. All these things, unlike spatial objects, have temporal parts, or *phases*: they do not exist in their entirety at any one time—they are known as *occurrents* rather than continuants.

Alongside these may be considered *temporal fields*, or *fluents*, which are mappings from times to values: the temperature or wind-speed at the summit of Mount Everest, the population of Canada, the number of fatalities in road traffic accidents each day in the UK, sunspot activity, and so on. In addition to standard numerical measures or qualitative descriptors, the value of a temporal field can be an ordinary ('spatial') object, e.g., the Prime Minister of the UK (a field which in recent years has successively taken the values Margaret Thatcher, John Major, and Tony Blair), or even a temporal object (event), such as 'the last general election', which maps each moment onto the latest general election preceding that moment.

With the introduction of time, the requirement to take heed of the manifold interconnections between spatial objects and fields must be extended in scope. Similar interconnections exist in the temporal case, but in addition, there are interconnections between the spatial and the temporal. For example, we may wish to pick out processes and events as salient patterns in an evolving spatial field, e.g., weather fronts or anticyclones. Or we may create spatial fields which encapsulate information relating to temporally changing data, e.g., the rate of population growth in a given area.

So far, then, we have a four-fold cross-classification: the phenomena of interest can be represented as spatial or temporal, and as object-like or field-like. I have indicated that these categories do not exist in isolation from one another, but beyond this, I shall be especially interested in phenomena which cut across these simple dichotomies and assume an ambiguous character, neither clearly spatial nor clearly temporal, neither clearly object-like nor clearly field-like. For this it may be necessary to embrace a more integrated approach to space and time—a four-dimensional view, if you will [1].

3   Locations

In addition to the phenomena themselves, it may be helpful to consider the more abstract elements which provide a locational framework within which the phenomena have their being. The next items on our inventory are thus *locations* of various kinds. There are purely spatial locations (regions, lines, and points), and purely temporal locations (intervals and instants), but for a fully spatio-temporal view we might need to consider also spatio-temporal locations: portions of space-time. Indeed, once these have been admitted, it is tempting to coerce purely spatial and temporal locations into this form too. We cannot speak of a spatial location *simpliciter*: we must specify whether we mean a spatial location at a particular point of time, or over some specified interval, or over all time; and when we speak of an instant in time, do we mean all of space at that instant, or some delineated region, or just a single point?

What kinds of spatio-temporal location should we consider? Some examples are

- *Region-histories*, i.e., $R \times I$, where R is a region of space and I is a time interval. This requires regions to have stable identities over time—e.g., defined by some set of spatial coordinates. So more exactly it is $\{\langle s, t\rangle \mid s \in R, t \in I\}$.
- *Object-histories*, i.e., for a given object a, its history occupies the space-time chunk $\{\langle s, t\rangle \mid s \in pos(a, t), a \text{ exists at } t\}$, where $pos(a, t)$ is the (spatial) position of object a at time t. The condition 'a exists at t' begs questions concerning the identity of objects over time—this is explored further below.
- *Event-locations.* An event may be a delineated episode in the life-history of an object, in which case the event-location could be a part of an object history; or it may be an instantaneous change in an object, so its location is not *part* of the life-history so much as an 'infinitely thin slice' through it. But an event may also involve more than one object, in which case its location must include parts of several individual object histories; and of course an event need not involve readily-individuated objects at all—e.g., a flooding event, whose spatio-temporal location might consist of a set of $\langle s, t\rangle$ such that the (normally dry) land at s is under water at time t.
- *Any* subset of $S \times T$—so long as we can specify it! The basis for specifying it will often be provided by one of the earlier items on this list, but it may be desirable to leave open a more flexible route to defining space-time chunks.

The ontology of space-time chunks must, of course, provide more than merely a means to identify them; it should describe their properties and interrelations as well as saying something about how other entities of the ontology stand in relation to them. Thus for example Muller [10] provides an axiomatic mereotopology for space-time chunks, and Hazarika and Cohn[11] investigate the notion of *continuity* as it applies to them.

4   Spatio-temporal Fields

If we are to embrace a fully four-dimensional view of things, we must revisit the object/field dichotomy. For as well as spatial objects and spatial fields, and temporal objects and temporal fields, we might also consider spatio-temporal objects and spatio-temporal fields. A spatio-temporal field cannot always be treated as just a spatial field with a history. To see this, contrast the following two data-sets relating to a field f:

1. For some sets of times $T_1 \subset T$ and spatial locations $S_1 \subset S$ (where T and S are the complete sets of times and locations), at each $t \in T_1$ we have a snapshot giving the values of f at the locations in S_1; thus we are given $\{W(t) \mid t \in T_1\}$, where $W(t) = \{(s, f(s,t)) \mid s \in S_1\}$.

2. We have a set of space-time locations $L \subset S \times T$, and at each location we have the value of f, so our data takes the form $\{((s,t), f(s,t)) \mid (s,t) \in L\}$.

In either case one might want to use some form of interpolation procedure to estimate values of f at places and times not in the data. In the first case the natural thing to do is to interpolate within snapshots first (purely spatial interpolation, e.g., kriging), and then to interpolate between snapshots (purely temporal interpolation). In the second case something more sophisticated is required: spatio-temporal interpolation. This requires us to compare distances in space with intervals in time, a problem for which there currently does not exist any 'routine' or 'standard' solution.

5   Identity and Change

As with fields, a spatio-temporal object is not just a spatial object with a history. Indeed, the four-dimensional view can provide a means to side-step some of the difficult problems of identity associated with normal spatial objects, since it gives us the freedom to consider a wider variety of portions of space-time, not just standard object-histories. To illustrate this, I first turn to a consideration of identity for ordinary spatial objects.

The foremost question concerning objects is how to individuate them: *When are two object-manifestations manifestations of the same object?*. By a 'manifestation' I mean the manner in which an object is presented to an observer on a particular occasion. Of course, this presupposes that a logically prior question has already been answered, at least in part: *When is a manifestation an object-manifestation?* An object-manifestation is, minimally, some observable feature-complex located at a particular place and time. When two object-manifestations are under consideration, these may be separated in space, in time, or in both space and time.

By what criteria is it to be determined whether two object-manifestations are manifestations of the same object? Minimally, they must be manifestations of the same *type* of object, which brings to the fore another aspect of manifestations: their *intensionality*. A stretch of the River Thames may be regarded as a manifestation of a river, of a river-system, or of a piece of territory. If I point to the Thames in London, and then point to the M1 motorway somewhere near Leeds, then there is no question of my having pointed out two manifestations of the same river, or of the same road, but I could say that I have pointed out two manifestations of the same *country*, England.

Even if we know the type under which we are considering an object-manifestation, there remain delicate issues concerning identity. A 'river-manifestation', for example, is a river-*stretch* observable at a particular time. A complete candidate river is a connected linear aggregate of river-stretches, from one river-landmark to another—where a river-landmark is a source, a confluence, or a river-mouth. In general, it is not given in the nature of things *which* such aggregates of river-stretches are to count as individual rivers. Perhaps the only criterion is one of *naming*—a river is a maximal connected linear aggregate of river-stretches all of which are referred to by the same river name. There is a strong element of arbitrariness about this in many cases [12, §3.5]. For some purposes it would be much more natural to take as the object under consideration a complete *river-system*, consisting of all those river stretches which ultimately feed into the same river mouth.

Roads are equally problematic. The A38 is the main road which runs from Bodmin in Cornwall through Plymouth, Exeter, Bristol, Gloucester, Birmingham, and Derby to Mansfield in Nottinghamshire. Of course this is an aggregate of road-stretches. At each junction there is, in principle, a choice as to which of the road-stretches meeting there should be designated as parts of the same road. The A38 counts as *one road* only because the transport authorities have decided to use the same label for all its substretches. There is an element of arbitrariness about the labelling of roads—though it is not completely arbitrary, there being in many places natural choices of continuation. In this case, there is no clear analogue to the river-system, unless it be the entire network of roads in a given (how specified?) area.

In both these examples what is considered to be a single object is an aggregate of parts for which there is an element of arbitrariness or conventionality about the particular selection of parts that is taken to constitute the object. They are, to that extent, *fiat objects* [13]. The selection is not totally arbitrary, however: there are some principles governing what is possible and what is not—e.g., that the selected totality should form a connected whole.

Questions of identity become more acute when the parts are discrete objects capable of independent motion. Flocks, herds, shoals, etc. are aggregates of individuals ('second-order objects'). The theory of identity-based change expounded by Hornsby and Egenhofer [14,15] can be adapted to apply to entities of this kind. Groups of individuals can come into existence, merge, split, or cease to be, but because the constituent individuals have their own identities there are distinctions we can make which may be important in some circumstances. For example, a group may cease to exist either as a result of the elimination of its members, or of their dispersal, the latter but not the former being a reversible change. Similarly, a group may grow in size either as a result of individuals joining the group from outside, or as a result of new individuals being generated from individuals already in the group; and it may become smaller either through members leaving or through their dying off. Likewise there are two ways for a group to remain the same size: either through nothing happening or through the loss of existing members being exactly matched by a compensating gain of new ones.

In practice we encounter complex mixtures of these types of change, placing severe strain on our concept of group identity. Suppose that every winter a flock of sparrows comes to feed in my garden, but then disperses to separate breeding territories in the spring. Next winter the sparrows return—but who are 'the sparrows'? Some will be the same individuals as last year, but some of last year's will have died or left the group, and there will be new individuals, either born within the group or joining it from outside. After a few years there may be wholesale replacement. What defines the group as a group, given that for half the year it does not exist? What if there is a regular exchange of membership with a neighbouring group? We could simply stipulate that the identity of the group is conferred by its location, so the group in my garden this year is the 'same' as the group in my garden last year regardless of possible shifts of allegiance of individuals. The answers to these questions are arbitrary in that there exist equally good alternatives between which there is no reason for preferring one over the other—there being, at least in many cases, no fact of the matter concerning group identity: it is something we confer on phenomena of our experience so that we can apply to them the ontological categories available to us. To some extent, a four-dimensional view could allow us to set such questions of identity on one side and instead consider the bundle of all individual sparrow-histories intersecting a given space-time region as the primary object of interest (analogous to considering the whole river-system and not worrying about individual rivers).

Much of the discussion of group identity applies, appropriately modified, to objects which are not aggregates of discrete macroscopic individuals but *masses of fluid*, e.g., oil slicks, lakes. In some cases, such as clouds, storms, floods, and fires, we are dealing with rapidly-changing phenomena which, while they *can* be considered as objects, in which case questions of identity arise, they can also quite naturally be considered as processes, with a concomitant shift in ontological properties (e.g., temporal phases instead of spatial parts). The object/process dichotomy is far from being an all-or-nothing issue, and some of my examples are clearly intermediate in regard to the naturalness with which we might accord them object or process status. What is needed here is an ontological framework which can accommodate this kind of slippage, while also adequately supporting those cases which are more clearly at either extreme.

6   Processes

Before considering this, I turn to the other extreme, namely *processes*. There is no consensus as to the exact meaning of this term, in part because it is used in various ways. Here I focus on two somewhat different notions of process, which I characterise as *process-as-activity* and *process-as-procedure*.

Process-as-activity is the temporal analogue of a *material* or *stuff*: processes, in this sense, are to events as materials are to objects. They are generic *kinds of going-on* which may be instantiated on particular space-time regions. The instantiation of a process on a well-delineated chunk of space-time may be thought of as a durative event. The boundaries of the space-time region are determined by the boundaries of the process-instantiation. The process/material analogy runs deep. Here 'material' stands for the referent of a mass noun and may include types of matter (e.g., stone, wood, water, oil, copper), more or less homogeneous aggregates of many small, relatively featureless objects (e.g., sand, rice, pollen, mud, foliage, grass, fur), and more or less heterogeneous aggregates of larger, structured objects (e.g., furniture, traffic, shopping, shipping). Likewise processes may include uniform states of change (e.g., gliding, falling, rolling, burning, blowing, getting warmer, colder, denser, darker), more or less homogeneous aggregates of many small, relatively featureless events (e.g., raining [many individual drops falling], walking [many steps], swimming [strokes], rattling, playing the piano, reading, a motor or engine running), and more or less heterogeneous aggregates of larger, structured events (e.g., playing football, lecturing, washing up, cooking, housecleaning, pub-crawling, revising, the tides). Note that many of the examples cited are human activities—and the class of processes, as here described, corresponds well to Vendler's class of *activities* [16][1].

A process-as-procedure is a characteristic sequence of events and/or activities leading to some outcome. It may be open-ended or closed. An example of the former is the process of walking, which in humans consists of alternately advancing each of the two feet to effect a forward displacement of the body which continues for as long as the process remains in operation—this kind of process fits in well with the view of process-as-activity, being the species of activity that has a specifiable outcome. A closed process has an instrinsic termination point beyond which it is not possible to continue[2], like the process of making a pot of tea, which consists of heating the water, warming the pot, putting in the tea-leaves, and pouring the boiling water on top of them—after which the process is complete and the only way to go on making tea is to start again and make another pot. An open-ended process description can be converted into a closed process description by adding an extra term to specify the end-point: instead of walking *simpliciter*, we can consider walking a mile, or walking to the post office[3]. But not every closed process can be described in this way—e.g., the process of making a pot of tea is not just the continuation of some activity until a specified end-point is reached, it is much more structured than that. This is also true of such things as industrial processes, the processes of the law, etc.

7 Spatio-temporal Objects

So far I have focussed on the temporal aspect of processes; but processes take place in space as well as time, and if we are to move to a fully four-dimensional view we must pay as much heed to the spatial aspects of temporal entities as to the temporal aspects of spatial ones—our goal in the end being to achieve an overarching theory of spatio-temporal entities which will encompass not only objects and events but also the awkward intermediate cases that pose such problems to the ontological enterprise.

On a fully four-dimensional view, objects and processes are remodelled as *four-dimensional material extents*. Thus instead of the table (an object) and its life-history (an event of sorts), we consider a four-dimensional 'object' occupying a determinate chunk of space-time whose cross-sections orthogonal to the time axis constitute the successive stages in the life of the table and whose extent parallel to the time-axis corresponds to the history of the table as normally understood. We can only do this, of course, if we are able to track the identity of the table over time—and even with so apparently straightforward an object as a table this is not necessarily unproblematic.

When we turn from objects to processes, we encounter even more difficulties. Some processes relate to individual objects, in which case the location of the process might be identifiable with the location of the object—or rather the space-time chunk built up out

---

[1] Mourelatos [17, p.201] points out that 'PROCESS ... is the topic-neutral counterpart of activity'—'topic-neutral' here implying no commitment to human (or other) agency.

[2] This is the distinction between 'telic' and 'atelic' verbs discussed by Dahl [18].

[3] In the terminology of Vendler [16], this converts an *activity* into an *accomplishment*.

of the successive locations of the object over the duration of the process. But even this is problematic: take the process which consists of someone's singing a song. Can we identify the location of the process in space-time with the location of the singer—including, say, her feet, her hair, and any other parts which play no part in the singing itself, while excluding the surrounding air through which the sound of the song is propagated? It is more usual to locate processes and events by reference to some larger space within which the process and its more immediate effects are confined. She sang the song in the drawing room, next to the piano. If the singing of the song is to be modelled as a four-dimensional object, it is not entirely straightforward to locate it in space-time.

Even more problematic are processes involving groups of individuals. Is the location of a group the sum of the locations of its members, or some region including that, or what? A performance of a string quartet involves four performers and their instruments (and their seats, music stands?). We can say it takes place in a certain room, or a part of that room—but the minimal location might be the very complex 'spiky' space-time chunk which is the sum of the locations of the participants over the interval of the performance. More generally we may wish to select *a* location from some set of regions predesignated as possible locations. It is a mistake always to look for an *exact* location for every event—a telephone conversation, for example.

Not all processes readily relate to discrete objects: consider the ebbing and flowing of the tide. What is the space-time location of this process, as realised on a particular beach over an interval I? Its maximal extent is the prism $B \times I$, where B is the beach's spatial location. The minimal extent is a surface whose successive slices at right-angles to the time axis are the positions of the water's edge at different times. No doubt for different purposes one might find one or the other the more convenient, so ideally we want a system with the flexibility to allow us to switch between them as we wish.

8   Dual Aspect Spatio-temporal Phenomena

While we may wish to remodel ordinary spatial objects and temporal processes as fully spatio-temporal phenomena in this way, the four-dimensional view only really comes into its own when we turn to the difficult intermediate cases. Consider a phenomenon from the human world: a *procession*. A typical procession involves a number of people marching in a more-or-less orderly way along a route through the streets of a town; but it may be viewed in various ways:

- The spectator's view. We are standing at one point along the route, waiting. We hear the sound of marching feet, the beat of drums, but see nothing. Then someone calls out 'Here they come!' and we see the first marchers rounding the corner. They come ever closer and are soon level with us; they pass by, then disappear from view round the next corner. Gradually the sound of marching dies away.
- The participant's view. We assemble near the railway station. The drums start beating, and we move off. We walk down the street, turning right near the post office, up the gentle hill to the park, then left, . . . ; all the while there is the constant beating of the drums, we pass many spectators lined up by the side of the route. Finally we arrive at the Town Hall.

– The view of the police helicopter surveillance team. The town is spread out like a map below us. We fly back and forth watching the events unfold. The streets are full of people. We see the snake-like procession working its way from the station, past the post office, all the way to the Town Hall.

What have these people described: an object, a process, or an event? Arguably, it is most natural to say that the spectator watched an event, with a beginning (when the marchers first came into view), a middle (as they were passing), and an end (when they disappeared); that the participant took part in a process, a process of marching along a predetermined route to a destination; and that the police in their helicopter watched an object, the procession, moving slowly along the streets from station to Town Hall. But beyond all these, is the notion that what we have here is a spatio-temporal entity of which the object-like and event-like aspects are both partial views.

Many geographical phenomena resemble the procession at least to some extent. Like the procession, phenomena such as storms, floods, wildfires, and traffic jams can be seen as having a *dual aspect*: they are both process-like and object-like. If there is a *continuum* of intermediate cases between 'hard objects' (such as a mountain) and 'raw flux' (such as the tides), then these procession-like phenomena fall somewhere in the middle. With all such phenomena it is important to have access not only to their entire spatio-temporal extent, but also to various different *views* of them, e.g., to trace the sequence of spatial locations affected by them, or the sequence of events associated with the phenomenon at one spatial location. The term 'views' encompasses not only various kinds of cross-sections (time-slices or space-slices) but also tracking along a trajectory, perhaps a winding path in space-time. What is needed—my third and final key desideratum—is therefore a flexible means of providing different views of arbitrary spatio-temporal extents and the phenomona that inhabit them.

9   Conclusions

To summarise, I have identified three key desiderata for a fully temporal geo-ontology:

1. to provide suitable forms of representation and manipulation to do justice to the rich network of interconnections between field-based and object-based views of the world;
2. to extend the field-based and object-based views, and the forms of representation developed to handle them, into the temporal domain;
3. to provide a means to develop different views of spatio-temporal extents and the phenomena that inhabit them, especially with reference to those phenomena which seem to present dual aspects as both object-like and process-like.

These desiderata have implications for how we represent our data and for the operations we define on it. In this paper I have merely identified the problems; the real work comes later, in fleshing them out with a technical infrastructure that can provide a basis for geo-information systems adequate to support the ever more sophisticated tasks that users want to perform.

References

1. Jonathan Raper. *Multidimensional Geographic Information Science*. Taylor and Francis, London and New York, 2000.
2. Brian Cantwell Smith. *On the Origin of Objects*. MIT Press, Cambridge MA, 1996.
3. Donna J. Peuquet. Making space for time: issues in space-time data representation. *GeoInformatica*, 5(1):11–32, 2001.
4. Nadine Schuurman. Critical GIS: Theorizing an emerging science. *Cartographica Monograph 53*, 36(4), 1999.
5. Donna Peuquet, Barry Smith, and Berit Brogaard. The ontology of fields. Technical report, NCGIA, University of California at Santa Barbara, 1998. Report of a specialist meeting held under the auspices of the Varenius Project, Bar Harbour, Maine, June 11-13, 1998, http://www.ncgia.ucsb.edu/~vanzuyle/varenius/ontologyoffields_rpt.html.
6. Antony P. Galton. A formal theory of objects and fields. In Daniel R. Montello, editor, *Spatial Information Theory: Foundations of Geographic Information Science*, pages 458–473, Berlin, 2001. Springer-Verlag. Proceedings of International Conference COSIT'01.
7. Barry Smith. Agglomerations. In Christian Freksa and David M. Mark, editors, *Spatial Information Theory: Cognitive and Computational Foundations of Geographic Science*, pages 267–82, New York, 1999. Springer-Verlag. Proceedings of International Conference COSIT'99.
8. Michael F. Worboys. Unifying the spatial and temporal components of geographic information. In T. C. Waugh and R. G. Healey, editors, *Advances in Geographic Information Systems: Proceedings of the Sixth International Symposium on Spatial Data Handling, Edinburgh*, pages 505–17, London, 1994. Taylor and Francis.
9. Antony P. Galton. Space, time, and the representation of geographic reality. *Topoi*, pages 173–187, 2001.
10. Philippe Muller. Space-time as a primitive for space and motion. In Nicola Guarino, editor, *Formal Ontology in Information Systems*, pages 63–76. IOS Press, Amsterdam, 1998.
11. Shyamanta M. Hazarika and Anthony G. Cohn. Qualitative spatio-temporal continuity. In Daniel R. Montello, editor, *Spatial Information Theory: Foundations of Geographic Information Science*, pages 92–107. Springer-Verlag, Berlin, 2001. Proceedings of International Conference COSIT'01.
12. Antony P. Galton. *Qualitative Spatial Change*. Oxford University Press, Oxford, 2000.
13. Barry Smith. Fiat objects. *Topoi*, 20:131–148, 2001.
14. Kathleen Hornsby and Max J. Egenhofer. Identity-based change operations for composite objects. In T. Poiker and N. Chrisman, editors, *Proceedings of the 8th International Symposium on Spatial Data Handling*, pages 202–213. International Geographical Union, 1998.
15. Kathleen Hornsby and Max J. Egenhofer. Identity-based change: a foundation for spatio-temporal knowledge representation. *International Journal of Geographical Information Science*, 14(3):207–224, 2000.
16. Zeno Vendler. *Linguistics and Philosophy*. Cornell University Press, Ithaca, 1967.
17. A. P. D. Mourelatos. Events, processes, and states. In Philip Tedeschi and Annie Zaenen, editors, *Tense and Aspect*, pages 191–212. Academic Press, New York, 1981.
18. Östen Dahl. On the definition of the telic-atelic (bounded-nonbounded) distinction. In Philip Tedeschi and Annie Zaenen, editors, *Tense and Aspect*, pages 79–90. Academic Press, New York, 1981.

Scale in Object and Process Ontologies

Femke Reitsma[1] and Thomas Bittner[2]

[1] Department of Geography, 2181 LeFrak Hall, University of Maryland, College Park
Maryland 20742, USA
femke@geog.umd.edu
[2] Institute for Formal Ontology and Medical Information Science, University of Leipzig
Härtelstraße 16-18, 04107 Leipzig, Germany
thomas.bittner@ifomis.uni-leipzig.de

Abstract. Scale is of great importance to the analysis of real world phenomena, be they enduring objects or perduring processes. This paper presents a new perspective on the concept of scale by considering it within two complementary ontological views. The first, called SNAP, recognizes enduring entities or objects, the other, called SPAN, perduring entities or processes. Within the meta-theory provided by the complementary SNAP and SPAN ontologies, we apply different theories of formal ontology such as mereology and granular partitions, and ideas derived from hierarchy theory. These theories are applied to objects and processes and form the framework within which we present tentative definitions of scale, which are found to differ between the two ontologies.

Keywords: scale, granularity, hierarchy, process, ontology, mereology

1 Introduction

For geography as an empirical science, the *observation* of enduring entities and processes at different levels of granularity is of critical importance. Given a certain range of resolution only processes and enduring entities at certain levels of granularity can be observed. In geography this kind of problem is referred to as the problem of scale of observation, which has long been recognized as a central problem of the analysis of geographical phenomena. The concept of scale is controversial in its definition and often applied without clarification of its meaning (Lam and Quattrochi 1992). The meaning of scale discussed in this paper is not the traditional cartographic notion of scale, but rather the level at which reality is sampled and observed (Goodchild and Proctor 1997). Thus, scale is defined by and incorporates both grain and extent, spatial and temporal (Ahl and Allen 1996; Albrecht and Car 1999; Pereira 2002).

In this paper we apply different theories of formal ontology such as mereology (Simons 1987) and granular partitions (Bittner and Smith 2001; Smith and Brogaard 2002), as well as hierarchy theory (Allen and Starr 1982, Ahl and Allen 1996), within the overarching framework of spatio-temporal ontology (Bittner and Smith 2003) in the context of considering the problem of scale. Our objective is to determine whether the combination of these theories provides new insights into the issue of scale. The aspect of scale that is focused upon is the linkage between scales, that is, the relationship between *objects* operating at large scales and small scales, and the

W. Kuhn, M.F. Worboys, and S. Timpf (Eds.): COSIT 2003, LNCS 2825, pp. 13–27, 2003.

relationship between *processes* operating at large scales and small scales.   We consider a definition of scale for both enduring entities, such as humans and mountains, and processes, such as tidal currents and gentrification, within their formalization in different kinds of ontologies (SNAP-ontologies for enduring entities and SPAN-ontologies for processes).

The paper is structured as follows, Section 2 describes the notions of grain and extent and their use in the paper. Section 3 introduces the SPAN and SNAP ontologies as formal ways of representing objects and processes. Section 4 clarifies the notion of scale with regards to objects in a SNAP ontology, and Section 5 explores scale with regards to processes in a SPAN ontology. Section 6 concludes the paper.

2   Grain and Extent

Extent is concerned with the spatial size of phenomena, in (x,y,z) dimensions, or the temporal length of duration over which those phenomena operate (Lam and Quattrochi 1992). For example, continental glaciers operate over a much larger extent, both spatial and temporal, than thunderstorms. Both spatial and temporal extent are referenced to an absolute space and time for the purpose of this paper. Grain, whether spatial or temporal, is a relative notion. That is, it has no absolute spatial or temporal location in the sense of being positioned on the earth at a certain longitude or latitude, or at a certain Greenwich Mean Time. Such positioning is defined by extent.

Grain refers to the fineness of distinctions recorded in the data, often referred to as resolution (Albrecht and Car 1999). For example, the spatial grain of a remotely sensed image is the size of each pixel in its relation to the patch of the earth it represents. Determining which spatial grain is appropriate for a certain observation is dependent on the spatial extent of the objects to be observed, that is, the spatial grain of our observation depends upon the granular structure of reality. Temporal grain refers to the frequency of behavior. For example, the temporal grain of a longitudinal study of the daily patterns of human movement is the frequency of observation or sampling. Determining which temporal grain is appropriate for a certain observation is dependent on the temporal extent of the objects to be observed, that is, the temporal grain of our observation depends upon the granular structure of reality in the sense that only a certain range of temporal grain is appropriate for the observation of temporal objects of a certain temporal extent. Frequency is traditionally defined as the number of cycles a phenomenon completes within a specified time interval. In the context of this paper, fast behavior is defined by high frequency and slow behavior by low frequency (Ahl and Allen 1996). For example, the movement of a glacier occurs at a much lower frequency than the ephemeral cusp formation at a beach.

The determination of the appropriate scale of analysis is fundamental to the analysis of all geographic phenomena. It is typically dependent on the objectives and resources of the research at hand. The effects of the scale of observation on geographic phenomena are well recognized and pertain to a common recognition that different processes are observed to operate at different scales, thus phenomena or processes may be obscured when research is concentrated at an inappropriate scale (Meyer *et al.* 1992). For example, an exclusive analysis of microclimates may lead to explanation based on local processes when there are processes of large extent, such as El Niño, that are influencing or controlling the microclimate and should be observed with coarser granularity. Although the problem of scale definition is well recognized, it continues as an unsolved problem, thus it remains a topical issue. Many questions continue unresolved from more recent initiatives such as NCGIA's (National Center for Geographic Information Analysis) Varanius project: Scale and Detail in the Cognition of Geographic Information, and the UCGIS (University Consortium for Geographic Information Science) research priority of scale (UCGIS 1998).

3   Two Ontological Views

Following Bittner and Smith (2003) we recognize two complementary views of the world, one directed towards enduring entities, or objects, and the other directed towards perduring entities, or processes. Taken together these two kinds of ontologies capture both enduring entities, such as political boundaries and humans, and spatiotemporal processes, such as erosion and urban sprawl. The former view gives raise to a series of snapshot-like ontologies called SNAP, each representing objects existing at a certain moment in time. The latter view presents an ontology called SPAN, which represents processes and other spatio-temporal entities as four-dimensional worms.

To describe geographic processes in an appropriate and complete manner we cannot have a SPAN ontology without the SNAP ontology and vice versa. For example, the processes of migration and gentrification cannot be understood without the enduring entities of humans, which are involved in these processes. Likewise, we cannot understand the interaction of sedimentary particles in the littoral zone of a coast without understanding processes such as longshore drift, rip currents, and mass transport. Thus the interaction between processes, enduring entities, and changes in enduring entities – which is related to participation in certain processes – is critical.

SNAP-ontologies are an intuitive approach to the world that is reflected in how we experience reality at every given moment in time, and corresponds to the way we typically model it: as a collection of interacting enduring entities and the relations between them. The temporal character of the world is reflected by a series of three-dimensional temporal slices or snap-shots of enduring entities. Considering sequences of 3D-SNAPshots is the common approach to modeling processes in Geographic Information Systems (GIS); for example, in trying to model the demographics of a region over time we use snapshots of the state of the region at specified time intervals such as with census data. An important aspect is that processes are not represented in ontologies of the SNAP kind, although they can be discovered indirectly in terms of the changes they cause. Change is hereby defined as the difference in some attribute in an *enduring* entity at different time indexes, be that spatial location or some other property of the entity.

SPAN-ontologies are derived from four-dimensionalism, a popular position in contemporary analytic metaphysics (see Sider 2001). The four-dimensionalist views the world from a God's eye perspective, as spread out in time and populated by space-time worms that represent processes. Time is considered as just another dimension in addition to the three spatial dimensions, as in the theory of relativity. Hence this is an atemporal view where things that are spread out over time are said to *perdure*. This does not deny the temporal extension of spatio-temporal entities, rather that their traditional temporal qualifiers of past, present, and future are traced over because all are simultaneously evident.

Another important aspect is that SPAN ontologies trace over enduring entities. For example, a flood event described by a SPAN ontology traces over the existence of enduring entities such as water bodies. However there exist complex interrelationships between SNAP and SPAN ontologies. Enduring entities at every moment in time have a corresponding three dimensional slice of the processes they participate in. For example, enduring entities such as humans or animals participate in migration processes. One cannot reduce one view to the other. Rather one must translate between SNAP and SPAN ontologies, requiring an analysis of cross-ontological relationships between those different ontological views on a meta-theoretical level in order to establish a relationship between them, which will not be discussed in this paper (see Bittner and Smith 2003 for details). This is evident in the difficulties that GIS have with temporal phenomena because the process view of the world cannot be reconstructed with the view of enduring entities without losing something.

Within each ontology, scale is defined differently. We give a provisional definition of scale within the SNAP ontology, in terms of classes of objects. Here scale is a set of levels of granularity composed of objects that can be distinguished at these levels of granularity with observation or measurement within a certain range of spatial grain.

4   SNAP Objects, Granularity, and Scale

We begin with a discussion of the granular structure of enduring entities, which is developed by applying the theories of mereology and granular partitions. Within this context the notion of scale is considered. The presented notions will then be extended to Section 5 in order to take into account the more complex granular structure of processes.

4.1   Mereology and Granularity Trees

Our basic tool for understanding the phenomenon of granularity is mereology (Simons, 1987). The most basic mereological concept is the relation of part to whole. The concept of part and whole applies in every domain, from ordinary objects such as people, chairs, and mangos, to processes such as frontal systems, information flows, and erosion. They may also apply to abstract entities such as classes and properties. In this section only enduring entities will be considered.

The simplest expression of this part-whole relation is given by $\leq$, where $x \leq y$ is read as 'x is a part of y'. For example, if a house (an enduring entity) is considered a whole, its doors, roof, and windows are parts of that whole. The relation $\leq$ includes both the case of proper parthood ($<$) and equality. The core axioms define the part-whole relation as reflexive, transitive, and antisymmetric, i.e., as a partial ordering (Simons 1987). In the context of this paper we are not interested in mereology in its whole generality but in hierarchical structures which represent mereological structure in a restricted form: as finite trees. Those structures are subject to the features of granularity.

Formally we define a granularity tree as a pair, $G = (R, \subseteq)$, where R is a set of objects with a binary relation $\subseteq$ which is a restricted form of the mereological part-of relation. Following Smith and Brogaard (2002), we call the objects forming a system of granularities, cells, and the relation $\subseteq$, the subcell relation. Using the subcell relation we define the relations of overlap, proper subcell, and immediate proper subcell.

D1 $O\,xy \equiv \exists z\,(z \subseteq x \wedge z \subseteq y)$
D2 $x \subset y \equiv x \subseteq y \wedge x \neq y$
D3 $x \subset_i y \equiv x \subset y \wedge \neg \exists z (x \subset z \wedge z \subset y)$

We here assume that quantification ranges over entities in R. Moreover we assume that leading quantifiers are understood. Two cells overlap if and only if they have a subcell in common (D1). The cell x is a proper subcell of the cell y if and only if x is a subcell of y and x and y are not identical (D2). The cell x is an immediate subcell of y if and only if x is a proper subcell of y and there does not exist a cell z such that x is a proper subcell of z and z is a proper subcell of y.
We continue by defining the predicates 'being a root element' and 'being an atom':

D4 $\text{Root } x \equiv \forall y\,(y \subseteq x)$
D5 $\text{Atom } x \equiv \neg \exists y\,(y \subset x)$

An entity is a root if and only if all entities are its subcells (D4). An entity is an atom if and only if it does not have a proper subcell.

Granularity trees are governed by the following axioms (Bittner and Smith (2003):

G1 $x \subseteq x$
G2 $x \subseteq y \wedge y \subseteq x \rightarrow x = y$
G3 $x \subseteq y \wedge y \subseteq z \rightarrow x \subseteq z$
G4 $\exists x\,(\text{Root } x)$
G5 $O\,xy \rightarrow x \subseteq y \vee y \subseteq x$
G6 $x \subset y \rightarrow \exists z\,(z \subset y \wedge \neg O\,zx)$
G7 $\exists y\,(\text{Atom } y \wedge y \subseteq x)$

(G1-G3) ensure that the subcell relation is reflexive, antisymmetric, and transitive, i.e., a partial ordering. That there is a root element is guaranteed by (G4). Axiom (G5) ensures that if two cells overlap then one is a subcell of the other. This excludes partial overlap. For this reason there cannot be circles and the resulting structure is a tree. (G6) is known in the literature as the weak supplementation principle (Simons, 1987). It ensures that if a cell has a proper subcell then it has at least two non-overlapping proper subcells. For this reason the resulting tree structure cannot degenerate into a list. (G7) ensures that every cell has an atom as subcell. Consequently, every cell is connected to the root cell by a finite chain of immediate subcells. This axiom might be disputable and is from a formal point of view not needed for the discussion that follows. However we believe that it is an important aspect of granularity that the subdivision into parts stops after finite steps.

Given those axioms one can prove that the root element is unique and that the strong supplementation principle holds, from which the extensionality of the overlap and the proper part relation follows (Simons, 1987).

It follows that every granularity tree can be represented using a tree structure in the mathematical sense, that is, as rooted directed graphs without circles, by taking the regions as nodes and by demanding that there is an edge from node a to node b if and only if $a \subseteq b$. For example, the parts of London may be referred to as 'boroughs', such as Westminster, Camden, Southwark, and Greenwich. The parts of these boroughs are referred to as 'suburb', for example, in the case of the borough of Camden, we find suburbs such as Kingcross, Hampstead Heath, Swiss Cottage, and Euston, which belong to such a system of granularities (Figure 1). Thus, we can describe London at varying levels of granularity, the composition of which defines our granularity tree.

Granularity trees have finite depth, where depth is defined as the maximal length of the finite chain from the root to a leaf. Therefore the granularity tree can be described as having finite grain, that is, the range of grain (from largest to smallest) of a granularity tree is defined. Thus parts of the structured domain that are below a certain grain are not recognized in the tree in the sense that they are not cells of the tree connected to the root. For example, the parts of Euston are not recognized by the granularity tree in Figure 1. However, this granularity tree is not full, that is, its subcells do not exhaustively describe each cell (Bittner and Smith 2001). For example, London is composed of 32 boroughs in total.

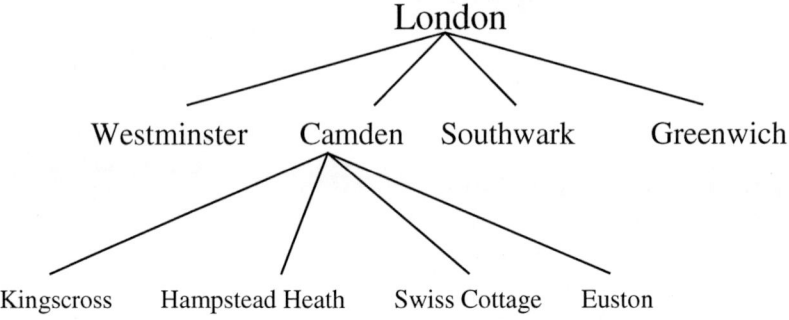

Fig. 1. London depicted as a granularity tree

Another example of granularity tree may be found in the classification of channel geomorphology (Montgomery and Buffington 1998). The watershed is the root of this granularity tree, which may be divided into sub-watersheds, sub-watersheds having valley segments as their parts, which are in turn composed of channel reaches, which are composed of channel units (Figure 2).

Take for example the Chesapeake Bay Watershed, the sub-watersheds in this watershed include the Potomac River, Susquehanna, York, and James. In the Potomac River sub-watershed, valley segments may include the north and south fork of the Shenandoah river, Monocacy River, Anacostia River, and the lower Potomac River (Figure 3). This granularity tree could be expanded to include channel reaches and channel units.

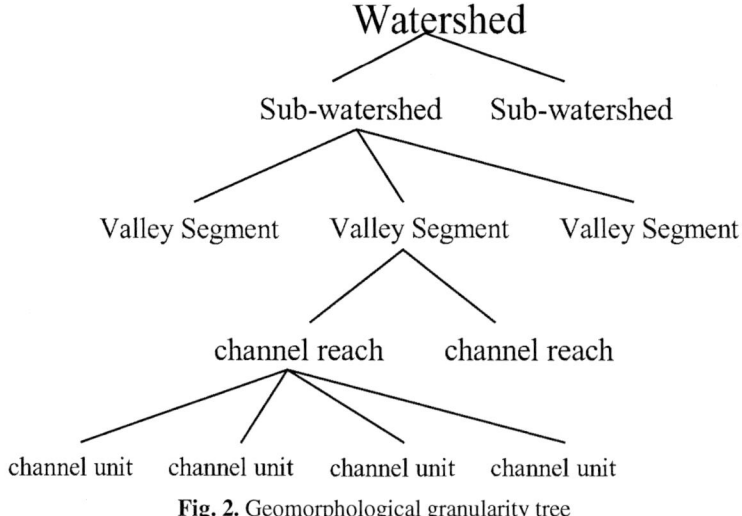

Fig. 2. Geomorphological granularity tree

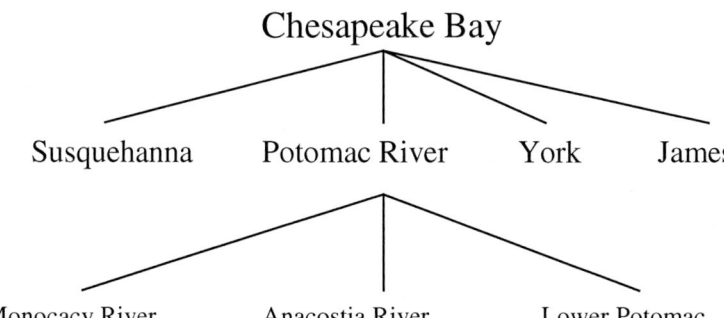

Fig. 3. Chesapeake Bay described as a granularity tree

The example of the Chesapeake Bay provides grounds for introducing the distinction between bona fide and fiat granularity trees. The objects within the granularity tree can be either bona fide or fiat (Smith 1995; Bittner and Smith 2001). Bona fide objects exist independently of human partitioning activity, take for example the Potomac River, which exists regardless of how it may be organized into a granularity tree. Fiat objects are created by our partitioning activity, for example the census units used to divide the population into spatial partitions. Orthogonal to this division between bona fide and fiat objects is the division of hierarchies into hierarchies of kinds (taxonomies) and hierarchies of parts (partonomies) (Tversky 1990). A taxonomy is the partitioning of reality into kinds, for example a poodle is a kind of dog, which is a kind of animal. A partonomy is the partitioning of reality into parts. The Geomorphologic granularity tree (Figure 2), instantiated by the Chesapeake Bay granularity tree (Figure 3), is an example of a partonomy.

4.2   Levels of Granularity

Let $G = (R_G, \subseteq)$ be a granularity tree and let R_G be non-empty. *Levels of granularity* in G are then defined as sets of cells inductively as follows (Rigaux and Scholl 1995):

1. The set containing only the root cell of the granularity tree is a level of granularity
2. Let C be a level of granularity consisting of the cells $z_1, ..., z_n$. Then we can replace every z_i by its immediate subcells (if it has any) and the result is another level of granularity.

Consider Figure 1; the levels of granularity, for example, are:

$$g_0 \quad \{London\}$$

$$g_1 \quad \{Westminster, Camden, Southwark, Greenwich\}$$

$$g_2 \quad \left\{\begin{array}{l} Westminster, Kingscross, Hampstead\ Heath, Swiss\ Cottage, Euston, Southwark, \\ Greenwich \end{array}\right\} \quad (1)$$

Our definition captures only certain necessary conditions that characterize levels of granularity, which are purely mereological in nature:

1. The elements forming a level of granularity are pair-wise disjoint
2. Levels of granularity are exhaustive in the sense that for every cell z that does not belong to the granularity-level $\delta = (z_1, ... z_k)$ there exists a cell $z_i \in \delta$ such that $z_i \subseteq z$ or $z \subseteq z_i$.

Our definition, so far is relatively rough, for example, the lengths of each finite chain that connect each leaf cell to the root cell in the granularity tree are not equal. This results in the problem of objects being repeated at a number of levels of granularity. In the London example, Westminster, Southwark, and Greenwich, are repeated in level g_2 after their initial appearance in level g_1.

For a more complete characterization of levels of granularity more domain-specific properties need to be taken into account. In SNAP domains metrical notions could also be included, for example, we might impose the requirement that objects forming a certain level of granularity have roughly the same size. In SPAN domains, as we shall see in Section 5 below, specific properties of the processes studied need to be considered.

4.3   Resolution of Observation and Scale

With the organization of objects into a granularity tree we return to the distinction between the granular structure of reality and the resolution of observation that was introduced in the introduction. Given a bona fide granularity tree, its ability to reflect the granular structure of reality that we wish to describe depends on the resolution of observation. Objects of human scale, such as books, filing cabinets, elephants, and anything else at the level of Zubin's *A-spaces* (Zubin 1989), are observable by humans without other instruments. Objects of larger scale, such as cities, watersheds, and islands, and of smaller scale, such as cells, bacteria, and atoms, require a different resolution of observation that depends on scale specific observation tools.

Scale, then, is determined by the relationship between the granular structure of reality and the resolution of observation. Returning to the provisional definition given at the end of Section 3, scale is a set of levels of granularity in the defined granularity tree that is partitioned into equivalence classes that are distinguished on the granular structure of reality through a certain range of resolution of observation. For example, human scale, as a class of levels of granularity, is composed of objects that can be distinguished on the granular structure of reality through a range of resolution defined by the limits of our perception.

The notion of scale discussed here only recognizes the relative extent and grain of objects based on their description as a granularity tree. The root cell in a system of cells is of greater relative scale than its children, as are any lower levels of granularity with respect to their descendents. Thus, going down the levels of granularity from the root cell, with coarsest relative granularity, to leaf-cells of finest relative granularity, the granularity tree is organized from phenomena of large relative extent to phenomena of small relative extent. For example, London is much larger than any of its descendents (Westminster, Camden, Southwark, Greenwich) taken individually. We may then define scale on our system of granularities as the partition of the set of levels of granularities into equivalence classes observable at a certain range of resolution, where from the root cell to depth x is scale 1, from depth x to y is scale 2. However, the definition of these depths is not specified in mereology, the observer defines them.

5   SPAN Processes, Granularity, and Scale

As with Section 4, we begin by discussing granular structures using the theories of mereology and granular partitions, however here we focus on perduring entities or processes. Within the framework provided by these theories of formal ontology applied to processes, the notion of scale is considered.

5.1   Process Mereology

In geography we are typically interested in perduring entities, or processes, rather than in enduring entities, or objects. Spatio-temporal entities, or processes, are best described by a SPAN-ontology. We may also distinguish between processes and classes of processes. For example, tropical cyclones are a class of processes and tropical cyclone Paka is an instance of this class.

As with the mereology of enduring entities, the relations of parthood and proper parthood also apply to the processes, or perduring entities, of a SPAN-ontology. However, the mereological part-of relation behaves differently in SNAP and SPAN ontologies in the sense that enduring entities have only spatial parts. Perduring entities, in contrast, also have temporal parts. For example, a temporal part of a process such as gentrification may be the temporal interval over which rent is increased by a certain amount; another temporal part of that same process may the temporal interval over which inhabitants are evicted.

This means that in the domain of enduring entities, granularity is a spatial notion, whereas it is a spatio-temporal notion in the domain of processes. These processes have spatial and temporal extent and spatial and temporal grain, and are structured mereologicaly. Therefore their organization into a tree of granularity must consider both their spatial and temporal extent and their spatial and temporal grain.

5.2   Hierarchy Theory and Rocesses

In descriptions of the process class of urban growth we can move from the level of granularity defined by processes at the neighborhood level to that level defined by processes observed at the metropolitan statistical area level. Or, in the case of weather phenomena, we can describe them at the granularity of microclimates or large-scale phenomena such as the El Niño weather pattern. The part-whole nature of these processes defines the hierarchy, or granularity tree, that they compose (note that hierarchy and granularity tree are taken to be synonymous). The organization of processes into a hierarchy is developed in hierarchy theory. Hierarchy theory, propounded by Ahl and Allen (1996) and Allen and Star (1982), is based on the recognition that processes, spatio-temporal phenomena such as urban growth or weather phenomena, can be described at different levels of granularity.

There is a strong consonance between hierarchy theory, mereology, and granular partitions, at least at the conceptual level. Simon (1973), one of the foundational thinkers of hierarchy theory, uses some of the mereological basics and the tree-like systems of granularities, defining a hierarchy as a partial ordering, a tree. The basic mereological axioms, noted above in Section 3.1, hold true for hierarchy theory, however, hierarchy theory does not formalize these relationships between parts and wholes in any form of explicit calculus. Its novelty lies in its rules for the organization of these parts and wholes into a hierarchy of spatio-temporal processes. We interpret hierarchy theory as an extension to mereology-based notions of parthood, scale, and levels of granularity with the specification of the relationship between levels, thereby extending our definition of scale above. It defines a number of ordering principles of upper levels relative to lower levels of granularity for processes (Ahl and Allen 1996). In this paper we will focus on two of these principles, that is, higher levels in the hierarchy behave at lower frequency than lower levels, and, as with mereological structure, higher levels contain processes at lower levels as parts. Our granular structure is defined by both spatial and temporal grain and extent.

5.3   Granularity Trees Formed by Processes

Processes have sub-processes as their parts. We organize processes and their parts into granularity trees using spatio-temporal grain and extent, extending beyond the granular structure of enduring entities by paying particular attention to temporal grain, or frequency. Frequency, as described in Section 2, is traditionally defined as the number of cycles a phenomenon completes within a specified time. Therefore, processes of high frequency are characterized by fast behavior where processes recur within a short duration of time, and processes of low frequency are characterized by slow behavior where processes recur within a long duration of time. For example, tides are a low frequency class of processes, oscillating over a 12 hour period, compared to the relative high frequency of a wave, which recurs over mere seconds.

Extending this notion of the frequency of a process hierarchically, each process is composed of a pattern of processes at a lower level of granularity, the parts of the process. A pattern of processes, then, is a sequence of different individual processes. The frequency of a process is the reoccurrence of the pattern of processes that it is composed of. Take, for example, a typhoon, which is a class of processes that have a certain lifecycle they undergo, which are characterized by a certain pattern of processes. The pattern of processes of a typhoon is likely to include a tropical depression and a tropical storm. The frequency of a typhoon is then the reoccurrence of the same pattern of processes.

Processes of low frequency, which recur over a long duration of time, are higher in the granularity tree than processes of high frequency, which recur over a short duration of time. This relationship between process at a higher level in the granularity tree and patterns of individual processes at a lower level is repeated down through the granularity tree. For example the temporal grain or frequency of atmospheric phenomena are typically classified as micro-scale: seconds to minutes, meso-scale: minutes to days, synoptic scale: days to weeks, and macro-scale: weeks and greater (Ahrens 1991). Another example of a system of granularities is defined on coastal processes. Consider the tidal cycle in the Bay of Fundy as a granularity tree (Figure 4). Note that this is a class of processes that can be applied to any instance of a tidal cycle in the Bay of Fundy. One tidal cycle occurs over almost a 13-hour period, the parts of which are its currents which have a higher frequency, followed by the subprocesses of waves which have a higher frequency again (Figure 5). The difficulty with finding an appropriate geographic example is that few researchers describe processes hierarchically, thus the example given above is very incomplete and provisional.

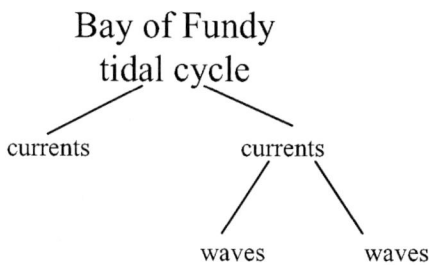

Fig. 4. The Bay of Fundy tidal cycle as a granularity tree

The temporal frequency of processes often reflects their spatial and temporal extent. Processes that are higher in our granularity tree are not only characterized by lower temporal frequency but also greater extent, thus processes that are lower in the system have higher frequency and smaller spatial and temporal extent. This reflects a well recognized principle whereby phenomena that are large in space, such as El

Niño, have a low frequency, whereas small phenomena, such as micro-climates, have a higher frequency (Meyer *et al.* 1992). For example, take the classification of channel geomorphology, which incorporates both the structural geomorphology and the functional geomorphic processes operating at each level. The processes operating in a channel unit are contained, spatially and temporally, in those of a channel reach, which are contained within those processes operating in a valley segment, and so forth within a watershed and a geomorphic province (Montgomery and Buffington 1998). Similarly, in the Bay of Fundy example it can be said that the tidal cycle contains all of the lower processes in the granularity tree within its temporal and spatial extent.

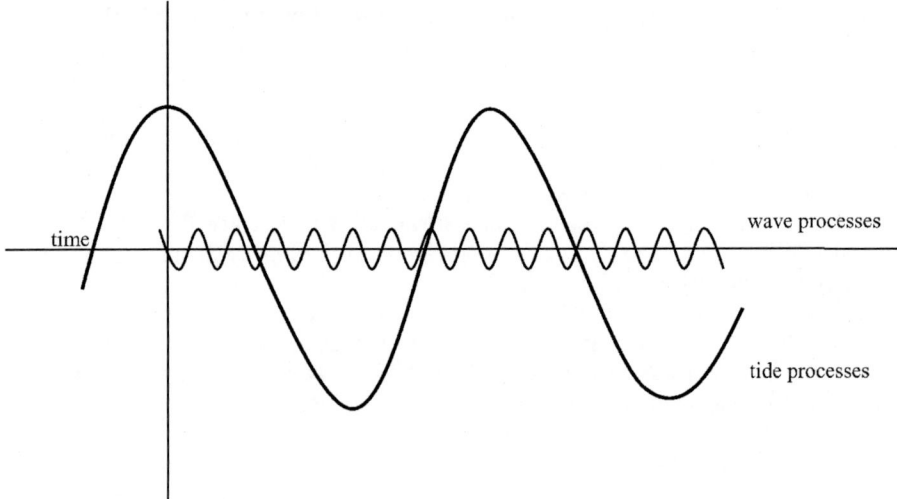

Fig. 5. A pattern of wave processes superimposed on the tide process

A system of process granularities is a pair, $G_p = (P, \subseteq)$, where P is a set of processes with a binary relation $\subseteq$ satisfying G1-G7 given above. As with enduring entities, we call the processes forming a system of granularities, cells, and the relation $\subseteq$, the subcell relation. However, for processes this subcell relation has to obey additional constraints which are defined in terms of spatial grain and extent and temporal grain and extent, where a subcell must:

1. have smaller temporal extent
2. have smaller spatial extent
3. have greater temporal frequency, or finer temporal grain
4. have greater spatial resolution, or finer spatial grain

5.4   Levels of Process Granularity

The nature of the part-whole relationship for processes, beyond the requirements of mereology, can then be summarized as: a part is contained within the spatial and temporal extent of the whole and has a higher frequency (temporal grain) and higher resolution (spatial grain) than the whole. In mereology, the parts exhaustively sum to the whole, such as the full granularity tree for London that would include all of its 32 Boroughs completely covering London's spatial extent. In contrast, the granularity tree that is formed by processes is not spatio-temporally exhaustive.

Within a granularity tree we can distinguish levels of granularity at which certain processes reside. Because of the added temporal dimensions of processes, our organization of processes into a granularity tree reflects this temporality. The sequence of processes that composes a process at a higher level has the requirement of having the same (or a similar range of) frequency or temporal granularity, which is the primary organizing construct of the granularity tree. Thus the levels in a granularity tree and the depth of each level is defined by the frequency of the process, that is, the recurrence of a sequence or pattern of sub-processes that compose a process. The approach for slicing a granularity tree into levels defined in Section 4.2 applies to processes also.

5.5   Process, Resolution of Observation, and Scale

As with objects, in Section 4.3, our definition of a system of spatio-temporal granularity that reflects the granular structure of reality depends on the resolution of observation. Processes at different scales require different resolutions of observation. However, unlike objects, with processes this notion of resolution is both spatial and temporal, that is, both spatial and temporal extent and spatial and temporal grain must be considered. For example, processes of human scale are observable without other instruments, such as the cusp formation at a beach, or pedestrian traffic along a limited stretch of pavement. Larger scale processes, such as oceanic circulation and urban sprawl, and smaller scale processes such as cancer or the migration of dust mites, require a different resolution of observation that depends on spatial and temporal scale specific observation tools. From the perspective of a SPAN-ontology, the length of the space-time worm that describes our process defines the temporal extent and implies a certain temporal resolution of observation to capture the temporal grain of the process. Likewise, the three dimensional geometry of that space-time worm defines the spatial extent and therefore implies a certain spatial resolution of observation to capture the spatial grain of the process. As noted earlier, the spatial and temporal dimensions are independent. This inclusion of temporal resolution extends to the definition of scale for processes. For example, human scale processes, as a class of levels of spatio-temporal granularity, are composed of processes that can be distinguished on the granular structure of reality through a range of spatial and temporal resolution that is defined by the spatial and temporal limits of our perception.

Given the definition of scale in Section 4.4, it may be extended by considering the temporality of processes and by defining the relationship between levels of granularity as the frequency of the process under observation. We may then define scale on our system of granularities as the partition of the set of levels of granularities into equivalence classes of processes observable within a certain range of frequency or temporal granularity. From the root cell to depth x is scale 1, from depth x to y is scale 2. Frequency classes determine the depths. More specifically, we may define a cut by the natural breaks in the observed frequency of a phenomena, where 'the natural breaking points in systems, the natural surfaces about which systems are nearly decomposed, have been identified as portions of the scale gradient which are so steep that they can be functionally considered as steps' (Allen and Star 1982). Alternatively, to define the depths of scale we might temporalise other classification measures such as Equal Interval or Standard Deviation.

6   Conclusions

Understanding the organization of processes into hierarchically structured granularity trees, where patterns of sub-processes at finer levels of granularity sum up to processes at coarser levels of granularity, helps us to understand the interrelationships between phenomena at different levels of granularity. These granularity trees form a framework for partitioning the world of enduring entities, viewed through the SNAP ontology, and processes, viewed through the SPAN ontology. Within both ontologies, the spatial and temporal grain of observation identifies the levels of the granularity tree. For enduring entities, their spatial extent and grain defines the structure of the granularity tree. For processes, their spatial and temporal grain and spatial and temporal extent defines the structure of the granularity tree. We defined scale as the depth between partitioned equivalence classes of either objects or processes described by the granularity tree.

It is, however, important to stress that our examples are somewhat simplified and completely ignore the issue of vagueness usually involved. This extends beyond defining the spatial bounds of objects to include the temporal dimension of defining the spatio-temporal bounds of processes. For example, what defines the beginning of a tropical cyclone? What are the pattern of processes that indicate the precise instant of a migration process?

Other questions that surface from this research include the exploration of bona fide and fiat hierarchies and the relationship between them. Furthermore, consideration the influence of varying projections on our hierarchy, not only varying in the sense of different views on the same subject matter, but also varying dynamically where scales change over time, that is, we have dynamic granularity trees, or dynamic hierarchies. Furthermore, the difficulty in finding hierarchical descriptions of processes presents the question of whether or not the world is organized hierarchically or whether our tools and views are limited.

Acknowledgements

We would like to thank Jochen Albrecht for many helpful comments and valuable insights throughout the development of this paper, and support from the Wolfgang Paul Program of the Alexander von Humboldt Foundation is gratefully acknowledged.

References

Ahl, V. and T. F. H. Allen (1996). *Hierarchy Theory: a vision, vocabulary, and epistemology.* New York, Columbia University Press.

Ahrens, C. D. (1991). *Meteorology Today: an introduction to weather, climate, and the environment.* New York, West Publishing Company.

Albrecht, J. and A. Car (1999). GIS analysis for scale-sensitive environmental modelling based on hierarchy theory. In: *GIS for Earth Surface Systems.* R. Dikau and H. Saurer. Berlin, Gebruder Borntraeger: 1-23.

Allen, T. F. H. and T. B. Starr (1982). *Hierarchy Theory: perspectives for ecological complexity.* Chicago, The University of Chicago Press.

Bittner, T. and B. Smith (2001). *A Taxonomy of Granular Partitions.* COSIT 2001, Morro Bay, Lecture Notes in Computer Science, Berlin-Heidelberg, Springer-Verlag.

Bittner, T. and B. Smith (2003). *Granular Spatio-Temporal Ontologies.* AAAI Spring Symposium on Foundations and Applications of
Spatio-Temporal Reasoning (FASTR), Palo Alto, California.

Goodchild, M. F. and J. Proctor (1997). Scale in a Digital Geographic World. *Geographical & Environmental Modelling* 1(1): 5-23.

Lam, N. S. and D. A. Quattrochi (1992). On the Issue of Scale, Resolution, and Fractal Analysis in the Mapping Sciences. *Professional Geographer* 44(1): 88-98.

Meyer, W. B., D. Gregory, B. L. T. II and P. F. McDowell (1992). The Local-global Continuum. In: *Geography's Inner Worlds.* J. M. Olson. New Jersey, Rutgers University Press: 255-279.

Montgomery, D. R. and J. M. Buffington (1998). Channel Processes, Classification, and Response. In: *River Ecology and Management: Lessons from the Pacific Coastal Ecoregion.* R. J. Naiman and R. E. Bilby. New York, Springer-Verlag.

Pereira, G. M. (2002). A Typology of Spatial and Temoral Scale Relations. *Geographical Analysis* 34(1): 21-33.

Rigaux, P. and M. Scholl (1995). *Multi-scale partitions: Application to spatial and statistical databases.* Advances in Spatial databases (SSD '95) - Lecture Notes in Computer Science, Berlin, Springer-Verlag.

Sider, T. (2001). *Four-Dimensionalism.* Oxford, Clarendon Press.

Simon, H. A. (1973). The Organization of Complex Systems. In: *Hierarchy Theory: the challenge of complex systems.* H. H. Pattee.

Simons, P. (1987). *Parts: a study in ontology.* Oxford, Clarendon Press.

Smith, B. (1995). *On Drawing Lines on a Map.* Spatial Information Theory, Proceedings of COSIT '95, Berlin, Springer Verlag.

Smith, B. and B. Brogaard (2002). Quantum Mereotopology. *Annals of Mathematics and Artificial Intelligence* 36(1-2): 153-175.

Tversky, B. (1990). Where Partonomies and Taxonomies Meet. In: *Meanings and Prototypes: studies in linguistic catogorization.* S. L. Tsohatzidis. London, Routledge.

UCGIS (1998). Scale. *Research Priorities: revised white papers.*

Zubin, D. (1989). Natural Language Understanding and Reference Frames. In: *Languages of Spatial Relations: Initiative 2 Specialist Meeting Report Technical Paper 89-2.* D. M. Mark, A. Frank, M. J. Egenhofer *et al.* Santa Barbara, CA, National Center for Geographic Information and Analysis: 13-16.

Landscape Categories in Yindjibarndi: Ontology, Environment, and Language

David M. Mark[1] and Andrew G. Turk[2]

[1] Department of Geography
National Center for Geographic Information and Analysis, and
Center for Cognitive Science
University at Buffalo, Buffalo, NY 14261, USA
dmark@geog.buffalo.edu
[2] School of Information Technology, Murdoch University
Perth, Western Australia 6150, Australia
a.turk@murdoch.edu.au

Abstract. This paper describes categories for landscape elements in the language of the Yindjibarndi people, a community of Indigenous Australians. Yindjibarndi terms for topographic features were obtained from dictionaries, and augmented and refined through discussions with local language experts in the Yindjibarndi community. In this paper, the Yindjibarndi terms for convex landforms and for water bodies are compared to English-language terms used to describe the Australian landscape, both in general terms and in the AUSLIG Gazetteer. The investigation found fundamental differences between the two conceptual systems at the basic level, supporting the notion that people from different places and cultures may use different categories for geographic features.

Keywords: Geographic categories, geographic ontology, landscape terms, natural language, cultural differences, Yindjibarndi, Indigenous Australians, spatial cognition, geographic information systems, GIS.

1 Introduction

Do all people, and all peoples, think about the landscapes and its elements in more or less the same way? Or are there significant cross-cultural and cross-linguistic differences in the ways human beings perceive and cognize their environments at geographic or landscape scales? These are important scientific questions, and also important challenges to designers of geographic information systems (GIS) and compilers of geographic databases and spatial data infrastructures. For the past several years, we and our colleagues have been approaching these questions from a variety of research perspectives, most recently the perspective of ontology.

W. Kuhn, M.F. Worboys, and S. Timpf (Eds.): COSIT 2003, LNCS 2825, pp. 28–45, 2003.
© Springer-Verlag Berlin Heidelberg 2003

In this paper, we attempt to gain perspective on such questions by examining landscape categories in Yindjibarndi[1], an Australian language spoken in the Pilbara region of northwestern Australia, approximately 1,600 km north of Perth. Yindjibarndi is a language that is very distant linguistically from English and other Indo-European languages, and is that spoken by people from an environment that is very different from northwestern Europe. Thus, if the comparison does not reveal significant differences, such a result would support the proposition that there are universal landscape concepts for the domains examined. Such universality can be regarded as the null hypothesis for the study.

2   Theoretical Background

2.1   Ontology

Ontology, in its long-established philosophical sense, seeks to identify the constituents of reality. In its more recent information systems sense, an ontology is "a logical theory which gives an explicit, partial account of a conceptualization" (Guarino and Giaretta, 1995, p. 32). The ontology stipulates the taxonomy that forms the basis of a data dictionary used in building an information system. At a relatively abstract level, ontology determines the kinds of entities that can exist—objects, fields, parts, solids, fluids, etc. Ontology also identifies observable properties or attributes, such as size, shape, or curvature. Another important aspect of the ontology of a domain is the categories to which entities can belong, and the relations among those categories. Recently, the geographic information science research community has devoted considerable attention to the ontology of the geospatial domain (Winter, 2001; Smith and Mark, 2001).

2.2   Categories

Categories are central to cognition (Rosch, 1973a, 1973b, 1978; Smith and Medin, 1981; Lakoff, 1987; Mark, 1993b). Some categories reflect groups of similar entities in the real world—natural kinds, if they exist, would be an example of this. Other categories exist by design—most artifacts belong to the categories that their manufacturers intended them to belong to. In contrast, there is more room for different people or groups to come up with different categorization schemes for natural inorganic domains—and geographic entities fall into this subdomain. For geographic entities,

---

[1] The name of this language and group has been spelled in various ways. Recently, local groups in the community have preferred "Indjibarndi". Von Brandenstein (1970) spelled the name as "Jindjiparndi", and Tindale (1974) used "Indjibandi". For conformance with current scholarly work on Aboriginal languages, In this paper we have spelled the language name as "Yindjibarndi", following the current spelling standard from AIATSIS, the Australian Institute of Aboriginal and Torres Strait Islander Studies (AIATSIS, 1994).

categories may in part reflect similarities and discontinuities in the landscape, but to some extent are projected onto the landscape by human cognition and language. This study explores the relative balance of such factors in the case of landforms and waterbodies.

2.3   Standards

Theoretical aspects of cognitive categories meet practical issues of geographic information systems and spatial databases in the area of standards (Mark, 1993a). Geospatial data exchange and data infrastructures depend on the use of standards for data formats. Semantics of spatial information often are expressed through some system of feature codes or entity types that indicate a real-world geographic category to which a feature on a map or in a database belongs. Feature codes also play a key role in gazetteers, which are important in digital map and image libraries. Government-endorsed feature categories also provide a baseline description of landscape categories according to the dominant culture in a society, against which category systems by other groups such as indigenous peoples may be compared.

2.4   Ontology of the Geographic Domain

As Smith and Mark (1998) noted, both geographic entities and their categories may differ in kind from entities and categories in other domains. Geographic entities are not simply large versions of their counterparts at smaller scales: "geographic objects are not merely located in space, but are tied intrinsically to space in a manner that implies that they inherit from space many of its structural (mereological, topological, geometrical) properties" (Smith and Mark, 1998, p. 592).

Smith and Mark (1998) speculated that categories of geographic entities might be organized differently than other categories studied by psychologists and cognitive scientists. However, subsequent empirical evidence appears to show that geographic categories have the same sorts of structures and internal organizations as do categories in other domains (Mark et al., 1999; Smith and Mark, 2001; Mark et al., 2001). In contrast, the distinctive nature of individual geographic entities, compared to entities in most other domains, remains apparent, especially in terms of their boundaries. For example, graded or transitional boundaries are common for geographic entities (Turk, 2000) but extremely rare in other domains. The existence of individual objects is a brute fact in the cases of organisms, fruits, or tools, but geographic entities such as mountains do not quite exist as objects to the same degree (Smith and Mark, 2003). Rather, most geographic entities are parts of the Earth's surface that are delimited from neighboring parts in a variety of ways, some of which may be contingent on the conceptual system of the delimiters. For example, believing that some region is a marsh may give it a different boundary than it would have if it were thought to be a lake. Such contingency provides more opportunity for cultural, linguistic, or individual differences in the delimitation of individual geographic entities. Even in the propensity to transform a continuous landscape into objects at all may vary across cultures.

A key issue then, is how do individual people, or the people in a speech community, divide the landscape into entities such as mountains or valleys? In addition, how are those entities categorized, and is there an interaction between the classification and delimitation processes? How important is the nature of the particular landscape that provides the environment for a speech community, and especially the range of forms in that landscape? How influential is the culture and lifestyle of the people, that is, the nature of human interaction with the landscape? How influential is the nature of the language itself, its grammar and lexicon?

Cross-linguistic comparisons might help tease apart these and other effects. In order to attempt this, we examined the terms for landscape entities in the Yindjibarndi language of northwestern Australia. We decided to study an Aboriginal Australian language because the Australian languages are only very distantly related to the Indo-European languages. We chose Yindjibarndi because one of us (AT) has worked with the Yindjibarndi-speaking community in Roebourne for several years, together with his collaborator, Dr. Kathryn Trees (Australian Indigenous Studies, Murdoch University) (Turk, 2000, 2002; Turk and Trees, 1998, 1999, 2000).

3   The Yindjibarndi People, Country, and Language

Before European colonization of Australia, the Yindjibarndi people lived mostly along the middle part of the valley of the Yarnda-Nyirra-na (Fortescue River) in northwestern Australia, and on adjacent uplands. To the south, their traditional country is bounded by the higher mountains of the Hammersley Range, occupied by the Banjima and Gurrama peoples, and to the north, their country ends approximately at the escarpment leading down to the coastal plain occupied by Ngarluma speakers (Tindale, 1974). Yindjibarndi, Ngarluma and Gurrama belong to the Coastal Ngayarda language group, and Banjima isclassified among Inland Ngayarda languages; all of these languages are in the South-West group of Pama-Nyungan languages (SIL, 2001).

There are no permanent or even seasonal rivers or creeks in Yindjibarndi country. Larger watercourses have running water in them only after major precipitation events, usually associated with cyclones (hurricanes). Between such major rain events, rivers continue to 'run', however, the water is underground, beneath the (usually sandy) surface. Permanent pools occur where the lie of the land and the geology cause the water table to break the surface of the ground. Permanent sources of water include permanent pools along the channels of the Yarnda-Nyirra-na (Fortescue) and other larger rivers, as well as some permanent small springs, and soaks where water can be obtained by digging. Unlike many areas of inland Australia, there are no significant intermittent or seasonal lakes in Yindjibarndi country. Local relief (elevation differences) within most of the traditional country of the Yindjibarndi is relatively low, with rolling hills and extensive flats.

Fig. 1. The **yinda** called "Jindawarrina", located at the place of the same name (Jindawarrina) also known as "Millstream".

As part of the European colonialization process, Yindjibarndi country was taken over by sheep and cattle stations (ranches) from the 1860s. The Yindjibarndi people were moved off their traditional territory into camps and settlements (Ieramugadu Group, 1995; Rijavec et al., 1995). Today, most of the Yindjibarndi speakers live in and around Roebourne, in what traditionally was Ngarluma country. Most of the surviving Ngarluma people now speak Yindjibarndi and English in addition to their own language. The Roebourne community is mostly Indigenous and people use their own languages and English to differing degrees, depending on the context, sometimes with terms from both languages occurring in the same sentence.

Several linguists have studied the Yindjibarndi language. Von Brandenstein (1970, 1992) studied Yindjibarndi and Ngarluma and collected stories. Wordick (1982) also collected stories and produced both a grammar and a Yindjibarndi-English dictionary. Anderson revised Wordick's system of phonetic spelling for Yindjibarndi, and produced both Yindjibarndi-English and English-Yindjibarndi versions (Anderson, 1986). Anderson also coded the words according to topic. A digital version of Anderson's compilation is available from the Aboriginal Studies Electronic Data Archive (ASEDA) of the Australian Institute of Aboriginal and Torres Strait Islander Studies (Anderson and Thieberger, no date); this electronic source with thematic coding was extremely useful in this study. In this paper, we will sometimes refer to Anderson and Thieberger's reworking of the Wordick (1982) and Anderson (1986) dictionaries, and the 1992 word lists from von Brandenstein (1970) collectively as "the dictionaries".

4   Landscape Categories in Yindjibarndi

4.1   Research Methods

Before our November 2002 visit to Yindjibarndi country, we compiled lists of all geographic terms that we could find in the dictionaries. Anderson (1986) coded 55 terms as "geographical features". These included all but one or two of the Yindjibarndi landscape terms that we could find in the dictionary, plus several terms for geologic and earth materials. We then classified the geographic terms into semantic groups according to the usual meanings of their English equivalents, using groups such as water features, land forms, land cover types, etc., to assist in organizing fieldwork.

In November 2002, the authors visited the Roebourne area for a week. We met with local language experts Allery Sandy, Trevor Soloman, and Nita Fishhook, and also toured the area to take photographs of landform examples. During these meetings, we asked our Yindjibarndi collaborators whether they agreed with the meanings of Yindjibarndi words given in the dictionaries that appeared to refer to kinds of geographical features. We also asked them to suggest additional words for kinds of features in the landscape. The elicitation aspects of these meetings were assisted by the use of color prints of photographs of parts of Yindjibarndi country and neighboring areas taken (by AT) on previous fieldtrips. At least initially, the discussion was structured in terms of particular superordinate classes of feature (e.g. water features; hills), and we asked questions seeking to clarify issues identified from analysis of the dictionaries. We also asked about entity types in English for which the dictionaries did not include Yindjibarndi equivalents, such as "island" or "waterfall". Terms were both written on a whiteboard and discussed verbally, and most of the sessions were recorded on digital audio tape. One of us (AT) returned to Roebourne for a week in late January 2003 and had further consultations with Trevor Soloman and with Marion Cheedy, using word lists and color photographic images from the November 2002 fieldtrip. This assisted in clarifying the meanings of some of the terms and established arrangements for more detailed and extensive assistance from Yindjibarndi elders, to be conducted during the first half of 2003. Both authors spent another week in Roebourne in May 2003, taking more photographs and measurements and discussing terms with Marion Cheedy, Jane Cheedy, and Trevor Soloman.

Spelling is somewhat of an issue in this research, since the Yindjibarndi had no written language before European contact, and since some of the phonemes used in Yindjibarndi are not used in English. In this paper, we have used Anderson's spelling for any word that he included in his dictionary, although these sometimes disagreed with the preferred spelling according to our Yindjibarndi colleagues. Inadequacies of the process of compilation of the dictionaries, differing linguistic approaches used, variation in usage of terms over space and time, and the influence of English, all make it impossible to be completely definitive regarding either the exact meaning or the most appropriate spelling for Yindjibarndi words. Our intention is not to make judgments regarding proper spelling, but merely to try to obtain an understanding of the Yindjibarndi landscape terms sufficient for the research project. Compilations of terms and photographs resulting from the research project will be provided to the community to assist our collaborators with teaching of Yindjibarndi language in schools, and at that time the standards for phonetic spelling must be re-visited.

4.2   Yindjibarndi Categories for Water in the Landscape

The Yindjibarndi language has several terms that refer to water bodies or watercourses. In Table 1, we compare these terms to the relevant set of water terms and categories from the Australian Gazetteer standard (AUSLIG, 2002).

Fig. 2. Part of the **yinda** called "Nangarnyungu" by the Yindjibarndi people, which is referred to in English as "Deepreach Pool". The pool is located at Jindawarrina.

One of the most important Yindjibarndi landscape concepts is **yinda**, a permanent pool. A **yinda** may be either large or small, but a body of water must be permanent to be a **yinda**. As noted above, in Yindjibarndi country, all rivers are dry at the surface almost all the time—thus the small number of **yinda** along the river beds take on great ecological and cultural significance. Every **yinda** has its own proper name. Most **yinda** are in the beds of the major rivers in Yindjibarndi country. Yindjibarndi believe that the river channels were formed "when the world was soft" by the river spirit (**warlu**) and that the **warlu** currently occupies and protects the **yinda** (Iera-mugadu Group Inc., 1995; Rijavec et al., 1995). Hence, proper behavior at a **yinda** incorporates respect for the **warlu**. Our collaborators said that during extended dry periods, a **yinda** may be reduced to a small pool, termed a **thula**. Anderson spells it *thurla* and says it means, among other things, "eye". Hence, a **thula** may be thought of as the eye of the **warlu**. Intermittent or temporary pools are not given a water body term at all, but are simply referred to by the general term for water as a substance, **bawa**.

Fig. 3. A **wundu**, referred to in English as "Dawson Creek", in Yindjibarndi country north of Jindawarrina.

Yindjibarndi has two terms that appear to refer to fluvial channels. **Wundu** is usually translated as "river", and refers to riverbed and channels. All the examples of **wundu** that we confirmed through photographs were broad, low-gradient channels at least several meters wide. Anderson (1986) states that **wundu** can also mean "gorge". The other term for channels was **garga**, which seems roughly equivalent to "gully" in English; it appears to refer both to the concave topographic feature and to the channel in it.

Yindjibarndi also has two words for water flow in nature. **Manggurdu** is the term for flood, or for other strong, deep water flow. **Yijirdi** is the Yindjibarndi word for a shallow, narrow flow or trickle of water. It appears that, unlike in English, the Yindjibandi treat the water flow and the channel as *different things*. If this is confirmed by further research, it would be a sharp difference from the conceptualization of watercourses in English.

In English and most other European languages water features are first divided into standing or flowing ones, and then the standing water bodies are divided into larger ones (such as lakes) and smaller ones (such as ponds and pools). Additional terms in English refer to water bodies with distinct origins, such as lagoons. Mark (1993b) discussed minor differences in water body categorization between French and English, as an example of the linguistic phenomenon that is the focus of this paper— French appears to distinguish *étangs* from *lacs* based on water quality and a lack of a surface outlet, rather than giving priority to the size difference that usually separates ponds from lakes in English. In contrast, the Yindjibarndi appear to give primary emphasis to permanence, a factor which is not encoded in the basic level water categories of English. Clearly, the conceptual organization of water body terms in Yindjibarndi contrasts sharply with the organization of terms and concepts in English.

Fig. 4. This flowing water near Jindawarrina would probably be referred to in Yindjibarndi by the term **yijirdi**.

Fig. 5. Another **yijirdi** (small stream) flowing into a **yinda** (permanent pool) at the place known in English as "Fortescue Falls" in Karijini National Park, which is in Banjima country.

4.3   Yindjibarndi Categories for Hills

Marnda is the common Yindjibarndi term for most hills and mountains. Even though most **marnda** would be called hills in English, Wordick and Anderson do not list "hill" as a possible translation of **marnda**, which according to those authors translates to "rock, mountain, metal, hard material, money" (Anderson, 1986). The word **marnda** was very familiar to our collaborators. There are several other Yindjibarndi terms for small hills or mounds, but **marnda** appears to include most things that would be called hills in English, as well as mountains or mountain ranges such as the Hammersley Range (at the Southern end of Yindjibarndi country). **Marnda** is also used for ridges. **Marnda** has other meanings—rock, metal, any hard material, and money (coins). **Marnda** is almost certainly a basic level term, and is one of the most common geographic terms in Yindjibarndi.

Table 1. Comparison of Water Terms and Categories.

AUSLIG category	Language	Terms
LAKE	English:	lake, tarn, loch, lough
	Yindjibarndi:	(some **yinda** are large enough to be considered to be lakes in English)
SOAK	English:	native well, soak, soakage
	Yindjibarndi:	**yurrama**
SPRG	English:	spring, pool spring, hotsprings, mineral spring
	Yindjibarndi:	**jinbi** (permanent spring)
STRM	English:	stream, brook, watercourse, anabranch, backwash, backwater, run, creek, river, gully, rivulet, beck, backwater, burn
	Yindjibarndi:	**Wundu** (riverbed), **yijirdi** (small stream of water), **garga** (gully)
WRFL	English:	waterfall, cascade, cataract, falls, rapids
	Yindjibarndi:	(no Yindjibarndi term for waterfall, however **yijirdi** is used for a small running stream of water over rocks)
WTRH	English:	waterhole, lagoon, hole, pool, washpool, billabong, oxbow
	Yindjibarndi:	**yinda**

Fig. 6. The tablelands between the northern edge of Yindjibarndi country and the Yarnda-Nyirra-na (Fortescue River) have scattered **marnda**, of which this is one of the larger ones near the road to Jindawarrina.

Fig. 7. This small feature in the Jindawarrina area, with a top about 2 meters above its base, would almost certainly be a **bargu** in Yindjibarndi.

Although many features that would be hills to an English speaker would be **marnda** in Yindjibarndi, there are at least three other terms in Yindjibarndi that refer to smaller hills or mounds. A **bargu** is a small hill or a sand hill—the key distinction between a **bargu** and a **marnda** appears to be size, rather than shape, steepness of slope, or material. A **burbaa** is a steep slope along a road, the sort of thing that is referred to as a "hill" in English. But **burbaa** also can refer to a mound, a small sandy hill, an incline, a slope on the side of any hill, or a vegetated sand ridge. Yet another term for a convex topographic feature is **bantha,** a mound or pile, banks, or a hump. At one point our collaborators suggested that a typical **bantha** is very small (e.g., a mound of earth covering a grave), but later it seemed that the main thing that distinguishes a **bantha** from a **bargu** or **marnda** is artificiality.

Again, we compared these terms to the convex topographic terms and categories from the Australian Gazetteer standard (Table 2).

Note that **marnda** appears under six of the seven English terms. A single basic-level term in Yindjibarndi appears to cover a range of topographic convexities described by several terms in English: mountain, hill, ridge, range, and others, while the meaning of the basic-level term "hill" in English is expressed by several terms in Yindjibarndi. For convex topographic features, it appears that the relation between Yindjibarndi terms and English terms is many-to-many. Thus, one would need reference to the exact form of the real-world referent in order to translate these terms correctly. This closely parallels the situation for the water body terms *pond* in English and *étang* in French (Mark, 1993b).

5   Discussion

In one sense, the conceptual systems for water features and for convex topographic features in English and in Yindjibarndi are very similar. The meanings of the Yindji-

Table 2. Comparison of Terms for Convex Topographic Features.

AUSLIG category	Language	Terms
HILL	English:	Hill, Knoll, Knob, Mesa, Sugarloaf, Lookout, Butte, Hillock, Kopje
	Yindjibarndi:	**marnda, bargu, burbaa**
MT	English:	Mountain, Peak
	Yindjibarndi:	**marnda**
PEAK	English:	Mountain Peak, Summit, Point (inland), Rock Column, Butte
	Yindjibarndi:	**marnda, gankala** (*)
RDGE	English:	Ridge, Saddle, Spur
	Yindjibarndi:	**marnda**
RNGE	English:	Range, Mountain Range, Hills, Mountains, Rock, Boulder, Pinnacle, Crag, Needle, Pillar,
	Yindjibarndi:	**marnda**
ROCK	English:	Rock Formation, Tor, Rocks (on land), Rocks (offshore)
	Yindjibarndi:	**marnda, jurrun** (D*), **thalungarn** (F*)
(Other)	Yindjibarndi:	**bantha**
	English:	pile, mound

(* = term not discussed in this paper; D = term only from dictionary, not recognized by our Yindjibarndi colleagues; F = term only from our fieldwork, not in the dictionaries)

barndi terms for such features can easily be expressed in English, and we had no problem communicating in English with our bilingual Yindjibarndi colleagues regarding the meanings of Yindjibarndi landscape terms. Of course, it is possible that there are subtleties of Yindjibarndi landscape concepts that cannot be expressed in English. On the other hand, at the basic level of category terms, the Yindjibarndi landscape vocabulary is completely different from the terms covering the equivalent domain in English. None of the Yindjibarndi terms discussed in this paper is exactly equivalent to one single term in English. Yindjibarndi terms divide up subdomains of geographic reality quite differently than do English terms. For example, permanent and temporary water features that otherwise are similar are considered to be different kinds of features in Yindjibarndi; English, in contrast, treats permanence of water bodies and water courses as only an attribute or property, and expresses it through adjectives such as "temporary", "seasonal", "intermittent", or "ephemeral". In addition, there are several kinds of small hills in Yindjibarndi, but this is not simply a refinement of terminology for convex terrain features, since, from the Yindjibarndi perspective, there are several kinds of **marnda** in English--the basic level terms simply do not match.

Indeed, this is exactly what we should have expected. The basic level categories in a language *must* be tuned to the variations in the particular environment in which a speech community lives, and to the ways in which that environment affords various activities essential to life, if it is to provide the common terms needed in every-day speech. The popular myth of the large number of Eskimo words for snow (Pullum, 1991) appears to be an exaggeration of a real tendency of environmental variation to influence vocabulary. The basic-level category system for environmental features *should* vary across environments. Of course, such a relation between categories and environment would not be deterministic, but would be probabilistic. Also, different cultures occupying the same landscape could have developed different concepts because of differences in lifestyle. For example, Indigenous Australians in their traditional lifestyle did not have the technology to store large quantities of water, and thus it is not surprising that permanent sources of water take on a vital significance.

6   Some Significant Issues for Further Research

There are many unresolved matters regarding the true nature of the Yindjibarndi landscape ontology (as revealed through their language), which require further research. Some of the more significant are as follows:

6.1   The Role of Compound Words and Phrases

A language might have a large number of words to refer to different kinds of geographic entities. Alternatively, speakers of a language might use a small number of general terms, and combine them with adjectives describing attributes, forming either phrases or compound words. However, in a language without a written tradition, the difference between a compound noun and a noun phrase is not always obvious. For hills of different size, it appears that in Yindjibarndi different terms are often used (**marnda; bargu**), however, at times our collaborators used the expression "**gubija marnda**" to mean a small hill. There does not seem to be a simple term for a flat-topped hill (mesa, butte) with the compound word **marndamarlirri** (literally: hill + flattened) used. Similarly, a type of hill in Yindjibarndi country and adjacent areas has a surface composed of slabs of loose iron-rich rock, which weathers to a very dark brown color. These are called **marndawarrura** (literally: hill + black, brown, dark). A similar, though somewhat different, etymology applies to the term for "mountain country" - **marndamirdayi** (mountain + place of, place where the ... is). Cognitive linguists often assume that the encoding of some concepts in monolexemic words, rather than as noun phrases, indicates that those concepts have in some sense a deeper importance to the speakers of the language in question. According to Berlin and Kay's classic work on color terms (Berlin and Kay, 1969), one characteristic of terms for basic level concepts is that they are monolexemic, and Wierzbicka (1996) also promotes this criterion, calling it "Morphological Structure" (Wierzbicka, 1996, p. 356). It would be very interesting to understand more clearly why some kinds of geographic features are denoted by monolexemic terms and why others are dealt with by compound words and phrases, and to try to establish whether this reflects some underlying cognitive salience or environmental importance, whether it is largely linguistic or historical effect, or whether it is due to chance.

6.2   The Role of Proper Names

During discussions with our Yindjibarndi collaborators, they frequently mentioned that significant geographic features are usually referred to by their individual (proper) names, rather than by generic terms. For instance, one of the authors (AT) was present (during an earlier fieldtrip) when a Ngarluma elder listed in order the first twenty **yinda** (permanent pools) that one would encounter when traveling inland from Roebourne along the Ngurin (Harding River). Knowing the names for pools, mountains, etc is an important part of Indigenous Australian culture, often passed on by 'singing' lists of names. In these cultures, one is expected to know the limits of one's own country and the cultural significance of places, and be able to demonstrate this by knowing the proper names of its features (Ieramugadu Group Inc., 1995). Malpass (1999, p. 3) notes that this is a key component of relationship to the land for Indigenous Australians: "So important is this tie of person to place that for Aboriginal peoples the land around them everywhere is filled with marks of individual and ancestral origins and is dense with story and myth".

The authors have not yet been able to establish the full extent of use of proper names for geographic features for Yindjibarndi, although it is clearly extensive. One of our collaborators said that all permanent features (of significant size) in the landscape had names - rivers and creeks, pools, hills, rocky outcrops, flat areas, etc - but that many of the names were not recorded and may now be lost. More fieldwork is needed before the way that this influences the form of geographic terms could be reasonably inferred.

6.3   Object vs. Field Conceptualizations of Landscape

Western conceptualizations of space, and the categories and data structures of GIS which arise from them, tend to treat geographic features in the landscape as objects. However, there is at least anecdotal evidence that Indigenous Australians (including the Yindjibarndi) tend to view landscape more as a continuous field. Parallel ideas have been suggested by Atran and Medin (1997), who claimed that "Westerners make much more use of categories for purposes of inductive inference than do members of other societies," and that members of other cultures are "more likely to organize on the basis of relationships and similarities". We have not yet been able to design and implement an experimental method to adequately explore this issue with respect to Yindjibarndi landscape categories. However, if the anecdotal evidence noted above is well founded, it would be of considerable significance to the design and usability of GIS.

6.4   The Role of Spirituality

For Indigenous Australians, including the Yindjibarndi people, spirituality and topography are inseparable (e.g. all **yinda** have **warlu**) (Ieramugadu Group Inc., 1995; Rijavec and Harrison, 1992; Turnbull, 1989). The significance of this issue for agency is highlighted by Malpass (1999, p. 95): "Understanding an agent, understanding oneself, as engaged in some activity is a matter both of understanding the agent as standing in certain causal and spatial relations to objects and of grasping the agent as having certain attributes - notably certain relevant beliefs and desires - about the objects concerned". Hence, in order to fully comprehend Yindjibarndi geographic concepts, it is necessary to adopt a method of inquiry that allows this possibility. Treating the spiritual as real is in conflict with prevailing Western philosophical assumptions underlying ontological investigation. Hence, a way of resolving this conflict needs to be found, especially, if the objective is to provide information systems suited to specific users (Remenyi et al, 1997; Wilson, 1998). A pluralist approach to knowledge systems would seem necessary (Watson-Verran and Turnbull, 1995). Robin Horton's efforts to reconcile African traditional thought with Western science may provide a viable approach for integrating spirituality into a comprehensive ontology of landscapes for information system design, as may Searle's (1995) ideas for characterizing the nature of social reality within a realist ontological framework. We plan to conduct further research that could lead to an integration of cross-cultural belief systems into geographic ontology and geographic information systems.

6.5   Ethnophysiography?

The research reported here appears to open a new research topic, which might best be called *ethnophysiography*. The Oxford English Dictionary gives one meaning of physiography as "physical geography", which captures the domain we are studying very well. We also came to realize during the research that the methods employed in this study parallel the methods used in ethnosciences such as ethnobiology (Berlin, 1992; Medin and Atran, 1999). Considering the importance of landscape to culture, it would be surprising if ethnographic methods have not been used to study common-sense categories for landscape elements, yet we have been unable to find examples of such work.

7   Conclusions

The results of this study support the hypothesis that people from different places and cultures use different conceptual categories for geographic features. Hence, if GIS are to be most effective, their design needs to take account of such matters. These research findings have practical implications. For instance, if the current Ngarluma-Yindjibarndi native title land claim is at least partially successful, it may well lead to joint management arrangements between the Yindjibarndi people and the State Government for large national parks in their country (Turk, 1996; Walsh and Mitchell, 2002). If a GIS were to be used to support this management, it would probably be based on the digital version of the relevant 1:100,000 maps, which incorporate the sorts of ontological assumptions and feature codes (AUSLIG, 2002) discussed above. The results of this study indicate that such an approach might not reflect Yindjibarndi landscape concepts, and hence a more complex inter-cultural approach would need to be adopted. To do otherwise would amount to ontological imperialism, and perhaps ontological assimilation.

This paper reports only some of the initial findings of an ongoing research project. As indicated above, much more research is needed before it is possible to arrive at a reasonably comprehensive understanding of the way geographic categories are expressed in the Yindjibarndi language.

Acknowledgments

Members of the Roebourne community, especially Allery Sandy, Trevor Soloman, Marion Cheedy, Nita Fishook, and Jane Cheedy provided invaluable assistance regarding the Yindjibarndi language. We also wish to thank the Australian Institute of Aboriginal and Torres Strait Islander Studies, especially David Nash, for providing material from the Aboriginal Studies Electronic Data Archive (ASEDA). Nicholas Thieberger, Barry Smith, and Werner Kuhn also contributed to the research process. This material is part of a project "Geographic Categories: An Ontological Investigation" supported by the U. S. National Science Foundation under Grant No. BCS-9975557. Support of the National Science Foundation is gratefully acknowledged.

References

AIATSIS, 1994. A List of Electronic Data Files Held in ASEDA. Aboriginal Studies Electronic Data Archive, Australian Institute of Aboriginal and Torres Strait Islander Studies, GPO Box 553, Canberra, ACT 2601, Australia.

Anderson, B., 1986.*Yindjibarndi dictionary*. Photocopy.

Anderson, B., and Thieberger, N. No date. *Yindjibarndi dictionary*. Document 0297 of the Aboriginal Studies Electronic Data Archive (ASEDA) Australian Institute of Aboriginal and Torres Strait Islander Studies, GPO Box 553, Canberra, ACT 2601, Australia.

Atran, S., and Medin, D., 1997. Knowledge and action: Cultural models of nature and resource management in Mesoamerica. In M. Bazerman et al. (ed.) *Environment, ethics, and behavior*. San Francisco: New Lexington Press.

AUSLIG, 2002. *Feature Codes used by the Gazetteer of Australia*. http://www.auslig.gov.au/mapping/names/featurecodes.htm (Downloaded December 2002).

Berlin, B., 1992. *Ethnobiological classification. Principles of categorization of plants and animals in traditional societies*. Princeton University Press, New Jersey.

Berlin, B., and Kay, P, 1969. *Basic Color Terms: Their Universality and Evolution*. Berkeley: University of California Press.

Von Brandenstein, C. G., 1970. *Narratives from the north-west of Western Australia in the Ngarluma and Jindjiparndi languages.* Canberra, AIAS 1970.

Von Brandenstein, C. G., 1992. *Wordlist from Narratives from the north-west of Western Australia in the Ngarluma and Jindjiparndi languages.* Canberra, ASEDA Document #0428.

Guarino N., and Giaretta P., 1995. Ontologies and Knowledge Bases: Towards a Term-terminological Clarification. In N. J. I. Mars (ed.), *Towards Very Large Knowledge Bases,* Amsterdam: IOS Press, pp. 25-32.

Ieramugadu Group Inc., 1995. *Know the Song, Know the Country: The Ngarda-Ngali story of culture and history in the Roebourne District.* Roebourne, Western Australia: Ieramugadu Group Inc.

Lakoff, G., 1987, *Women, Fire, and Dangerous Things: What Categories Reveal About the Mind.* Chicago: University of Chicago Press.

Lowe, P., 1990. *Jilji, Life in the Great Sandy Desert.* Broome, Western Australia: Magabala Books.

Malpass, J. E., 1999. *Place and Experience: A Philosophical Topography.* Cambridge University Press.

Mark, D. M., 1993a. A Theoretical Framework for Extending the Set of Geographic Entity Types in the U.S. Spatial Data Transfer Standard (SDTS). *Proceedings, GIS/LIS'93,* Minneapolis, November 1993, v. 2, pp. 475-483.

Mark, D. M., 1993b. Toward a Theoretical Framework for Geographic Entity Types. In Frank, A. U., and Campari, I., editors, *Spatial Information Theory: A Theoretical Basis for GIS.* Berlin: Springer-Verlag, Lecture Notes in Computer Sciences No. 716, pp. 270-283.

Mark, D. M., Skupin, A., and Smith, B., 2001. Features, Objects, and other Things: Ontological Distinctions in the Geographic Domain. In Montello, D., (ed.), *Spatial Information Theory: Foundations of Geographic Information Science,* Berlin: Springer-Verlag, Lecture Notes in Computer Science No. 2205, pp. 489-502.

Mark, D. M., and Smith, B., 1999. *Geographic Categories: An Ontological Investigation.* NSF Award BCS 9975557.

Mark, D. M., Smith, B., and Tversky, B., 1999. Ontology and Geographic Objects: An Empirical Study of Cognitive Categorization. In Freksa, C., and Mark, D. M., Editors, *Spatial Information Theory: A Theoretical Basis for GIS,* Berlin: Springer-Verlag, Lecture Notes in Computer Science No. 1661, pp. 283-298.

Medin, D.L., and Atran, S., eds. 1999. *Folkbiology.* Cambridge, MA: MIT Press

Pullum, G. K., 1991. The great Eskimo vocabulary hoax. In Pullum, G. K., (ed.), *The Great Eskimo Vocabulary Hoax and Other Irreverent Essays on the Study of Language,* pp. 159-171.

Rijavec, F., Harrison, N., and Soloman, R., 1995. *Exile and the Kingdom* [documentary film]. Roebourne, Western Australia: Ieramugadu Group Inc. and Film Australia.

Remenyi, D., Sherwood-Smith, M. and White, T. (1997) *Achieving Maximum Value From Information Systems: A Process Approach.* Chapter 2. John Wiley and Sons, New York.

Rosch, E., 1973a. Natural categories. *Cognitive Psychology* 4, 328-350.

Rosch, E., 1973b. On the internal structure of perceptual and semantic categories. In T. E. Moore (ed.), *Cognitive Development and the Acquisition of Language,* New York, Academic Press.

Rosch, E., 1978. Principles of categorization. In E. Rosch and B. B. Lloyd (eds.) *Cognition and Categorization.* Hillsdale, NJ: Erlbaum.

Searle, J. R., 1995. *The Construction of Social Reality.* New York: The Free Press.

SIL, 2001. *Ethnologue: Languages of the World,* 14th Edition.
http://www.ethnologue.com/show_lang_family.asp?code=YIJ

Smith, B., and Mark, D. M., 1998. Ontology and Geographic Kinds. in T. K. Poiker and N. Chrisman (eds.), *Proceedings. 8th International Symposium on Spatial Data Handling (SDH'98)*, Vancouver: International Geographical Union, 1998, 308–320.

Smith, B., and Mark, D. M., 2001. Geographic categories: An ontological investigation. *International Journal of Geographical Information Science*, 15 (7), 591-612.

Smith, B., and Mark, D. M., 2003. Do Mountains Exist? Towards an Ontology of Landforms. *Environment and Planning, B*, accepted, in press.

Smith, E. E., and Medin, D. L., 1981. *Categories and Concepts*. Cambridge, Massachusetts: Harvard University Press.

Tindale, N. B., 1974. *Aboriginal Tribes of Australia. Their Terrain, Environmental Controls, Distribution, Limits, and Proper Names*. Canberra: Australian National University Press.

Turk, A. G., 1996. Presenting Aboriginal knowledge: Using technology to progress native title claims. *Alternative Law Journal*, Vol. 21, No. 1, 6-9.

Turk, A. G., 2000. Tribal Boundaries of Australian Indigenous Peoples. Paper presented at *Geographical Domain and Geographical Information Systems: EuroConference on Ontology and Epistemology for Spatial Data Standards*. La Londe-les-Maures, France, 22-27 September 2000, unpublished manuscript.

Turk, A. G., 2002. A Critique of Government Grant Based Approaches to Addressing Digital Divide Issues in an Australian Indigenous Community. *Proceedings of the International Conference: The Digital Divide: Technology and Politics in the Information Age - 22-24 August, 2002 - Hong Kong Baptist University*.

Turk, A. G. and Trees, K. A., 1998. Ethical Issues Concerning the Development of an Indigenous Cultural Heritage Information System. *Systemist* Volume 20, Special Issue, 229-242.

Turk, A. G. and Trees, K. A., 1999. Culturally Appropriate Computer Mediated Communication: An Australian Indigenous Information System Case Study. *AI and Society*, 13, 377-388.

Turk, A. G. and Trees, K. A., 2000. Facilitating Community Processes Through Culturally Appropriate Informatics: An Australian Indigenous Community Information System Case Study. In: Gurstein, M. (ed.) *Community Informatics: Enabling Communities with Information and Communication Technologies*, Idea Group Publishing, 339-358.

Turnbull, D., 1989. *Maps are Territories: Science is an Atlas*. Deakin University Press, Geelong, Australia

Walsh, F. and Mitchell, P. (eds.), 2002. *Planning for Country: Cross-cultural Approaches to Decision-making on Aboriginal Lands*. Central Land Council, Alice Springs, Australia: Jukurrpa Books.

Watson-Verran, H. and Turnbull, D., 1995. Science and Other Indigenous Knowledge Systems. In: Jasanoff, S., Markle, G.E., Petersen, J.C. and Pinch, T. (eds.) *Handbook of Science and Technology Studies*. Sage Publications. Pp. 115-139.

Wierzbicka, A., 1996. *Semantics - Primes and Universals*. Oxford, England: Oxford University Press.

Wilson, D., 1998. Ontological Pluralism and Information Systems Research. In: *Proceedings of PAIS II, the Second Symposium and Workshop on Philosophical Aspects of Information Systems: Methodology, Theory, Practice and Critique*, University of the West of England, Bristol, UK, July 27-29, 1998.

Winter, S., 2001. Ontology: buzzword or paradigm shift in GI science? *International Journal of Geographical Information Science*, 15 (7), 587-590.

Wordick, F. J. F., 1982. *The Yindjibarndi language*. Pacific linguistics. Series C., no. 71. Canberra: Dept. of Linguistics, Research School of Pacific Studies, Australian University, 1982.

Layers: A New Approach to Locating Objects in Space

Maureen Donnelly[1] and Barry Smith[1,2]

[1]Institute for Formal Ontology and Medical Information Science, University of Leipzig
{maureen.donnelly,bsmith}@ifomis.uni-leipzig.de
[2]Department of Philosophy, University at Buffalo, NY

Abstract. Standard theories in mereotopology focus on relations of parthood and connection among spatial or spatio-temporal regions. Objects or processes which might be located in such regions are not normally directly treated in such theories. At best, they are simulated via appeal to distributions of attributes across the regions occupied or by functions from times to regions. The present paper offers a richer framework, in which it is possible to represent directly the relations between entities of various types at different levels, including both objects and the regions they occupy. What results is a layered mereotopology, a theory which can handle multiple layers (analogous to the layers of a lasagna) of spatially or spatiotemporally coincident but mereologically non-overlapping entities.

Keywords: Ontology, mereology, mereotopology, qualitative spatial reasoning, map layers, dynamic GIS

1 The Problem

Suppose you are a conservation biologist whose job it is to keep track of vulnerable species in a large national park. You develop a system of *map layers* representing not only the topography of the park region, the soil-types and vegetation zones, the system of roads and other man-made geographic objects, but also the most important habitats for each species. Your GISystem then needs to be able to use the results of your layer-making efforts in order to answer questions like: *Which of the four most vulnerable species live in riparian zones? Do Chiricahua leopard frogs live in the same vegetation zones as shovel-nosed snakes? How do climate changes in the elevated regions of the park affect long-term patterns of declining populations of scimitar-horned oryx?* And so forth.

Standard methods for dealing with map overlays in GIS were developed in the 1970s and have been incorporated into commercial GISystems software for more than two decades. Such systems can comfortably represent objects with crisp boundaries by describing the positions of their boundaries, and they can handle multiple objects which share the same footprint wherever the objects in question occupy the same regions of space. But such approaches are static in nature, and they have no direct facility for dealing with moving objects in space. With the increasing availability of temporally indexed geospatial information comes the ever more pressing need on the part of GISystems to solve the problems which arise when we need to deal with those complex interactions among different sorts of dynamic phenomena which arise in

W. Kuhn, M.F. Worboys, and S. Timpf (Eds.): COSIT 2003, LNCS 2825, pp. 46–60, 2003.
© Springer-Verlag Berlin Heidelberg 2003

areas such as conservation biology, or predator tracking, or in the building of early warning systems for chemical hazards. For such purposes, however, we need a framework within which we can distinguish objects from the regions they occupy at successive points in time.

2   Region-Based Approaches

In the literature on spatial reasoning one dominant approach is in terms of region-based calculi of the sort that have been developed by Cohn and his associates. Such calculi have found favor among some GIScientists, not least because they correspond well not only to the field-based representations of geospatial phenomena used in scientific geography but also to our understanding of maps and map layers as consisting, in effect, of regions coloured differently according to the attributes which they instantiate. (Casati, Smith and Varzi 1998)

Region-based approaches are of interest to us here because they have served of late as the basis for serious attempts to create frameworks for reasoning about moving objects. The simplest such approach treats moving objects indirectly, in terms of the regions they occupy at successive intervals of time. Not only space itself is understood in terms of regions, but all the different kinds of entities located in space. We have not: *the talus snail spawns in the area of needle-spined pineapple cactus* or: *the tumamoc globeberry grows in the habitat of the Mexican long-tongued bat*, but rather, in each case: *(connected or scattered) region A is included as subregion within (connected or scattered) region B.*

How this region-based framework leads to problems becomes clear when we need to formulate a qualitative theory of motion. If we are to be able even to attempt to characterize movement, something more, for example temporal indexing, must be added to the mereotopology. But even then, regions-plus-attributes representations of organisms in their habitats must necessarily obscure what is involved when an enduring object is registered at different spatial locations in successive instants of time. For an adequate account of such registration data requires at least two independent sorts of spatial entities: one, the locations, which remain fixed, and the other, the objects, which move relative to them. Since the region-based approach admits only the first type of entity – the locations or regions – it must somehow simulate motion, for example via successive assignments of attributes to a fixed frame of locations.

That *a rufous-winged sparrow moves from one location (region A) to another (region B)* must then be cashed out, non-intuitively, as: *each member of this continuous sequence of sparrow-shaped regions, starting with region A and ending with region B, has at successive times, rufous-winged (etc.) attributes.* That is, instead of talking about sparrows flying about in the sky, we talk rather of mappings of the form: Sparrow$_{152}$: time $\rightarrow$ regular closed subsets of $\mathbf{R}^3$.

Notice that, besides failing to match our intuitions about what it is for birds to move through space and to inhabit different locations at different times, this picture also does less than adequate justice to regions themselves. For it leaves unexplained what it is for attributes to be correctly assigned to given region at given times. Intuitively, of course, this turns on the fact that there are corresponding *objects* which occupy those regions at the times in question. Appeal to this separate layer of objects is however precisely what is ruled out by the region-based approach, which, for reasons of mathematico-logical simplicity, attempts to explain what is cognitively more salient (objects) in terms of what is cognitively less salient (regions).

One side-effect of such reductionist treatments of physical objects and of their independent spatial properties – as well as of other spatial entities such as epidemics, wildfires, hurricanes and the like – is that we are unable to distinguish cases of true mereological overlap (i.e. the sharing of parts) from mere spatial co-location. We cannot, for example, distinguish the relation of a fish to the lake it inhabits (but is not a part of) from the relation of a genuine part of a lake (a bay, an inlet) to the lake as a whole. Both are represented in the regions-plus-attributes picture as the inclusion of a smaller in a larger region. Similarly, we cannot distinguish between genuine parts of the human body such as the heart or lungs, and foreign occupants such as parasites or shrapnel. This weakness in the region-based account is clear in the representation of phagocytosis and exocytosis offered in (Cui et al., 1992), which offers no means of distinguishing between the relation of an amoeba to a portion of food which it has recently ingested and the relation of the amoeba to a genuine part, such as its nucleus.

A slightly different type of mereotopology-based analysis of motion is given in (Cohn and Hazarika, 2001). Here, the domain of the mereotopology is again restricted to regions, but this time to regions of a four-dimensional, spatial-temporal sort. This sort of account incurs the same sorts of problems as the three-dimensional region-based approach. Again, the relation between regions and the material entities (here: four-dimensional *processes*) that are supposed to inhabit them is left unexplained. Also, as in the three-dimensional region-based approach, this theory does not allow us to distinguish true overlap from spatio-temporal coincidence. Moreover, this approach is marked by an additional counter-intuitive feature in that we are forced to reduce all spatial objects to four-dimensional processes. Individuals, such as you and me, are thereby identified with their *lives* or with some totality of *histories* in which they are involved.

Bennett (2001) introduces a more sophisticated analysis, in which individuals are interpreted as distributions of matter with which count nouns are associated in a way which yields an account of motion of individuals in terms of the continuous changes in distributions of matter. Bennett thus distinguishes individuals from their spatial extensions; but he still defines individuals in terms of phenomena – distributions of matter – which are cognitively less salient than individuals themselves. Moreover, he does not take the further step of developing a framework in which entities on a plurality of levels might be recognized as coinciding spatially.

The vector methods commonly used in GIS for the manipulation of data about objects solve some of the problems at issue. From the vector point of view objects are identified with sets of points referenced to a spatial coordinate system. Moving objects are identified with sequences of sets of points indexed by times. Unfortunately however the vector approach involves a highly unrealistic understanding of what objects are. Moreover, embracing this approach means giving up many of the benefits for our understanding of spatial reasoning and of the structures of the spatial continuum which have accrued in recent years as a result of applications of mereotopology. In what follows, therefore, we seek a new sort of framework, building on mereotopology by adding a new conception of objects and their locations in space that is encapsulated in the notion of layer.

3   Mereotopology

The fruitfulness of mereotopology rests precisely on the fact that it can admit *extended* individuals such as regions, material objects, chunks of stuff, or spatio-temporally extended processes. In this way it can yield a more direct and realistic representation of the qualitative space of common sense than is available under standard reconstructions of the spatial continuum in terms of sets of points or vectors.

A mereotopology is a formal theory of parthood and connection relations. Several different mereotopologies have been proposed, including not only those of Cohn *et al.*, but also those of Asher and Vieu, Smith, and others. These theories are, it is clear, intended to be used for reasoning about spatial relations among material objects. When they are examined more closely, however, it becomes evident that – in keeping with what was said above – they assume that their immediate domains of application will be restricted to regions. The axioms formulated in (Smith 1996) are, it is true, neutral as between objects and regions; but even there no resources are provided for giving an account of the distinction between objects and the regions in which they are located. Because distinct *location* relations are not introduced into these mereotopologies, coincidence of spatial location collapses onto overlap. Where, as in (Cohn 2001), material objects are explicitly introduced, the mereotopological relations are still restricted to associated regions. Each object's spatial properties are determined by those of the region at which it is at any given time located.

Terminological clarity is important here. Note, first, that we are using 'overlap' to mean what some might prefer to call 'partial overlap'. That is, two entities will be said to overlap (share parts in common) even when they are identical. We shall say that two entities *coincide* when they occupy overlapping regions of space. Coincidence will then in general fall far short of overlap. We shall say that objects *coincide completely* with the spatial regions which, at any given instant of time, they occupy. The relation of coincidence holds not only between objects and their regions; it holds also between objects themselves. This first of all in a trivial sense: my hand and my arm (partially) coincide, and so also do the British Commonwealth and the European Union. In fact we have to deal in cases such as this with objects which do not merely coincide but also overlap (i.e. have parts in common). My hand is a part of both itself and my arm. Cyprus and Malta are both parts of the British Commonwealth and (will shortly be) parts of the European Union. If all entities (and thus all parts of entities) are spatial, then any two mereologically overlapping entities are, trivially, coincident in our sense. That is: their locations are identical at their intersections.

The relation of coincidence is however strictly broader than that of overlap. For there are pairs of coincident objects that do not share parts. The food that I am currently digesting coincides with, but does not overlap, my stomach cavity. The brain is located in the cranial cavity; but it is not a part of the cranial cavity. (Schulz and Hahn, 2001) The Great Plague of 1664 coincides with, but does not overlap, Holland.

The goal of this paper is to sketch a mereotopological framework that has the resources to deal with all types of coincident but non-overlapping entities, including not only material objects and their regions but also other types of entities such as *qualities*, *processes* and *holes*.

Qualities may coincide with material objects in the way in which, for example, the individual redness of a cube of coloured glass coincides with, but does not overlap, the glass itself. Qualities may also coincide with each other: the qualities of temperature and pressure of a given mass of air will coincide in this sense, and both will coincide with the mass of air itself.

Processes, too, may coincide in similar ways with the spatiotemporal regions they occupy. A process of deforestation may coincide with, but does not overlap, a specific geopolitical region over a given time. Processes may also coincide with each other, as when a process of absorption of a drug in a patient's body coincides with, but does not share parts with, the disease processes which the drug is designed to alleviate.

Holes may coincide with material objects in the way in which, for example, the chamber of a revolver coincides completely with, but does not overlap, the bullet which fills it (Casati and Varzi, 1994). If the revolver is moved around inside a moving railway carriage, then we can distinguish two levels of holes which are, in each temporal instant, coincident with each other but yet moving relative to each other and also relative to the territory through which the train is moving. The theory here presented allows reasoning about such multi-layered structures of coincident entities as they arise in domains such as mechanics (valves, pathways formed by piping) and medicine (body cavities and orifices). Here, however, we are concerned with layered structures in the domain of the geographic sciences.

4   Examples of Layers

Example 1
Suppose that you wish to represent the relations holding between a lake, the water in the lake, the fish swimming in the lake, and the mercury in the fish tissue. You might distinguish here four coincident three-dimensional layers:

   L1. a *region* layer, consisting of a regular spatial volume including in its interior the spatial region occupied by the lake,

   L2. a *lake* layer, consisting of a certain concave portion of the earth's surface together with a body of water,

   L3. a *fish* layer, consisting of a certain aggregate of fish,

   L4. a *mercury* (or *chemical contaminant*) layer, consisting of tiny deposits of organic mercury scattered through the lake and through the tissue of the fish.

L1 serves as underlying reference-system for the edifice of layers taken as a whole. The mapping which takes every object to its spatial region is a projection of the whole domain onto the region layer L1.

Objects from the separate layers never overlap. Clearly, however, the corresponding *regions* may overlap or be parts of one another. For example, the spatial region occupied by the mercury in the fish is properly included within the spatial region occupied by the fish, which is in turn properly included within the spatial region occupied by the whole lake.

Summation may occur within layers. The mereological sum of any aggregate of fish in the lake is itself an aggregate of fish in the lake and thus included in layer L3. But there are no mixed sums, i.e. entities which have members of different layers as their parts. Thus there is no mereological sum of the fish, the mercury, and the lake. In tackling the relations between these different entities we must rather find ways to do justice to the fact that they exist on different levels.

Example 2
Suppose that you wish to represent the relations holding between topographical features, weather phenomena such as storms and winds, and the interactions of these with each other and with wind-borne nuclear, chemical or biological agents which have been released into the atmosphere. Here again you might distinguish four layers, as follows:

L1. a region layer, consisting of a certain collection of spatial regions defined in relation to the relevant part of the surface of the earth,

L2. a topographical layer, consisting of mountains, valleys, deserts, gullies, and other geographic features,

L3. a storm-system and all its parts occupying sub-regions of this layer,

L4. an airborne cloud of smallpox virus particles.

Here again the different layers are to be understood as mereologically exclusive. That is, while there can be (physical, chemical, biological) relations between the entities on neighboring layers, we stipulate that such entities never stand to each other in relations of part to whole.

Example 3
Suppose that you wish to represent the political systems of the United States, including the political systems of the constituent towns, counties and states, as well as the overarching system of the federal government, including also the governments of the other US territories (Guam, Puerto Rico, Palua, etc.). You might distinguish here five layers:

L1. a *region* layer, defined in the obvious way,

L2. a *local government* layer, consisting of collections of administrative geospatial units dealing with water, sewers, sanitation services, schools, etc.,

L3. a *county government* layer, consisting of collections of administrative geospatial units dealing with property assessment and record keeping, administration of elections, consumer protection, planning and zoning, etc.,

L4. a *state government* layer, consisting of collections of administrative geospatial units dealing with judicial functions, welfare provision, public higher education, etc.,

L5. a *federal government* layer, consisting of collections of administrative geospatial units dealing with foreign relations, defense and the other functions specified in the Constitution.

Here again L1 serves as supporting framework for the edifice as a whole, and if we examine the results of projecting onto L1 the individual units in each successive layer then we see that they are organized in a system of nested honeycomb structures, with the largest cell in the structure comprehending the totality of the relevant territory, with successively finer spatial divisions as we move down through states and counties to individual towns. Some higher layers will when taken as a whole be spatially perfectly coincident with the region layer L1. Others will fall short of perfect coincidence (there is no layer of functioning county governments, for example, in Connecticut and Rhode Island).

When we construct an administrative layered hierarchy of the given sort – a hierarchy of fiat objects (Smith 2001) – then we make it true by stipulation that the objects from the separate layers do not overlap. We declare at the same time that sumformation is admissible (or admissible where certain conditions are satisfied) within any given layer. Thus for example when two county government bodies in L3 collaborate, e.g. in dealing with mercury in a lake, then the resultant joint administrative unit belongs also to this layer.

Example 4
Imagine that we wish to compare distinct modes of segmentation of topographic reality in a given area, for example the distinct segmentations of landforms effected by European and non-European peoples in rural Australia. Here (following Mark and Turk 2003) we might distinguish:

L1. a region layer, consisting of the middle part of the valley of the Fortescue River in northwestern Australia and of its adjacent uplands,

L2. a layer consisting of the segmentations effected by the Yindjibarndi people, employing terms such as *marnda, bargu, burbaa, bantha, gankala, thalungarn* and the various combinations thereof,

L3. a layer consisting of the segmentations effected by English speakers, employing terms such as *mound, hill, mountain, ridge, saddle, spur*, etc.

That what English-speakers call 'Pyramid Hill' corresponds to what the Yindjibarndi people call 'Googarana' signifies, in the terms of layered mereology, that the corresponding segments in L2 and L3 are centered on the peak at 21° 10' S latitude; 117° 22' E longitude in L1.

Example 5
Examples 1–4 are synchronic; they represent snapshots of reality at given times; the relative positions of the entities on certain layers, especially in the first two examples, may however vary greatly from one moment (snapshot) to the next. Layer ontologies can however be developed also on the basis of spatiotemporal region layers, where we talk not of change but rather of an entity's having different qualities in different temporal parts (in just the way that they might have different qualities in different spatial parts). An example might be:

L1. a four-dimensional region layer, defined as the product of a spatial component: the territory of Iraq, and a temporal component (shared by all the other layers): the interval between March 19 and April 14 2003,

L2. a four-dimensional topographical layer, whose spatial component is defined as the sum of all topographical features (mountains, valleys, rivers, deserts, ...) within the locus of L1,

L3. a four-dimensional geographical artifact layer, whose spatial component is defined as the sum of all geographical-scale artifacts (cities, oilwells, mine-fields, dams, ...) within the locus of L1,

L4. a process layer, consisting of all movements of troops, materiel, and civilians within the locus of L1.

5   General Extensional Mereology

The formal theory presented here is taken from (Donnelly 2003), which also includes a semantics for the theory in terms of structures called Layered Models. The theory is formulated in standard first-order logic. It takes its starting point from mereology, the formal theory of the binary parthood relation P, and uses the following additional mereological relations, which are defined in terms of P:

D1. $PPxy =: Pxy \ \& \sim Pyx$ (x is a proper part of y)
D2. $Oxy =: \exists z(Pzx \ \& \ Pzy)$ (x and y overlap)

For the sake of simplicity we take P to be axiomatized via the familiar axioms of General Extensional Mereology (GEM) (Simons 1987), but with two small modifications, to be addressed below. The adoption of GEM axioms means that the relation of parthood is a partial ordering: it is reflexive (each entity is a part of itself), anti-symmetric (if two entities are part of each other then they are identical), and transitive. In addition it satisfies a remainder axiom, which states that, if x is not a part of y, then there is some part, z, of x that does not overlap y. Finally GEM employs an unrestricted comprehension or summation axiom (or better: axiom schema), which is formulated as follows:

$$\exists x\phi \rightarrow \exists z \forall w \ (Owz \leftrightarrow \exists x \ (\phi \ \& \ Owx))$$

Here ϕ stands for any first-order formula in which z does not occur free. The axiom schema states that, if any member of the domain satisfies the formula ϕ, then there must be in the domain a sum of all objects satisfying ϕ. More simply put: if any entity satisfies ϕ, then there is a sum of all entities satisfying ϕ. This unrestricted comprehension principle represents for many application purposes an unacceptable simplification. Modified versions of the comprehension principle can be formulated, for example by restricting admissible sums to those consisting of topologically connected parts, and corresponding modified versions of layered mereology can then be constructed to fit.

Axioms and theorems for the base mereology will be labeled in what follows with 'P' for 'parthood'. (Here, as throughout this paper, initial universal quantifiers are suppressed unless confusion will otherwise result).

P1. Pxx
P2. $Pxy \ \& \ Pyx \rightarrow x = y$
P3. $Pxy \ \& \ Pyz \rightarrow Pxz$
P4. $\sim Pxy \rightarrow \exists z(Pzx \ \& \sim Ozy)$

The first three axioms require that P is a partial ordering. P4. is the remainder principle.

It follows from P1.–4. that overlap is extensional. This means that any two members of the domain that overlap the same entities are identical:

P-T1. $x = y \leftrightarrow \forall z(Oxz \leftrightarrow Oyz)$

Because O is extensional, for any formula ϕ in which z does not occur free, if we can assign z to a member of the domain that satisfies

$$\forall w\,(Owz \leftrightarrow \exists x\,(\phi\ \&\ Owx)),$$

then this object is unique. Thus, the comprehesion axiom of GEM tells us that, if any member of the domain satisfies ϕ, then there is a *unique* sum of all ϕ-ers.

6   Adding Layers

The summation axiom schema included in Layered Mereology must be weaker than GEM's comprehension schema, since we do not allow sums to cross layers. For example, let ϕ be the formula: $x = x$, then entities on different layers would satisfy ϕ, so there can in this case be no sum of all the ϕ-ers. What we need is a restricted version of the comprehension schema that requires sums to exist only if all summands are part of the same layer. To this end we introduce a new notion, that of *underlapping*, defined as follows:

D5. $Uxy =: \exists z\,(Pxz\ \&\ Pyz)$ (x and y underlap)

Two entities underlap whenever there exists some whole in which both are included as parts.

Our restricted version of GEM's comprehension schema then reads:

P5. $(\exists x\phi\ \&\ \forall xy(\phi\ \&\ \phi/y \rightarrow Uxy)) \rightarrow \exists z\forall w\,(Owz \leftrightarrow \exists x\,(\phi\ \&\ Owx))$

where ϕ/y is the formula which results from ϕ when all free instances of x are replaced by instances of y (and using variable substitution as necessary so that y is free in ϕ/y where x is free in ϕ).

P5. says that if there is some object that satisfies ϕ, and any two entities that satisfy ϕ underlap, then there is a sum of all entities that satisfy ϕ.

So far, we can say only that two entities share the same layer. We would like to be able to say that a certain entity *is a layer* or *is the layer of a particular object*. To this end we first of all define the relation that holds between z and y when z is y's layer, as follows:

D6. $Lyz =: \forall w(Owz \leftrightarrow \exists x\,(Uxy\ \&\ Owx))$ (z is y's layer)

D6 tells us that z is y's layer if and only if z is the unique sum of all entities underlapping y.

Does it hold that every entity has some layer to which it belongs? This would follow from P5., but only if we knew that any two entities that underlap y must underlap each other (i.e. that U is transitive). The transitivity of U is not, however, derivable from P1.–P5., and so one final axiom needs to be added to layered mereology:

P6. $(Uxy\ \&\ Uyz) \rightarrow Uxz$ (underlap is transitive)

Thus if x and y are each parts of a single whole, and if the same holds for y and z, then there is some whole of which x and z are parts. From this it follows that every object has a layer, and that every object is part of its layer. We can prove further that two entities underlap if and only if they have the same layer, and we can introduce the unary predicate, L, which distinguishes those special entities called layers:

D7. $Lz =: \exists x\,Lxz$ (z is a layer)

It is easily proved that an entity is a layer iff it is its own layer. In other words, the layers are just those members of the domain which are the mereological sums of all entities they underlap.

7   The Region Layer

Our framework thus far has been that of General Extensional Mereology with just two modifications: (i) we have restricted the comprehension principle by allowing sums to exist only if their constiuents underlap; (ii) we have added an axiom to the effect that underlap is transitive. Since both layers and the relation of underlap are defined in terms of the usual mereological predicates, no new mereological primitives are needed.

For some applications, however, we may want a theory that is weaker than layered mereology in that the separate layers need not satisfy the full comprehension schema of GEM. In other words, for some applications we may not want to require arbitrary collections of entities from the same layer to have a mereological sum. Consider a modification of *Example 2*, above, which allows on the topographical layer only contiguous features to form wholes (including that maximal contiguous whole which is the given layer in its entirety). In cases such as this we can use a subtheory of layered mereology called *weak layered mereology*, in which axiom schmema P5 is replaced by the following axiom, which merely requires that every entity has a layer:

P5′. $\forall x \exists z\, Lxz$ (every entity has a layer)

P5′ tells us that for any x, there is a mereological sum of *all* entities underlapping x. But there need not be mereological sums corresponding to smaller collections of entities on x's layer.

In what follows we shall presuppose the full resources of layered mereology, showing how these resources can be extended to account for the special role of the region layer and also for topological relations. It is then a simple matter to formulate analogous extensions within the framework of weak layered mereology.

We move beyond the strictly mereological framework first of all by adding the facility to specify the spatial locations of entities in a way that allows us to formulate relations between entities on different layers. We introduce a unary function r which assigns each object x to the region, r(x), at which x is exactly located. (Compare Casati and Varzi 1999.)

Using r, we can define a one-place predicate R that distinguishes the subdomain of regions:

D11. $Ry =: \exists x(\mathrm{r}(x) = y)$ (y is a region)

Appropriate axioms can now be appended to the axioms P1.–6. of layered mereology, as follows:

R1. $Ry\ \&\ Rz \rightarrow Uyz$ (all regions underlap)
R2. $Ry\ \&\ Uyz \rightarrow \mathrm{r}(z) = z$ (every member of the region
                                                  layer is its own region)

We can then prove theorems for example to the effect that arbitrary sums of regions are themselves regions.

The next axiom relates the region-function to parthood:

R3. $Pxy \rightarrow Pr(x)r(y)$ (if x is part of y, then x's region
                                                          is part of y's region)

From this we can infer that the mereological relations among objects on a given layer are mirrored in the mereological relations among their regions. Clearly, however, mereological relations among regions are not of necessity inherited by corresponding objects located at those regions – not every region is occupied by an object, and objects which do occupy regions may lie on distinct layers. We can, however, add the axiom:

R4. $Uxy \& O(r(x), r(y)) \rightarrow Oxy$

which states that objects on the same layer which are located at overlapping regions themselves overlap.

We can finally define the relation of complete coincidence, which holds between two entities – for example a lake and the interior of its exact container – that occupy exactly the same region of space. (One might then explore the question of whether it is possible for layered mereotopology to make sense of philosophically problematic candidate examples of exact coincidence such as you and your body, *Hamburg Stadt* and *Hamburg Land*, or Michelangelo's *David* and a certain lump of stone.)

Given axioms R1.–4. we can prove theorems to the effect that any object can completely coincide with at most one object in any given layer and that if two objects completely coincide and are on the same layer, then they are identical.

The relation of complete coincidence is reflexive, symmetric and transitive, which means that it partitions the domain of objects into equivalence classes of co-located entities. At the same time the equivalence relation of underlap partitions the domain, orthogonally, into layers, yielding a simple structure as in Figure 1.

8   Layered Mereotopology

As is shown in (Donnelly, 2003), we can extend the theory of layered mereology by adding a binary relation of connectedness in the standard way, but with the addition of an axiom which restricts connection to a single layer.

Additional axioms are needed to tie the connection relation to the region function, namely axioms to the effect that connected objects occupy connected regions and objects on the same layer with connected regions are themselves connected objects.

In *Example 3* above, New York City cannot be connected to the state of New Jersey, because New York City belongs to the city government layer, while New Jersey belongs to the state government layer. They do however stand in the weaker relation of *abutment*, which holds among objects whenever their spatial regions are connected and do not overlap.

An entity is self-connected whenever all ways of dividing the entity into two parts yield parts which are themselves connected. For some purposes it will be useful to impose on the region layer the condition that it be self-connected. The idea here is that, in contrast to what is the case regarding layers housing other types of objects (for example the fish layer in *Example 1*), the region layer is characteristically a self-connected whole that extends beyond the scattered objects occupying other layers. We can impose further conditions on the region layer for other application purposes. Thus we may want to insist that the region layer be fixed across time. In the realm of cadastral ontology we may want to demand that region layers be not only self-connected but also subject to a Euclidean metric, so that they can serve as the coordinate-systems in relation to which parcels can be defined. To do this, we would need to extend layered mereotopology to include relations strong enough to specify the geometric properties of the region layer. (See Bennett 2001a)

co-located entities

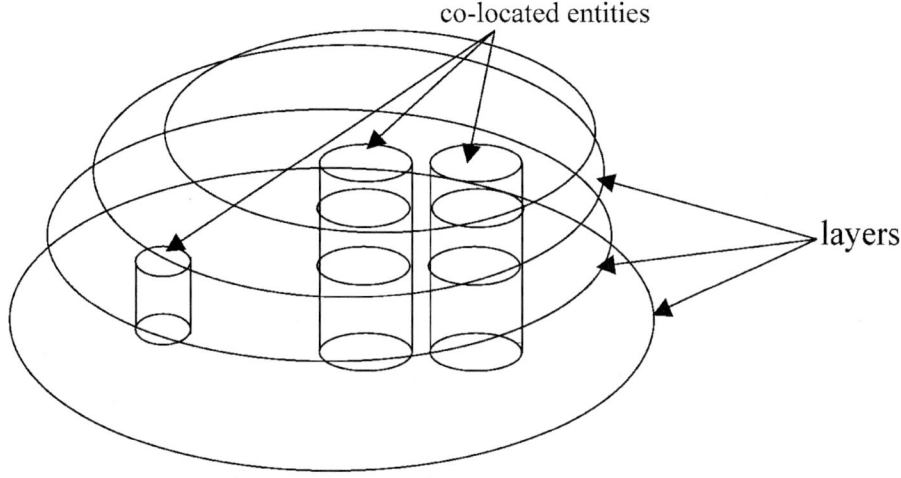

layers

Fig. 1.

Allowing direct descriptions of the spatial properties of material objects yields more generally the possibility of attributing different spatial structures to material objects and to the regions at which they are located at different times. For example, we may wish to represent material objects as having only closed, regular, divisible parts, but represent spatial region as sums of points; or we may wish to represent the fact that geospatial parcels and regions may possess *fiat* boundaries while the geographical features and artifacts located in them are marked by bona fide boundaries of a physical sort (Smith 1995, Smith and Varzi 2000, Bittner 2001).

9   Time and Change

There are two mutually complementary ways of introducing the factors of time and tense into layered mereotopology. The first is associated with what we shall call a 'SNAP' (for: snapshot) perspective on reality (Grenon and Smith 2003, Smith and Grenon 2003). This is designed to give us the facility to represent a series of inventories of what exists on a given system of layers at successive times. The conservation biologist takes samples at regular intervals of the populations of different species in given regions of the area for which he is responsible. At the same time he samples air and water quality, records meteorological data, and so forth. The machinery of the SNAP layered ontology then facilitates synchronic reasoning about those relations between populations and topographical and other features on different layers which are of ecological significance. In addition it supports the tracking of specific individuals and groups of organisms over time. This is because the SNAP ontological perspective recognizes *continuant entities* – which means objects together with their qualities, functions, roles, and powers –which endure identically through time even while undergoing certain sorts of changes.

On the SNAP perspective not only the region function but also all mereotopological and location relations are indexed by times. This allows us to represent in a natural way the results of our samplings of the positions of moving objects or populations over time. Movement is understood as change in an object's position within a fixed location-structure from one time to the next. Hence:

rufous-winged sparrow #152 moves from region A to region B

can be represented quite naturally as:

(1) $r(\text{Sparrow}_{152}, t_1) = A \ \& \ r(\text{Sparrow}_{152}, t_2) = B.$

We would like, however, to be able to draw on axioms which allow us to move from statements such as (1) to statements concerning the location of the sparrow during the intervening period, for example a statement to the effect that during the interval (t_1, t_2) there occurred a continuous process of spatial dislocation of Sparrow_{152}. To this end we need to appeal to something like the resources of the spatiotemporal process ontology ('SPAN'). Here the region layer consists of spatiotemporal volumes, and the entities on the higher layers include the occurrent processes which unfold themselves therein. When the conservation biologist needs to represent the properly dynamic features of the territory under his charge, he will want to reason (diachronically) about *processes* on different layers – about patterns of change and development. But such entities are invisible when all we have at our disposal are the successive inventories of the SNAP ontology. Such processes belong, rather, to the four-dimensional world of SPAN. According to the temporal granularity selected by the researcher the phenomena in the higher layers may be, for example, meteorological processes occurring over the span of a single day, or processes of climatic change occurring over periods of centuries. They may be cyclical processes of seasonal change in water quantity in lakes and rivers or irreversible processes of erosion stretching over millennia – and again, one of the virtues of layered mereology is that it allows us to represent such processes as spatio-temporally coincident even where they do not overlap.

10   Conclusions

The framework presented in this paper allows mereotopological relations to apply directly to all spatial entities, including spatial regions, material objects, and holes. We saw that the framework can be extended to yield a mereotopological treatment of four-dimensional entities such as motions of sparrows as well as of the spatio-temporal regions in which such processes occur. Our framework thus allows us to do justice to the expressive possibilities encapsulated in the idea of map layers while at the same time incorporating the resources of a rich and naturalistic ontology of geo-spatial dynamics.

Acknowledgement

This work was supported by the Alexander von Humboldt Foundation under the auspices of its Wolfgang Paul Program, and also by the National Science Foundation under grant BCS-9975557: Geographic Categories: An Ontological Investigation. Our thanks go to David Mark for helpful comments.

Bibliography

(Asher and Vieu, 1995) N. Asher and L. Vieu. Towards a Geometry of Commonsense: A Semantics and a Complete Axiomatization of Mereotopology. *Proceedings of IJCAI '95*, 846–852, San Mateo, CA: Morgan Kaufmann.

(Bennett, 2001) B. Bennett. Space, Time, Matter, and Things. In (Welty and Smith), 105–116.

(Bennett, 2001a) B. Bennett. A Categorical Axiomatization of Region-Based Geometry. *Fundamenta Informaticae* 46, 1–2, 145–158.

(Bittner 2001) T. Bittner. The Qualitative Structure of Built Environments. *Fundamenta Informaticae*, 46, 97–128.

(Bittner and Smith, 2002) T. Bittner and B. Smith. A Theory of Granular Partitions. *Foundations of Geographic Information Science*, M. Duckham, M. F. Goodchild and M. F. Worboys, eds., London: Taylor & Francis, 117–151

(Casati and Varzi, 1994) R. Casati and A. C. Varzi. *Holes and Other Superficialities*. Cambridge, MA: MIT Press.

(Casati and Varzi, 1999) R. Casati and A. C. Varzi. *Parts and Places. The Structures of Spatial Representation*, Cambridge, MA: MIT Press.

(Casati, Smith and Varzi 1998) R. Casati, B. Smith and A. C. Varzi. Ontological Tools for Geographic Representation. In: N. Guarino (ed.), *Formal Ontology in Information Systems*, Amsterdam: IOS Press, 77–85.

(Cohn, 2001) A. G. Cohn. Formalizing Bio-Spatial Knowledge. In (Welty and Smith) 198–209.

(Cohn and Hazarika 2001) A. G. Cohn and S. M. Hazarika. Continuous Transitions in Mereotopology. In *Commonsense 2001: 5th Symposium on Logical Formalizations of Commonsense Reasoning* (to appear).

(Cohn and Varzi, 1998) A. G. Cohn and A. C. Varzi. Connection Relations in Mereotopology. In H. Prade (Ed.), *Proceedings of the 13th European Conference on Artificial Intelligence* (ECAI '98), 150–154, New York: Wiley.

(Cui et al., 1992) Z. Cui, A. G. Cohn, D. A. Randell. Qualitative Simulation Based on a Logical Formalism of Space and Time. *Proceedings of AAAI-92*.

(Donnelly, 2003) M. Donnelly. Layered Mereotopology, forthcoming in *IJCAI 2003*.

(Gotts *et al.*, 1996) N. M. Gotts, J. M. Gooday, and A. G. Cohn. A Connection Based Approach to Commonsense Topological Description and Reasoning. *The Monist*, 79: 51–75, 1996.

(Grenon and Smith, 2003) P. Grenon and B. Smith. SNAP and SPAN. A Prolegomena to Geodynamic Ontology. IFOMIS Technical Report, University of Leipzig.

(Mark and Turk, 2003) D. M. Mark and A. Turk. Landscape Categories in Yindjibarndi: Ontology, Environment, and Language (in this volume).

(Schulz and Hahn, 2001) S. Schulz and U. Hahn. Mereotopological Reasoning about Parts and (W)Holes in Bio-Ontologies. In (Welty and Smith), 210–221.

(Simons, 1987) P. M. Simons. *Parts: A Study in Ontology*. Oxford: Oxford University Press.

(Smith, 1995) B. Smith. On Drawing Lines on a Map. In: A. U. Frank and W. Kuhn (eds.), *Spatial Information Theory. A Theoretical Basis for GIS* (Lecture Notes in Computer Science 988), Berlin/Heidelberg/New York, etc.: Springer, 475–484.

(Smith, 1996) B. Smith. Mereotopology: A Theory of Parts and Boundaries. *Data & Knowl-edge Engineering*. 20: 287–303.
(Smith and Grenon, 2003) B. Smith and P. Grenon. The Cornucopia of Formal-Ontological Relations. Forthcoming in *Dialectica*.
(Smith and Varzi, 2000) B. Smith and A. C. Varzi. Fiat and Bona Fide Boundaries. *Philosophy and Phenomenological Research*, 60: 2, 401–420.
(Smith, 2001) B. Smith, Fiat Objects, *Topoi*, 20: 2, 131–148.
(Welty and Smith, 2001) *Proceedings of the 2nd International Conference on Formal Ontology in Information Systems* (FOIS 2001), New York: ACM Press.

Spatial Reasoning about Relative Orientation and Distance for Robot Exploration

Reinhard Moratz and Jan Oliver Wallgrün

Universität Bremen, Department of Mathematics and Informatics
Bibliothekstr. 1, 28359 Bremen, Germany
{moratz,wallgruen}@informatik.uni-bremen.de

Abstract. Spatial agents often have to infer global knowledge from local knowledge about orientations and distances. Thereby, the local knowledge based on sensor data is typically imprecise. We propose an approach that propagates orientation and distance intervals to produce global knowledge and compare it with qualitative reasoning calculi in the context of indoor mobile robot exploration. We combine this propagation method with a path-based and predominantly topological mapping approach and demonstrate how it can be utilized to solve the cycle detection problem.

Keywords: Spatial Reasoning, Robot Navigation

1 Introduction

Spatial knowledge acquired from the environment by a mobile agent is typically incomplete and imprecise. This poses a serious problem for the integration of local spatial knowledge into survey knowledge as required e.g. for a robot exploring an unknown environment. To represent and reason about imprecise spatial information different qualitative spatial calculi have been proposed that deal with certain aspects of space (e.g. topology [4,3,12], orientation [5,18,10,7,11], orientation and distance [1,8]).

To solve navigation tasks topological information is not sufficient and calculi that handle orientation and distance information are required [9,13]. With regard to orientation information two general ways can be distinguished to describe a perceived configuration from the robot's point of view. The direction towards an object relative to the agent's current position can be specified with regard to a global reference direction (e.g. the cardinal direction north), or the direction towards the object can be specified relatively to a reference object and relative to the direction from the agent's position to this reference object.

In this paper we will apply existing calculi for both cases to indoor mobile robot exploration and test how well they are suited to support the integration of local knowledge into survey kowledge. As expected it turns out that in both cases the results become too coarse after a small number of inference steps to be of much use for the given task. Therefore, we propose a method called distance/orientation-interval (DOI) propagation which uses continuous interval borders for modelling imprecision and which can be seen as an extension of the qualitative approaches. We show that DOI propagation outperforms classical qualitative approaches when applied to mobile robot exploration.

W. Kuhn, M.F. Worboys, and S. Timpf (Eds.): COSIT 2003, LNCS 2825, pp. 61–74, 2003.

We then combine DOI propagation with a path-based representation scheme described in [17]. The representation is based on the generalized Voronoi graph of the robot's free space and describes qualitatively different paths through the robot's environment. DOIs are used to specify the relative positions of decision points (corresponding to nodes in the graph). DOI propagation is utilized to help solve the notoriously hard problem of cycle detection.

2   Qualitative Spatial Reasoning about Point Orientations

Qualitative spatial reasoning (QSR) abstracts from the metrical details of the physical world and enables computers to make predictions about spatial relations, even when precise quantitative information is not available [2]. Using QSR imprecise measurements can be represented in a straightforward manner. From a practical viewpoint QSR is an abstraction that summarizes similar quantitative states into one qualitative characterization. A complementary view from the cognitive perspective is that the qualitative method *compares* features within the object domain rather than by *measuring* them in terms of some artificial external scale [6]. This is the reason why qualitative descriptions are quite natural for humans.

Based on previous work by Frank [5] a qualitative calculus about cardinal directions was developed by Ligozat [10]. The cardinal direction calculus splits the plane relative to the reference point (relatum) and the reference direction into nine regions (see figure 1).

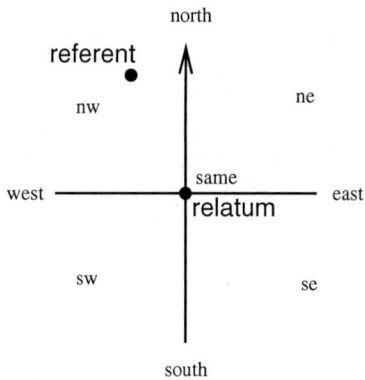

Fig. 1. The cardinal directions calculus.

The spatial relation between the reference system and the *referent* is then described by naming the part of the partition in which the referent lies. Given points A and B where A is the relatum and B the referent the expression

$$A \text{ nw } B$$

would correspond to the proposition: "B is northwest from A".

With the composition operation local observations can be combined into more global knowledge. For example A north B and B north C can be composed into A north C. The corresponding composition rules are expressed as composition tables [5,10]. The result of a composition can also be a disjunction of base relations.

In a relative reference system there is no absolute reference direction. Instead the reference axis is defined by two local points, origin and relatum. Freksa's double cross calculus [6] (see figure 2) also uses right angles as the granularity of orientation expressions. For the notation we adopt the scheme of Scivos and Nebel [14].

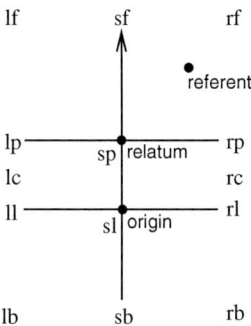

Fig. 2. The double cross calculus.

A relation in the double cross calculus comprises a proposition about the configuration of three points on the plane. As an example a motion from point A straight to B and then oblique right to C would generate a relation $A,\ B$ rf C. If the motion is continued to point D again by a relative oblique right turn, the configuration can be represented by $B,\ C$ rf D. Both propositions share two common points and therefore can be composed. The result of the composition is the relation $A,\ B$ rf, rp, rc, rl, rb D (this notation expresses a disjunction of five base relations). The composition table of the double cross calculus can be found in [6].

In the next section the fixed, coarse intervals for imprecise orientation expressions are generalized with continuous interval borders. Thereby, cognitive adequacy is sacrificed for higher technical performance.

3   Finer Grained Spatial Reasoning about Relative Orientation and Distance

The goal of our approach is to combine benefits from the qualitative approach with benefits from the metrical approach. Qualitative calculi can represent imprecise spatial knowledge. Metric representations are good at distinguishing different spatial enitities. A combination then uses the intervals from the qualitative approach but uses continuous, metric borders rather than fixed salient borders.

Our approach is based on continuous distance/orientation-intervals. A distance/orientation-interval (DOI) uses a point and a reference direction as anchor and has four additional parameters r^{min}, r^{max}, ϕ^{min} and ϕ^{max} (see figure 3). A DOI d is a set of polar vectors (r^i, ϕ^j) with:

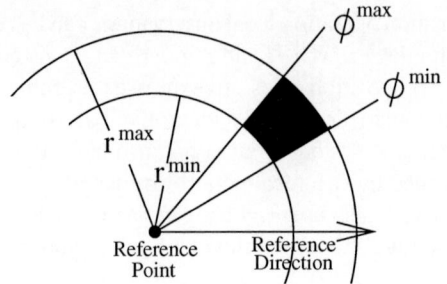

Fig. 3. A distance/orientation-interval and its parameters.

$$\mathbf{d} = \left\{ \left(r^i, \phi^l \right) \quad | \quad r^{min} \leq r^i \leq r^{max} \wedge \phi^{min} \leq \phi^j \leq \phi^{max} \right\}$$

Also $\phi^{max} - \phi^{min} \leq \pi$ holds with the exception of the special case having $\phi^{max} = \pi$ and $\phi^{min} = -\pi$ and $r^{min} = 0$. This special case represents the spatial arrangement in which the location which is expressed by the DOI can be the same as the reference point. For convenience it is assumed that $-2\pi \leq \phi^{min} \leq \pi$ and $-\pi \leq \phi^{max} \leq \pi$.

These quantitative intervals can be propagated along paths. The respective reference directions then are determined by the direction through adjacent points on the path. The composition operation between two DOIs is the basic step for the propagation (see figure 4).

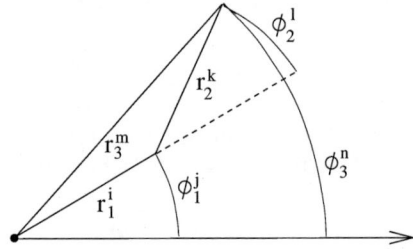

Fig. 4. Composition of adjacent path segments.

The composition of two DOIs $\mathbf{d}_1 = (r_1^{min}, r_1^{max}, \phi_1^{min}, \phi_1^{max})$ and $\mathbf{d}_2 = (r_2^{min}, r_2^{max}, \phi_2^{min}, \phi_2^{max})$ results in a third DOI $\mathbf{d}_3 = (r_3^{min}, r_3^{max}, \phi_3^{min}, \phi_3^{max})$. It holds:

$$\mathbf{d}_3 = \left(\min_{r_1^i, r_2^k, \phi_1^j, \phi_2^l} r_{\Sigma}\left(r_1^i, r_2^k, \phi_1^j, \phi_2^l \right), \max_{r_1^i, r_2^k, \phi_1^j, \phi_2^l} r_{\Sigma}\left(r_1^i, r_2^k, \phi_1^j, \phi_2^l \right), \right.$$

$$\left. \min_{r_1^i, r_2^k, \phi_1^j, \phi_2^l} \phi_{\Sigma}\left(r_1^i, r_2^k, \phi_1^j, \phi_2^l \right), \max_{r_1^i, r_2^k, \phi_1^j, \phi_2^l} \phi_{\Sigma}\left(r_1^i, r_2^k, \phi_1^j, \phi_2^l \right) \right)$$

with

$$r_1^{min} \le r_1^i \le r_1^{max}, \phi_1^{min} \le \phi_1^j \le \phi_1^{max}, r_2^{min} \le r_2^k \le r_2^{max}, \phi_2^{min} \le \phi_2^l \le \phi_2^{max}.$$

The functions r_Σ and ϕ_Σ have the following meaning:

$$r_\Sigma\left(r_1^i, r_2^k, \phi_1^j, \phi_2^l\right) = \left(\left(r_1^i \sin \phi_1^j + r_2^k \sin\left(\phi_1^j + \phi_2^l\right)\right)^2 + \right.$$
$$\left. \left(r_1^i \cos \phi_1^j + r_2^k \cos\left(\phi_1^j + \phi_2^l\right)\right)^2\right)^{\frac{1}{2}}$$

and:

$$\phi_\Sigma\left(r_1^i, r_2^k, \phi_1^j, \phi_2^l\right) = \tan^{-1} \frac{r_1^i \sin \phi_1^j + r_2^k \sin\left(\phi_1^j + \phi_2^l\right)}{r_1^i \cos \phi_1^j + r_2^k \cos\left(\phi_1^j + \phi_2^l\right)}$$

The cases for which r_3^m and ϕ_3^n have their minimum or maximum can be listed by a simple geometric analysis. The results of this analysis are listed in the appendix. In the following we refer to this composition operation by the symbol $\diamond$. This composition is only an approximation in form of an upper bound of the area consisting of the vectors (r_3^m, ϕ_3^n) which can be directly composed by vectors (r_1^i, ϕ_1^j) and (r_2^k, ϕ_2^l) from $\mathbf{d}_1$ and $\mathbf{d}_2$, respectively. A typical spatial layout of these areas is depicted in figure 5.

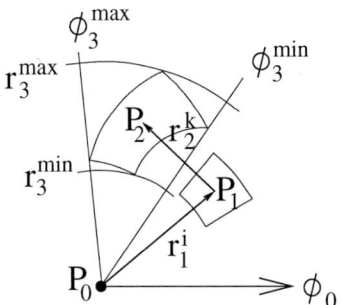

Fig. 5. Resulting DOI as approximation of the composed area.

A DOI can be viewed as a set of difference vectors between two points. These difference vectors are then expressed in polar coordinates relative to a reference direction. A function Δ maps two points and a reference direction to the corresponding difference vector. A function ϕ maps two points to the reference direction from the first point to the second point. Now we can express the relation between points in imprecise relative position and the respective DOIs:

$$\Delta\left(\phi_0, P_0, P_1\right) \in \mathbf{d}_1 \wedge \Delta\left(\phi\left(P_0, P_1\right), P_1, P_2\right) \in \mathbf{d}_2 \implies$$
$$\Delta\left(\phi_0, P_0, P_2\right) \in \mathbf{d}_1 \diamond \mathbf{d}_2$$

We can apply the DOI composition to the propagation of local, relative spatial knowledge along a path. We have an anchor point P_0, a reference direction ϕ_0 and a sequence of points which determine the path segments P_1, P_2, ... P_i, ... P_n. Each point P_i on the path has an associated DOI $\mathbf{d}_i$. Then a stepwise composition recursively beginning with the end of the path yields the relative position of the end point with respect to the anchor point P_0 and reference direction ϕ_0:

$$\Delta(\phi_0, P_0, P_n) \in \mathbf{d}_1 \diamond (\mathbf{d}_2 \diamond (\dots (\mathbf{d}_{n-1} \diamond \mathbf{d}_n)\dots))$$

In the next section we apply the DOI propagation to the spatial unification of landmarks.

4   Spatial Unification of Landmarks with Spatial Reasoning Calculi

A commonly used method of navigation in living beings is that of navigation by landmarks. Landmarks are visual structures already in the environment which are easy to recognize from different positions [15]. The problem with robot navigation using landmarks is to recognize previously registered landmarks especially when coming back via a cycle.

In our experiments with a mobile robot we use a laser based distance sensor (see figure 6). In a simple first test environment (see figure 7) we use the corners as landmarks. The robot's centroid is used as its point position. The task for spatial reasoning is to support the unification of landmarks for cyclic paths.

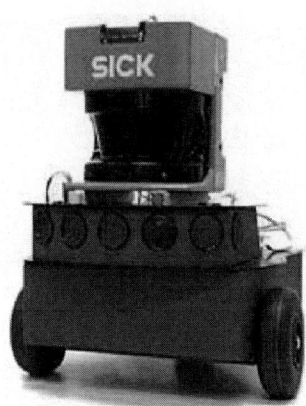

Fig. 6. Our pioneer robot with mounted laser range finder.

In our first experiment the mobile robot drives in four steps from P_1 to P_5. At each position it can perceive three of the four landmarks. Since the landmarks cannot be distinguished visually they have to be separated by their spatial position. The measurement tolerance for the corner detection is 1 degree and 1 cm. The measurement tolerance for the robot's straight movement segments is 2 degree and 2.5 cm. This relative high precision (compared to human judgements) can be better used by the DOI propagation approach then by classical qualitative spatial reasoning. Landmark A seen from position P_i is referred to by A_i. Using the cardinal direction calculus we have the following observations:

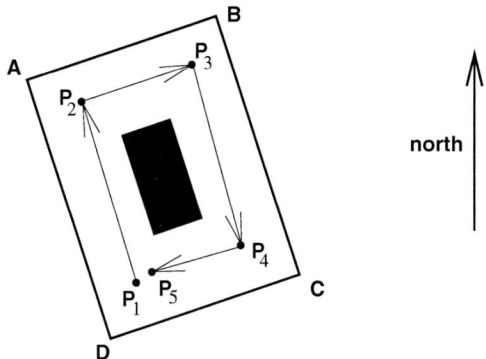

Fig. 7. The environment for the landmark unification experiment.

$$P_1 \text{ sw } D_1 \tag{1}$$
$$P_2 \text{ se } P_1 \tag{2}$$
$$P_2 \text{ nw } A_2 \tag{3}$$

From equation 1, 2 and 3 we can infer A_2 sw, south, se D_1. Then A_2 and D_1 have to be different landmarks. The same observations represented by the double cross calculus are:

$$P_2, \ P_1 \text{ rf } D_1 \tag{4}$$
$$P_1, \ P_2 \text{ lf } A_2 \tag{5}$$

From 4 we can deduce P_1, P_2 lb D_1. From this and from 5 we can again infer that D_1 and A_2 are different.

As we have seen spatial reasoning can distinguish landmarks. The identity of landmarks cannot be decided based on point based spatial reasoning with imprecise data only. But the information that two landmarks have overlapping position ranges can support the hypothesis that these landmarks are identical. However, for longer paths the QSR approaches cannot distinguish landmarks. E.g. in the case of the cardinal direction calculus moving to P_3 and adding the information P_2 ne P_3 and P_3 se C_3, only a disjunction of all relations can be inferred for D_1 and C_3.

As a first test for the DOI propagation we look at a cycle from P_1 to P_5. We compare the expressions for the relative position of A_1 and A_5. The initial view direction (also symmetry axis) of the robot is used as reference direction. The measurements are:

$$\Delta(\phi_1, P_1, A_1) \ \in \ \mathbf{d}_0 = (172.6cm, 174.6cm, 107.25°, 109.25°) \tag{6}$$
$$\Delta(\phi_1, P_1, P_2) \ \in \ \mathbf{d}_1 = (117.5cm, 122.5cm, -2°, 2°) \tag{7}$$
$$\Delta(\phi(P_1, P_2), P_2, P_3) \ \in \ \mathbf{d}_2 = (97.5cm, 102.5cm, -92°, -88°) \tag{8}$$
$$\Delta(\phi(P_2, P_3), P_3, P_4) \ \in \ \mathbf{d}_3 = (117.5cm, 122.5cm, -2°, 2°) \tag{9}$$
$$\Delta(\phi(P_3, P_4), P_4, P_5) \ \in \ \mathbf{d}_4 = (97.5cm, 102.5cm, -92°, -88°) \tag{10}$$
$$\Delta(\phi(P_4, P_5), P_5, A_5) \ \in \ \mathbf{d}_5 = (180.4cm, 182.4cm, 11.75°, 13.75°) \tag{11}$$

Using DOI propagation we can infer from these observations that:

$$\Delta\left(\phi_1, P_1, A_5\right) \quad \in \quad \mathbf{d}_6 = (149.3cm, 214.7cm, 88.10°, 118.42°) \qquad (12)$$

The relative position expressions 6 and 12 have the same reference system. Since they have a non-empty intersection we can infer that landmarks A_1 and A_5 are potentially identical. We have detected a potential cycle.

The above setting has 225 landmark pairs to be compared (the cross product of all the 15 perceived landmark entities), 57 identical pairs, 168 separations. The cardinal direction calculus detects 55 of the 168 separations. The double cross calculus detects 65 separations (the reason might be that the double cross calculus has more base relations). However, the DOI propagation can infer all 168 separations correctly. We conclude from these results that the qualitative approach cannot make enough use of fine-grained data, in contrast. But the DOI approach is promising for these settings. In the next section we transfer DOI propagation to more complex robot exploration settings.

5   Evaluation of DOI Propagation in the Context of Mobile Robot Exploration

For the simple setting of the last section the path planning of the robot was done manually. To evaluate the usefulness of the DOI propagation approach for mobile robot navigation in general indoor environments we combine it with our path-based representation approach that automatically handles path planning [17]. There we developed a representation consisting of a generalized Voronoi graph (GVG), the graph corresponding to the generalized Voronoi diagram of the robot's free space, annotated with additional information. A focus of our work was to develop an incremental construction scheme for the GVG that merges locally computed GVGs and thus makes optimal use of the sensor information available from a single position in the environment, an idea that has also been employed in [16].

This compact predominantly topological representation brings together advantages stemming from the use of topological maps and from the use of Voronoi diagrams in mobile robot applications. Path planning can be done directly by using graph search to search through the GVG. Since only qualitatively different paths are represented in this graph path planning becomes very efficient. The representation also allows for a systematic exploration since having explored the complete GVG means having explored the complete environment.

After describing the representation and navigation based on it in more detail, we will employ DOI propagation to deal with the uncertainty in sensor information and robot position by helping to decide if positions corresponding to nodes of the GVG perceived at different times during the exploration can be identical.

5.1   The Path-Based Representation

The underlying representation is based on the generalized Voronoi graph of the environment's free space, which is the graph corresponding to the generalized Voronoi diagram (GVD) with nodes corresponding to meet points and edges corresponding to Voronoi curves (see figure 8 for an example of a GVD and its corresponding GVG in which – for visualization purposes – the positions of the meet points have been used to place the nodes of the GVG, though this metric information is not contained in the basic GVG). This graph that only represents the topology of the GVD is annotated with additional information (including DOI information between neighbouring nodes) which is required to enable the robot to localize itself within the graph, for path planning, to drive along a planned path, and also for the construction scheme and systematic exploration. The clockwise order of leaving edges is stored at each node to unambigiously describe the embedding of the graph in the plane.

 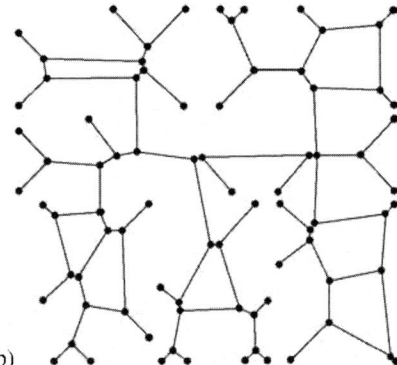

Fig. 8. a) The generalized Voronoi diagram (fine lines) of a 2D environment, b) the corresponding generalized Voronoi graph with nodes placed at positions corresponding to the meet points for visualization.

Labels attached to the edges of the GVG are used by the robot to keep track of which parts of the environments are still unexplored and to mark impassable paths that would lead the robot too close to an obstacle.

The construction scheme to build up the representation of an unknown environment incrementally includes the comparison of nodes in different GVGs based on their *signature*, which contains the following information:

- the type of the node (inner node, corner node, border node)
- the angle intervals between the lines that connect the meet point corresponding to the node with its generating points on the obstacle borders (the points that all have minimal distance to the meet point)
- the interval for the distance from the meet point to its generating points

The distance and the angles are included because they are relatively stable features of a Voronoi node and thus well-suited to recognize nodes. However, the signature alone is not sufficient to unambiguously distinguish nodes under realistic conditions (noise, discreteness of the sensor, etc.). Therefore, the comparison also takes into account the configuration of nodes classified as similar.

Figure 9 shows a partially constructed global GVG together with the additional information annotated to nodes and edges as it has just been described.

To simplify matters, we will use the term "GVG" also for this more complex representation consisting of a GVG with all the annotations listed here in the remainder of the text.

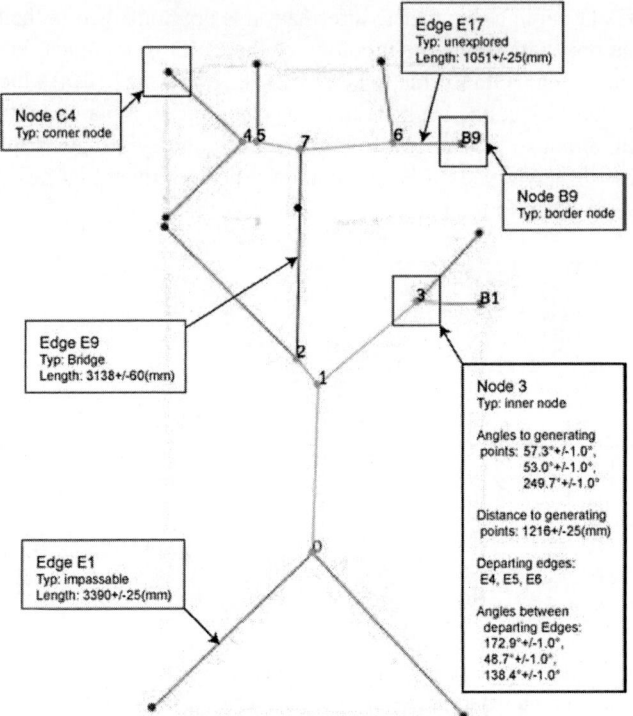

Fig. 9. The used spatial representation consisting of a partially constructed GVG, annotated with additional information at the nodes and edges.

5.2   Incremental Construction

The incremental construction of the global GVG works by computing a so-called *local GVG* (again annotated with the additional information described above) that is based on a single 360° scan of the environment taken from the robot's current position. The local GVG describes how the global GVG "looks" in the neighborhood of the robot as far as this can be derived from the current scan. The current local GVG is then compared to a subgraph of the partially constructed global GVG that represents the region around the place that the robot was supposed go to, called the *comparison GVG*.

In the next step, pairs of nodes from the local GVG and the comparison GVG are identified, which have similar signatures, which configurations in terms of their relative positions sufficiently match, and which are plausible when the last robot movement is taken into account. Based on the best hypothesis the robot can localize itself within the comparison GVG and by that within the global GVG as well. Furthermore, parts contained in the local GVG that are not yet represented in the global GVG can be appended.

5.3   Navigation

Navigating with this representation is rather simple. After a path through the GVG leading to the goal node has been planned using standard graph search techniques, following this path is done in a way similar to the construction scheme described in the previous section. The robot repeatedly performs a movement that should bring him to the next intermediate goal, computes a local GVG, localizes itself within the graph by comparing this local GVG with the global one and selects a new intermediate goal.

6   Cycle Detection Using DOI Propagation

Since localization is only done locally by comparing the local GVG to a subgraph of the global representation based on the last position of the robot and the movement executed afterwards, cycles in the environment will not be detected this way. Therefore, we extended the mapping scheme with the DOI propagation approach to determine possible candidates for identical nodes based on the DOI information stored between nodes of the graph. Thereby, the DOIs are propagated along the path through the GVG that connects the nodes that are checked to be identical. If the DOI propagation indicates that the nodes may be identical, the matching scheme described in section 5.2 is applied for verification.

We carried out first experiments in a medium sized indoor environment comprising hallways, offices, and laboratories that were not especially prepared. In 12 exploration runs the robot explored different parts of the environment including cyclic and U-shaped sections. The average length of the path travelled was 21.8 meters corresponding to 10.2 steps. We compared the results of the DOI propagation based classification for all 624 pairs of decision points (inner nodes) registered during the exploration with the manually identified true cycles. Due to the worst-case approximation principle of the DOI propagation all true cycles were identified correctly. Altogether, the DOI propagation classified 96 percent correctly. Cycles were erroneously reported when the part of the environment explored had a tight U-shape or when it was indeed cyclic but the DOI propagation also classified additional pairs in the neighbourhood of the nodes forming the true cycle as potentially identical. In this cases the accuracy of discrimination was not sufficient anymore and the matching scheme had to rule out the wrong hypotheses.

7   Conclusion and Perspective

Spatial reasoning can be used to distinguish salient positions for robot exploration. The typical precision of robot systems cannot be used by classical qualitative approaches. Our DOI propagation approach which is inspired by qualitative spatial reasoning but uses continuous interval borders can distinguish salient points by their spatial position in medium sized environments as the experiments in combination with our GVG-based mapping approach have shown. The worst-case approximation of a spatial position allows DOI propagation to be used as a preprocessing step to reduce the set of potential identifications that can then be verified using a computationally more expensive method.

We plan to improve the accuracy of discrimination by using refined scan matching for better measurement of the position of the nodes. Thus, the number of hypotheses for identifications generated by the DOI propagation should be reduced further and DOI propagation should also become applicable in larger environments.

Acknowledgement

We would like to thank Christian Freksa and Diedrich Wolter for fruitful discussions and Thora Tenbrink for valuable comments. The work reported here is funded by the German Research Foundation (DFG) grant SFB/TR 8 Spatial Cognition.

References

1. E. Clementini, P. Di Felice, and D. Hernandez. Qualitative represenation of positional information. *Artificial Intelligence*, 95:317–356, 1997.
2. A.G. Cohn. Qualitative spatial representation and reasoning techniques. In G. Brewka, C. Habel, and B. Nebel, editors, *KI-97: Advances in Artificial Intelligence*, Lecture Notes in Artificial Intelligence, pages 1–30. Springer-Verlag, Berlin, 1997.
3. Z. Cui, A.G. Cohn, and D.A. Randell. Qualitative and topological relationships in spatial databases. In D. Abel, editor, *Advances in spatial databases*, Lecture Notes in Computer Science 692, pages 296–315. Springer-Verlag, Berlin, 1993.
4. M. Egenhofer. Reasoning about Binary Topological Relations. In O. Gunther and H.-J. Schek, editors, *Second Symposium on Large Spatial Databases*, Lecture Notes in Computer Science 525, pages 143–160. Springer-Verlag, Berlin, 1991.
5. A. Frank. Qualitative spatial reasoning with cardinal directions. In *Proceedings of 7th Österreichische Artificial-Intelligence-Tagung*, pages 157–167, Berlin, 1991. Springer.
6. C. Freksa. Using Orientation Information for Qualitative Spatial Reasoning . In A. U. Frank, I. Campari, and U. Formentini, editors, *Theories and Methods of Spatial-Temporal Reasoning in Geographic Space*, pages 162–178. Springer, Berlin, 1992.
7. A. Isli and A. Cohn. Qualitative spatial reasoning: A new approach to cyclic ordering of 2d orientation. *Artificial Intelligence*, 122:137–187, 2000.
8. A. Isli and R. Moratz. *Qualitative Spatial Representation and Reasoning: Algebraic Models for Relative Position*. Universität Hamburg, FB Informatik, Technical Report FBI-HH-M-284/99, Hamburg, 1999.
9. J-C Latombe. *Robot Motion Planning*. Kluwer, 1991.
10. G. Ligozat. Reasoning about cardinal directions. *Journal of Visual Languages and Computing*, 9:23–44, 1998.
11. R. Moratz, J. Renz, and D. Wolter. Qualitative spatial reasoning about line segments. In Horn. W., editor, *ECAI 2000. Proceedings of the 14th European Conference on Artifical Intelligence*, Amsterdam, 2000. IOS Press.
12. J. Renz and B. Nebel. On the complexity of qualitative spatial reasoning: A maximal tractable fragment of the region connection calculus. *Artificial Intelligence*, 108(1-2):69–123, 1999.
13. T. Röfer. Route Navigation Using Motion Analysis. In C. Freksa and D. Mark, editors, *COSIT 1999*, pages 21–36, Berlin, 1999. Springer.
14. A. Scivos and B. Nebel. Double-crossing: Decidability and computational complexity of a qualitative calculus for navigation. In *COSIT 2001*, Berlin, 2001. Springer.
15. S. Thrun. Finding landmarks for mobile robot navigation. In *Proceedings IEEE International Conference on Robotics and Automation (ICRA) 1998*, pages 958–963, 1998.
16. D. van Zwynsvoorde, T. Simeon, and R. Alami. Incremental topological modeling using local voronoi-like graphs. In *IEEE/RSJ Int. Conf. on Intelligent Robots and System (IROS 2000)*, pages 897–902, 2000.
17. J. O. Wallgrün. *Exploration und Pfadplanung für mobile Roboter basierend auf Generalisierten Voronoi-Graphen*. Diplomarbeit, Fachbereich Informatik, Universität Hamburg, 2002.
18. K. Zimmermann and C. Freksa. Qualitative spatial reasoning using orientation, distance, path knowledge. *Applied Intelligence*, 6:49–58, 1996.

A Appendix: Computing the Composition Result

For the composition we have to compute the following minima and maxima: $\min_{r_1^i, r_2^k, \phi_1^j, \phi_2^l} r_\Sigma \left(r_1^i, r_2^k, \phi_1^j, \phi_2^l \right)$, $\max_{r_1^i, r_2^k, \phi_1^j, \phi_2^l} r_\Sigma \left(r_1^i, r_2^k, \phi_1^j, \phi_2^l \right)$, $\min_{r_1^i, r_2^k, \phi_1^j, \phi_2^l} \phi_\Sigma \left(r_1^i, r_2^k, \phi_1^j, \phi_2^l \right)$, and $\max_{r_1^i, r_2^k, \phi_1^j, \phi_2^l} \phi_\Sigma \left(r_1^i, r_2^k, \phi_1^j, \phi_2^l \right)$.

For $\min_{r_1^i, r_2^k, \phi_1^j, \phi_2^l} r_\Sigma \left(r_1^i, r_2^k, \phi_1^j, \phi_2^l \right)$ we have to consider 12 geometric cases $r_3^1, \ldots, r_3^{12}$. Then $\min_{r_1^i, r_2^k, \phi_1^j, \phi_2^l} r_\Sigma \left(r_1^i, r_2^k, \phi_1^j, \phi_2^l \right) = \min(r_3^1, \ldots, r_3^{12})$.

$$r_3^1 = r_\Sigma \left(r_1^{min}, r_2^{min}, 0, \phi_2^{min} \right)$$
$$r_3^2 = r_\Sigma \left(r_1^{min}, r_2^{min}, 0, \phi_2^{max} \right)$$
$$r_3^3 = r_\Sigma \left(r_1^{min}, r_2^{max}, 0, \phi_2^{min} \right)$$
$$r_3^4 = r_\Sigma \left(r_1^{min}, r_2^{max}, 0, \phi_2^{max} \right)$$
$$r_3^5 = r_\Sigma \left(r_1^{max}, r_2^{min}, 0, \phi_2^{min} \right)$$
$$r_3^6 = r_\Sigma \left(r_1^{max}, r_2^{min}, 0, \phi_2^{max} \right)$$
$$r_3^7 = r_1^{min} - r_2^{max} \quad \Longleftarrow \quad \left(\phi_2^{min} \leq -\pi \leq \phi_2^{max} \wedge r_1^{min} > r_2^{max} \right)$$
$$r_3^8 = r_2^{min} - r_1^{max} \quad \Longleftarrow \quad \left(\phi_2^{min} \leq -\pi \leq \phi_2^{max} \wedge r_2^{min} > r_1^{max} \right)$$
$$r_3^9 = r_\Sigma \left(r_1^{min}, r_1^{min} \cos(\pi - \phi_2^{max}), 0, \phi_2^{max} \right) \quad \Longleftarrow$$
$$\left(\phi_2^{max} > \frac{\pi}{2} \quad \wedge \quad r_2^{min} < r_1^{min} \cos(\pi - \phi_2^{max}) < r_2^{max} \right)$$
$$r_3^{10} = r_\Sigma \left(r_1^{min}, r_1^{min} \cos(\pi + \phi_2^{min}), 0, \phi_2^{min} \right) \quad \Longleftarrow$$
$$\left(\phi_2^{min} < -\frac{\pi}{2} \quad \wedge \quad r_2^{min} < r_1^{min} \cos(\pi + \phi_2^{min}) < r_2^{max} \right)$$
$$r_3^{11} = r_\Sigma \left(-\cos \phi_2^{max} r_2^{min}, r_2^{min}, 0, \phi_2^{max} \right) \quad \Longleftarrow$$
$$\cos \phi_2^{max} r_2^{min} + r_1^{min} < 0 < \cos \phi_2^{max} r_2^{min} + r_1^{max}$$
$$r_3^{12} = r_\Sigma \left(-\cos \phi_2^{min} r_2^{min}, r_2^{min}, 0, \phi_2^{min} \right) \quad \Longleftarrow$$
$$\cos \phi_2^{min} r_2^{min} + r_1^{min} < 0 < \cos \phi_2^{min} r_2^{min} + r_1^{max}$$

For $\max_{r_1^i, r_2^k, \phi_1^j, \phi_2^l} r_\Sigma \left(r_1^i, r_2^k, \phi_1^j, \phi_2^l \right)$ the geometric analysis shows seven distinct cases over which the maximum $\max(r_3^{13}, \ldots, r_3^{19})$ has to be computed:

$$r_3^{13} = r_\Sigma \left(r_1^{max}, r_2^{min}, 0, \phi_2^{min} \right)$$
$$r_3^{14} = r_\Sigma \left(r_1^{max}, r_2^{min}, 0, \phi_2^{max} \right)$$
$$r_3^{15} = r_\Sigma \left(r_1^{max}, r_2^{max}, 0, \phi_2^{min} \right)$$
$$r_3^{16} = r_\Sigma \left(r_1^{max}, r_2^{max}, 0, \phi_2^{max} \right)$$
$$r_3^{17} = r_\Sigma \left(r_1^{min}, r_2^{max}, 0, \phi_2^{min} \right)$$
$$r_3^{18} = r_\Sigma \left(r_1^{min}, r_2^{max}, 0, \phi_2^{max} \right)$$
$$r_3^{19} = r_\Sigma \left(r_1^{max}, r_2^{max}, 0, 0 \right) \quad \Longleftarrow \quad \phi_2^{min} < 0 < \phi_2^{max}$$

The six cases to be considered for $\min_{r_1^i, r_2^k, \phi_1^j, \phi_2^l} \phi_\Sigma \left(r_1^i, r_2^k, \phi_1^j, \phi_2^l \right)$ are:

$$\phi_3^1 = \phi_\Sigma \left(r_1^{min}, r_2^{min}, \phi_1^{min}, \phi_2^{min} \right)$$

$$\phi_3^2 = \phi_\Sigma \left(r_1^{min}, r_2^{max}, \phi_1^{min}, \phi_2^{min} \right)$$

$$\phi_3^3 = \phi_\Sigma \left(r_1^{max}, r_2^{min}, \phi_1^{min}, \phi_2^{min} \right)$$

$$\phi_3^4 = \phi_\Sigma \left(r_1^{max}, r_2^{max}, \phi_1^{min}, \phi_2^{min} \right)$$

$$\phi_3^5 = \phi_\Sigma \left(r_1^{min}, r_2^{max}, \phi_1^{min}, \phi_2^{max} \right)$$

$$\phi_3^6 = \phi_\Sigma \left(r_1^{min}, r_2^{max}, \phi_1^{min}, -\frac{\pi}{2} - \sin^{-1} \frac{r_2^{max}}{r_1^{min}} \right) \quad \Longleftarrow$$

$$\phi_2^{min} < -\frac{\pi}{2} - \sin^{-1} \frac{r_2^{max}}{r_1^{min}} < \phi_2^{max}$$

The cases for $\max_{r_1^i, r_2^k, \phi_1^j, \phi_2^l} \phi_\Sigma \left(r_1^i, r_2^k, \phi_1^j, \phi_2^l \right)$ are symmetric to the cases for $\min_{r_1^i, r_2^k, \phi_1^j, \phi_2^l} \phi_\Sigma \left(r_1^i, r_2^k, \phi_1^j, \phi_2^l \right)$:

$$\phi_3^7 = \phi_\Sigma \left(r_1^{min}, r_2^{min}, \phi_1^{max}, \phi_2^{max} \right)$$

$$\phi_3^8 = \phi_\Sigma \left(r_1^{min}, r_2^{max}, \phi_1^{max}, \phi_2^{max} \right)$$

$$\phi_3^9 = \phi_\Sigma \left(r_1^{max}, r_2^{min}, \phi_1^{max}, \phi_2^{max} \right)$$

$$\phi_3^{10} = \phi_\Sigma \left(r_1^{max}, r_2^{max}, \phi_1^{max}, \phi_2^{max} \right)$$

$$\phi_3^{11} = \phi_\Sigma \left(r_1^{min}, r_2^{max}, \phi_1^{max}, \phi_2^{min} \right)$$

$$\phi_3^{12} = \phi_\Sigma \left(r_1^{min}, r_2^{max}, \phi_1^{max}, \frac{\pi}{2} + \sin^{-1} \frac{r_2^{max}}{r_1^{min}} \right) \quad \Longleftarrow$$

$$\phi_2^{min} < \frac{\pi}{2} + \sin^{-1} \frac{r_2^{max}}{r_1^{min}} < \phi_2^{max}$$

In addition to the above cases the case with $\phi_2^{min} \leq -\pi \leq \phi_2^{max}$ and $r_1^{min} \leq r_2^{min} \leq r_2^{max} \leq r_1^{max} \quad \vee \quad r_1^{min} \leq r_2^{min} \leq r_1^{max}$ results in the special case that the resulting DOI d_3 includes the reference point. Then $\phi_3^{max} = \pi$ and $\phi_3^{min} = -\pi$ and $r_3^{min} = 0$.

Structuring a Wayfinder's
Dynamic Space-Time Environment

Michael D. Hendricks[1,2], Max J. Egenhofer[1,2,3], and Kathleen Hornsby[1]

[1] National Center for Geographic Information and Analysis
[2] Department of Spatial Information Science and Engineering
[3] Department of Computer Science
University of Maine
Orono, ME 04469-5711, USA
{hendrick,max,khornsby}@spatial.maine.edu

Abstract. To travel successfully in a dynamic space-time setting, wayfinders must project the impact of a changing environment onto future travel choices. When making decisions, however, people often fail to consider the impact of future changes. They instead overly rely on current system states. In addition, spatial information systems designed for wayfinders typically focus on current or historic travel information. To address these limitations, this paper presents an approach to structure the dynamic space-time environment of a wayfinder. With this structure, improved spatial information systems can be designed to support wayfinders in dynamic environments. To create this structure, four primitives of space-time wayfinding are presented: maximum travel speed, a starting point, barriers, and compulsions. Combining the speed limitation with each of the remaining three primitives creates distinctive partitions of space-time. To integrate all four primitives, a method of sequentially partitioning space-time is described which results in four partition categories that account for the different constraints of wayfinding. These partitions are described in a cognitively plausible manner using modal verbs *can*, *may*, *must*, and *should*. The creation of this structure along with these descriptive semantics creates a rich representation of the wayfinder's space-time environment and allows for reasoning about space-time decision points and their impact on future possibilities.

1 Introduction

Wayfinding in a spatio-temporal environment is a complex task (Golledge 1999; Davies and Pederson 2001; Sholl 2001), resulting from the dynamic existence and location of objects over time. For example, a route may be available to a wayfinder early in the day, but a scheduled bridge closure later creates a barrier. A travel requirement may be in place for a specific time interval, such as meeting a friend for lunch. Barriers and requirements may also move or change shape over time, as in the case of a forest fire (Yuan 1997). In addition, the wayfinder's location also changes throughout this complex task.

Failing to account for the dynamic nature of the spatio-temporal environment may result in unsuccessful wayfinding. Arriving at the bridge after it closes, or going to the restaurant at the wrong time, may make the wayfinding task impossible. The unidirec-

W. Kuhn, M.F. Worboys, and S. Timpf (Eds.): COSIT 2003, LNCS 2825, pp. 75–92, 2003.
© Springer-Verlag Berlin Heidelberg 2003

tional nature of time (Galton 1997; Boroditsky 2000), along with the wayfinder's maximum travel speed, establishes a structure in space-time that defines available choices. For example, choosing not to cross the bridge before it closes precludes travel options on the other side of the river. Other choices may be contingent on previous ones, for example, waiting for the bridge to open may be irrelevant if you have already crossed and do not wish to return.

To account for these changes when planning for and traveling in dynamic spatio-temporal environments, an integrated space-time model is required. In addition, there must be a method for determining the location and time for making decisions, and a means to realize the impacts of these decisions on future possibilities.

A typical strategy used by people in day-to-day decision making is to project possible future states from known patterns and trends (Barlow 1998). When crossing the street, people project both their location and the location of any cars into the future to ensure that they do not intersect in space and time. This process of perceiving and understanding elements in a dynamic spatio-temporal environment and projecting their status into the future is referred to as *situational awareness* (Endsley 1988). In the professional world, success of air-traffic controllers (Andre *et al.* 1998; Durso *et al.* 1999; Azumea *et al.* 2000), pilots (Endsley 1995; Zhang and Hill 2000), and military personnel (NRC 1997; Ellis and Johnston 1999) relies heavily on situational awareness in order to make effective decisions in these spatio-temporal environments.

It has been found, however, that people often fail to consider effectively the impact of future changes when making decisions, but overly rely on current system states (Kerstholt and Raajimakers 1997). A wayfinder may fail to project into the future the impacts a scheduled bridge closure will have on her travel options, and as a result may be unsuccessful in reaching her destination on time. Since wayfinding in a dynamic spatio-temporal environment is particularly complex, it is expected that wayfinders often fail to project future states when determining space-time decision points and reasoning about the impacts of these decisions on future possibilities.

Spatial information systems typically support wayfinders with reference maps and more recently location information derived from GPS data. Route guidance systems are becoming more popular in automobiles and are beginning to account for dynamic traffic conditions (Eby and Kostyniuk 1999). Though increasing the information available to the wayfinder, these systems fail to address the the impact of a changing environment on future travel possibilities.

To address these limitations, this paper introduces an approach that structures and describes the space-time environment of a wayfinder in a manner that will support people's qualitative reasoning capability with complex tasks (Egenhofer and Mark 1995) and provide a framework for the development of more effective spatial information systems. The space-time structure is created by extending into the temporal domain space partitioning symbolic projection techniques for route planning (Holmes and Jungert 1992; Jungert 1992; Chang and Jungert 1996). The resulting space-time structure extends the approaches of space-time prisms (Miller 1991) and geospatial lifelines (Hariharan and Hornsby 2000; Hornsby and Egenhofer 2002) by providing a richer and more complete representation of the wayfinder's space-time environment.

Four primitives of space-time wayfinding are presented: maximum travel speed, a starting point, barriers, and compulsions. The maximum travel speed constrains the wayfinder's rate of movement. When projected through space-time, the maximum travel speed creates a structure that defines travel possibilities. The start point represents the location and time where wayfinding begins. Barriers are the locations in space and time where travel is not allowed, while compulsions represent the space-time requirements of the wayfinder, such as having lunch with a friend.

Combining the speed limitation with each of the remaining three primitives creates distinct partitions of space-time. Combining the maximum speed with the starting point partitions space-time into accessible and inaccessible space based on the wayfinder's capability. The combination of the maximum speed and barriers creates a partition of space-time into accessible and inaccessible spaces based on the barrier's travel restrictions. The last combination considered—the speed limitation with the wayfinder's compulsions—creates a partition that defines where and when the wayfinder should or should not travel in order to meet the requirements. To integrate all four primitives, a method of sequentially partitioning space-time is described, which results in four partition categories that account for the different constraints of wayfinding.

To describe the resulting partitions of this space-time structure in a cognitively plausible manner, a method of employing the modal verbs *can, may, must,* and *should* is presented. Though each modal verb is used in many ways during normal conversations, *can* is related to a positive ability (capability); *must* denotes obligation or compelling force; and *may* is associated with permission or lack of a potential barrier (Sweetser 1990). In addition, we include one other modal verb, *should,* as a weaker form of must. In an inherently physical and dynamic act, such as wayfinding, describing spatial and temporal constraints modally provides insights for argument and reasoning.

The result of creating this space-time structure and linking modal verbs to its partitions allows a wayfinder to reason about space-time decision points and their impact on future possibilities. For example, the wayfinder can consider the following queries: Where *can* I go given some time limit? Where *may* I go given barriers in my environment? Where *must* I go? Where *should* I go and still meet my requirements? When is it possible to go to a location given my other requirements? If I go to some space-time point, is it still possible to meet my requirements? Developing a framework for queries such as these provides additional decision support tools for wayfinders traveling in dynamic space-time environments.

The remainder of this paper continues with a summary of approaches for structuring a wayfinder's environment, to include symbolic projections and space-time prisms (Section 2). In Section 3, the four wayfinding primitives used in this approach are introduced. Section 4 describes the three distinct partitions of space-time, resulting from combining the wayfinder's maximum speed with each of the remaining primitives. Section 5 describes the sequential partitioning technique that combines all four partitions and creates the integrated structure of the wayfinder's space-time. Section 6 presents an approach for using modal verbs to describe these partitions and Section 7 provides conclusions and suggestions for future work.

2   Structuring a Wayfinder's Environment

Numerous methods of structuring a wayfinder's environment have been proposed in the past. Many focus on the spatial structure of the environment alone, while others attempt to integrate the temporal component of wayfinding. This section reviews in particular those spatial (Section 2.1) and spatio-temporal (Section 2.2) approaches that form the basis for later discussions in this paper.

2.1   Spatial Structure for Wayfinding

Spatial approaches to wayfinding typically consider a flat, 2-dimensional space in which they aim at identifying optimal paths. A popular structure of a wayfinder's environment identifies barriers and calculates the Euclidean shortest path around them (Lee and Preparata 1984; Stefanakis and Kavouras 1995; Hershergery and Suri 1999). More complex approaches classify the environment's objects as obstructers or facilitators, which allows for pragmatic navigation through the environment (Epstein 1997). Other approaches organize space hierarchically based on different functionalities (Timpf 2002). When emphasizing human perceptive and cognitive lines, image schemata and affordances have been used to structure the space of wayfinders in built-up environments (Raubal *et al.* 1997; Raubal and Egenhofer 1998).

A method based on *symbolic projections* (Holmes and Jungert 1992; Jungert 1992) partitions the traveler's 2-dimensional space around barriers, enabling qualitative spatial reasoning about routes. These symbolic projections record objects' extents along the horizontal and vertical axes (Figure 1), resulting in strings of object locations associated with each axis (Chang and Jungert 1986; Jungert 1988). When specifically applied to wayfinding, the focus of the symbolic projection approach is the partitioning of space by projecting the extents of barriers.

2.2   Spatio-temporal Structure for Wayfinding

Since wayfinding is often time-related, a comprehensive approach to reasoning in this environment requires an integrated space-time model. A foundation of such a space-time model considers an n-dimensional space, denoted by S, and one orthogonal dimension of time, denoted by T, on $\Re^{n+1}$. The case of two spatial dimensions (x-y plane) and an orthogonal dimension for time (ascending z-axis) yields a three-dimensional space-time cube, analogous to the foundation of Hägerstrand's Time Geography (Hägerstrand 1967). In this setting, the wayfinder is modeled as a 0-dimensional point object, whose location in space and time is referenced by (x_i, y_i, t_i). As time progresses, this point object traces a space-time path (Miller 1991) from the

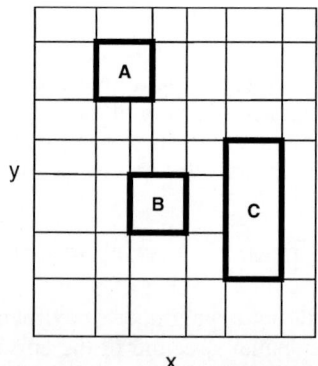

Fig. 1. Structuring the space of a wayfinder through symbolic projection techniques. Barriers A, B, and C are projected onto each axis, creating a grid representing travel possibilities.

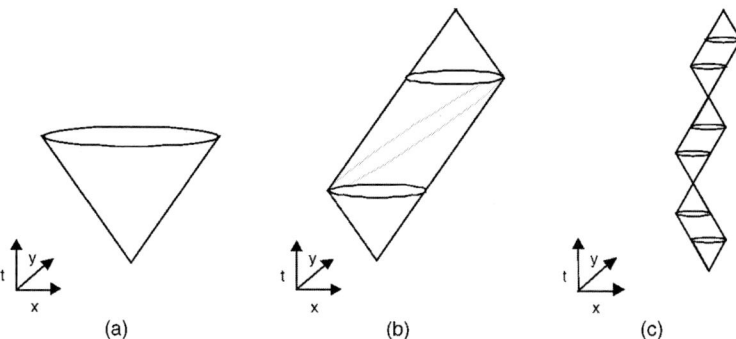

Fig. 2. Possible travel locations in a space-time volume: (a) a cone, (b) space-time prism or lifeline bead, and (c) a lifeline necklace.

origin (x_0, y_0, t_0) to the destination (x_1, y_1, t_1), also modeled as a geospatial lifeline thread (Hornsby and Egenhofer 2002). An immobile object would trace a vertical line in the space-time cube, whereas an object moving at a constant speed creates a sloped space-time path, with flatter lines representing faster travel and steeper lines standing for slower travel. Projecting such a space-time path onto the x-y plane creates the *route* traveled through space (Miller 1991).

Most often, the exact path through space-time is unknown. One method of handling this uncertainty is to determine the set of all possible locations that an object can travel to between time intervals. With a given start point and a constant maximum speed, a half cone is created in space-time, which represents the set of all possible locations the traveler can reach (Figure 2a). If a destination point is added to this scenario, a second half cone extending back in time is created and whose intersection with the first half cone creates a potential path space (Figure 2b), referred to as a space-time prism (Miller 1991) or lifeline bead (Hornsby and Egenhofer 2002). This lifeline bead represents the set of all possible space-time points that the object may have occupied between the origin and the destination, based on heuristics about the object's maximum travel speed. The aggregate of simply connected beads (Figure 2c) forms a lifeline necklace (Hornsby and Egenhofer 2002).

Time Geography considers three classes of constraints that restrict individual movement and shape space-time prisms (Pred 1977). *Capability constraints* restrict movement based on physiological needs, such as sleeping or eating, and the speed limitations of available transportation. *Coupling constraints* limit travel by the requirements to meet other people or objects in space and time. *Authority constraints* restrict travel as a result of certain activities being only available at certain times. This paper presents an alternative classification of travel constraints modeled as four wayfinding primitives which are described in the next section.

The space-time prisms are projected onto the spatial surface to represent accessibility, a qualitative spatial measure used by trsavelers (Weibull 1980; Miller 1991; Kwan 1998; Miller 1999). While these concepts support the analysis as to whether two or more individuals could have met, they say little about space-time inside or outside the prisms, as well as the effects of barriers on their shapes.

3   Integrated Space-Time Structure for Wayfinding

To provide a richer and more complete structure to the space-time environment of the wayfinder, a systematic approach is presented, which builds on a set of wayfinding primitives. The framework for this space-time environment is that space and time are bound, creating an $n+1$ dimensional space-time *container*, denoted by $ST_{(n+1)D}$. This container creates a closed world of the wayfinder's environment with the following assumptions:

- The wayfinder experiences time as being continuous and unidirectional (Boroditsky 2000).

- All objects related to the wayfinder's task must be contained in ST.

In addition, to simplify this discussion and presentation we consider a space-time environment with one spatial dimension, such as a road or rail segment. This simplification creates a 2-dimensional space-time *container* ST_{2D} that allows us to represent these concepts in planar maps and diagrams.

Within this framework, the following primitives structure the wayfinder's space:

- the maximum speed limitation of the wayfinder (L);

- space-time information of the start point (O);

- space-time information of compulsions through which the wayfinder *must* pass (M); and

- space-time information about barriers that the wayfinder *must not* travel through (¬M).

Each of these primitives adds a set of constraints to the space-time environment of the wayfinder. The maximum speed limitation constrains the rate of travel through the environment. This speed limitation provides a method of projecting through space-time the impacts of the remaining three primitives and allows the wayfinder to reason about space-time decision points and their impact on future possibilities. The second primitive, the wayfinder's start point, establishes the origin in space and time, thereby constraining future possibilities. The remaining two primitives, compulsions and barriers, are more complex, and each is discussed below.

3.1   Compulsions: The *M-Space* of Wayfinding

A compulsion represents the space-time task requirements of the wayfinder, or the *where* and *when* the wayfinder must be in the future. This approach assumes independence of compulsions. A compulsion cannot be contingent upon another, and disjunctive compulsions, as in one or the other, are not allowed. The—possibly empty—set of compulsions, in combination with the start point, is referred to as *M-space*.

Compulsions come in various shapes and sizes. The following list gives examples:

- A 0-dimensional space-time point compulsion represents a requirement at a specific location and time (M_1 in Figure 3). For example, "you *must* be at my office at 10:00 am." This is the type of compulsion typically modeled in the space-time prism and geospatial lifeline approaches.

- A compulsion may also occur over larger areas of space. In these cases, since the wayfinder is modeled as a point object, the requirement is to travel to some point in the compulsion space. A compulsion with a spatial extent at an instance of time is modeled as a horizontal line in ST_2 (M_2 in Figure 3).

- Often a compulsion occurs over a temporal interval. The extreme case is a compulsion that requires a wayfinder to occupy a location continuously. Scenarios that are more interesting occur when the compulsion is active for only a subset of the available time. Consider the case where a compulsion lasts only until a certain time. For example, "you *must* stay home until 8:00 am." In this case, the wayfinder's travel possibilities are completely restricted until after the compulsion disappears (M_3 in Figure 3).

- An alternate compulsion arises when you *must* occupy some location after a specified time. For example, "be home by 10:00 pm" (M_4 in Figure 3).

- A combination of these two types models the compulsion that begins and ends inside the space-time container (M_5 in Figure 3). For example, "you *must* be at work from 8:00 am to 5:00 pm."

- Compulsions may also occur over both a spatial extent and a temporal interval (M_6 in Figure 3).

- In addition, compulsions may themselves travel through space-time. A requirement to intercept a moving object would be a compulsion of this type. Moving point compulsions are modeled as a sloped line through space-time, representing the speed of the compulsion object (M_7 in Figure 3).

- A compulsion also may change shape and size over time (M_8 in Figure 3).

An additional characteristic of *M-space* is that there is a distinction between compulsions that are required to be occupied the entire time and compulsions that only require occupation at some instance of the temporal interval. The compulsion, "Go to the post office after 4 pm," does not compel the wayfinder to spend his entire evening at the post office, whereas the compulsion to have lunch with a friend requires occupation of the compulsion space for the entire time.

3.2 Barriers: The ¬*M-Space* of Wayfinding

Barriers constrain the wayfinder by disallowing travel through them. These spaces are referred to as ¬*M-spaces*. *M-space* and ¬*M-space* do not themselves partition space-time in that there are points in the wayfinder's space-time container that are neither compulsions nor barriers. Barriers are classified into those that partition space and those that do not. For example, a river running entirely through the wayfinder's space will create a partition, whereas a lake will not.

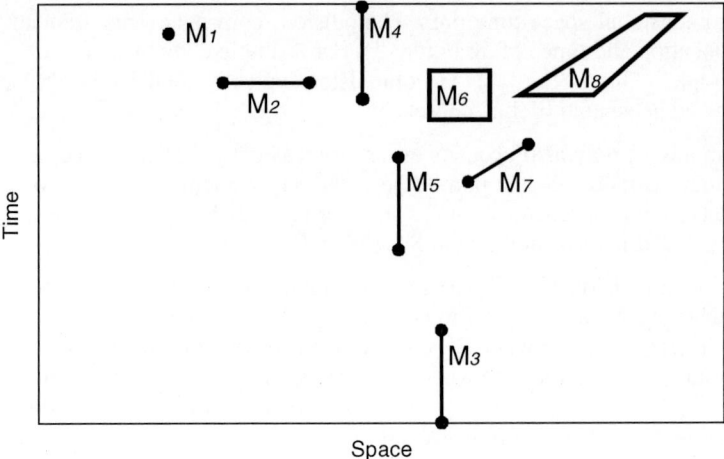

Fig. 3. Various compulsion shapes: a space-time point compulsion (M_1); a compulsion with a spatial extent at an instance of time (M_2); spatial point compulsions at various temporal intervals (M_3 M_4 M_5); a compulsion over both a spatial and temporal interval (M_6); a moving point compulsion (M_7); and a compulsion that changes shape and size over time (M_8).

Barriers, as with compulsions, vary in size and shape. The following list gives examples:

- A typical temporary barrier has a spatial extent, and begins and ends within the wayfinder's space-time environment. For example, closing a section of road for 2 hours ($\neg M_1$ in Figure 4).

- Partitioning barriers may themselves have minimal area, but by partitioning space, they may constrain movement over a large spatial extent. For example, closing a bridge creates a point barrier that constrains movement to one side of the river ($\neg M_2$ in Figure 4).

- Barriers may also change size over time. For example, a forest fire modeled in this way would appear, increase in size, and then decrease until it goes out ($\neg M_3$ in Figure 4).

4   Combining Wayfinding Primitives

Combining the maximum travel speed with each of the remaining three primitives creates a distinctive structure that partitions space-time and defines the accessibility related to that combination. These structures are created by projecting the maximum travel speed limitation through space-time from the start point, and each barrier and compulsion. This procedure is similar to that used in the symbolic projection technique's partitioning of a wayfinder's space. The partitions created from these three combinations are assigned a symbol and its complement to indicate the accessible and inaccessible spaces created from each combination (Figure 5). This section continues by introducing in more detail *C-space* (Section 4.1), *Y-space* (Section 4.2), and *H-space* (Section 4.3)

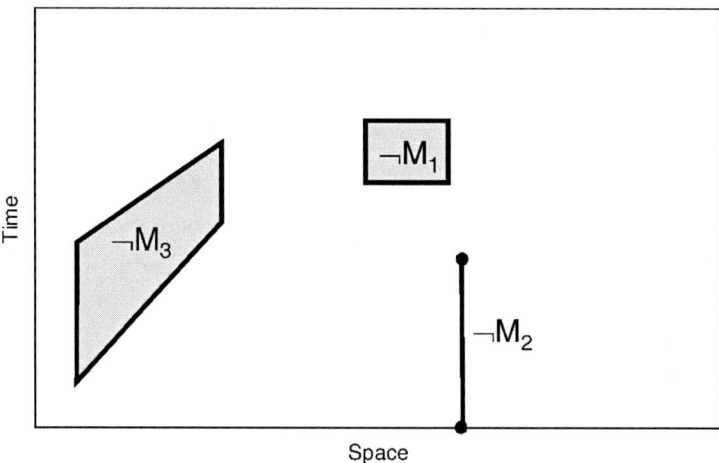

Fig. 4. Various barrier shapes: a barrier within a spatial extent beginning and ending within the wayfinder's space-time environment ($\neg M_1$); a spatial point barrier that lasts until a specified time ($\neg M_2$); and a barrier appearing, increasing in size and then decreasing until disappearing ($\neg M_3$).

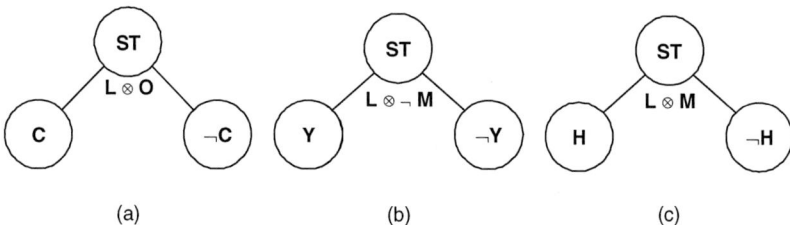

Fig. 5. Three accessibility partitions of space-time, ST, by wayfinding primitive combinations: (a) speed limitation, L, and start point, O, create *C-space* and $\neg C$*-space*; (b) speed limitation and barriers, $\neg M$, create *H-space* and $\neg H$*-space*; and (c) speed limitation and compulsions, M, create *Y-space* and $\neg Y$*-space*.

4.1   The *C-Space* of Wayfinding

Combining a wayfinder's maximum travel speed and start point partitions the space-time container into two spaces: *C-space* as the set of locations that can be accessed from the start point and $\neg C$*-space* as the inaccessible space. This process is analogous to generating the half cone used to create a lifeline bead. *C-space* and $\neg C$*-space* are simply connected; the separation of $\neg C$*-space* into two areas (Figure 6) is only a side effect of the planar graphical presentation. The following observations are made about this structure:

• Any valid space-time path must be fully contained within *C-space*.

• If the wayfinder's maximum speed increases, the *C-space* also grows in space and time.

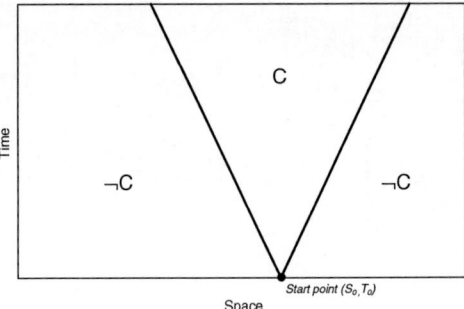

Fig. 6. Partitioning of the space-time container into *C-space* and ¬*C-space* as an implication of the wayfinder's start point and maximum travel speed.

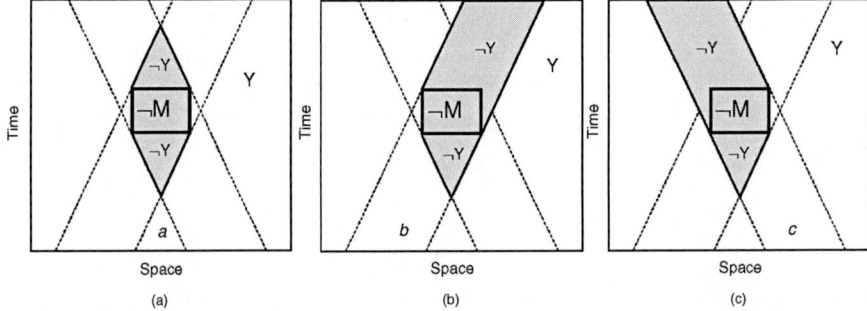

Fig. 7. A barrier combined with the maximum travel speed limitation partitions space-time: (a) when the wayfinder is in *region a*, ¬*Y-space* is created before and after the barrier; (b) when the wayfinder is in *region b*, additional ¬*Y-space* is created after the barrier; and (c) when the wayfinder is in *region c*, a different ¬*Y-space* is created.

- The reverse holds true as well, that is, decreasing the maximum travel speed shrinks the *C-space*.

4.2   The *Y-Space* of Wayfinding

A wayfinder's maximum travel speed and a set of barriers partition space-time into additional accessible and inaccessible spaces. The inaccessible spaces, referred to as ¬*Y-spaces,* result from the absolute travel restriction of these barriers. The accessible spaces that remain are referred to as *Y-spaces.*

The maximum travel speed projections through space-time create a ¬*Y-space* before and after each temporary barrier with a spatial extent (Figure 7a). The ¬*Y-space* before each barrier is a danger area where, if a wayfinder enters, it is impossible not to encounter the barrier. For example, if an explosion occurs when the wayfinder is in this region then injury will occur. Another way to describe this situation is that the inaccessible space (i.e. ¬*Y-space)* is beyond the reaction time of the wayfinder.

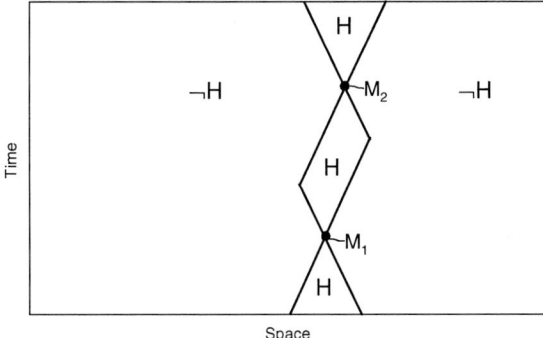

Fig. 8. Partitioning space-time into *H-space* and *¬H-space* as a result of the wayfinder's maximum travel speed and space-time point compulsions M_1 and M_2.

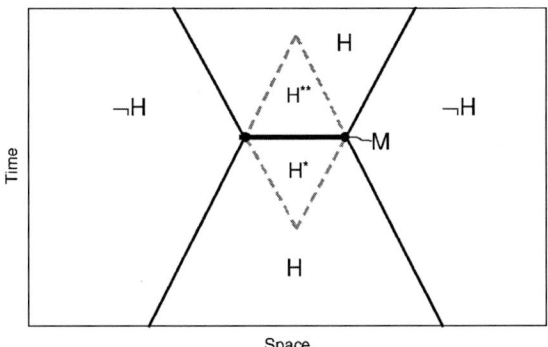

Fig. 9. A compulsion with a spatial extent creates a larger *H-space* than that created by a space-time point compulsion. In addition, spaces can be identified that indicate that the compulsion absolutely will be met (H^*) or absolutely has been met (H^{**}).

Barriers that partition space create additional *¬Y-spaces* that are dependent on the location of the wayfinder. For example, as shown in Figure 7b, a wayfinder in *region b* finds that the barrier creates a *space-time shadow* of *¬Y-space*. This *shadow* differs if the wayfinder is located in *region c* (Figure 7c). An example of this type of barrier is a closed bridge over a space-partitioning river. The inaccessible space resulting from the barrier, the blocked bridge, is dependent on what side of the barrier the wayfinder is located. As opposed to the *¬C-space*, the partitioning nature of the barrier creates *¬Y-spaces* that are not simply connected and in fact represent disconnect spaces.

4.3   The *H-Space* of Wayfinding

The combination of maximum travel speed and the set of compulsions create a third partition of space-time into *H-space* and *¬H-space*. The *H-space* represents *where* and *when* the wayfinder should travel while still meeting the required compulsions.

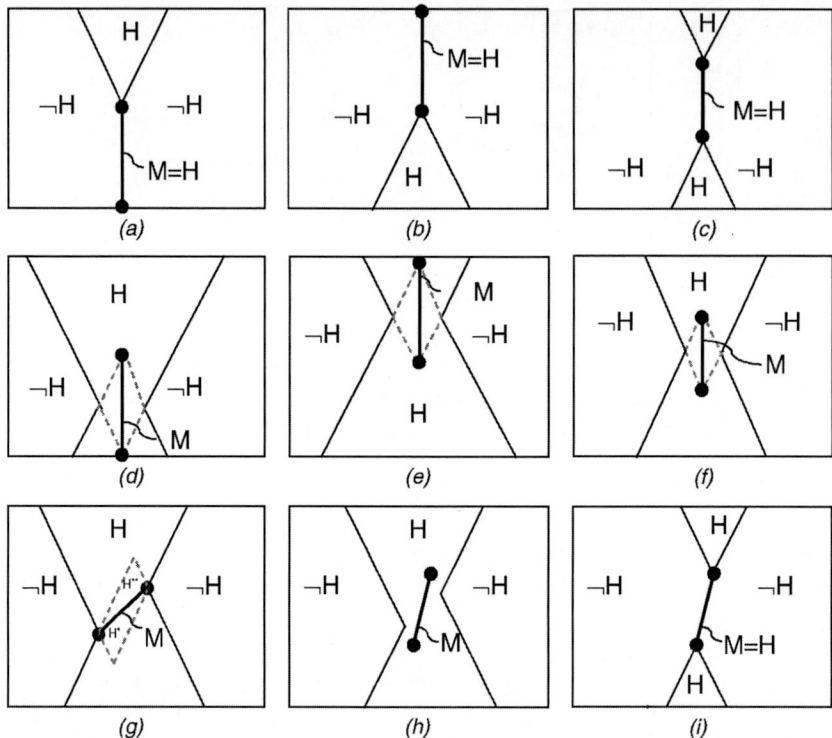

Fig. 10. Various *H-space* shapes resulting from different compulsion types: (a) compulsion to continually occupy until a specified time; (b) compulsion to continually occupy after a specified time; (c) compulsion to continually occupy during a temporal interval; (d) compulsion to occupy for an instant until a specified time; (e) compulsion to occupy for an instant after a specified time; (f) compulsion to occupy for an instant during a temporal interval; (g) compulsion to occupy for an instant moving faster than wayfinder; (h) compulsion to occupy for an instant moving slower than the wayfinder; and (i) compulsion to occupy continuously moving slower than the wayfinder.

The remainder of space-time that is inaccessible as a result of meeting these compulsions is *¬H-space*. The *H-space* in a 2-dimensional spatial world is a lifeline bead and a sequence of valid compulsion objects in space-time create a connected sequence of *H-spaces* as in a lifeline necklace (Figure 8, Hornsby and Egenhofer 2002).

The various compulsion shapes (Figure 3) create different *H-spaces* and *¬H-spaces*. The implication of a compulsion over a spatial region is a much wider *H-space* than a space-time point compulsion, indicating the additional flexibility afforded to the wayfinder (Figure 9). An additional characteristic of a compulsion occupying a spatial extent is—due to the maximum speed restrictions of the wayfinder—the creation of spaces before and after each compulsion that allow us to reason that if the wayfinder travels through these spaces the compulsion absolutely will be met, denoted by H^*, or absolutely was met, denoted by H^{**} (Figure 9).

The different compulsions occurring over a subset of the available time in ST create distinctive *H-spaces*. Consider the case where a compulsion lasts only until a certain time. For example, "you *must* stay home until 8:00 am." In this case, the way-

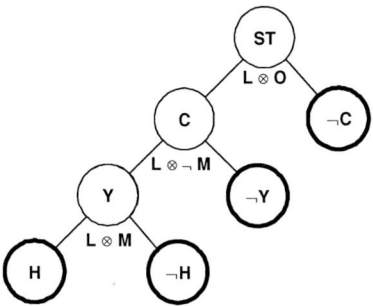

Fig. 11. Sequential partitioning of space-time with the four wayfinding primitives. Space-time is partitioned by the speed limitation L and start point O into *C-space* and *¬C-space*. *C-space* then is partitioned by the speed limitation and the barriers *¬M* into *Y-space* and *¬Y-space*. Finally *Y-space* is partitioned by the speed limitation and compulsions *M* into *H-space* and *¬H-space*. The four highlighted spaces occupying the leaf nodes of the graph indicate the results of a combined sequential structuring of wayfinder's space-time environment.

finder's travel possibilities are completely restricted until after the compulsion ends and *H-space* equals the *M-space*. After the compulsion ends, the travel possibilities expand, as shown in the increasing size of the *H-space* (Figure 10a). The alternate compulsion that the traveler *must* occupy some location after a specified time creates a similar *H-space*, but oriented temporally in the opposite direction. For example, "be home by 10:00 pm." In this case, the wayfinder's *H-space* continually decreases as the compulsion nears until it equals the *M-space* (Figure 10b). The *during-compulsion* over a temporal interval creates an *H-space* shape that is the union of the two previous examples (Figure 10c). The implications of compulsions that do not require the wayfinder to occupy the compulsion space during the entire temporal interval create different *H-spaces* than those that do. These *H-spaces* are larger, indicating additional flexibility (Figure 10d-f).

The *H-space* created by a moving compulsion varies depending on whether its speed is greater than the wayfinder's maximum travel speed. A compulsion that moves faster than the wayfinder's maximum speed cannot be a valid compulsion to occupy for its entire time, but may be a valid compulsion to occupy for an instant. The *H-space* created is similar to a compulsion over a spatial extent (Figure 10g) and, therefore, spaces are created that allow us to reason about whether or not the wayfinder absolutely will meet or absolutely did meet the compulsion. Moving compulsions slower than the wayfinder may be of either type, creating *H-spaces* similar to non-moving compulsions (Figure 10h and 10i).

5   Sequential Partitioning
of the Wayfinder's Space-Time Environment

The wayfinding primitives may instead of separately partitioning space-time, sequentially partition each other in a hierarchical manner. The first partition of space-time is created by combining the maximum speed limitation and the start point, yielding *C-space* and *¬C-space*. The *C-space* is then partitioned by a set of barriers (¬M) into *Y-space* and *¬Y-space*. This process is run once more by partitioning *Y-space* with the

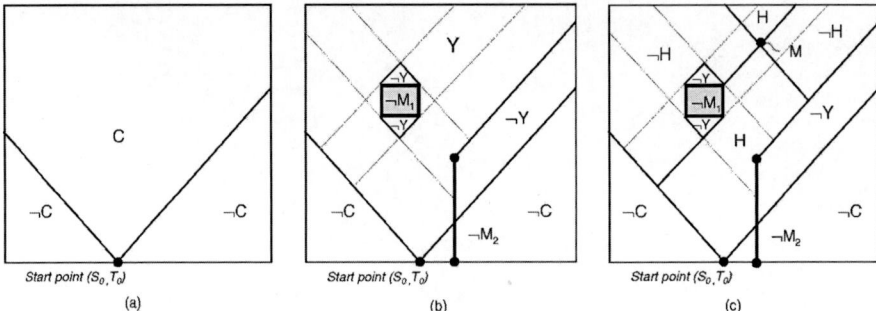

Fig. 12. The structuring of a wayfinder's space-time environment by sequential partitioning: (a) combining the speed limitation and start point to partition the space time container into *C-space* and *¬C space*; (b) partitioning *C-space*, by combining the barriers and maximum travel speed into *Y space* and *¬Y-space*; and (c) partitioning *Y-space* by the compulsion into *H space* and *¬H-space*. Spaces influencing potential *¬Y-spaces* are delineated by dashed lines.

compulsions (M) into *H-space* and *¬H-space*. This sequential partitioning of space-time is represented graphically as shown in Figure 11. The leaf nodes of this graph indicate the resulting spaces from a sequential partitioning of space-time: *¬C-space*, *¬Y-space*, *¬H-space*, and *H-space*.

To illustrate the sequential partitioning technique, consider a wayfinding environment consisting of a maximum travel speed, a start point (S_0, T_0), a compulsion M, and two barriers $¬M_1$ and $¬M_2$. The maximum travel speed, combined with the wayfinder's start point, partitions space-time into *C-space* and *¬C-space* (Figure 12a). The *C-space* is then partitioned by combining the barriers with the maximum travel speed to yield *Y-spaces* and *¬Y-spaces* (Figure 12b). Finally, *Y-space* is partitioned by combining the wayfinder's compulsion with the travel speed limitation creating *H-spaces* and *¬H-spaces* (Figure 12c).

6   Describing Partitions with Modal Verbs

To describe the partitions of the structure created with this approach in a cognitively plausible manner, the modal verbs *can, may, must,* and *should* are used. Modal verbs indicate whether things, events, or relations are actual, possible, or necessary (Johnson 1987). Sweetser (1990) argues that the meaning of modal verbs as used in the physical or social realm are similarly used for argument and reasoning. In an inherently physical act, such as wayfinding, describing spatial and temporal constraints modally provides insights for argument and reasoning of the wayfinding task.

Though each modal verb is used in various ways during normal conversations, in this approach *can* is related to a positive ability (capability); *must* denotes obligation or compelling force; and *may* is roughly associated with permission or lack of a potential barrier (Sweetser 1990). Johnson relates each of these modal verbs to various image schemata, which structure knowledge through abstract high-level experiential gestalts of common situations (Johnson 1987). For example, *must* is a compulsion; *may* is the removal of restraint; and *can* is enablement. In addition, we include one

Table 1. Assignment of modal verbs to the primitives and partition spaces of the wayfinder's space-time environment.

Space	Modal Verb	Example Usage
M	*must*	"You *must* be home by 5:00 pm."
$\neg M$	*must not*	"You *must not* cross the railroad tracks."
C	*can*	"I *can* be at the restaurant at noon."
$\neg C$	*cannot*	"I *cannot* get to the post office by 2:00 pm."
Y	*may*	"You *may* cross the river, because the bridge is open."
$\neg Y$	*may not*	"You *may not* be in the library at 9:00 pm, because it closes at 8:00 pm."
H	*should*	"To get to the restaurant by noon, I *should* cross the bridge before it closes."
$\neg H$	*should not*	"To get to the post office by 5:00 pm, I *should not* go first to the grocery."

other modal verb to describe a wayfinder's space-time structure, *should*, indicating a weaker form of *must*. Defining a wayfinder's compulsions (*must*), cabilities (*can*), and permissions (*may*) while traveling through a volume of space-time provides a concise, yet simple description of a wayfinding scenario.

Modal verbs are associated with the various wayfinding primitives and the partitions of space-time. We begin by assigning the modal verb *must* (M) to compulsions and *must not* ($\neg$M) to barriers. The modal verb *can* (C) effectively describes the accessible partition of space-time created by the combination of the start point and maximum travel speed, or the space that the wayfinder's capability allows travel to. Its complement *cannot* ($\neg$C), describes the inaccessible spaces of this partition. The modal verb *may* (Y), effectively describes the partition of *C-space* that the wayfinder has access to based on temporary barriers in space-time. Again, its complement *may not* ($\neg$Y) is employed to describe the spaces made inaccessible as a result of barriers in space-time. Finally, the modal verbs *should* and *should not* are employed to describe *Y-space* and $\neg$*Y-space,* respectively. A summary of these modal verb assignments, along with example usages, is shown in Table 1.

7   Conclusions

This paper introduced an approach to structure and describe the dynamic space-time environment of a wayfinder. This structure allows for better situational awareness by providing a mechanism to reason about space-time decision points and the impacts of these decisions on future possibilities. Four primitives of spatio-temporal wayfinding were identified: the maximum travel speed, the start point in space and time, a set of barriers, and a set of compulsions. Leveraging existing space-time prism concepts with symbolic projection techniques provides a framework for combining the speed limitation with the remaining wayfinding primitives to create various forms of accessible and not accessible spaces. It was shown that the combination of the speed limitation with the start point partitioned space-time into accessible and inaccessible spaces based on the wayfinder's capability. Combining the speed limitation and the barriers created accessibility partitions based on the constraints of these barriers. The combination of speed limitation with the wayfinder's compulsions partitioned space-time into those areas that the wayfinder should and should not travel through to meet the requirements. These three partitioning mechanisms were then integrated into a sequential partitioning scheme that creates four partition categories of a wayfinder's space-time environment: $\neg C$-space, $\neg Y$-space, H-space, and $\neg H$-space. To describe these spaces in a cognitively plausible manner, the modal verbs *can, must, may,* and *should* along with their complements were assigned to appropriate primitives and partition spaces.

The structure created with this approach could be represented as a directed graph. The partitions would be the nodes of the graph, and the edges would represent the ability of the wayfinder to travel between these partitions in space and time. By employing a graph to represent this structure, a rich set of theoretical and applied techniques can be leveraged to develop the structure and answer queries. For example, a simple least cost algorithm could be used to determine if a proposed compulsion is located in *H-space (should)* and, therefore, can be added to the wayfinder's travel plan. In addition, a search of the directed graph would indicate what travel options are available to the wayfinder.

Acknowledgments

This work was partially supported by the National Imagery and Mapping Agency under grant numbers NMA201-00-1-2009, NMA201-01-1-2003, and NMA401-02-1-2009, The National Science Foundation under grant numbers 11S-9970123 and EPS-9983432, and the National Institute of Environmental Health Sciences, NIH, under grant number 1 R 01 ES09816-01

References

Andre, A., B. Hooey, D. Foyle and R. McCann. 1998. Field Evaluation of T-NASA: Taxi Navigation and Situation Awareness System. In: *Proceedings of the AIAA/IEEE/SAE 17th Digital Avionics System Conference.* 1-8.

Azumea, R., H. Neely, M. Daily and R. Geiss. 2000. Visualization Tools for Free Flight Air-Traffic Management. *IEEE Computer Graphics and Applications.* 20(5): 32-36.

Barlow, H. 1998. Cerebral Predictions. *Perception.* 27: 885-888.

Boroditsky, L. 2000. Metaphoric Structuring: Understanding Time through Spatial Metaphors. *Cognition.* 75(1): 1-28.

Chang, S. and E. Jungert. 1986. A Spatial Knowledge Structure for Image Information Systems Using Symbolic Projections. In: *Proceedings of 1986 Fall Joint Computer Conference*, Dallas, Texas, IEEE Computer Society Press. 79-86.

Chang, S. K. and E. Jungert. 1996. *Symbolic Projection for Image Information Retrieval and Spatial Reasoning.* London: Academic Press.

Davies, C. and E. Pederson. 2001. Grid Patterns and Cultural Expectations in Urban Wayfinding. In: D. Montello, (Ed.) *Spatial Information Theory - Foundations of Geographic Information Science, COSIT '01*, Morro Bay, CA, Springer. 400-414.

Durso, F., C. Hackworth, T. Truitt, J. Crutchfield, D. Nikolic and C. Manning. 1999. *Situation Awareness as a Predictor of Performance in Enroute Air Traffic Controllers*. Technical Report DOT/FAA/AM-99/3. U.S. Department of Transportation, Federal Aviation Administration, Office of Aviation Medicine.

Eby, D. and L. Kostyniuk. 1999. An On-the-Road Comparison of In-Vehicle Navigation Assistance Systems. *Human Factors*. 41(2): 295-311.

Egenhofer, M. and D. Mark. 1995. Naive Geography. In: A. Frank and W. Kuhn, (Eds.) *Spatial Information Theory: A Theoretical Basis for GIS (COSIT '95)*, Semmering Austria, Springer-Verlag. 1-15.

Ellis, C. and D. Johnston. 1999. Qualitative Spatial Representation for Situational Awareness and Spatial Decision Support. In: C. Freksa and D. Mark, (Eds.) *Spatial Information Theory - Foundations of Geographic Information Science, (COSIT 1999)*, Stade, Germany, Springer-Verlag. 449-460.

Endsley, M. 1988. Design and Evaluation for Situation Awareness Enhancement. In: *Human Factors Society 32nd Annual Meeting*, Santa Monica, CA, Human Factors and Ergonomic Society. 97-101.

Endsley, M. 1995. Toward a Theory of Situational Awareness in Dynamic Systems. *Human Factors*. 37(1): 32-64.

Epstein, S. 1997. Spatial Representation for Pragmatic Navigation. In: S. Hirtle and A. Frank, (Eds.) *Spatial Information Theory: A Theoretical Basis for GIS (COSIT '97)*, Laurel Highlands, PA, Springer. 373-388.

Galton, A. 1997. Space, Time, and Movement. In: *Spatial and Temporal Reasoning*. O. Stock. Dordrecht, Kluwer Academic: 321-352.

Golledge, R. 1999. *Wayfinding Behavior*. Baltimore: The Johns Hopkins University Press.

Hägerstrand, T. 1967. *Innovation Diffusion as a Spatial Process*. Chicago: University of Chicago Press.

Hariharan, R. and K. Hornsby. 2000. Modeling Intersections of Geospatial Lifelines. In: *First International Conference on Geographic Information Science, GIScience 2000*, Savannah, GA. 208-210.

Hershergery, J. and S. Suri. 1999. An Optimal Algorithm for Euclidean Shortest Paths in the Plane. *Society of Industrial and Applied Mathematics (SIAM) Journal of Computing*. 28(6): 2215-2256.

Holmes, P. and E. Jungert. 1992. Symbolic and Geometric Connectivity Graph Methods for Route Planning in Digitized Maps. *IEEE Transactions on Pattern Analysis and Machine Intelligence*. 14(5): 549-565.

Hornsby, K. and M. Egenhofer. 2002. Modeling Moving Objects over Multiple Granularities. *Annals of Mathematics and Artificial Intelligence*. 36: 177-194.

Johnson, M. 1987. *The Body in the Mind, the Bodily Basis of Meaning, Imagination, and Reason*. Chicago: The University of Chicago Press.

Jungert, E. 1988. Extended Symbolic Projections as a Knowledge Structure for Spatial Reasoning and Planning. In: *Pattern Recognition*. J. Kittler. New York, Springer Verlag.

Jungert, E. 1992. The Observers Point of View: An Extension of Symbolic Projections. In: A. Frank, I. Campari and U. Formentini, (Eds.) *Theory and Methods of Spatio-Temporal Reasoning in Geographic Space*, Pisa, Italy, Springer-Verlag. 179-195.

Kerstholt, J. and J. Raajimakers. 1997. Dynamic Task Environments. In: *Decision Making, Cognitive Models and Explanations*. R. Ranyard, R. Crozier and O. Svenson. New York, Routledge: 205-217.

Kwan, M.-P. 1998. Space-Time and Integral Measures of Individual Accessibility: A Comparative Analysis Using a Point-Based Framework. *Geographical Analysis*. 30(3): 191-216.

Lee, D. and F. Preparata. 1984. Euclidean Shortest Paths in the Presence of Rectilinear Barriers. *Networks*. 14: 393-410.

Miller, H. 1991. Modeling Accessibility Using Space-Time Prism Concepts within Geographical Information Systems. *International Journal of Geographical Information Systems.* 5(3): 287-301.

Miller, H. 1999. Measuring Space-Time Accessibility Benefits within Transportation Networks: Basic Theory and Computational Procedures. *Geographical Analysis.* 31(2): 187-212.

NRC. 1997. *Tactical Display for Soldiers: Human Factors Considerations.* Washington, D.C.: National Academy Press.

Pred, A. 1977. The Choreography of Existence: Comments on Hägerstrand's Time-Geography and Its Usefulness. *Economic Geography.* 53: 207-221.

Raubal, M. and M. Egenhofer. 1998. Comparing the Complexity of Wayfinding Tasks in Built Environments. *Environment & Planning B.* 25(6): 895-913.

Raubal, M., M. Egenhofer, D. Pfoser and N. Tryfona. 1997. Structuring Space with Image Schemata: Wayfinding in Airports as a Case Study. In: S. Hirtle and A. Frank, (Eds.) *Spatial Information Theory: A Theoretical Basis for GIS, (COSIT '97),* Laurel Highlands, PA, Springer-Verlag. 85-102.

Sholl, J. 2001. The Role of a Self-Reference System in Spatial Navigation. In: D. R. Montello, (Ed.) *Spatial Information Theory, Foundations of Geographic Information Science, (COSIT 2001),* Morro Bay, CA, Springer. 217-232.

Stefanakis, E. and M. Kavouras. 1995. On the Determination of the Optimal Path in Space. In: A. Frank and W. Kuhn, (Eds.) *Spatial Information Theory: A Theoretical Basis for GIS (COSIT '95),* Semmering Austria, Springer. 241-257.

Sweetser, E. 1990. *From Etymology to Pragmatics.* Cambridge: Cambridge University Press.

Timpf, S. 2002. Ontologies of Wayfinding: A Traveler's Perspective. *Networks and Spatial Economics.* 2(1): 9-33.

Weibull, J. 1980. On the Numerical Measurement of Accessibility. *Environment & Planning A.* 12: 53-67.

Yuan, M. 1997. Use of Knowledge Acquisition to Build Wildfire Representation in Geographical Information Systems. *International Journal of Geographic Information Science.* 11(8): 723-745.

Zhang, W. and R. Hill. 2000. A Template-Based and Pattern-Driven Approach to Situation Awareness and Assessment in Virtual Humans. In: *Fourth International Conference on Autonomous Agents,* Barcelona, Spain. 116-123.

Systematic Distortions in Cognitive Maps: The North American West Coast vs. the (West) Coast of Israel

Juval Portugali and Itzhak Omer

ESLab (Environmental Simulation Lab)
Department of Geography and the Human Environment
Tel Aviv University, Tel Aviv 69978, Israel
{juval,omery}@post.tau.ac.il

Abstract. This article suggests a second thought on two papers published in *Cognitive psychology* in 1978 and 1981. Both articles deal with systematic distortions in cognitive mapping and both are based on experiments conducted along the North American West Coast. The first, by Stevens and Coupe, deals with distortions due to hierarchical organization while the second, by Tversky, with distortions due to rotation. Our second thought follows a set empirical results from a study conducted along the (West) coast of Israel. These results suggest that the experiments, on the basis of which the above two forms of systematic distortions were determined, could have resulted from another form of systematic distortion that we term the *edge effect*.

1   Introduction

This paper suggests a second thought on two types of 'systematic distortions in cognitive mapping' [1] – distortions due to 'hierarchical organization' and 'rotation'. The motivation to reexamine these distortions followed the analysis of a set of experiments conducted in Israel on the phenomenon of systematic distortions. On the one hand, these experiments have exposed systematic distortions that remind one of the above effects of hierarchy and rotation. On the other hand, they have also raised some doubts about the validity and generality of systematic distortions in cognitive mapping due to hierarchy and rotations as they currently appear in the literature.

The first two sections of the paper introduce distortions due to hierarchy then examine the phenomenon critically in light of the results from Israel. The next pair of sections does the same to distortions due to rotation. The section that follows introduces a new form of systematic distortion termed the *edge* effect. It shows that the edge effect might explain several equivocal cases of rotation. The paper concludes by noting that in certain cases the edge effect might provide a better interpretive framework for both hierarchy and rotation.

2   From San Diego California to Reno Nevada

In a paper entitled "Distortions in judged spatial relations" [2], Stevens and Coupe report on an experiment in which subjects in San Diego, California were asked to

W. Kuhn, M.F. Worboys, and S. Timpf (Eds.): COSIT 2003, LNCS 2825, pp. 93–100, 2003.
© Springer-Verlag Berlin Heidelberg 2003

indicate from memory the direction to Reno Nevada (as well as to, and between, several other American cities). This, by drawing a line in the proper orientation on a circle with north noted at the top. Most subjects have indicated that Reno is northeast of San Diego [2], while from Fig. 1 it can be seen that it is, in fact, northwest.

Fig. 1. California and Nevada – The geographical context for Stevens & Coupe's [2] and Tversky's [4] studies

Stevens and Coupe's interpretation is that this distortion is due to *hierarchical organization* of spatial knowledge. That is, people tend to store in memory not the exact, or relative, location of all cities, but rather the relative location of states. Thus, when asked to make judgment about directions between cities, subjects infer the direction between cities from the spatial relations between the states. In our case such an inference can go like this: Since Reno is in Nevada and since Nevada is generally east of California, then every city in Nevada must be east of every city in California. Reno must, therefore, be east of San Diego. In terms of memory space, this way of hierarchical organization of spatial information is evidently more efficient and economical, but it has a price – systematic distortion in cognitive mapping due to hierarchy.

Since its publication in 1978 several subsequent studies have found similar results, have elaborated various aspects and have established the working of systematic distortions in cognitive maps due to hierarchy. In all these studies, as well in several textbooks that have since appeared, Stevens and Coupe's San Diego-Reno experiment stands as the seminal, paradigmatic case study. But is it really so? Does their experiment really indicate the working of hierarchical organization in cognitive mapping?

3   From Haifa to Be'er Sheva

Our doubts regarding Stevens and Coupe's interpretation of the San Diego–Reno experiment followed a systematic distortion we've found in the relations between the cities Haifa and Be'er Sheva in Israel (Fig. 2), as described in Experiment 1.

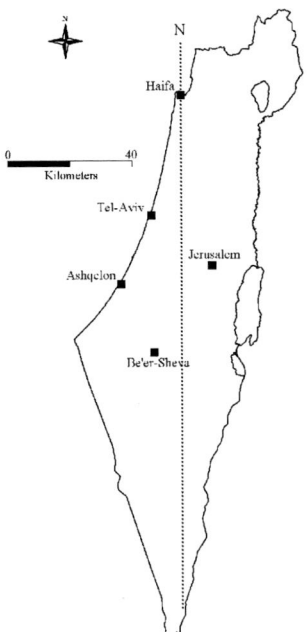

Fig. 2. Map of Israel – The geographical context for experiments conducted in the present study

Experiment 1: Two groups – 28 undergraduate geography students and 25 under-graduate architecture students at Tel Aviv University – participated in this experiment, that was conducted within the frame of a "cognitive geography" class. The subjects of the first group were asked to judge the direction from Be'er Sheva to Haifa, while those from the second group from Haifa to Be'er Sheva. The subjects were asked to draw a line from the center of a 2.7 cm radius circle according to the following instructions:

Imagine that Beer Sheva is at the center of the circle and that you are in Be'er Sheve. Draw a straight line from <u>*Be'er Sheva*</u> *to the direction of* <u>*Haifa*</u>*. Make sure that the line reaches the edge of the circle.*

At the outset of the experiment the experimenter made sure that the subjects understand the task. At the end of the experiment the responses were screened and distortions of over 60° from the true direction were taken out.

As can be seen, our experimental set-up was *almost* identical to Stevens and Coupe's. Comparing Fig. 1 with Fig. 2 one can see that both the North American West Coast and the Israeli coastline run in a general north-south direction and that both bend somewhat from the exact north-south axis: The West Coast towards SE, while the Israeli coastline towards SW. Furthermore, Haifa in Israel has a geographical position similar to San Diego in California – both are located "on the sea"; while Be'er Sheva in Israel is the parallel of Reno in the USA – both are inland, far from the coastline.

The two experimental setups differ, however, in two respects. First in scale; the Israeli coast is some 250 km long, while the North American one some 2,200 km long.

Second, and this is of major significance in the present context, the Israeli study area is not subdivided into states – there are no parallels to California and Nevada here. True, one can still claim that Be'er Sheva is in the Negev – Israel's southernmost region. But all that can be said about the Negev is that it is south of any other region in Israel, not east nor west of it.

The results from experiment 1 are presented in Fig. 3. The average discrepancy from the true direction of the two sets of judgment is 21°. As can be seen, the systematic distortions in Israel are identical to those found by Stevens and Coupe in the US West Coast. But in Israel, as noted, there is no California and Nevada and the systematic distortion in the cognitive maps of Israelis cannot be a consequence of hierarchical organization of spatial geographical relations. What is it then?

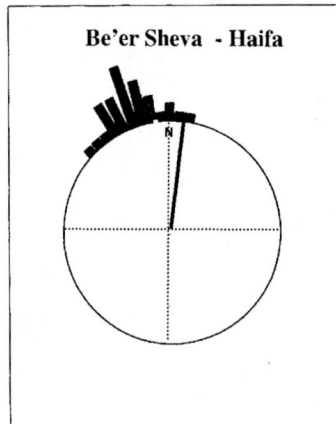 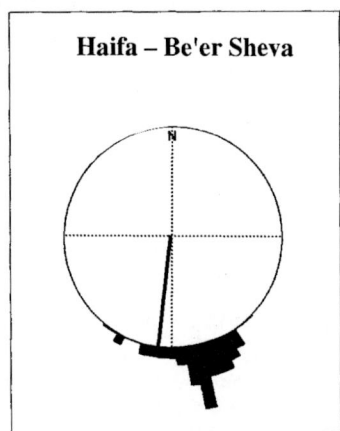

Fig. 3. Experiment 1: distribution of responses. Bold lines indicate the true directions

In a seminar that took place in Israel in 1988, Barbara Tversky suggested that since the coastline of Israel bends from north to SW, it is rather likely that it might cause systematic distortions in cognitive mapping due to *rotation* (See report on that seminar in [3]). The notion of rotation effect was first introduced by Tversky [4] in a study where she asked subjects to judge the direction between several cities located along the Bay Area coastline. Most subjects correctly indicated that Berkeley, for example, is north of Palo Alto, but they incorrectly indicated that Berkeley is east of Palo Alto. Similar systematic distortions were identifies by Tversky with respect to other cities along the Bay Area. Tversky's interpretation to the above distortions was that, in their minds, people tend to rotate geographical objects so that they will correspond the exact N-S frame of reference. As a consequence, in cases where the orientation of a geographical object is somewhat in "conflict" with the "natural" 90 degrees N-S axis, the N-S frame of reference tends to "win" and people wrongly judge spatial relations.

Since in the Haifa – Be'er Sheva case 'hierarchy' as a cause for systematic distortion must be ruled out, we thought that Tversky's 'rotation effect' might provide an alternative and more appropriate interpretational framework. The logic of this interpretation might be as follows: The Israeli coastline runs from north to SW, but people tend to rotate it, in their mind, to the exact N-S axis. And since Haifa is on the coastline itself and Be'er Sheva some 60 km east of it, people wrongly infer or imagine that it must be SE of Haifa. And vice versa: When considering the task from the point of view of Be'er Sheva, they imagine/infer that Haifa's direction is NW.

Cases of systematic distortion due to rotation were typically identified between pairs of cities located along, or close to, the coastline – along the San Francisco Bay Area, for example. Now, Haifa in Israel is indeed on the coastline, but Be'er Sheva is some 60 km from the sea. The latter case study thus indicates that the effect of rotation might extend beyond the area of the coastline. Is such an extended rotation effect active only in the Israeli case or is it a general property active also in the west coast of California and its Bay Area were the systematic distortions due to hierarchy and rotation were originally identified? The goal of experiment 2 was to answer this question, concerning the spatial relations between San Diego and two sites in north California: the city of Sacramento and Lake Tahoe. Both places are located far from the coastline; Sacramento is within California while Lake Tahoe is divided between Californian and Nevada. No hierarchy affect could be at work with respect to Sacramento.

Experiment 2: This experiment was conducted with the help of Ms. N. Taylor – a graduate student at Tel Aviv University and a former undergraduate student at San Diego. 22 undergraduate students, living in San Diego, participated voluntarily in the experiment. They were asked to judge the direction from San Diego to Sacramento and Lake Tahoe. The procedure and method employed was identical to experiment 1. The subjects were asked the following:

Imagine that San Diego is at the center of the circle:
1. Draw a straight line from <u>San Diego</u> to the direction of <u>Sacramento</u>. Make sure that the line reaches the edge of the circle.
2. Draw a straight line from <u>San Diego</u> to the direction of <u>Lake Tahoe</u>. Make sure that the line reaches the edge of the circle.

The results from experiment 2 are presented in Fig. 4. Although the pairs of locations represent different principle geographical positions, the results show the same systematic error: The subjects judge the direction between the locations as running southwest to northeast instead of southeast to northwest.

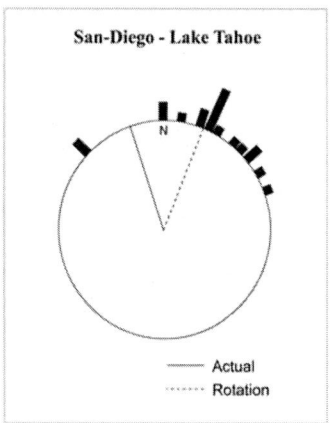
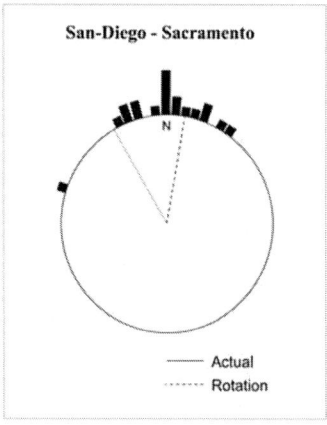

Fig. 4. Experiment 2: Distribution of responses.

4   Systematic Distortion due to Rotation – a Second Thought

The above results about the extended rotation effect entailed two questions: Firstly, how far from the reference line the rotation effect is active? Secondly, does it effect also the perceived spatial relations between cities that are far from the reference line? To answer these questions we've conducted Experiment 3.

Experiment 3: 27 students from Tel Aviv University participated voluntarily in this experiment in the context of an urban geography class. The procedure and methodology employed here are identical to the previous two experiments. The subjects were asked to draw a line between the following pairs of cities in Israel (see map in Fig. 2), when each of the four tasks was implemented separately:

a. From Haifa (on the sea) to Tel-Aviv (on the sea).
b. From Tel-Aviv (on the sea) to Ashkelon (on the sea).
c. From Haifa (on the sea) to Jerusalem (inland – further east of Be'er Sheva).
d. From Be'er Sheva (inland) to Jerusalem (inland).

The results of experiment 3 are presented in Fig. 5. Firstly, as we've already seen above (Fig. 3), there are significant systematic distortions in the pair Haifa-Be'er Sheva (The average distortion from the true direction is 47 °). Secondly, as can be seen, there are also significant distortions with respect to the pair Haifa–Jerusalem. Thirdly, no distortions were found when subjects were asked to judge the direction from Be'er Sheva to Jerusalem and vice versa. Fourthly, and this is a rather surprising finding, no significant distortions were found between the pairs Haifa–Tel Aviv and Tel Aviv–Ashkelon.

The intermediate conclusions from the above are straightforward: What we've considered so far as the rotation effect is active when one site/city of the pair in question is on the reference line and the other is inland; it is not active when both sites are not related to the reference line – people do not rotate the entire country, in their mind, in line with the exact N-S axis. Furthermore, it is not effective either when the pairs of sites are on the reference line itself – the Israeli coastline in our case.

5   Rotation or Edge?

The above findings, in particular the fourth point, put a question mark on the whole idea of systematic distortion due to rotation: If people really tend to rotate the coastline in line with the N-S axis, then in the above tests they should have responded that Tel Aviv is south of Haifa and Ashkelon is directly south of Tel Aviv. The analysis of their responses indicates that they make no such mistakes; they know that the Israeli coastline bends toward SW and that as a consequence Tel Aviv is SW of Haifa and Ashkelon is SW of Tel Aviv. In other words, the effect of the Israeli coastline on systematic distortions is not unequivocal. In some cases it indeed causes distortions, but in other cases it "helps" subjects to judge spatial relations accurately. We suggest calling this equivocal effect of the coast as a reference line *the edge effect*.

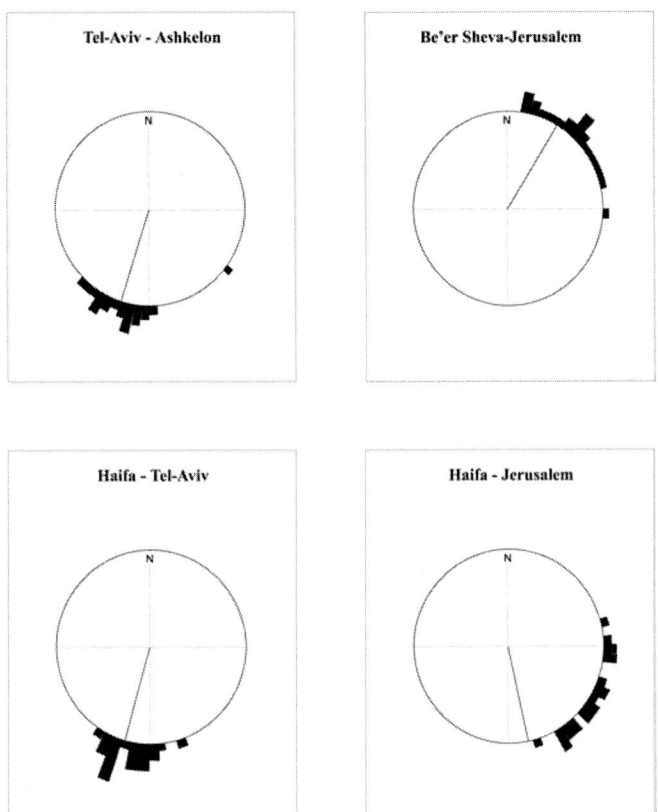

Fig. 5. Experiment 3: Distribution of responses. Bold lines indicate the true directions

The notion of 'edge', as is well known, goes back to Lynch's [5] *The Image of the City*. According to him edges are very useful elements when dealing with perceived spatial relations in large-scale extended environments such as cities – they provide a very prominent reference *line*. Thus, unlike Tversky's notion of 'rotation,' that has the "negative" effect as a cause of distortions, Lynch's notion of edge is associated with the "positive" effect of improving *legibility*. Edges belong to a set of five elements (edge, path, landmark, node, district) that according to Lynch are specifically significant in making the city and its image legible.

As we've just seen, due to the edge effect of the bending Israeli coastline, our subjects were rather accurate when judging spatial relations between the three 'edge cities': Haifa, Tel Aviv and Ashkelon. The difficulty started when they were asked to judge the relations between the edge city Haifa and the "floating point" cities Be'er Sheva and Jerusalem. Most subjects simply underestimated the cumulative effect of a bending edge. That is to say, that what to the eye looks as a slight inclination westward, off the N-S axis, near Haifa, might accumulate to some 60 km at the south part of the country, close to 200 km down the coastline south of Haifa. On the Israeli coastline this is particularly prominent because as can be seen from the map (Fig. 2), the Israeli coastline bends westward "exponentially" in a nonlinear fashion; and nonlinear relations are counter-intuitive and hard to perceive and judge.

6   Concluding Notes

The above results from Israel and California have direct implications to the original Steven and Coupe interpretation of the San Diego–Reno systematic distortions: The actual direction of the Californian coast is NW-SE. San Diego is on the coastline itself, while Reno is inland, far away from the reference line of the west coast; exactly like Be'er Sheva in Israel and Sacramento and Lake Tahoe in California. In light of the above results, the edge effect might explain the San Diego–Reno distortion without the need to employ the hierarchical organization effect, nor the rotation effect.

There is no way to tell, of course, whether the systematic distortions made by Stevens and Coupe's subjects were due to hierarchical organization, as it is commonly assumed, or due to the other effects discussed above. That is to day, due to the rotation effect, or due to the edge effect – that to our view explains the systematic distortions we've identified between cities in Israel, as well as in relation to Sacramento and probably also Lake Tahoe in California – or possibly due to several effects. But, it can clearly be stated that from the point of view of Popper's principle of falsification, the experimental results from Israel and California falsify the interpretation suggested by Stevens and Coupe.

Certainly it would be hasty to claim that our results falsify the very possibility of the hierarchical organization effect, the rotation effect and the very possibility of systematic distortions due to them. In fact, several other studies have demonstrated the effect of hierarchy in different circumstances, even when the area concerned had no prior hierarchical subdivision [6]. Our results falsify only what has become a canonical interpretation of the Stevens and Coupe's experiment – that the distortions of their subjects were due to hierarchy only. The lesson from the above is rather interesting; it shows that a fallacious interpretation might entail true results – but this is an issue that goes beyond the short research note we are engaged with here.

References

1. Tversky, B.: Distortions in Cognitive maps. *Geoforum,* **23** (1992) 131-138.
2. Stevens, A., & Coupe, P.: Distortions in Judged Spatial Relation. *Cognitive Psychology* **10** (1978) 422-437
3. Glicksohn, J.: Rotation, Orientation and Cognitive Mapping. *American Journal of Psychology* **107** (1994) 39-51
4. Tversky, B.: Distortions in Memory for Maps. *Cognitive Psychology* **13** (1981) 407-433
5. Lynch, K.: *The Image of the City.* Cambridge: Mit Press (1960)
6. McNamara, T.: Spatial Representation. *Geoforum 23*, 39-150

Tripartite Line Tracks
Qualitative Curvature Information

Björn Gottfried

Artificial Intelligence Group, TZI, University of Bremen
Universitätsallee 21-23, 28359 Bremen, Germany
bg@tzi.de

Abstract. We present a qualitative shape description which has previously been proven to be useful for object categorisation. The description is based on a set of shape primitives which we will restrict to a new subset of relations representing stylised curvature information. In contrast to other qualitative shape theories, this description enables us to distinguish different convex shapes. This is especially interesting from a cognitive point of view since these shapes show salient visual differences. It turns out that the distinction between two sides of a line together with the distinction between acute and obtuse angles make up a powerful concept of orientation information for shapes in two dimensions.

1 Introduction

In recent years, a number of qualitative shape descriptions has been introduced. Among other things, theories of shapes can be classified with regard to the entities on which they are based. Schlieder, 1996, introduces a *point*-orientated approach, describing polygonal shapes by triangle orientations of the polygons' vertices. Galton & Meathrel, 1999, propose a representation of *outlines* by means of strings over an alphabet of seven qualitative curvature types. A *region*-orientated technique has been investigated by Cohn, 1995, who distinguishes different concave shapes by considering the notion of the connection of regions and their convex hulls. Different as these approaches may be, common to all of them is their consideration of substructures in order to describe qualitative shape properties. The first approach considers properties which are defined by the curvature extrema of boundaries, while the other two refer to structures such as *a line segment followed by an inward pointing cusp*, or *a region consisting of two connected concavities which themselves are convex*. In this way, these approaches resemble theories pertaining to shape decompositions. Shape decompositions into parts have been mainly established in the psychological community by the work of Biederman, 1987, and from a computer science point of view by Marr & Nishihara, 1978. Newer theories of the kind mentioned above are more related to the natural structure of shapes than to artificial parts such as Biederman's *geons* or Marr's *generalised cylinders*, which have been devised to approximate shapes.

W. Kuhn, M.F. Worboys, and S. Timpf (Eds.): COSIT 2003, LNCS 2825, pp. 101–117, 2003.

In this paper, we are emphasising methods which take into account the boundaries of shapes. Boundaries are particularly important in performing grasping movements, and in navigating efficiently around obstacles. The former task requires knowledge about the interface between an object's region and its surroundings, and the latter requires knowledge about the curve progression along obstacles. Boundaries are also related to object recognition, in that visual perception is particularly sensitive to discontinuities in space which arise at the boundaries of objects. It is a perpetual problem in performing tasks such as those mentioned above that boundaries are in general imprecisely determined. We will demonstrate a reasonable way of coping with imprecise boundaries, in particular by introducing a qualitative concept of curvature information.

We will concentrate on the outlines of boundaries in two dimensions, since some interesting shape properties can already be investigated in two dimensions. Moreover, properties which are essential to our approach have to be investigated by considering polygons, and in addition many algorithms exist for polygonal approximations of shapes, as for instance Horng & Li, 2002. Accordingly, we shall confine ourselves to analysing two-dimensional polygonal outlines and dealing with the simplest extended entities, namely lines.

2   Shape and Curvature

When dealing with shapes, the question arises as to what kind of information is particularly relevant to perceiving and describing them. We motivate one possible answer to this question from a cognitive point of view. In order to do this, we shall make use of the correlation between vision and haptics. For instance, in James et al., 2002, it is shown that the haptic exploration of novel objects activates both the somatosensory cortex, relevant to processing haptic information, and areas of the occipital cortex, which is associated with visual processing. Furthermore, earlier haptic experiences of an object enhance activations in visual areas of the cortex when the same object is viewed subsequently. Such cross-modal interactions have been investigated in several studies, and for our purpose it is important to be aware of the close relationship between vision and haptics, in order to appreciate the following notion of shape. The curvature of an arbitrary shape becomes apparent when one traces its boundary with a forefinger. The two opposite extremes of the *curvature-spectrum* may be considered to be a straight line and a circle, and any shape may be viewed as made up of a mixture of different curvatures between these two extremes. On the basis of these considerations we place great importance on curvature information.

In the field of *Qualitative Spatial Reasoning*, the question is how to represent information about curvature qualitatively. When tracing a contour, we must bear in mind that curvature is related to change in orientation. Orientation is often considered in relation to reasoning about positions and it is worth analysing how adequate any orientation calculus is for a qualitative shape description. Schlieder, 1996, adapted his approach to reasoning about positions in the context of robot navigation to the description of polygons. By contrast, Cohn, 1995, and Galton & Meathrel, 1999, describe relationships between parts. Galton & Meathrel explicitly propose various different curvature types and give a grammar for their combination. The parts described by Cohn concern the concavities and convexities of regions which can either be connected or disconnected, and which may themselves consist of further concave or convex regions which can be described recursively. Curvature is treated implicitly by means of recursive descriptions of concavities. We conclude that qualitative shape descriptions either emphasise the characterisation of parts or orientation information, or treat both equally.

We propose an approach which is based on orientation information. But it can also be considered as a part-based approach, in the sense that parts may be characterised by their description of change in orientation. In this way, we represent qualitative curvature information.

The next section introduces *Tripartite Line Tracks*. In the following paragraph, the proposed set of relations is restricted to a subset of orientation relations. The expressiveness of the proposed shape description is shown by comparison with another similar approach. Finally, we define a stylised concept of curvature on the basis of the new relations and demonstrate how to reason about change in curvature.

3   Tripartite Line Tracks

Gottfried, 2002, introduces a qualitative description of the two-dimensional boundaries of polygonal shapes. This description adopts the *orientation grid* from Freksa & Zimmermann, 1992, which was originally used to reason about orientation information, to represent shape primitives. These primitives are tracks of three lines which are described by the orientation grid, as shown in Fig. 1. In section 3.1 we will see why three lines are particularly suitable in order to describe two-dimensional shape primitives.

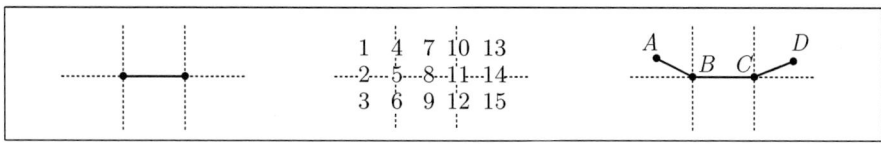

Fig. 1. Left: The orientation grid introduced by one continuously drawn line, distinguishes fifteen positions (Middle); Right: A line track $(\overline{AB}, \overline{BC}, \overline{CD})$ which consists of three connected lines

In order to construct shape primitives only general positions are considered, i.e. the six positions which do not lie on the orientation grid: $1, 3, 7, 9, 13$, and 15 in Fig. 1. The other nine positions which directly lie on the orientation grid are called singular positions. As both endpoints of one line track can be in six different general positions, there exist $6^2 = 36$ different relations as outlined in Fig. 2. These line tracks are called *Tripartite Line Tracks* ($\mathcal{TLT}$). The i-th relation is accessed by $\mathcal{TLT}(i)$. Specific $\mathcal{TLT}$-subsets are distinguishable by indices.

$\mathcal{TLT}_{36}(1)$	$\mathcal{TLT}_{36}(2)$	$\mathcal{TLT}_{36}(3)$	$\mathcal{TLT}_{36}(4)$	$\mathcal{TLT}_{36}(5)$	$\mathcal{TLT}_{36}(6)$
$\mathcal{TLT}_{36}(7)$	$\mathcal{TLT}_{36}(8)$	$\mathcal{TLT}_{36}(9)$	$\mathcal{TLT}_{36}(10)$	$\mathcal{TLT}_{36}(11)$	$\mathcal{TLT}_{36}(12)$
$\mathcal{TLT}_{36}(13)$	$\mathcal{TLT}_{36}(14)$	$\mathcal{TLT}_{36}(15)$	$\mathcal{TLT}_{36}(16)$	$\mathcal{TLT}_{36}(17)$	$\mathcal{TLT}_{36}(18)$
$\mathcal{TLT}_{36}(19)$	$\mathcal{TLT}_{36}(20)$	$\mathcal{TLT}_{36}(21)$	$\mathcal{TLT}_{36}(22)$	$\mathcal{TLT}_{36}(23)$	$\mathcal{TLT}_{36}(24)$
$\mathcal{TLT}_{36}(25)$	$\mathcal{TLT}_{36}(26)$	$\mathcal{TLT}_{36}(27)$	$\mathcal{TLT}_{36}(28)$	$\mathcal{TLT}_{36}(29)$	$\mathcal{TLT}_{36}(30)$
$\mathcal{TLT}_{36}(31)$	$\mathcal{TLT}_{36}(32)$	$\mathcal{TLT}_{36}(33)$	$\mathcal{TLT}_{36}(34)$	$\mathcal{TLT}_{36}(35)$	$\mathcal{TLT}_{36}(36)$

Fig. 2. 36 distinguishable classes of line track arrangements with three connected lines

The orientation of a shape can be considered with respect to any global frame of reference. The same holds for the parts of any shape. But for shape descriptions it is sometimes useful to consider the relationships between parts first, rather than the orientation of a single part with regard to a global frame of reference. Therefore, it is more expedient to consider parts which are invariant with respect to rotation and reflection. Such symmetries are primarily relevant to the whole shape. After removing all symmetrical relations of $\mathcal{TLT}_{36}$ there remain twelve distinguishable arrangements, as depicted in Fig. 3. The dotted lines outline the areas where the endpoints are allowed to lie in order to satisfy the denoted relation.

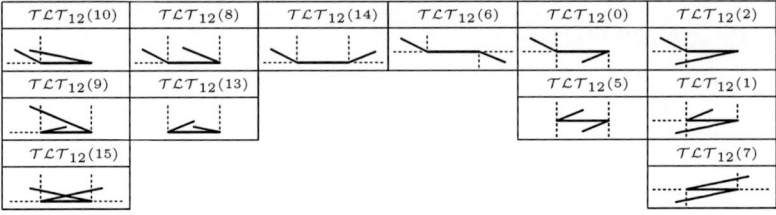

$\mathcal{TLT}_{12}(10)$	$\mathcal{TLT}_{12}(8)$	$\mathcal{TLT}_{12}(14)$	$\mathcal{TLT}_{12}(6)$	$\mathcal{TLT}_{12}(0)$	$\mathcal{TLT}_{12}(2)$
$\mathcal{TLT}_{12}(9)$	$\mathcal{TLT}_{12}(13)$			$\mathcal{TLT}_{12}(5)$	$\mathcal{TLT}_{12}(1)$
$\mathcal{TLT}_{12}(15)$					$\mathcal{TLT}_{12}(7)$

Fig. 3. Twelve boundary primitives distinguished by $\mathcal{TLT}_{12}$

Fig. 4 shows the neighbourhood graph adopted from Freksa, 1992, with two relations connected if they are only separated by one line of the orientation grid. Two relations are neighbours if one relation can be transformed into the other by continuously moving one endpoint to another position crossing the orientation grid once. Gottfried, 2002, shows how to distinguish different object categories by describing salient contour parts with the aid of $\mathcal{TLT}_{12}$. He uses the $\mathcal{TLT}_{12}$-neighbourhood graph as a measure of similarity between an object-instance and possible categories, since neighbouring relations correspond to similar shapes.

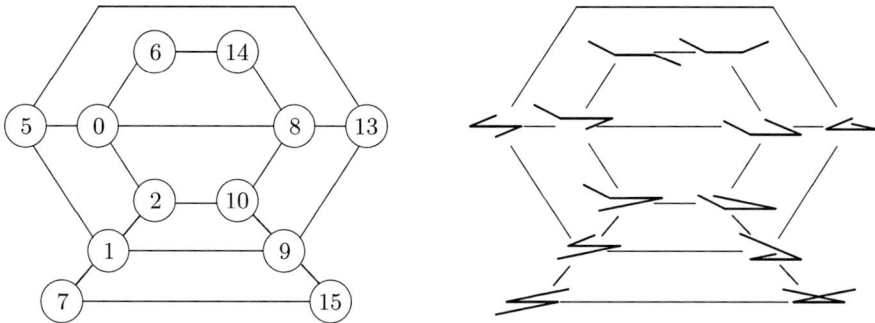

Fig. 4. Left: the neighbourhood graph; Right: example instantiations; the numbers refer to the $\mathcal{TLT}_{12}$-relations

3.1 $\mathcal{TLT}_6$

We will now consider a new subset of $\mathcal{TLT}$-relations. By examining $\mathcal{TLT}_{12}$ more precisely, we learn that some relations implicitly encode length information: $\mathcal{TLT}_{12}(0)$ and $\mathcal{TLT}_{12}(2)$, for example, are only different with respect to the length of the sideline forming an acute angle with the medial line. Consider an arbitrary instance of $\mathcal{TLT}_{12}(0)$. By lengthening the sideline a point instant can be found where the relation changes to $\mathcal{TLT}_{12}(2)$, that is, the corresponding edge between the nodes 0 and 2 of the neighbourhood graph in Fig. 4 is visited. A similar process can be applied to other $\mathcal{TLT}_{12}$-relations, and thus we subsume different $\mathcal{TLT}_{12}$-relations into single $\mathcal{TLT}_6$-relations as follows:

$$\mathcal{TLT}_6(0) := \mathcal{TLT}_{12}(0) \vee \mathcal{TLT}_{12}(2)$$

$$\mathcal{TLT}_6(8) := \mathcal{TLT}_{12}(8) \vee \mathcal{TLT}_{12}(10)$$

$$\mathcal{TLT}_6(1) := \mathcal{TLT}_{12}(1) \vee \mathcal{TLT}_{12}(5) \vee \mathcal{TLT}_{12}(7)$$

$$\mathcal{TLT}_6(9) := \mathcal{TLT}_{12}(9) \vee \mathcal{TLT}_{12}(13) \vee \mathcal{TLT}_{12}(15)$$

Furthermore, it holds that $\mathcal{TLT}_6(6) = \mathcal{TLT}_{12}(6)$ and $\mathcal{TLT}_6(14) = \mathcal{TLT}_{12}(14)$.

The relations of $\mathcal{TLT}_6$ are depicted in Fig. 5. Whereas $\mathcal{TLT}_{12}$ encodes orientation as well as some length information, $\mathcal{TLT}_6$ encodes only information about orientation. It can be argued that $\mathcal{TLT}_{12}$ also encodes only orientation information, but with respect to the orientation grid (see Fig. 1). In a way the concept of orientation is different for $\mathcal{TLT}_{12}$ and $\mathcal{TLT}_6$. The latter is simpler since it only distinguishes acute angles, obtuse angles, and the two different sides to which a sideline can point.

A great deal of work concerning shape descriptions is based on Attneave, 1954. Attneave stated that significant shape information is concentrated at those points on a contour where its direction changes most rapidly. Attneave himself used these contour points as vertices for polygonal shape approximations. We

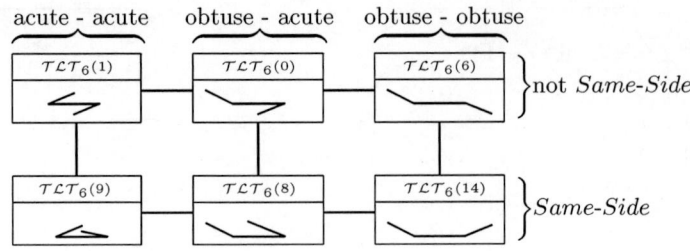

Fig. 5. Six primitives distinguished by $\mathcal{TLT}_6$ arranged in a neighbourhood graph

consider not only single contour points, but two curvature extrema, to be relevant together. This means, for example, that in a track of three lines two successive acute angles, i.e. two curvature extrema, could either be on the same or on different sides with respect to the medial line, as represented by $\mathcal{TLT}_6(1)$ and $\mathcal{TLT}_6(9)$. Interestingly, this *Same-Side* property does not arise until three connected lines are considered, and the *Same-Side* property cannot be reduced to a vector of angles without considering the relationship between adjacent angles. With the *Same-Side* property we are able to distinguish $\mathcal{TLT}_6(1)$ from $\mathcal{TLT}_6(9)$, both made up of two acute angles but differing with respect to *Same-Side*. Corresponding circumstances hold for $\mathcal{TLT}_6(0)$ and $\mathcal{TLT}_6(8)$ as well as $\mathcal{TLT}_6(6)$ and $\mathcal{TLT}_6(14)$. Configurations of more than two adjacent curvature extrema do not allow for more information, since no further direction exists in the two-dimensional Euclidean plane to which our description is restricted. Such configurations can always be distinguished by $\mathcal{TLT}_6$-combinations. Otherwise, further curvature extrema have to be considered in higher dimensions, since the number of possible directions increases when considering further dimensions.

To summarise, tracks of three lines are described in two dimensions. The medial line determines a local reference system, and both dimensions are taken into account in quite a simple manner: *Same-Side* is a distinction in one dimension, and the dichotomy between acute and obtuse angles is a distinction in the other perpendicular dimension. Note that the latter is related to two different ways of changing orientation on any contour segment. Thus, $\mathcal{TLT}$s represent all possible bipartite combinations of such orientation differences, and $\mathcal{TLT}$s can be described in a way solely based on orientation information.

Though we have been considering only general positions, we are able to cope with singular positions as well, i.e. those positions which lie on the orientation grid (Fig. 1).

3.2   Singular Line Tracks

To be more precise, the $\mathcal{TLT}$s which have been considered so far would have been better denominated as *General Tripartite Line Tracks*, because we have disregarded singular positions. Line tracks which involve singular positions as

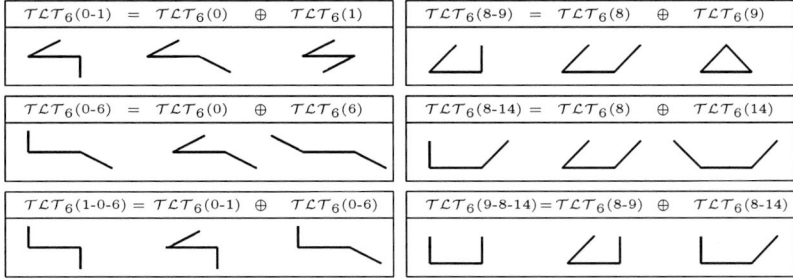

Fig. 6. Upper two rows: *Singular Tripartite Line Tracks* with one sideline in singular position and the other in general position; Lower row: both sidelines in singular position; $\oplus$ denotes what line tracks are related to each other for describing singular line tracks

well are of special interest with respect to the boundaries of artificial objects. That is to say, man-made objects often have perpendicular sides.

A *Singular Tripartite Line Track* is referred to as $\mathcal{TLT}(a\text{-}b)$, with one sideline in general position and the other sideline in singular position. Thus, $\mathcal{TLT}(a\text{-}b)$ refers to an edge in the $\mathcal{TLT}_6$-neighbourhood graph (see Fig. 5), where a and b are the nodes which are connected by this edge. For example, consider the relations $\mathcal{TLT}(0)$ and $\mathcal{TLT}(1)$. They are neighbours and as such they are separated by a singular relation which we refer to as $\mathcal{TLT}(0\text{-}1)$. Fig. 6 shows how singular line tracks are related to general line tracks. Furthermore, singular line tracks with both endpoints in singular position are related to two other singular line tracks each with one endpoint in singular position. Singular line tracks with the endpoints lying on the positions 2, 5, 8, 11, or 14 (see Fig. 1) correspond to degenerated cases[1]. We consider only visually distinguishable lines.

3.3   Combinations of $\mathcal{TLT}_6$

A closed polygon with $k \geq 3$ lines is described as a vector of k $\mathcal{TLT}$s, an open polygon as a vector of $k-2$ $\mathcal{TLT}$s: $\mathcal{TLT}_6(i, j, ...), i, j \in \{0, 1, 6, 8, 9, 14\}$. In order to be able to treat each arbitrary polygon with k lines, and not only those which consist of a multiple of three lines, a vector of $\mathcal{TLT}$s describes polygons in such a way that two consecutive $\mathcal{TLT}$s share two lines. In this way, there exist $\mathcal{TLT}$s which are incompatible. For example, $\mathcal{TLT}_6(1)$ and $\mathcal{TLT}_6(6)$ cannot be combined, since for a combination to be possible they would have to share two adjacent lines, or rather one angle. But the angles of $\mathcal{TLT}_6(1)$ are both acute, whereas the angles of $\mathcal{TLT}_6(6)$ are both obtuse. A compatible combination consists of four lines, or rather of two entwined $\mathcal{TLT}$s. Each such combination is referred to as a *Quadripartite Line Track* ($\mathcal{QLT}$). All $\mathcal{QLT}$s are given in Fig. 7.

---

[1] At least one endpoint and the medial points are rectilinear.

The $\mathcal{QLT}$s in Fig. 7 are arranged in their neighbourhood graph. This graph allows us to reason about change in shape. Consider, for instance, $\mathcal{TLT}_6(0,0)$ and $\mathcal{TLT}_6(6,6)$. The shortest possible way to transform $\mathcal{TLT}_6(0,0)$ into $\mathcal{TLT}_6(6,6)$ is via $\mathcal{TLT}_6(0,6)$. If special intermediate states are to be visited or to be avoided when transforming one $\mathcal{QLT}$ into another, an appropriate way can be found using the neighbourhood graph.

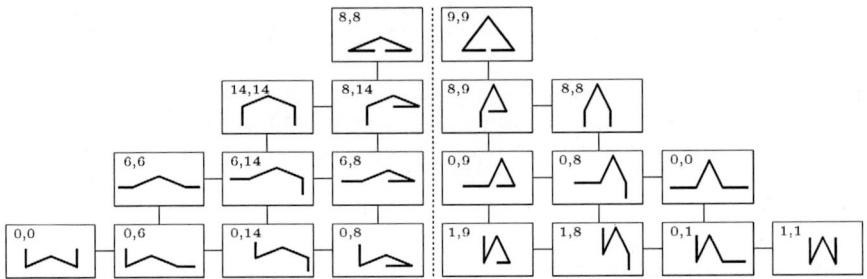

Fig. 7. All General Quadripartite Line Track classes ($\mathcal{QLT}$) distinguished by $\mathcal{TLT}_6$ and arranged in their neighbourhood graph; one half can be mirrored at the dotted line, thereby the medial angles switch between obtuse and acute, thus each $\mathcal{QLT}$ is also neighbour of its counterpart on the other side, for example $\mathcal{TLT}_6(14,14)$ is a neighbour of $\mathcal{TLT}_6(8,8)$

Attention should be paid to those relations where one sideline crosses one of the two medial lines when transforming a relation into its neighbouring relation. These self-crossing relationships have been omitted, and thus this graph allows only for non-intersecting transformations. In the complete neighbourhood graph each symmetrical $\mathcal{QLT}$, i.e. each $\mathcal{TLT}_6(i,i)$, $i \in \{0,1,6,8,9,14\}$, has three neighbours, and each asymmetrical $\mathcal{QLT}$ has five neighbours. The eight missing relationships are as follows:

$$[\mathcal{TLT}_6(0,0) - \mathcal{TLT}_6(0,8)], \ [\mathcal{TLT}_6(0,8) - \mathcal{TLT}_6(8,8)],$$

$$[\mathcal{TLT}_6(0,14) - \mathcal{TLT}_6(8,14)], \ [\mathcal{TLT}_6(0,6) - \mathcal{TLT}_6(6,8)],$$

and on the other side of the graph,

$$[\mathcal{TLT}_6(1,1) - \mathcal{TLT}_6(1,9)], \ [\mathcal{TLT}_6(1,9) - \mathcal{TLT}_6(9,9)],$$

$$[\mathcal{TLT}_6(1,8) - \mathcal{TLT}_6(8,9)], \ [\mathcal{TLT}_6(0,1) - \mathcal{TLT}_6(0,9)].$$

For closed polygons, different $\mathcal{TLT}$ descriptions will be obtained depending on where one starts to enumerate the polygon, and on whether one is tracing the contour clockwise or anticlockwise. Any description can be converted into another equivalent description by means of a cyclic permutation of the $\mathcal{TLT}$s involved. For a given polygon, we choose that description which comes first in

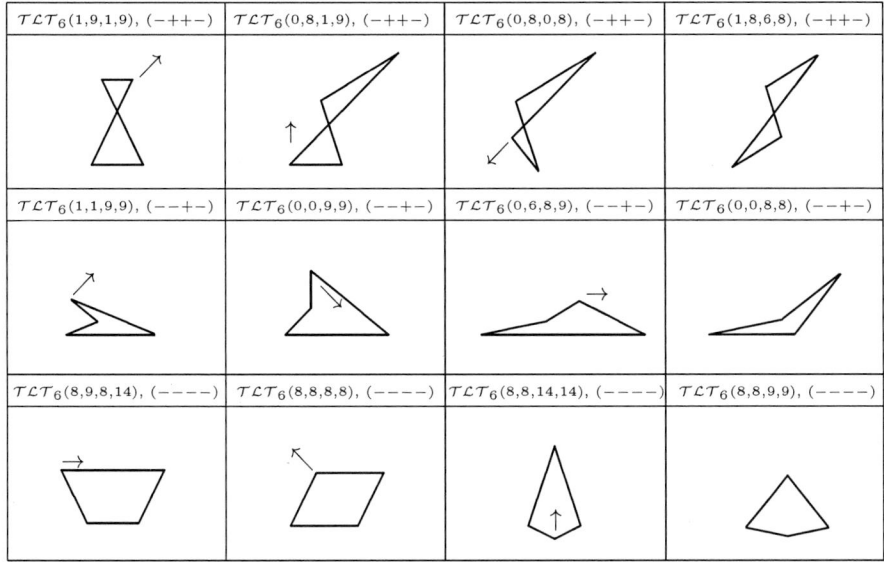

Fig. 8. All twelve general quadrilaterals of $\mathcal{TLT}_6$; pulling or pressing a figure's vertex which is marked by the arrow deforms it to the next one to the right; the clockwise triangle orientations starting in the upper left of the polygons are also depicted

the ordering with respect to the $\mathcal{TLT}$ numbers. Fig. 8 shows all distinguishable quadrilaterals of $\mathcal{TLT}_6$ by examples: all non-simple polygons are in the upper row, the simple-concave polygons are in the middle row, and all convex polygons are in the lower row.

4 $\mathcal{TLT}$ and the Triangle Orientation

The notion of $\mathcal{TLT}$s is closely related to the approach of Schlieder, 1996, which describes polygons by considering the triangle orientation of vertices. Both approaches are based on orientation information, and both describe polygons. We should therefore compare these two approaches more thoroughly.

A triangle is defined by three points in the oriented plane. Its orientation is defined as "+" if the path of the three points follows the mathematically positive orientation, "0" if it is rectilinear, and "-" if it follows the negative orientation. Let us consider the four hexagons of Fig. 9, which are adopted from Schlieder, 1996. These hexagons can be distinguished neither by the triangle orientation of adjacent vertices nor by $\mathcal{TLT}$s taking adjacent lines. Hence, both descriptions treat these four shapes as though they were identical when tracing their contours. In order to distinguish these shapes, Schlieder proposes to consider additionally non-adjacent vertices. In this way, the four shapes can be distinguished using either approach. Below the shapes in Fig. 9 there are descriptions of substructures, using which it is possible to distinguish the hexagons. The consideration of such substructures comprehends non-local shape information, which is not available either from adjacent contour vertices or adjacent contour lines. By *non-local* $\mathcal{TLT}s$ we mean $\mathcal{TLT}s$ which are made up of non-adjacent vertices with respect to the given polygon.

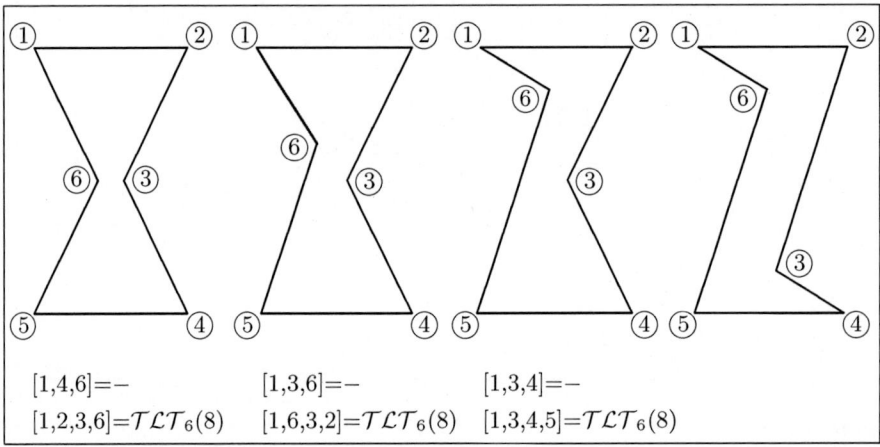

Fig. 9. Four hexagons which cannot be distinguished by their clockwise contour description beginning by vertex number 1: The triangle orientation $(-+--+-)$ and $\mathcal{TLT}_6(9,0,0,9,0,0)$ are equal for all shapes; but it is possible to distinguish these shapes by the non-local substructures below the shapes

The question arises whether there are any differences between the two approaches. As shown in the lower row of Fig. 8, it is possible to distinguish different convex quadrilaterals using $\mathcal{TLT}$. Convexity can be defined as follows: circulating around a polygon and each time taking the triangle orientation of three adjacent vertices, the sign of the orientation never changes. In this way it is possible to discriminate concave and convex shapes by the triangle orientation, but the triangle orientation cannot be used to distinguish different convex quadrilaterals. Thus, $\mathcal{TLT}$ is more expressive concerning the concept of convexity, and also concerning non-simple and concave polygons as demonstrated in Fig. 8 and as already shown by tripartite polygons, as for example $\mathcal{TLT}_6(8)$ and $\mathcal{TLT}_6(14)$ in Fig. 10. But in order to prove that the triangle orientation is less expressive, it must be demonstrated that every conceivable pair of point configurations which differ with respect to their triangle orientation can be distinguished using $\mathcal{TLT}$, too.

For this purpose, let us consider configurations with at least four points, as depicted in Fig. 11.a. Two point configurations, C_1 and C_2, are different with respect to the triangle orientation if there exists a line l defined by two points, a and b, of C_1 and C_2, so that a third point p lies on one side of l regarding C_1 and on the other side regarding C_2 (Fig. 11.a/b). In order to distinguish two such configurations, we will construct one $\mathcal{TLT}$ for each configuration, and

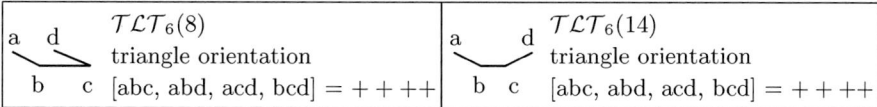

Fig. 10. $\mathcal{TLT}_6(8)$ and $\mathcal{TLT}_6(14)$ are equal regarding the triangle orientation

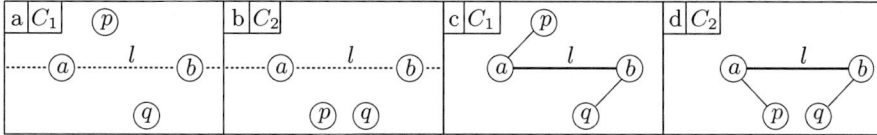

Fig. 11. Two example configurations C_1 and C_2, and their discrimination by $\mathcal{TLT}$

these $\mathcal{TLT}$s will be different although they are based on the same points. In each configuration we take a fourth point, q, and we assume that q is not coincident with l. Then q lies either on the same side as p with respect to l (Fig. 11.b), or it lies on the other side (Fig. 11.a). However, this is different for C_1 and C_2 since only p lies on a different side in C_1 than in C_2. Therefore, we can distinguish C_1 and C_2 using $\mathcal{TLT}$s defined by the points p-a-b-q, because *Same-Side* is satisfied for only one of the two configurations (Fig. 11.c/d). That is, for one configuration it holds p-a-b-$q \in \{\mathcal{TLT}_6(0), \mathcal{TLT}_6(1), \mathcal{TLT}_6(6)\}$, and for the other configuration it holds p-a-b-$q \in \{\mathcal{TLT}_6(8), \mathcal{TLT}_6(9), \mathcal{TLT}_6(14)\}$.

Note that in contrast to the triangle orientation, $\mathcal{TLT}_6$ does not distinguish configurations which are mirror symmetrical. In order to allow for this distinctions we have to use $\mathcal{TLT}_{36}$.

We have shown that $\mathcal{TLT}$ is more expressive than the triangle orientation. Finally, we will contrast these two approaches by considering the reference frames on which they are based. The triangle orientation requires the orientation of the plane to be defined, and thus it is possible to distinguish the position of any point in relation to the two sides of an oriented line. In contrast, $\mathcal{TLT}$ is not based on any globally determined orientation, but defines a local frame of reference with the medial line determining the first dimension and an imaginary line determining the second dimension. The latter is defined as perpendicular to the medial line. Relative to this local frame of reference the positions of points are considered in two different ways. First, the endpoint of one sideline can be positioned on either side of the medial line. Secondly, it can be positioned on either side of the imaginary line in the second dimension. For this dimension we define two half-planes by the imaginary line which orthogonally intersects the endpoint of the medial line where the sideline under consideration meets the medial line. This is just another way of stating that a sideline forms either an acute angle or an obtuse angle with the medial line. By this means, both dimensions in the two-dimensional plane are considered equally. It is not possible to distinguish, for example, different convex polygons until the second dimension is considered in this way.

5   Stylised Curvature

One important shape property is the curvature of the boundary, as we have already pointed out in section 2. Reasoning about shape curvature is relevant to shape comparisons and to shape deformations. We will consider these issues by defining a concept of curvature on the basis of $\mathcal{TLT}$s. Before this, we will examine some shape examples to illustrate the relationship between curvature and $\mathcal{TLT}$s. But first, we will show that polygons with almost no curvature, in the sense of smooth curves, are particularly suitable for object categorisation.

5.1   Object Categorisation

Shapes can be approximated by polygons with different levels of accuracy. Thus, the shape of an object-instance may be represented by different $\mathcal{TLT}$ descriptions depending on the chosen level of granularity. Gottfried, 2002, concentrates on coarse shape approximations, in order to demonstrate the applicability of $\mathcal{TLT}$ to the problem of object categorisation.

Object categorisation is concerned with the characterisation of properties common to most of the objects belonging to the same category. Such salient properties are typically found in stylised shapes which only contain a minimum of shape information. However, this is frequently specific enough to categorise objects.

Salient properties are not related to the details of specific object-instances, and different object-instances of the same category sometimes have only these properties in common. Coarse approximations normally maintain salient, category-specific features, while instance-specific details are omitted. In the case of two-dimensional polygons, salient properties are based on tracks of lines connected via combinations of acute and obtuse angles, that is, $\mathcal{TLT}$ combinations describe salient parts of coarse polygonal approximations.

Subtle polygonal approximations do not represent salient properties so directly, but reflect curvature information more precisely.

5.2   Accurate Curvature versus Stylised Curvature

At first sight, curvature does not seem to be related in the slightest to polygonal approximations - regardless of the level of granularity. Instead, the curvature at an arbitrary point on a contour is defined, for example, as the reciprocal of the radius of the circle approximating the contour of the shape at that point. By this definition, polygons have no curvature, in consequence of the fact that they are only made up of straight lines. Therefore, the question arises as to how $\mathcal{TLT}$s are related to curvature.

We argue that $\mathcal{TLT}$ defines a stylised concept of curvature, and that curvature information is represented by two adjacent curvature extrema, i.e. two successive angles on the polygonal contour. Instead of using single metrical curvature-values at each contour point, we obtain *spaciously curvature information*, expanded over more or less long contour segments. This becomes particularly important when dealing with shapes in large-scale spaces where one

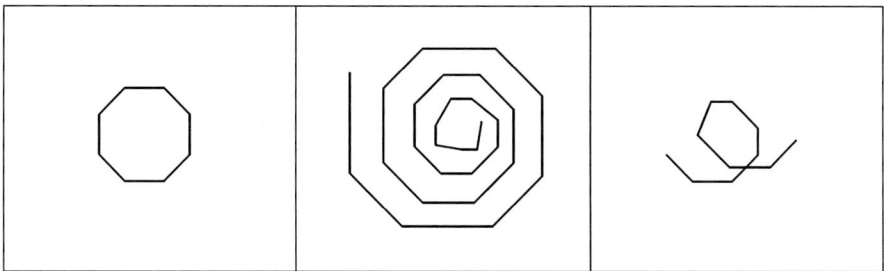

Fig. 12. Some examples of $\mathcal{TLT}_6(14)$-contours

may lose track of things. In such cases, it is helpful to obtain an idea of the overall shape or of single contour parts in a more comprehensible way.

5.3   Curvature by Examples

Just as stylised shapes are restricted to a minimum of shape information specific enough to categorise objects, a stylised concept of curvature restricts itself to a minimum of information specific enough to distinguish different curves. Any $\mathcal{TLT}$ description represents stylised curvature information, and sometimes this description allows us to comprehend the curve progression very well indeed.

For instance, consider polygons that consist only of successive $\mathcal{TLT}_6(14)$-segments. Such polygons are always bent to one side, and form circles or spirals. In order to understand these tendencies of $\mathcal{TLT}_6(14)$-line tracks, we must take a closer look at $\mathcal{TLT}_6(14)$-relations. The endpoints of such line tracks lie on the same side with respect to the medial line, and both angles are obtuse. Therefore, entwined $\mathcal{TLT}_6(14)$-relations always make up a curved line which never changes its local orientation. This constancy in orientation can be comprehended if we imagine tracing a $\mathcal{TLT}_6(14)$-contour with one finger without the need for wriggling. The circle-like figures described by our finger may get larger or smaller, but they always remain approximately like a circle or an ellipse. Fig. 12 shows three examples: depending on precise length and angle information, we either obtain circle-like figures, or snails where the lines become consistently shorter, or loops where the lines vary in length and angle.

As a second example, we will consider polygons that consist only of $\mathcal{TLT}_6(6)$, as shown in Fig. 13. With the exception of *Same-Side*, $\mathcal{TLT}(6)$ equals $\mathcal{TLT}_6(14)$. As *Same-Side* does not hold for $\mathcal{TLT}_6(6)$, when tracing the contour our finger constantly changes in orientation, if only slightly, and hence this is not as easy as tracing a $\mathcal{TLT}_6(14)$-contour. But the difference with $\mathcal{TLT}$s which consist of acute angles is far more considerable. Two examples of $\mathcal{TLT}_6(1)$-contours are shown in Fig. 14.

5.4   Local Curvature versus Global Curvature

As exemplified in Figs. 12 to 14, sometimes two shapes look quite different though they are based on the same $\mathcal{TLT}$s. Locally viewed, these shapes are absolutely

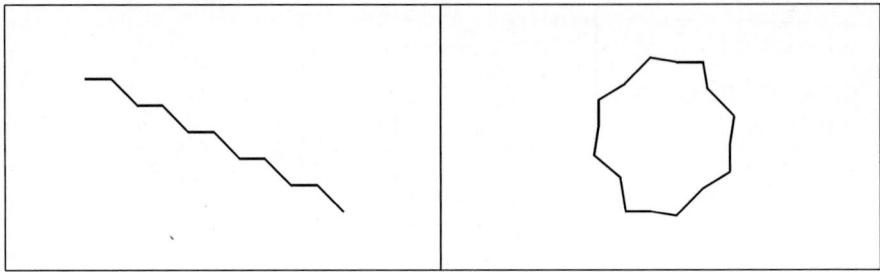

Fig. 13. Two globally very different examples of $\mathcal{TLT}_6(6)$-contours

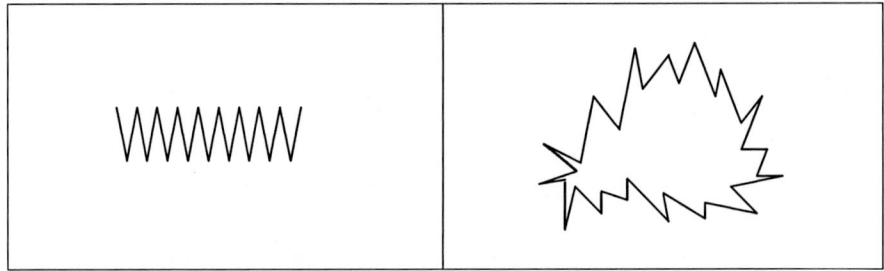

Fig. 14. Two examples of jagged $\mathcal{TLT}_6(1)$-contours

similar. Two curves may only differ a little locally. But the longer the line tracks are, the more such local differences accumulate into globally salient differences. These global differences are the consequence of exploiting the $\mathcal{TLT}$s' degree of freedom in different ways. Nevertheless, the overall curvature remains similar.

Locally similar but globally different shapes can be distinguished by using further non-local $\mathcal{TLT}$s, such as those applied in order to distinguish the hexagons in Fig. 9. For example, $\mathcal{TLT}_6(9)$ is only obtainable in the closed figure on the right in Fig. 13. Only those non-local $\mathcal{TLT}$s which are made up of points that adhere strictly to the order of vertices as given by the polygon clockwise or anti-clockwise are allowed. Another possible way to distinguish the two shapes is to relate each $\mathcal{TLT}$ to a global frame of reference, as we discussed when introducing $\mathcal{TLT}_{12}$.

The analysis of global shape properties is a rather complex topic. For the time being, we will restrict the exploration of shape curvature to Quadripartite Line Tracks.

5.5   Curvature of Quadripartite Line Tracks

So far, we have only considered the curvature of a number of special cases. But we are interested in the general case of arbitrary $\mathcal{TLT}$-combinations and their curvature. The distinction between $\mathcal{TLT}_6(1)$-shapes and $\mathcal{TLT}_6(6)$-shapes was made clear through comparisons, and we will now compare all conceivable Quadripartite Line Tracks ($\mathcal{QLT}$s). First, we define an order of curvature for single $\mathcal{TLT}$s, and then the curvature for $\mathcal{TLT}$-combinations will be inferable. By means of this concept of curvature it is decidable which of two given $\mathcal{QLT}$s is more curved.

A possible order of curvature can be defined on the assumption that obtuse angles are less curved than acute angles, and that secondarily $\mathcal{TLT}$s which satisfy *Same-Side* are less curved than $\mathcal{TLT}$s for which *Same-Side* is not satisfied. Starting from the least curved $\mathcal{TLT}$, we obtain the following curvature-order: $\mathcal{TLT}_6(14)$, $\mathcal{TLT}_6(6)$, $\mathcal{TLT}_6(8)$, $\mathcal{TLT}_6(0)$, $\mathcal{TLT}_6(9)$, and $\mathcal{TLT}_6(1)$. This order determines the order of all $\mathcal{QLT}$s which are depicted in Fig. 7. $\mathcal{TLT}_6(14, 14)$ on the left in the second row is the least curved $\mathcal{QLT}$, and $\mathcal{TLT}_6(1, 1)$ on the right of the bottom row is the most curved.

5.6   Reasoning about Change in Curvature

In paragraph 3.3 we considered change in shape with the aid of the neighbourhood graph in Fig. 7. When moving from one $\mathcal{QLT}$ to a neighbouring $\mathcal{QLT}$ in this graph, the curvature increases or decreases depending on the relations involved. Note that transitions in the neighbourhood graph correspond to change in shape, i.e. to single distortion steps.

Change in curvature can be different for the same number of distortion steps. For example, moving from $\mathcal{TLT}_6(14, 14)$ to $\mathcal{TLT}_6(6, 14)$ a single distortion step is taken, such as when moving from $\mathcal{TLT}_6(14, 14)$ to $\mathcal{TLT}_6(8, 14)$. But on the basis of our curvature-order, $\mathcal{TLT}_6(6, 14)$ is less curved than $\mathcal{TLT}_6(8, 14)$. Therefore, we can search for a path in the neighbourhood graph which is composed of the least curved transitions in order to transform one $\mathcal{QLT}$ into another one.

For instance, we may want to transform $\mathcal{TLT}_6(14, 14)$ into $\mathcal{TLT}_6(6, 8)$. In this case, we would prefer the path via $\mathcal{TLT}_6(6, 14)$, since the path via $\mathcal{TLT}_6(8, 14)$ is more curved. Note that we need both the neighbourhood graph, in order to determine possible transitions, and the curvature-order, in order to select from the set of possible transitions the one which optimally contributes to the least curved path. In a similar manner, any two shapes can be compared in terms of their curvature by analysing which shape describes the more curved path in the neighbourhood graph.

5.7   Summary

We have established a stylised concept of curvature which can be used to characterise shapes. This is especially interesting in the case of large-scale shapes for which metrical characterisations are too precise and become unmanageable. The concept of curvature presented here provides a concise and more manageable characterisation. Sometimes computations have to rely on coarse approximations because of a lack of precise information. The proposed concept of stylised curvature information allows us to reason about such coarse curve approximations.

6 Discussion

As the approach described here only considers orientation information, similar shapes such as squares and rectangles cannot be distinguished since they exhibit the same configuration in orientation. They are only different with respect to the relative length of their sides. Nevertheless, the *similarity* between these shapes may be represented by information about orientation, and since length cannot be reduced to orientation information we assume it to be another fundamental concept.

The question arises, to what extent applications are in need of concepts based on polygonal structures. From a technical point of view, in most instances curves have to be approximated using polygons. Cordella & Vento, 2000, reviewed almost one hundred papers in the field of computer vision, and found that 58% of the reviewed systems applied thinning techniques and polygonal approximations to shapes. Furthermore, polygonal data are relevant to geographical information systems and for representing trajectories, as in the context of navigation.

The use of polygons rather than smooth curves is less crucial than the determination of the right level of granularity. It remains to be investigated what kinds of relationships can be stated between different levels of granularity. Moreover, generalisations about arbitrary arrangements of lines have to be considered in order to cope with disconnected shape parts as efficiently as with tracks of lines. The application of $\mathcal{TLT}$s will eventually show its usefulness. It has to be investigated how appropriate $\mathcal{TLT}$s are to solve tasks such as those mentioned at the beginning of this paper. When performing grasping movements, for example, precise boundary values are of no importance. But we have to get an idea of the overall curvature of an object in order to be able to grasp it easily. The same holds for circumventing obstacles. We only aim at finding a convenient way to move around obstacles, and qualitative curvature information may alone be sufficient to solve such tasks. In the context of vision, precise boundary information about objects are not necessary either. Normally, we can recognise objects even in bad environmental conditions such as twilight. Obviously, our recognition abilities effortlessly manage coarse curvature information and it remains to be shown how descriptions restricted to qualitative curvature information are sufficient to solve such tasks.

To sum up, we have proposed a qualitative description of polygons in two dimensions. In comparison with other approaches, $\mathcal{TLT}_6$ defines a completely local frame of reference and is more expressive in terms of distinguishable polygons, since it considers two different sides for each dimension to be distinguished. In this way, $\mathcal{TLT}_6$ establishes orientation information which is especially suitable to describing a stylised concept of shape curvature.

References

Attneave, F. (1954). Some informational aspects of visual perception. *Psychological Review, 61*, 183-193.

Biederman, I. (1987). Recognition-by-components: A theory of human image understanding. *Psychological Review, 94*, 115-147.

Cohn, A. G. (1995). A hierarchical representation of qualitative shape based on connection and convexity. In A. M. Frank & W. Kuhn (Eds.), *Cosit 1995, Spatial Information Theory* (p. 311-326). Springer-Verlag.

Cordella, L. P. & Vento, M. (2000). Symbol recognition in documents: a collection of techniques? *International Journal on Document Analysis and Recognition, 3*, 73-88.

Freksa, C. (1992). Temporal reasoning based on semi-intervals. *Artificial Intelligence, 94*, 199-227.

Freksa, C. & Zimmermann, K. (1992). On the utilization of spatial structures for cognitively plausible and efficient reasoning. In *IEEE International Conference on Systems, Man and Cybernetics* (p. 261-266). Chicago.

Galton, A. & Meathrel, R. C. (1999). Qualitative outline theory. In T. Dean (Ed.), *Proc. of 16th IJCAI-99* (p. 1061-1066). Stockholm, Sweden: Morgan Kaufmann, Inc.

Gottfried, B. (2002). Tripartite Line Tracks. In K. Wojciechowski (Ed.), *International Conference on Computer Vision and Graphics* (p. 288-293). Zakopane.

Horng, J. & Li, J. T. (2002). An automatic and efficient dynamic programming algorithm for polygonal approximation of digital curves. *Pattern Recognition Letters, 23*, 171-182.

James, T. W., Humphrey, G. K., Gati, J. S., Servos, P., Menon, R. S. & Goodale, M. A. (2002). Haptic study of three-dimensional objects activates extrastriate visual areas. *Neuropsychologia, 40*, 1706-1714.

Marr, D. & Nishihara, H. K. (1978). Representation and recognition of the spatial organization of three-dimensional shapes. *Proc. R. Soc. of Lond. B, 200*, 269-294.

Schlieder, C. (1996). Qualitative shape representation. In P. Burrough & A. M. Frank (Eds.), *Geographic objects with indeterminate boundaries* (p. 123-140). London: Taylor & Francis.

Linearized Terrain:
Languages for Silhouette Representations

Lars Kulik and Max J. Egenhofer

National Center for Geographic Information and Analysis
348 Boardman Hall, University of Maine, Orono, ME 04469-5711, USA
{kulik,max}@spatial.maine.edu

Abstract. The scope of this paper is a qualitative description of terrain features
that can be characterized using the silhouette of a terrain. The silhouette is a
profile of a landform seen from a particular observer's perspective. We develop
a terrain language as a formal framework to capture terrain features. The horizon of a terrain silhouette is represented as a string. The alphabet of the terrain
language comprises straight-line segments. These line primitives are classified
according to three criteria: (1) the alignment of their slope, (2) their relative
lengths, characterized by orders of magnitude, and (3) their differences in
elevation, described by an order relation. We employ term rewriting rules to
identify terrain features at different granularity levels. There are three kinds of
rules: aggregation, generalization, and simplification rules. The aggregation
rules generate a description of the terrain features at a given granularity level.
For a terrain description the generalization and simplification rules specify the
transition from a finer granularity level to a coarser one. An example shows
how the three kinds of rules lead to a terrain description at different granularity
levels.

Keywords: Formal languages, granularity, qualitative spatial reasoning, terrain

1   Introduction

Terrain refers to the topographic surface of the Earth. As a type of *field* it captures the
distribution of *elevation* throughout space. Terrain is commonly represented by a real
bivariate function over a domain in Euclidean space, whose discretization leads to
various *digital terrain models* (De Floriani and Magillo 1999). Examples include
triangulated irregular networks (TINs) and regular square grids. A complementary
terrain representation is given by cognitively salient *terrain features*, such as valleys
and ridges. People commonly extract these features visually and use them to communicate spatial properties concisely. For example, a visitor standing at Jordan Pond in
Acadia National Park will see two characteristic terrain features, the North Bubble to
the left and South Bubble to the right (Fig. 1). This qualitative description of a terrain
scenery complements a digital terrain model of Mount Desert Island, which captures
terrain information in a quantitative way.

   The goal of this paper is to identify methods that provide a qualitative description
of terrain. A quantitative description of a terrain silhouette employs numerical values
to determine lengths and angles of the horizon segments, whereas a qualitative de-

W. Kuhn, M.F. Worboys, and S. Timpf (Eds.): COSIT 2003, LNCS 2825, pp. 118–135, 2003.

scription uses only a small set of line features. Languages for terrain features are typically domain-dependent. Geomorphologists have developed a detailed vocabulary for terrain features such as u-shaped vs. v-shaped valleys or horns vs. arêtes (Strahler and Strahler 1992). Military applications distinguish only five major terrain features—ridge, hill, valley, saddle, and depression—and three minor features—spur, draw, and cliff (Department of the Army 2001). Rather than defining another set of terrain features or identifying a comprehensive terrain feature ontology, we start with fundamental elements of a qualitative terrain representation and explore how these elements can be combined into meaningful terrain descriptions. This approach leads to the formulation of a *qualitative terrain language* as a framework within which certain terrain features can be captured.

Fig. 1. North and South Bubble in Acadia National Park

The scope of this paper is the modeling of a *terrain silhouette* as a projection of a particular terrain onto an observer's visual field. The same method applies to cross-sections through terrain and, therefore, forms a foundation for constructing qualitative terrain descriptions of higher-dimensional spaces as combinations of terrain silhouettes. Such an approach supports the egocentric view that an observer has about terrain in the field, and complements the panoptic perspective that underlies digital terrain models.

Applications of languages for terrain silhouettes are in intelligent image retrieval, where a user queries a database for such features as "a volcano next to a mountain with a cliff," and in mobile field computing, where a user interacts with the landscape with an intelligent geographic pointer, a *Magic Wand* (Egenhofer and Kuhn 1998), to identify terrain features and to obtain a natural-language description of the terrain.

The remainder of this paper is structured as follows: After a brief overview of shape languages, granularity issues, and orders of magnitude (Section 2), the formal language is presented to describe terrain features (Section 3). Horizons of a terrain silhouette are represented by finite strings of line segments of different slopes. In Section 4 three kinds of rules are introduced to rewrite a terrain string: (1) aggregation rules to identify terrain features, (2) generalization rules to neglect smaller terrain features, and (3) simplification rules to enable a terrain description at a coarser granularity. In Section 5 these rules are applied in an example terrain silhouette to illustrate how a terrain description is derived at different granularities.

2   Formal Languages for Terrain

One way to represent shape features of spatial objects is to specify the features in a formal language. Such a formal language is built up from shape primitives such as convex lines and straight lines. It allows us to identify classes of objects that can be characterized by the same representation of shape primitives. A shape language enables a qualitative description of shape (Clementini and Di Felice 1997) that abstracts from numerical details like the curvature of a boundary line at a certain point. Typically, a shape language is specified by a shape grammar. Leyton (1988) developed a shape grammar that considers the history that leads to the particular shape of an object. Since it assumes that an object's outline is the result of continuous deformations, it does not apply to terrain descriptions. Cinque and Lombardi (1995) describe object shapes at different levels of resolution, using rather a quantitative than a qualitative technique. Galton and Meathrel (1999) proposed a grammar that captures qualitative shape features of the boundary of two-dimensional objects. Since they do not consider the impact of the vertical dimension, their grammar is too general for a qualitative terrain description.

Terrain can be described at different levels of granularity: a detailed description might account for every small terrain feature, whereas a coarse description might only mention major terrain features. Hobbs (1985) proposed a general framework of granularity that enables the construction of simpler, less-detailed descriptions from more complex and detailed descriptions. He employed an indistinguishability relation to model the concept of simplification. We pursue this idea for terrain silhouettes to generate a characterization of their features at different granularity levels by using different orders of magnitude. At a given granularity, terrain features of smaller orders of magnitude are neglected with respect to features of greater order of magnitude. The coarsest granularity level provides an overview of a terrain silhouette.

Reasoning with orders of magnitude allows us to draw inferences for entities, which differ by orders of magnitude. Two entities could be similar but not necessarily equal in size or one entity can be neglected with respect to another entity. In the first case, the entities have the same order of magnitude (denoted by $\approx$); in the second case, the entities differ by orders of magnitude and the first entity has a smaller order than the second entity (denoted by $\ll$). Raiman (1986) proposed an axiomatic system to reason with orders of magnitude. Raiman used the symbols CO for $\approx$ and NE for $\ll$, respectively. Approaches formalizing the concept of orders of magnitude have been proposed by Raiman (1986), Nayak (1992), and Davis (1999).

Most closely related to this approach is earlier work in the area of terrain feature formalization (Frank et al. 1986). Their work, however, uses predicate calculus rules to derive from a TIN local, point-based terrain features such as peaks, pits, and saddle points. By contrast, our work focuses on the identification of complex terrain features from a terrain silhouette such as flat-floored valleys, plateaus, or canyons. Generalizations deriving coarser representations are not part of the predicate calculus and have to be performed in the TIN domain. De Berg and Dobrindt (1998) proposed an approach to generate a representation of a terrain at different levels of details, if the terrain is modeled as triangular irregular network. Overviews of multiresolution representations are given in Garland (1999) or De Floriani and Magillo (2002).

3   Primitives for Terrain Silhouettes

A *terrain silhouette* results from the projection of a terrain onto a 2-dimensional plane whose y-axis is parallel to gravity and the x-axis is perpendicular to the viewer's line of sight. The *horizon* (De Floriani and Magillo 1995) of a silhouette is the line that separates the terrain from the sky. As such, it is a 1-dimensional object embedded in a 2-dimensional space. Horizons have no self-intersections and, ignoring any occlusions in the line of sight, they are simply connected. Since the view of a terrain is bounded, a horizon has two endpoints, p_1 and p_2. Although a horizon may represent an arbitrary, connected subset of a terrain, it is reasonable to restrict horizons to $x(p_1) \neq x(p_2)$. A consistent, agreed-upon orientation (from left to right) allows us to further distinguish a horizon's start point from its endpoint (i.e., $x(p_1) < x(p_2)$). A horizon can be discretized by breaking it into an arbitrary number of connected line segments. The overarching 1-dimensional nature of a horizon implies a *linearization* of terrain.

This section focuses on formalisms of qualitative terrain descriptions. Since a formal language is determined by its strings over a finite alphabet, capturing horizons through a formal language encodes terrain properties by finite strings of symbols. For horizons, the finite alphabet is given by line segments of different slopes. A horizon is described by its corresponding string, from which terrain features can be identified as substrings. A mountain, for instance, is described by an ascending line followed by a descending line. The identification of terrain features can be seen as a concatenation of simple terrain primitives, such as ascending and descending lines, into terrain features such as mountains and plateaus. We model aggregation with term rewriting rules (Dershowitz 1993). The analysis of a terrain is characterized as a two-stage process: (1) the description of a terrain by simple terrain primitives and (2) the extraction of terrain features by aggregating line primitives into higher concepts.

3.1   The Slope of Terrain Silhouettes

The profile of the silhouette results from changes in the horizon's slope. Oriented straight line segments are used as *line primitives*. Similar to the flat earth in *Naive Geography* (Egenhofer and Mark 1995), which ignores the Earth's curvature, the representation by line primitives assumes that the terrain has a negligible curvature.

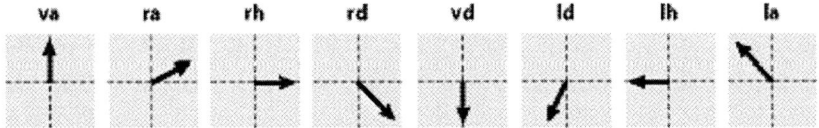

Fig. 2. Eight line primitives describing different qualitative slope values

Oriented straight line segments have eight different qualitative slope values, which can be derived from the combination of the directions vertical, left, and right with the directions horizontal, ascending, and descending—the combination of vertical with horizontal is excluded as it would yield a point rather than a line. These slope values correspond to the qualitative vectors that can be distinguished with the quadrants of a

cardinal direction system (Frank 1996). Exactly one type of line segment falls into each of the four quadrants and four line segments coincide with the separations between neighboring quadrants, leading to eight slope values in total (Fig. 2). The following notation is used for the eight line primitives:
- a vertical ascending (va), a right ascending (ra), a right horizontal (rh), a right descending (rd), a vertical descending (vd), a left descending (ld), a left horizontal (lh), and a left ascending (la) line.

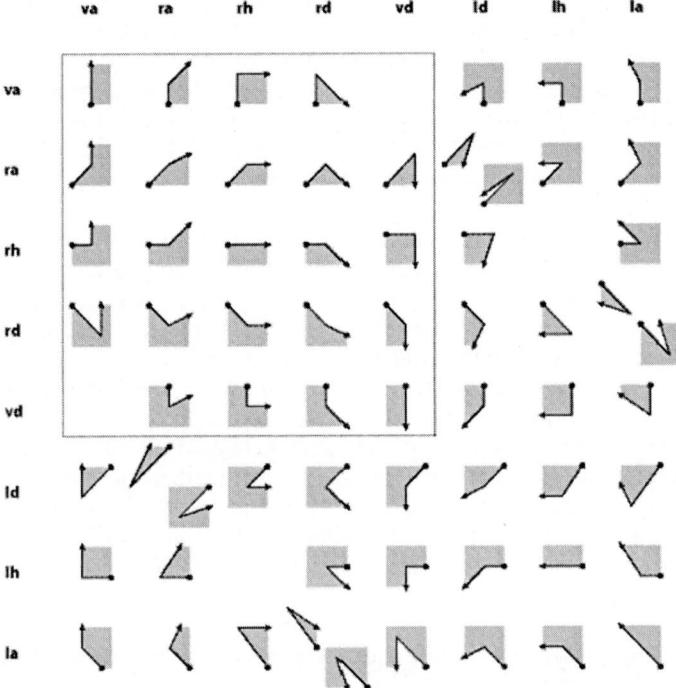

Fig. 3. Eight different line primitives, leading to 64 different terrain elements—grey areas indicate terrain. The 23 terrain elements in the top-left corner comprise those that do not point to the left

Among the eight-by-eight combinations of line primitives there are 60 line-line sequences that are fully constrained and describe unique terrain configurations. Four line-line sequences have three realizations each, as they result from the combination of two line primitives that are located in the same quadrant, which imposes no restriction to place them differently to each other. For example, the line sequence ldra could mean that the second segment is, in clockwise orientation, either to the left, to the right, or coincident with the first segment. The other pairs located in a single quadrant are rald, lard, and rdla, resulting in twelve additional line-line sequences. Among these 72 candidates, however, some do not represent valid terrain. For instance, the sequence lhrh would capture terrain above *and* below the line, with no space in between. There are eight illegal line sequences, where lines coincide: the four vertical and horizontal pairs with opposite directions (vavd, vdva, rhlh, and lhrh) and another four among the triples extending in the same quadrant. Thus, the combination of the eight line primitives (Fig. 3) yields 64 realizable, legal terrain elements.

Subsequently, we study a sublanguage that considers only the combination of primitives va, ra, rh, rd, and vd, where none of them points to the left. While this choice excludes the description of overhangs as terrain features, the overall approach becomes clearer as the alphabet is 64% smaller. The methods, however, scale up to larger alphabets.

A formal language describing terrain features on the basis of the five line primitives yields a single granularity level. Fig. 4 shows two horizons that are described by the same string dadadada. Since the line primitives are not distinguished according to their lengths or their differences in elevation, no further inferences about the terrain can be drawn. A small hill has the same representation as a big mountain. To capture such differences, we extend the qualitative line segment descriptions with two concepts: orders of magnitude for the length of segments (Section 3.2) and orders for differences in elevation (Section 3.3). The concepts slope direction, difference in elevation, and length of line segments are independent of each other:

- One cannot derive the slope direction (i.e., whether a line segment is left, vertical, or right ascending) from the difference in elevation, and vice versa.
- One cannot derive the slope direction from the lengths of the segments, and vice versa.
- One cannot derive the difference in elevation from the lengths of the segments, and vice versa.

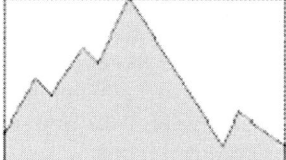 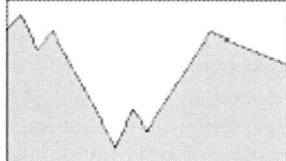

Fig. 4. Two terrains that are described by the same sequence of ascending and descending lines

3.2   Orders of Magnitude for Line Primitives

To capture differences in length qualitatively, we introduce *orders of magnitude*. We classify the line primitives of a terrain string according to their relative sizes, and assign to each line primitive an order of magnitude. This assignment can be done, for example, through a cluster analysis. In general, the number of orders of magnitude is small, typically up to five values. With the help of order-of-magnitude reasoning, smaller line primitives can be neglected with respect to longer ones.

Orders of magnitude are denoted by o, o', o'', o_1, o_2, ..., o_k, ω. They do not have to be natural numbers. It is sufficient that they are a linearly ordered list of symbols $\{o_1, ..., o_k\}$ with a total order relation $<$. These numbers roughly correspond to the number of granularity levels (see below). We denote the order of magnitude of a line primitive by a superscript, for instance, ra^{o_2} denotes a right ascending line with an order of magnitude of o_2.

3.3   Order of Differences in Elevation

A third component that determines a line primitive relative to a terrain silhouette is its difference in elevation. Two line primitives can have the same slope, for instance right ascending, but one could still be steeper than the other. We introduce an order relation $<_v$ that characterizes differences in elevation for line primitives. The relation $<_v$ enables the simplification of line primitives pointing in opposite directions. For example, for an ascending and a descending line there are three possibilities: the simplified line can be ascending, horizontal, or descending (Fig. 5).

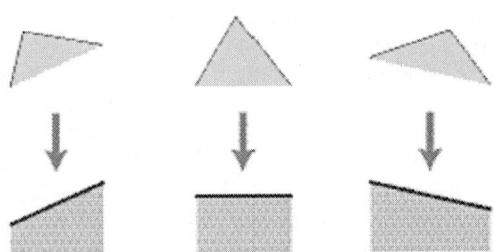

Fig. 5. Three different line primitives result from a right ascending and descending line

An order relation $<_v$ can be introduced in different ways. One way is to order the line primitives' endpoints according to their heights. If the points are denoted as a string then one can count the distance of the endpoints within this string. A second way is to introduce an order for the line primitives themselves: we take the common endpoint of two line primitives and determine which of the two lines has a greater difference in elevation by comparing the height of the two remaining endpoints. The difference in elevation induces an order for the line primitives. Both approaches are equivalent for the term rewriting rules in the succeeding section.

4   Levels of Terrain Granularity from Term Rewriting

The introduction of orders of magnitude enables a terrain analysis at different granularity levels. Hobbs' (1985) approach emphasizes the importance of different levels of detail. On each granularity level we can generate a description of terrain features from the string of line primitives that represent the horizon of a terrain silhouette. The rules to extract terrain features from a string of line primitives are given in Section 4.1.

In the case of a terrain description, the change from a finer granularity level to a coarser granularity level requires two steps: (1) generalization and (2) simplification. The generalization neglects line primitives with a smaller order of magnitude, whereas simplification deals with line primitives that are either of different types but the same order of magnitude, or of the same type but of different orders.

Since we restrict the analysis to the five line primitives va, ra, rh, rd, and vd, there are 23 valid combinations for two consecutive line primitives (Fig. 2). A further reduction is made if the combination of two line primitives of the same type is simplified into a single primitive (e.g., vava $\to$ va), which eliminates five combinations, leaving 18 combinations that are subject to refinements with orders of magnitude.

Either two consecutive line primitives have the same orders of magnitude, or one of them has a greater order of magnitude than the other one. This leads to $3 \times 18 = 54$ term rewriting rules. Rather than providing all rules, we focus on the process that systematically derives combinations of line primitives.

4.1   Aggregation of Line Primitives

This section presents term rewriting rules to aggregate line primitives into terrain features. These rules form just one possible set of rules for describing a terrain. The purpose of the rules is to demonstrate the general procedure to identify terrain features from a terrain string, not to provide the right way to determine terrain features.

S denotes a string representing a horizon in terms of line primitives. To generate terrain features from S, we process it sequentially. For two consecutive line primitives of S we evaluate whether they can be aggregated into a terrain feature. The string S is processed word by word. After evaluating a substring, we proceed with the next substring, which contains the second word of the previous substring now as the first word. Suppose, the string is rardrard. If we assume that a mountain is described by the substring rard and a valley by rdra, then the resulting terrain description comprises two mountains that share a valley. A procedure that moved two words at a time could not infer that there is a valley between two mountains. The word-by-word technique, however, ensures a complete enumeration of all identifiable terrain features.

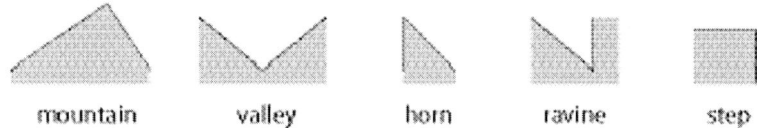

mountain   valley   horn   ravine   step

Fig. 6. Five different terrain features: a mountain, a valley, a horn, a ravine, and a step

We assume in the following a terrain description with a level of detail of order o'. In the first step, we present the aggregation rules for two line primitives of the same order of magnitude o, with $o \geq o'$. An ascending and a descending line represent a mountain (A1) or a valley (A2), depending on the sequence of the ascending and the descending line (Fig. 6).

(A1) $ra^o rd^o \to mt^o$

(A2) $rd^o ra^o \to vl^o$

A horn is a particular type of mountain, where one side is a sheer face (Fig. 6). Thus, a horn is described by either a right ascending and a vertical descending line, or a vertical ascending and a right descending line (A3a-b).

(A3a) $ra^o vd^o \to hn^o$

(A3b) $va^o rd^o \to hn^o$

Opposite to a horn, a ravine is a valley with a steep side (Fig. 6). Formally, a ravine is represented by a right descending and a vertical ascending line primitive, or a vertical descending and a right ascending line primitive (A4a-b).

(A4a) $rd^{\circ}va^{\circ} \rightarrow rv^{\circ}$

(A4b) $vd^{\circ}ra^{\circ} \rightarrow rv^{\circ}$

A steep difference in altitude starting from a horizontal level is considered a downward step (A5a), if it is described by a right horizontal and a right descending line primitive (Fig. 6). It is an upward step, if it is a sequence of a right horizontal and a right ascending line primitive (A5b).

(A5a) $rh^{\circ}vd^{\circ} \rightarrow ds^{\circ}$

(A5b) $rh^{\circ}va^{\circ} \rightarrow us^{\circ}$

The remaining cases are not aggregated into a single terrain entity, although geomorphology might offer further insights into aggregating more line primitives of the same order of magnitude. The main purpose of these rules is to show how a system with a set of aggregation rules is able to generate a terrain description for common terrain features. This systematic combination of line primitives establishes the option for an ontological terrain classification for all 18 possibilities.

In the second step, we demonstrate how to aggregate two line primitives with different orders of magnitude. We present two rules for ascending and descending lines. A right ascending line followed by a right descending line, both of different orders of magnitude, can be subsumed to a small *hillock* (A6a-b). The analog aggregation of a right descending with a right ascending line leads to a *notch* (A7a-b). This does not mean that the line primitives themselves form a hillock or a notch, but they indicate their existence. With o and ω denoting orders of magnitude such that o $\geq$ o' and o $> \omega$, the order of the hillock or notch is always the order of the right line primitive.

(A6a) $ra^{\circ}rd^{\omega} \rightarrow hk^{\omega}$

(A6b) $ra^{\omega}rd^{\circ} \rightarrow hk^{\circ}$

(A7a) $rd^{\circ}ra^{\omega} \rightarrow nt^{\omega}$

(A7b) $rd^{\omega}ra^{\circ} \rightarrow nt^{\circ}$

Terrain features, such as a plateau or a u-shaped valley, require more than two line primitives. The incorporation of orders of magnitude allows us to distinguish different types of plateaus: the butte (bt), the mesa (ms), and the plateau (pt). These landforms are elevations with little internal relief and at least one cliff at one side (Fig. 7).

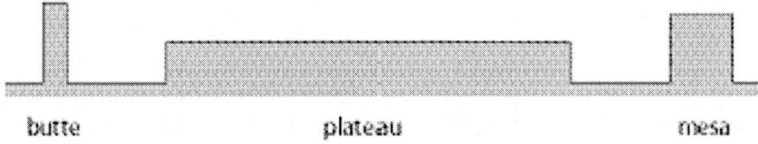

butte   plateau   mesa

Fig. 7. Butte, plateau, and mesa with the same line primitives, but different orders of magnitude

A butte (A8) has a small summit area that is bounded by cliffs. Using line primitives, a butte is described by a right horizontal line that is between a vertical ascending and descending line of a higher order of magnitude.

(A8) $va^{\circ}rh^{\omega}vd^{\circ} \rightarrow bt^{\circ}$

A plateau (A9) has a large uplifted area that is relatively flat. Thus, the geometric characterization is a right horizontal line that is surrounded by a vertical ascending and descending line, both of a smaller order of magnitude.

(A9) $va^{\omega}rh^{\circ}vd^{\omega} \rightarrow pl^{\circ}$

A mesa (A10) has a much larger elevated area than a butte. The summit area is of the same order of magnitude as the vertical ascending and descending line.

(A10) $va^{\circ}rh^{\circ}vd^{\circ} \rightarrow ms^{\circ}$

A terrace, similar to a plateau, is a large, flat area embedded on a relatively flat hillside. Geometrically, it can be described as right horizontal line that is bounded either by two small vertical ascending lines or two descending lines (A11a-b).

(A11a) $va^{\omega}rh^{\circ}va^{\omega} \rightarrow tr^{\circ}$

(A11b) $vd^{\omega}rh^{\circ}vd^{\omega} \rightarrow tr^{\circ}$

flat-floored valley   u-shaped valley

Fig. 8. A flat-floored valley and a u-shaped valley described by three line primitives

We can distinguish two further types of valleys using orders of magnitude: a flat-floored valley and a u-shaped valley (Fig. 8). A right horizontal line between a right descending and ascending line of the same order of magnitude describes a flat-floored valley (A12), and a right horizontal line of a smaller order a u-shaped valley (A13).

(A12) $rd^{\circ}rh^{\circ}ra^{\circ} \rightarrow fv^{\circ}$

(A13) $rd^{\circ}rh^{\omega}ra^{\circ} \rightarrow uv^{\circ}$

depression   canyon

Fig. 9. Depression and canyon described by five line primitives

The last two examples show that there are cases that require more line primitives to identify a terrain feature, the depression and the canyon (Fig. 9). A depression can be characterized as a right horizontal line that has a higher order of magnitude than its limiting vertical descending and ascending line (A14). The depression is embedded in a flat land, that is, there are two horizontal lines one both sides of the depression.

(A14) $rh^{\circ"}vd^{\omega}rh^{\circ}va^{\omega}rh^{\circ"} \rightarrow dp^{\circ}$ $(\circ" \geq \circ)$

A canyon may occur between two plateaus. It can be described by two horizontal lines and two vertical lines of the same order of magnitude in between (A15), such that the vertical lines are connected by a small horizontal line.

(A15) $rh^{o"}vd^{o}rh^{\omega}va^{o}rh^{o"} \rightarrow cy^{o}$ $(o" \geq o)$

If no further combinations of line primitives are possible, the primitive lines are directly mapped to terrain features: a right horizontal line describes a plain (A16), a descending line a descent (A17a), an ascending line an ascent (A17b), a vertical descending line a falling cliff (A18a), and a vertical ascending line a rising cliff (A18b).

(A16) $rh^{o} \rightarrow pn^{o}$

(A17a) $rd^{o} \rightarrow dc^{o}$

(A17b) $ra^{o} \rightarrow ac^{o}$

(A18a) $vd^{o} \rightarrow fc^{o}$

(A18b) $va^{o} \rightarrow rc^{o}$

4.2   Generalization of Line Primitives

In order to generate terrain descriptions at coarser levels of detail, two line primitives of different orders of magnitude are generalized to a single line primitive. Under order-of-magnitude reasoning, pairs of line segments are generalized only if the line segment with the higher order of magnitude has the order o' of the given granularity level. The reason is the following: suppose, a line primitive p_1 has the order $o_1 > o'$ followed by a line primitive p_2 of order o_2 with $o_1 > o_2 > o'$. If we were to generalize the substring p_1p_2 into the line primitive p, a terrain feature relying on p_2 could not be extracted at a coarser granularity level of order o_2. All generalization rules are commutative, that is, if $p_1^{o}p_2^{\omega} \rightarrow p^{o}$ holds, then it also follows $p_2^{\omega}p_1^{o} \rightarrow p^{o}$.

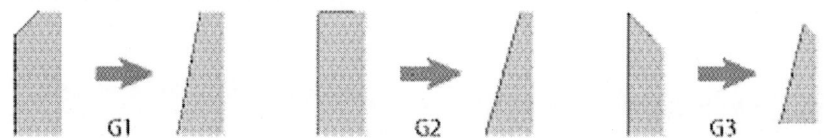

Fig. 10. Three generalizations: a vertical ascending line followed by a right ascending line, a right horizontal line, or a right descending line of a smaller order of magnitude (G1–3)

A pair with a vertical ascending and a right ascending, horizontal, or descending line primitive is generalized to a right ascending line (Fig. 10), with $o > \omega$ and $o = o'$ (G1–3). The corresponding generalization applies to a combination that starts with a vertical descending line (G4–6).

(G1) $va^{o}ra^{\omega} \rightarrow ra^{o}$

(G2) $va^{o}rh^{\omega} \rightarrow ra^{o}$

(G3) $va^{o}rd^{\omega} \rightarrow ra^{o}$

(G4) $vd^{o}ra^{\omega} \rightarrow rd^{o}$

(G5) $vd^{o}rh^{\omega} \rightarrow rd^{o}$

(G6) $vd^{o}rd^{\omega} \rightarrow rd^{o}$

A pair with a right horizontal line primitive of a higher order of magnitude ($o > \omega$) and an ascending line primitive is generalized to a right ascending line primitive (G7–8), with the corresponding generalizations for a sequence that ends in a vertical descending line (G9–10).

(G7) $\mathrm{rh^{o}va^{\omega}} \to \mathrm{ra^{o}}$

(G8) $\mathrm{rh^{o}ra^{\omega}} \to \mathrm{ra^{o}}$

(G9) $\mathrm{rh^{o}vd^{\omega}} \to \mathrm{rd^{o}}$

(G10) $\mathrm{rh^{o}rd^{\omega}} \to \mathrm{rd^{o}}$

A pair with a right ascending line and a right horizontal (G11) or vertical ascending line (G12), both of a smaller order of magnitude, is generalized to a right ascending line. Corresponding generalizations apply to a right descending line followed by either a right horizontal (G13) or a vertical descendent line (G14) if they are of a smaller order of magnitude.

(G11) $\mathrm{ra^{o}rh^{\omega}} \to \mathrm{ra^{o}}$

(G12) $\mathrm{ra^{o}va^{\omega}} \to \mathrm{ra^{o}}$

(G13) $\mathrm{rd^{o}rh^{\omega}} \to \mathrm{rd^{o}}$

(G14) $\mathrm{rd^{o}vd^{\omega}} \to \mathrm{rd^{o}}$

Rules G1–14 do not require any knowledge about the differences in elevation. Ascending and descending lines, however, can have small inclinations. To determine the combination of a right ascending or descending line with a line primitive pointing in the opposite vertical direction, we have to regard the differences in elevation of the line primitives. For a pair of a right ascending line and a descending line of a smaller order of magnitude (G15–16), there are three ways in which this pair can be generalized. Depending on the relative height difference of the line primitives, the resulting line primitive is either ascending, or horizontal, or descending. In all cases, the generalized line primitive has the order of magnitude of the right ascending line. The rules (G17–18) for a pair of a right ascending line and descending line are analogous.

(G15a) $\mathrm{rd^{\omega}} <_{v} \mathrm{ra^{o}} \wedge \mathrm{ra^{o}rd^{\omega}} \to \mathrm{ra^{o}}$

(G15b) $\mathrm{rd^{\omega}} =_{v} \mathrm{ra^{o}} \wedge \mathrm{ra^{o}rd^{\omega}} \to \mathrm{rh^{o}}$

(G15c) $\mathrm{rd^{\omega}} >_{v} \mathrm{ra^{o}} \wedge \mathrm{ra^{o}rd^{\omega}} \to \mathrm{rd^{o}}$

(G16a) $\mathrm{vd^{\omega}} <_{v} \mathrm{ra^{o}} \wedge \mathrm{ra^{o}vd^{\omega}} \to \mathrm{ra^{o}}$

(G16b) $\mathrm{vd^{\omega}} =_{v} \mathrm{ra^{o}} \wedge \mathrm{ra^{o}vd^{\omega}} \to \mathrm{rh^{o}}$

(G16c) $\mathrm{vd^{\omega}} >_{v} \mathrm{ra^{o}} \wedge \mathrm{ra^{o}vd^{\omega}} \to \mathrm{rd^{o}}$

(G17a) $\mathrm{ra^{\omega}} >_{v} \mathrm{rd^{o}} \wedge \mathrm{rd^{o}ra^{\omega}} \to \mathrm{ra^{o}}$

(G17b) $\mathrm{ra^{\omega}} =_{v} \mathrm{rd^{o}} \wedge \mathrm{rd^{o}ra^{\omega}} \to \mathrm{rh^{o}}$

(G17c) $\mathrm{ra^{\omega}} <_{v} \mathrm{rd^{o}} \wedge \mathrm{rd^{o}ra^{\omega}} \to \mathrm{rd^{o}}$

(G18a) $\mathrm{va^{\omega}} >_{v} \mathrm{rd^{o}} \wedge \mathrm{rd^{o}va^{\omega}} \to \mathrm{ra^{o}}$

(G18b) $\mathrm{va^{\omega}} =_{v} \mathrm{rd^{o}} \wedge \mathrm{rd^{o}va^{\omega}} \to \mathrm{rh^{o}}$

(G18c) $\mathrm{va^{\omega}} <_{v} \mathrm{rd^{o}} \wedge \mathrm{rd^{o}va^{\omega}} \to \mathrm{rd^{o}}$

The rules G1–18 reflect the 18 combinations for two line primitives of different orders of magnitude (see the introduction of this section). The generalization rules for lines primitives rewrite a terrain string such that it is no longer guaranteed that a line primitive cannot be followed by a line primitive of the same type. Moreover, a sequence of line primitives of the same type might accumulate to a line primitive of a higher order of magnitude. The rules in the next section simplify a generalized terrain string before it is reevaluated on a coarser granularity level.

4.3   Simplification of Line Primitives

After generalizing a terrain string two *simplifications* (i.e., re-assignments of line primitives and orders of magnitude) are required. First, without a higher-order neighbor primitive, a sequence of line primitives of the same order of magnitude cannot be generalized. For instance, the substring $\text{rd}^\circ\text{ra}^\circ\text{rd}^\circ$ cannot be further generalized if $\text{rd}^\circ\text{ra}^\circ$ and $\text{ra}^\circ\text{rd}^\circ$ are left and right neighbors. We only simplify line primitives of the same order if their orders are equal to the previous granularity level. Simplification of all line primitives of the same order would loose major terrain features. Since the 18 simplification rules are commutative, only 9 cases need to be considered. A pair of a right horizontal line primitive with a vertical ascending or a right ascending line primitive is simplified to a right ascending line (S1–2). The corresponding simplification applies to descending lines (S3–4). A natural number $k \in \{1, 2\}$, which captures the length of the simplified line primitive, is obtained from the comparison of the resulting line primitive with the existing line primitives of the same order (Fig. 11).

(S1) $\text{rh}^\circ\text{va}^\circ \rightarrow k\cdot\text{ra}^\circ,$

(S2) $\text{rh}^\circ\text{ra}^\circ \rightarrow k\cdot\text{ra}^\circ,$

(S3) $\text{rh}^\circ\text{vd}^\circ \rightarrow k\cdot\text{rd}^\circ,$

(S4) $\text{rh}^\circ\text{rd}^\circ \rightarrow k\cdot\text{rd}^\circ,$

$S2 \{k = 1\}$ $S2 \{k = 2\}$

Fig. 11. Two simplifications of a right horizontal and a right ascending line (S2) lead to different length multipliers k = 1 and k = 2, respectively, for the simplified right ascending line

Likewise, a pair of a right ascending and a vertical ascending line is simplified to a right ascending line (S5), with equivalent simplifications for descending lines (S6).

(S5) $\text{ra}^\circ\text{va}^\circ \rightarrow k\cdot\text{ra}^\circ,$

(S6) $\text{rd}^\circ\text{vd}^\circ \rightarrow k\cdot\text{rd}^\circ,$

If two line primitives point in opposite directions, one needs to incorporate the lines' elevation differences. Similar to the case with generalization, there are three different resulting line primitives: an ascending (S7a, S8a, S9a), a horizontal (S7b, S8b, S9b), or a descending line (S7c, S8c, S9c). The simplified line primitive can be very small and, therefore, of a smaller order of magnitude x than the order o of the two primitives before the simplification (Fig. 12). The length multiplier k is 1 if the simplified line primitive has a smaller order of magnitude, and 1 or 2 if the simplified line has the same order o.

(S7a) $rd^o <_v ra^o \wedge ra^o rd^o \rightarrow k \cdot ra^x$

(S7b) $rd^o =_v ra^o \wedge ra^o rd^o \rightarrow k \cdot rh^x$

(S7c) $rd^o >_v ra^o \wedge ra^o rd^o \rightarrow k \cdot rd^x$

(S8a) $vd^o <_v ra^o \wedge ra^o vd^o \rightarrow k \cdot ra^x$

(S8b) $vd^o =_v ra^o \wedge ra^o vd^o \rightarrow k \cdot rh^x$

(S8c) $vd^o >_v ra^o \wedge ra^o vd^o \rightarrow k \cdot rd^x$

(S9a) $va^o >_v rd^o \wedge rd^o va^o \rightarrow k \cdot ra^x$

(S9b) $va^o =_v rd^o \wedge rd^o va^o \rightarrow k \cdot rh^x$

(S9c) $va^o <_v rd^o \wedge rd^o va^o \rightarrow k \cdot rd^x$

Fig. 12. An illustration of the rule S7a: on the left, the simplified right ascending line has the same order of magnitude as the initial right ascending and right descending line; on the right, the simplified right ascending line has a smaller order of magnitude

In the second simplification scenario, the generalization can lead to sequences of line primitives of the same type (e.g., $ra^{o_2} ra^{o_2} ra^{o_2} ra^{o_4}$). Thus, one needs to assign orders of magnitude to generalized sequences of the same type. We accumulate all line primitives of the same type. Let $p \in \{va, ra, rh, rd, vd\}$ denote a single line primitive and n a natural number. First, we subsume all line primitives of the same order (S10).

(S10) $n \cdot p^o \, p^o \rightarrow (n+1) \cdot p^o$.

Second, according to rule S11 we assign for all line primitives of the same type a new order of magnitude (Fig. 13).

(S11) $S := n_{o_1} \cdot p^{o_1} \dots n_{o_i} \cdot p^{o_i} \rightarrow p^o$.

Fig. 13. An illustration of the application of rule S11 to right ascending and right descending line primitives

The order of magnitude o is chosen by the following three rules:
• If there is a line primitive p^{ro} that has the same order of magnitude as S (i.e., $S \approx p^{ro}$), then the order of magnitude of S is o.

- If there are two line primitives p' and p" with different orders of magnitude o' and o" and S satisfies $p^{o'} \ll S \ll p^{o''}$, then we add a new order of magnitude o_{k+1} to the list $\{o_1, ..., o_k\}$ with $o' < o_{k+1} < o''$.
- If S has a greater order of magnitude than any of the line primitives, we add a new order of magnitude o_{k+1} to the list $\{o_1, ..., o_k\}$ with $o_i < o_{k+1} \; \forall i \in \{1, ..., k\}$.

To obtain a coarser terrain description, the aggregation rules (A1–18) apply again. Because the simplification of line primitives can lead to higher orders of magnitude than the ones initially obtained from the original terrain, there might be more granularity levels than the initial number of orders of magnitude. In general, the number of granularity levels is determined by the highest order of magnitude for the line primitives that can be obtained by repeated application of the simplification rules.

5   Terrain Language Example

A terrain silhouette (Fig. 14) is used to demonstrate the utility of the terrain language. On a fine granularity level, the terrain is described as a sequence of two smaller mountains, followed by a large mountain with a jagged slope, a mountain where the right slope transitions into a plateau, and two terraces. The two largest mountains are separated by a u-shaped valley. At the coarsest granularity level, the terrain comprises two mountains.

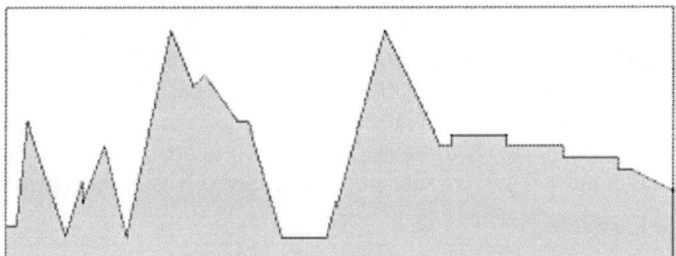

Fig. 14. Silhouette of a terrain

Line primitives are expressed by p_i^o, where o stands for the order of magnitude—a number between 1 and 4—and i is the index of the particular line primitive. The following string S is a description of the horizon of the terrain silhouette.

$$S := \quad rh_1^1 \; ra_2^3 \; rd_3^3 \; ra_4^2 \; rd_5^1 \; ra_6^2 \; rd_7^3 \; ra_8^4 \; rd_9^2 \; ra_{10}^1 \; rd_{11}^2 \; rh_{12}^1 \; rd_{13}^3 \; rh_{14}^2 \; ra_{15}^4 \; rd_{16}^3 \; rh_{17}^1 \; va_{18}^1$$
$$rh_{19}^2 vd_{20}^1 \; rh_{21}^2 \; vd_{22}^1 \; rh_{23}^2 \; vd_{24}^1 \; rh_{25}^1 \; rd_{26}^2$$

The finest granularity level generates a description of the terrain features for all line primitives. This level is skipped in the following discussion. The application of the generalization (G1–18) and simplification rules (S1–11) does not change the terrain string, since there are no line primitives of order 0. Thus, we start with a terrain description that describes the terrain features if the line primitives are at least of order 2. The string S is parsed word-by-word using the aggregation rules (A1–18).

- The first step considers terrain features that can be characterized by pairs of line primitives. The substring $rh_1^1 ra_2^3$ is not a terrain feature. The substring $ra_2^3 rd_3^3$ describes a mountain of order 3, while $rd_3^3 ra_4^2$ derives a notch of order 2. This analysis applies to the rest of all pairs.
- The second step aggregates three line primitives into a terrain feature, obtaining a plateau ($va_{18}^1 rh_{19}^2 vd_{20}^1$), followed by a terrace ($vd_{20}^1 rh_{21}^2 vd_{22}^1$), and another terrace ($vd_{22}^1 rh_{23}^2 vd_{24}^1$). Since there are no canyons and depression, the rules A1–18 lead to a terrain description that captures the sequence of a mountain, a notch, a hillock, a notch, a hillock, a notch, a hillock, two descents, a plain, a hillock, a descent, a plateau, two terraces, and a descent.

$$S_T(2) := mt^3 nt^2 hl^1 nt^2 hl^3 nt^4 hl^2 nt^1 hl^2 dc^2 dc^3 pn^2 hl^3 dc^3 pl^2 tr^2 tr^2 dc^2$$

A coarser terrain description is obtained from the generalization rules (G1–18), eliminating in S all line primitives of order 1. Rule G11 generalizes the substring $rh^1 ra^3$ to ra^3. Substring $ra^3 rd^3$ cannot be generalized, because ra^3 and rd^3 have the same order of magnitude. Likewise, substring $rd^3 ra^2$ cannot be generalized, because both orders are greater than 1. The repeated application of rules G1–18 on S leads to:

$$S_G(2) := ra_1^3 rd_2^3 ra_3^2 ra_4^2 rd_5^3 ra_6^4 rd_7^2 rd_8^2 rd_9^3 rh_{10}^2 ra_{11}^4 rd_{12}^3 ra_{13}^2 rd_{14}^2 rd_{15}^3 rd_{16}^2$$

This generalized string can be simplified by rules S1–11. For instance, $ra_3^2 ra_4^2$ get simplified to a right ascending line of order 3. Application of rules S1–11 yields:

$$S_S(2) := ra_1^3 rd_2^3 ra_3^3 rd_4^3 ra_5^3 rd_6^4 rh_7^2 ra_8^4 rd_9^3 ra_{10}^2 rd_{11}^3$$

The simplified string allows us to identify terrain features at the next coarser level, leading to a sequence of a mountain, a valley, a mountain, a notch, a mountain, a u-shaped valley, a hillock, a notch, and a hillock.

$$S_T(3) := mt_1^3 vl_2^3 mt_3^3 nt_4^4 mt_5^3 uv_6^4 hl_7^2 nt_8^2 hl_9^3$$

Generalizing and simplifying the string $S_T(3)$ again, we obtain $rd_1^3 ra_2^4 rd_3^3 ra_4^4 rd_5^4$. Aggregation rules A1 and A2 generate the coarsest terrain description $S_T(4) = mt_1^3 vl_2^3 mt_3^3$ (i.e. two mountains with a valley in between). The string $S_T(4)$ is the coarsest possible terrain description, because rules G1-18 do not rewrite the terrain string any further and thus no further generalization applies.

6   Conclusions and Future Work

This paper introduces an approach to identifying terrain features of a terrain silhouette at different levels of granularity. The horizon of a terrain silhouette is represented as a string that captures the changes in slope on the basis of line primitives. The line primitives are classified according to their inclination (ascending, horizontal, descending), and their vertical alignment (left, vertical, right). Orders of magnitude reflect the impact of the relative size of a line primitive for the overall terrain. Differences in elevation are represented by an order relation for the line primitives.

Three different classes of term rewriting rules—aggregation, generalization, and simplification—evaluate a terrain string. Aggregation rules enable a terrain description for a given granularity level, whereas generalization and simplification rules specify the principles for moving up to a coarser granularity level. The rules can be adapted to different audiences: less specific ones for tourists and more specific ones for experts in geomorphology. A terrain string encodes the horizon in such a way that it is possible to reconstruct a prototypical terrain for each granularity level.

The aggregation rules provide a framework for classifying a variety of terrain features. Cliffs, flat-floored valleys, as well as canyons can be identified by these rules. Since we restrict the terrain features mainly to five line primitives (i.e., dropping line primitives that point to the left), there are no rules for overhangs or caves. However, the presented approach can be easily enriched with these terrain features. This is a major advantage compared to other approaches for identifying terrain features.

There are two cases where a description based on orders of magnitude has its limitations. First, an assignment of orders of magnitude is difficult, if the line primitives representing a horizon cannot be combined into clusters such as a sequence of lines, where the succeeding line is exactly one unit greater than the previous one. Second, the simplification rules might fail to assign the right order of magnitude, if there are numerous small line primitives. This behavior is expected, since a qualitative description is appropriate only, if the data can be grouped into sensible classes and do not require a detailed numerical description. If the horizon is represented by a detailed polygonal curve, a shape-preserving algorithm (Latecki and Lakämper 2000) can be used to reduce the number of line segments, before applying term rewriting rules.

The approach describes linear terrain features that can be extracted from a horizon. Features such as saddles, ridges, and draws can be identified only for terrains represented as surfaces. Within our framework, it is impossible to distinguish a valley from a saddle. Two strategies are possible to apply the framework to a terrain represented as a surface: the construction of orthogonal cross-sections of the terrain silhouette at critical points to resolve ambiguities, or a terrain analysis with a sweep line algorithm (O'Rourke 1998). Both approaches will be investigated and compared in future work.

Acknowledgments

This research was supported by the National Imagery and Mapping Agency under grant number NMA201-01-1-2003. We thank Matt Duckham, Jim Farrugia, Mike Hendricks, Mike Worboys, and the reviewers for their comments and suggestions.

References

L. Cinque and L. Lombardi (1995) Shape description and recognition by a multiresolution approach. *Image and Vision Computing* 13(8): 599–607.
E. Clementini and P. Di Felice (1997) A global framework for qualitative shape description. *GeoInformatica* 1: 11–27.
E. Davis (1999) Order of magnitude comparisons of distance. *Journal of Artificial Intelligence Research* 10: 1–38.
M. T. de Berg and K. T. G. Dobrindt (1998) On levels of detail in terrains. *Graphical Models & Image Processing* 60(1): 1–12.

L. De Floriani and P. Magillo (1995) Horizon computation on a hierarchical triangulated terrain model. *The Visual Computer* 11(3): 134–149.

L. De Floriani and P. Magillo (1999) Intervisibility on terrains. In: P. Longley, M. Goodchild, D. Maguire, and D. Rhind (Eds). *Geographical information systems*: 543–556. New York, Wiley.

L. De Floriani and P. Magillo (2002) Multiresolution mesh representation: Models and data structures. In: A. Iske, Q. Ewald, and M. S. Floater (Eds). *Tutorials on Multiresolution in Geometric Modelling*: 363–418. Berlin, Springer.

Department of the Army (2001) *Field Manual No. 3-25.26: Map reading and land navigation.* Washington, DC.

N. Dershowitz (1993) A taste of rewrite systems. In: P. E. Lauer (Ed.), *Functional Programming, Concurrency, Simulation and Automated Reasoning: International Lecture Series 1991-1992*, McMaster University, Hamilton, Ontario, Canada, pp. 199–228, Springer.

M. J. Egenhofer and W. Kuhn (1998) Beyond desktop GIS, *GIS PlaNET*, Lisbon, Portugal.

M. J. Egenhofer and D. M. Mark (1995) Naive geography. In: A. U. Frank and W. Kuhn (Eds), *Spatial Information Theory: A Theoretical Basis for GIS, International Conference COSIT '95*, Semmering, Austria, pp. 1–15, Springer.

A. Frank, B. Palmer, and V. Robinson (1986) Formal methods of the accurate definition of some fundamental terms in physical geography, *Proceedings, Second International Symposium on Spatial Data Handling*, Seattle, Washington, pp. 583–599.

A. U. Frank (1996) Qualitative spatial reasoning: Cardinal directions as an example. *International Journal of Geographical Information Science* 10(3): 269–290.

A. P. Galton and R. C. Meathrel (1999) Qualitative outline theory. In: T. Dean (Ed.), *Proceedings of the 16th International Joint Conference on Artificial Intelligence, IJCAI 99*, Stockholm, Sweden, Vol. 2, pp. 1061–1066, Morgan Kaufmann.

M. Garland (1999) Multiresolution Modeling: Survey & Future Opportunities, *Eurographics '99 – State of the Art Reports*, Milan, Italy, pp. 111–131.

J. R. Hobbs (1985) Granularity. In: A. K. Joshi (Ed.), *Proceedings of the 9th International Joint Conference on Artificial Intelligence*, Los Angeles, CA, USA, Vol. 1, pp. 432–435, Morgan Kaufmann.

L. J. Latecki and R. Lakämper (2000) Shape similarity measure based on correspondence of visual parts. *IEEE Transaction Pattern Analysis and Machine Intelligence* 22(10): 1185–1190.

M. Leyton (1988) A process-grammar for shape. *Artificial Intelligence* 34(2): 213–247.

P. P. Nayak (1992) Order of magnitude reasoning using logarithms. In: B. Nebel, C. Rich, and W. R. Swartout (Eds), *Proceedings of the 3rd International Conference on Principles of Knowledge Representation and Reasoning (KR'92)*, Cambridge, MA, USA, pp. 201–210, Morgan Kaufmann.

J. O'Rourke (1998) *Computational Geometry in C*. New York, Cambridge University Press.

O. Raiman (1986) Order of magnitude reasoning, *Proceedings of AAAI-88*, pp. 100–104.

A. H. Strahler and A. N. Strahler (1992) *Modern physical geography*. New York, John Wiley & Sons.

Maintaining Spatial Relations
in an Incremental Diagrammatic Reasoner

Ronald W. Ferguson, Joseph L. Bokor, Rudolph L. Mappus IV, and Adam Feldman

College of Computing, Georgia Institute of Technology
Atlanta, GA 30332 USA
{rwf,jlbokor,cmappus,storm}@cc.gatech.edu

Abstract. This paper describes an architecture for dynamically handling spatial relations in an incremental, nonmonotonic diagrammatic reasoning system. The architecture represents jointly exhaustive and pairwise disjoint (JEPD) spatial relation sets as nodes in a dependency network. These spatial relation sets include interval relations, relative orientation relations, and connectivity relations, but in theory could include any JEPD spatial relation sets. This network then caches dependencies between low-level spatial relations, allowing those relations to be easily assumed or retracted as visual elements are added or removed from a diagram. For example, in the architecture's Undo mechanism, the dependency network can quickly reactivate cached spatial relations when a previously-deleted element is restored. As part of this work, we describe how the system supports higher-level reasoning, including support for creating default assumptions. We also describe how this system was integrated with an existing drawing program and discuss its possible use in diagrammatic and geographic reasoning.

1   Introduction

Diagrams are useful in a wide variety of reasoning tasks. Because a single diagram can convey many spatial relations at a glance, diagrams provide rich sources of information for a number of domains, including geographic, architectural, and engineering domains.

This capability is more interesting when we consider that the spatial relations in a diagram need not be static. Diagrams frequently change over time. The additions, removals, and modifications of elements also change the set of spatial relations. Handling such incremental changes without significant reprocessing of previously established spatial relations is key to making spatial and diagrammatic reasoning efficient. Incremental processing is also key to many geographic reasoning tasks. Maps indicating changing conditions, such as those for crisis planning, may be edited as conditions and needs change.

An incremental processing approach may also be driven by other factors. For example, problem solving with diagrams often requires many changes to a particular diagram over time as new ideas or subtasks emerge. The ability to understand the consequences of a change in a set of elements in conceptual terms may be affected by low-level changes. Finally, a more practical benefit of incremental processing is that it distributes the processing burden more evenly over the extent of the task, which is useful on low-end devices, such as personal digital assistants.

W. Kuhn, M.F. Worboys, and S. Timpf (Eds.): COSIT 2003, LNCS 2825, pp. 136–150, 2003.

In this paper, we describe an architecture for maintaining the set of qualitative spatial relations during incremental, nonmonotonic changes to a diagram. This work builds on earlier work on the GeoRep diagrammatic reasoner [1], which is described in the next section (after reviewing related research). After describing GeoRep, we discuss how GeoRep was modified to allow incremental processing, and cover a number of key theoretical and architectural issues: how to handle composite objects, the interface between low-level and high-level reasoning, and why such a system requires a new approach to handling default visual interpretations. We then discuss related work and future challenges for this system.

2   Related Work

A number of other researchers have looked at how incremental changes impact spatial reasoning.

In qualitative spatial reasoning, researchers have explored how to process qualitative spatial vocabularies incrementally. Notably, Hernández [2] proposed a mechanism for maintaining a consistent spatial description consisting of a combined set of orientational and topological spatial relations. Through careful analysis, separate orientation and topological relations sets are combined into a single vocabulary. These mechanisms also use a dependency network, similar to that in a truth-maintenance system, to allow relations to be added or deleted incrementally. At the same time, this mechanism does not address visual descriptions that contain multiple heterogeneous relation sets.

A number of researchers have explored the use of conceptual neighborhoods [3] and topological distances [4] to understand gradual change in the context of qualitative spatial vocabularies. Egenhofer and Al-Taha, for example, show how an analysis of topological distance between members of a relation set can be used to construct a graph that links the closest qualitative spatial relations. This graph can be used to show the set of necessary intervening qualitative states that must occur between two given states.

Along with this research in qualitative spatial reasoning, there are a number of sketching systems that must maintain knowledge of the links between individual visual elements and the inferred relations, although few of them have explicit frameworks for handling dependencies between spatial relations. A family of sketch interpretation systems by Davis and colleagues [5] use blackboard systems to integrate a low-level reasoner with a high-level description language, as with GeoRep, but can handle sketched shapes as well as vector graphics. Similarly, the Geometer's Sketchpad [6] uses a constraint network to enforce a set of constraints over a set of visual elements, which include line segments, rays, and circles. These constraints, however, are not discovered by the system, but must be entered by hand.

3   The GeoRep Diagrammatic Reasoner

This work extends an existing diagrammatic reasoner, GeoRep[1], to make processing incremental. GeoRep [1] has been used in a number of diagrammatic reasoning do-

---

[1] GeoRep is short for GEOmetric REPresenter, although it has also been used for geographic domains, such as reasoning about battlefield movements [7].

mains, including map-based military diagrams [7], simple physical diagrams [8], and logic circuits [9]. In addition, it has been used as the visual representation system for several cognitive modeling simulations [10, 11]. We first describe the GeoRep reasoner and then turn to the incremental processing mechanism.

GeoRep's input is a line-drawn figure, given as a vector graphics file. The vector graphics file contains a number of visual element types, including line segments, circles, ellipses, arcs, spline curves, and positioned text. As output, GeoRep produces a predicate calculus representation. This representation has three parts: the low-level spatial relations, the high-level interpretation of the figure, and intermediate spatial and conceptual relations that are produced in the process of interpretation.

To generate this description, GeoRep uses a two-stage architecture (Figure 1). We describe each stage in turn.

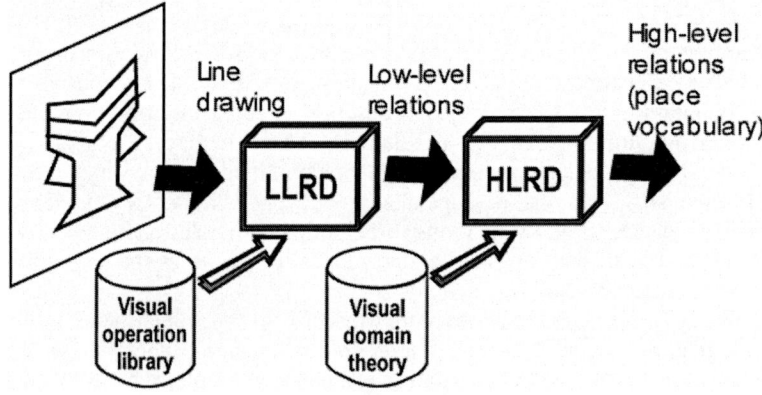

Fig. 1. Simplified GeoRep architecture, containing stages for low-level and high-level visual reasoning

The first stage, the *Low-Level Relational Describer* (LLRD) represents a set of low-level spatial relations in the figure (Figure 2). These low-level relations are generated in a pipelined architecture, starting with simple proximity calculations and ending with more sophisticated relations, such as interval relations between parallel line segments.

These particular spatial relations are also designed to model those qualitative spatial relationships that are detected in early vision. For example, it is well-known that humans are sensitive to relative angles (such as perpendicular lines), indentations in figure boundaries [12], and to vertical and horizontal orientations in the assumed frame of reference [13]. Interestingly, one relation set used that has not been tested for early vision is interval relations [14] between parallel lines. In practice, these relations are extremely useful in modeling aspects of visual perception such as the detection of qualitative symmetry [15]. A simple attention model uses a proximity detector to limit visual relation tests to proximate visual elements. This acts as a limited focusing mechanism that keeps processing tractable.

The system also models some aspects of attention and perceptual organization, though in a domain-dependent fashion. "Grouping rules" can be used to simulate similarity-based grouping. Similarly, multiple LLRDs can be used to simulate visual separation based on factors such as color [9].

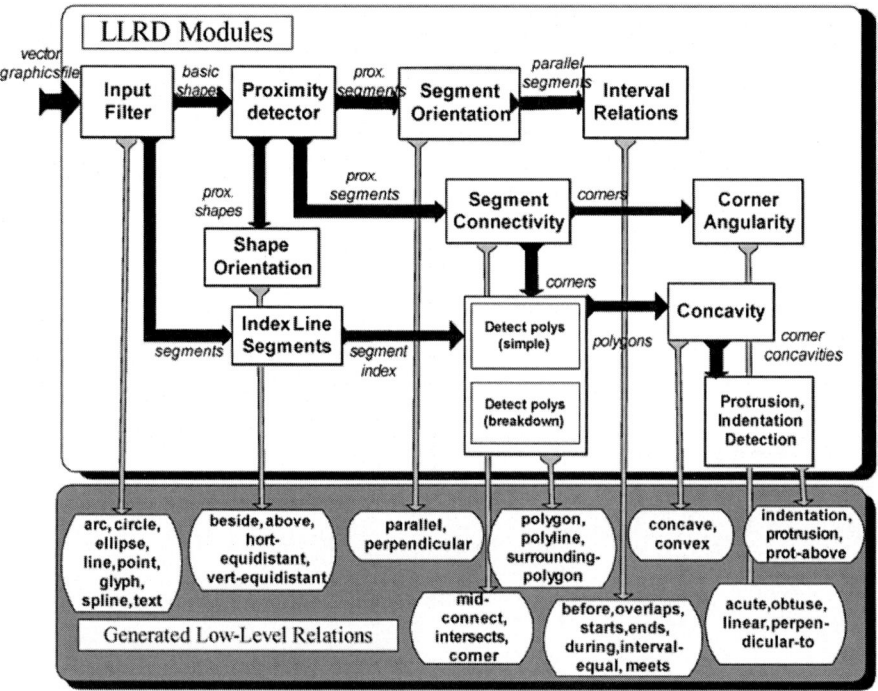

Fig. 2. Data flow diagram of GeoRep's Low-Level Relation Describer (LLRD). The top portion shows the collections of visual operators and the visual elements they process. The lower portion shows the set of spatial relations produced by each set of visual operators

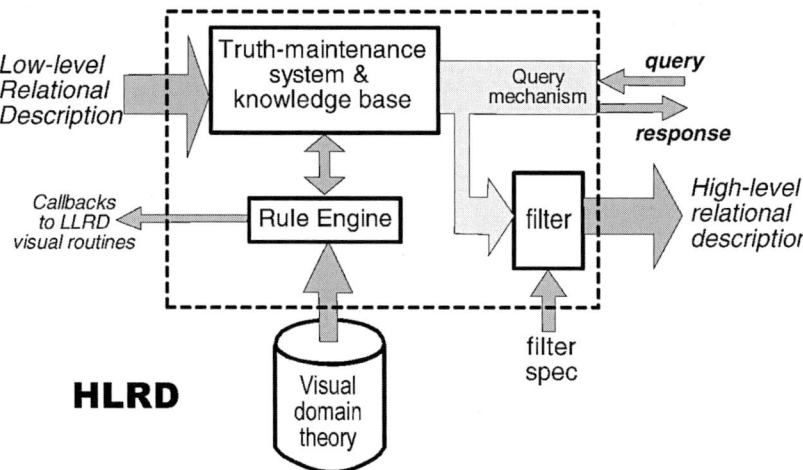

Fig. 3. The HLRD architecture, which reasons over the low-level description to produce a domain-specific interpretation

The second stage, the *High-Level Relational Describer* (HLRD; Figure 3), uses these low-level relations and a rule-based *visual domain theory* to produce a description of the diagram in question. The output of the HLRD is a description of the figure expressed in a domain-dependent high-level representation. For example, using domain-dependent rules, the HLRD produces the high-level representation of the logic circuit in Figure 4, describing the gates, the inputs and outputs, and the input and output labels. It has also been used in map-based military diagrams, called Course of Action (COA) diagrams (Figure 5, with a portion of the generated representation in Figure 6).

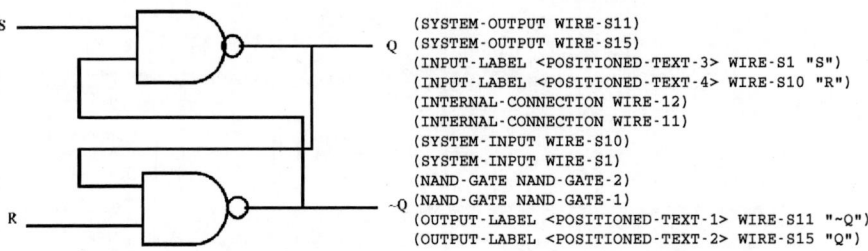

Fig. 4. SR-Latch logic diagram and the representation produced by GeoRep

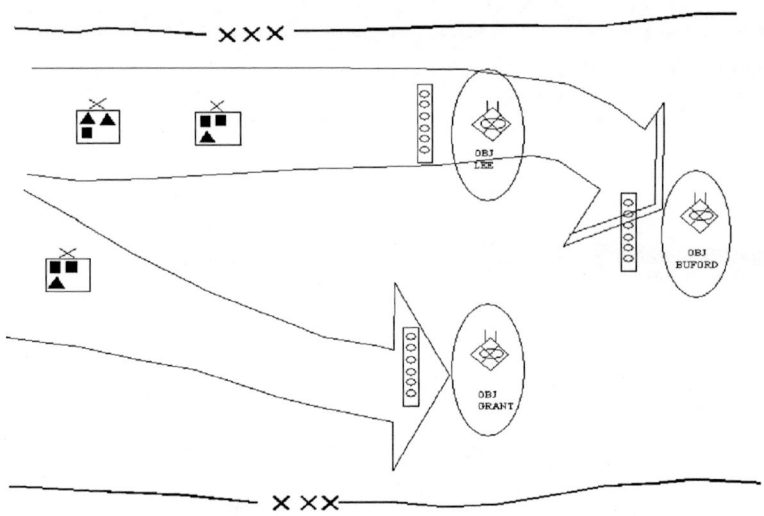

Fig. 5. A Course of Action diagram handled by GeoRep. This diagram depicts friendly units attacking three enemy positions

The HLRD rules utilize a pattern-directed inference system that is supported by a *logic-based truth maintenance system* (LTMS) [16]. HLRD rules can match against low-level spatial relations produced by the LLRD, as well as domain knowledge about the diagram (such as a small ontology of logic gate or military unit types). HLRD rules are specialized for spatial domains. They typically are constrained to run only on proximate visual elements, and can call visual operations within the LLRD.

Enemy unit inside area "Grant"	Subordinate units for the task force	Other
(armor-unit unit-6)	(subordinate-maneuver-unit unit-10 unit-17)	(completed-minefield minefield320)
(enemy-unit unit-6)	(mechanized-infantry-unit unit-10)	**Representation links**
(military-unit unit-6)	(battalion unit-10)	*Bottom arrow*
(motorized-rifle-unit unit-6)	(friendly-unit unit-10)	(represents (composite <polyline:4>)
(unit-at unit-6		(axis-of-advance-support attack4))
(objective-area area19 "Grant"))	(subordinate-maneuver-unit unit-13 unit-17)	*Echelon marker*
	(mechanized-infantry-unit unit-13)	(represents
Attack on area "Grant"	(battalion unit-13)	(x-mark <segment:71> <segment:72>)
(objective-area area19 "Grant")	(friendly-unit unit-13)	(marker marker317 brigade))
(area area19)		*Mech unit inside unit-17*
(attack-on attack4 area19)	(subordinate-maneuver-unit unit-15 unit-17)	(represents <polygon:10>
	(armor-unit unit-15)	(mechanized-infantry-unit unit-10))
Southernmost friendly task force	(battalion unit-15)	*Armor unit*
(composite-unit unit-17)	(friendly-unit unit-15)	(represents
(brigade unit-17)		(composite <polygon:6> <ellipse:23>)
(friendly-unit unit-17)		(armor-unit unit-6))

Fig. 6. Sample subset of representations generated for Figure 5

One limitation of GeoRep is its model of processing. GeoRep processes figures only in batch mode, and as a result, can run only on static diagrams. This is due to limitations of the LLRD rather than the HLRD. The HLRD is inherently incremental because it is based on an LTMS. The HLRD's relations can be assumed or retracted at any time, and the consequences of these relations will also be assumed or retracted accordingly. However, once the LLRD detects a visual relation between elements, it cannot retract it. While the use of visual tests is pipelined, the system does not store information about which visual elements are used in particular spatial relations.

Therefore, to make GeoRep incremental, it suffices to make the LLRD incremental. Our solution is to treat low-level elements and relations as assumptions that can be added or retracted as the diagram is modified by a user.

4   Making GeoRep Incremental

Creating an incremental LLRD requires a number of modifications to the existing system. First, a dependency network for spatial relations is needed. Along with handling low-level relations, the network must also handle nonmonotonic inferences created by particular composite objects, such as polygons. The second part of making the LLRD incremental is to ensure that incremental changes generated by the LLRD are properly handled by the HLRD. This in turn requires a new mechanism for handling default object interpretations, which cannot be handled with the HLRD's existing truth-maintenance system.

4.1   Creating a Dependency Network within the LLRD

The dependency network we developed for the incremental LLRD draws on previous research on the mathematical character of qualitative spatial vocabularies that are jointly exhaustive and pairwise disjoint (JEPD) [17]. By ensuring the JEPD character of each node's relation set, this network can take advantage of a number of such vocabularies shown to be JEPD, such as interval relations [14] and RCC [18, 19]. The logical properties of these JEPD sets are important because they allow the dependency network to isolate a set of spatial relations relative to other possible relations.

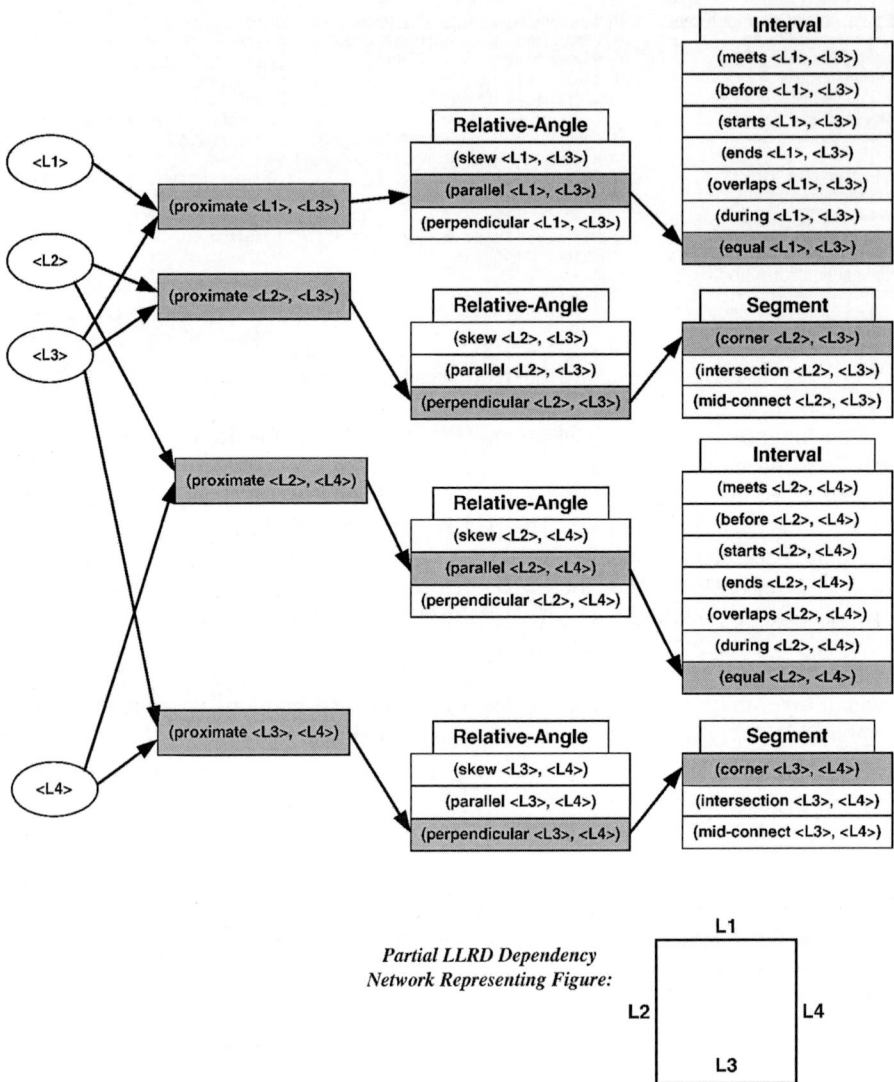

Fig. 7. Subsection of the dependency network. The selected internal relation for each node is shaded

To handle the incremental processing of input, we have re-designed the LLRD as a dependency network (Figure 7). This dependency network is responsible for several tasks. It tracks the low-level relations supported by each visual element, allows visual elements to be added or removed, and maintains the information needed to allow re-evaluation when elements are modified.

Each node in the network represents a set of alternative spatial relations – a single relation set that is JEPD. The node may be IN or OUT. If the node is IN, then one internal relation in the set is true. If the node is OUT, none of the internal relations are true. For example, each *relative-angle* node in Figure 7 must select one of three possible internal relations – *perpendicular, parallel,* or *skew* – to describe an angle relationship. Similarly, the *interval* node indicates that two line segments have one of seven interval relations [14].

The links between nodes allow antecedent relations to support consequent nodes. In this network, antecedent nodes represent how some spatial relations support the existence of one of a set of mutually exclusive alternatives.

Each internal relation in a node supports zero or more successor nodes. For example, a *parallel* internal relation supports the construction of an *interval* node. In other words, the internal relation combined with a visual test supports the whole truth value and labeling of the successor node.

Let M be the set of nodes in the LLRD dependency network. Each node $m \in M$ has a truth value (IN or OUT). If m is IN, it also has an internal relation, which is one of n possible relations in a JEPD set.

It is important to note that the truth value of the network is not a relationship between the truth values of the nodes, but between the nodes, their antecedents, and the visual tests that are performed for each node's set of internal relations. For example, for the *interval-equal* relation over segments L1 and L2, we can decompose the set of relations using the dependency network as follows:

interval- equal (L1,L2) $\equiv$ *parallel*(L1,L2) ^ *interval-test*(L1,L2,interval-equal).
parallel(L1,L2) $\equiv$ *proximate*(L1,L2) ^ *relative-angle-test*(L1,L2,parallel).
proximate(L1,L2) $\equiv$ L1 ^ L2 ^ *proximate-test*(L1,L2).

Therefore:
interval-equal(L1,L2) $\leftrightarrow$
    L1 ^ L2 ^ *proximate-test*(L1,L2) ^ *relative-angle-test*(L1,L2,parallel)
    ^ *interval-test*(L1,L2,interval-equal).

This is equivalent to the set of tests applied to segments in the original pipelined version of the LLRD architecture.

Other types of relations are handled somewhat differently. Boolean relations are handled as JEPD sets with one element. *Proximate* is one relation handled in this fashion.

4.2   Handling Composite Objects

Composite objects (objects composed of multiple visual elements, such as polygons used to depict regions) are a special problem in incremental spatial reasoning [25][2]. Composite objects are shapes, but can also be treated as relations, since composite objects are detected based on their constituent elements. As a result, they are the only visual objects that can be retracted due to the retraction of other elements. In addition,

---

[2] It is, of course, possible to take an opposite strategy, treating composite elements as primary and non-decomposable. This avoids the problem with incrementality described here. GeoRep has a *glyph* element that works in this fashion, and which has been used to reason over figures with complex but self-contained shape descriptions [8]. However, composability is a hard problem to avoid entirely, and so we are using this simpler form of composability to delineate the challenges of more complex composability types.

the retraction of a visual element can lead to the detection or re-assumption of other composite objects. For example, removing a single line segment from a square will lead to its interpretation as a polyline. The LLRD currently handles two forms of composite objects: polylines and polygons.

For this reason, the incremental addition or removal of visual elements introduces new situations where polyline and polygon detection must be reapplied to existing visual elements. For example, when line segments are removed, the polylines and polygons that contained them must be reconsidered. Removing a line segment from a polyline or polygon can result in several possible combinations of polylines.

The LLRD performs polyline and polygon detection by using a *vertex index table*. As new line segments are added, the LLRD maintains the table of added segments indexed by endpoint. This table is then used to determine which line segments share endpoints with others. Groups of line segments that are not closed form polylines. Once a polyline is detected, it is added to the dependency network, and its vertices are removed from the table. Polygon detection uses the remaining entries in the table. If a set of vertices is closed, then a polygon is added to the dependency network and its vertices are removed from the table.

A node in the network for a composite object is not a relation node in our implementation. Instead, it represents the object by storing geometric information about the shape as well as linking the node to its constituent (subsumed) elements.

The LLRD uses different methods to determine which elements to reconsider for composite objects depending on whether an element is added, removed, or restored. In the first case, when a line segment is created, a new node is added and set to IN. Existing elements that are proximate to the added segment are added to the vertex index table in order to discover new polylines or polygons or changes to existing polylines. In contrast, when a line segment is retracted, the dependency network is used to retract any affected polylines and polygons. Lines that are part of the affected polyline or polygon, yet remain IN (i.e., have not been removed) are re-analyzed, and new polylines may be assumed. Finally, when an element is restored, the dependency network reactivates composite objects containing the restored element if all their subsumed elements are IN.

An example of how the LLRD handles composite objects is given in Figure 8. In step 1, line segments are added to form a polyline. In step 2, another segment is added, closing this polyline to form a new polygon. The polyline composite element is now OUT, and the new polygon is IN. In step 3, L4 is removed (OUT), and the polygon becomes OUT. The polyline formed by {L1 L2 L3} is IN. If L1, L2 or L3 became OUT (instead of L4), the polygon would still become OUT and a polyline formed by the remaining segments would become IN. In step 4, L4 is restored, and the polygon that contained L4 becomes IN again. The previous polyline containing L1, L2 and L3 becomes OUT because it is a subset of the polygon.

4.3   Supporting High-Level Inferences

Along with maintaining the set of spatial relations, the LLRD also supports the HLRD's high-level reasoning. The LLRD continually provides a correct set of spatial relations for the HLRD. In addition, when relations change in the LLRD's dependency network, these changes are propagated directly to the HLRD. The HLRD then changes its diagram interpretation accordingly.

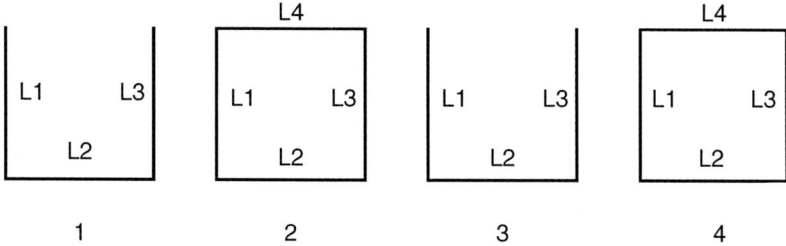

Polyline/Polygon Detection Example:
Step 1 A polyline is detected (polyline <L1> <L2> <L3>).
Step 2 A new line <L4> is added, and it as well as its proximate neighbors <L1>, <L2>, <L3> are added to the list of elements to process. The polyline element is marked OUT and a new polygon element (polygon <L1> <L2> <L3> <L4>) is added.
Step 3 The most recently added line <L4> is removed, the polygon is marked OUT and the remaining lines are processed. The previous polyline is detected and marked IN.
Step 4 The removed line <L4> is restored and marked IN, and its component elements are processed, marking the polyline OUT and the polygon IN.

Fig. 8. Handling polylines and polygons in the incremental LLRD

Figure 9 shows an example of how small visual changes can dynamically change the diagram interpretation. Here, gradual additions and deletions to a diagram change the diagram from an uninterpreted figure (A), to a NAND gate (B), then to an AND gate (C), back to a NAND gate (D), and then to a NOT gate (F). At each point, the LLRD's dependency network manages the set of spatial relations that are available to the HLRD. The HLRD, in turn, modifies its representation automatically as these changes are made to the figure.

To make this work, the internal relations of nodes in the LLRD's dependency network are linked with logic nodes (each representing a specific visual relation) in the HLRD's LTMS. If a node in the LLRD is retracted or if the selected internal relation is changed, these changes are propagated to the LTMS. Changes to an LTMS node's truth value automatically triggers the Boolean Constraint Propagation algorithm [20] to update the LTMS's belief state.

The LTMS is a more powerful reasoner than the LLRD's dependency network, but the network is still an adequate foundation for the LTMS given the constraints of spatial domains. An LTMS make inferences based on both true and false nodes, while the LLRD's dependency network is roughly equivalent to the nodes in a Justification-based TMS (JTMS). Such nodes represent only Horn clauses, and do not distinguish between facts that are false and those that are unknown. However, for any JEPD set represented by a LLRD node, we can make a closed-world assumption over the set of internal relations that allows us to treat the selected internal relation as true, and the rest as false. In addition, due to the nature of visual relation detection, when an LLRD node is OUT, all of its internal relations can be treated as false, and not simply unknown. This is because an LLRD node becomes OUT only when a necessary visual precondition becomes invalid.

146   Ronald W. Ferguson et al.

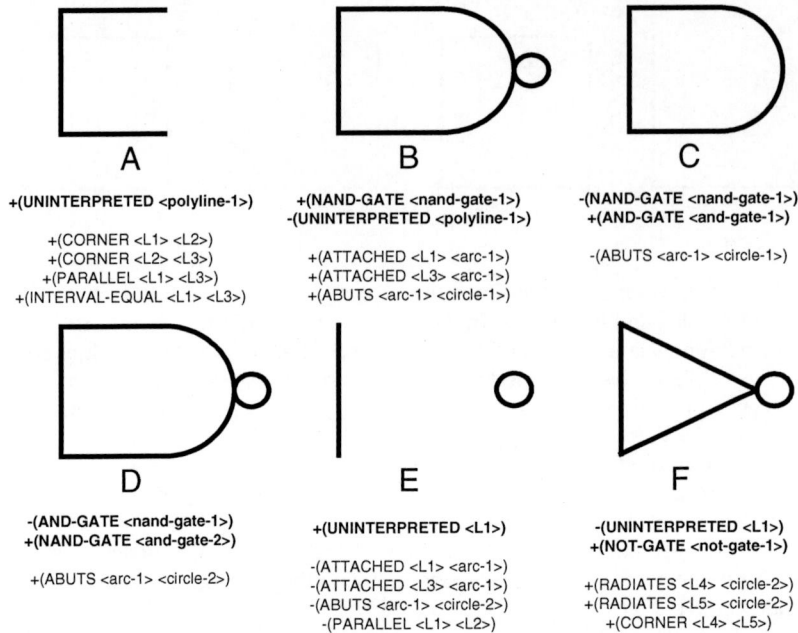

+(UNINTERPRETED <polyline-1>)

+(CORNER <L1> <L2>)
+(CORNER <L2> <L3>)
+(PARALLEL <L1> <L3>)
+(INTERVAL-EQUAL <L1> <L3>)

+(NAND-GATE <nand-gate-1>)
-(UNINTERPRETED <polyline-1>)

+(ATTACHED <L1> <arc-1>)
+(ATTACHED <L3> <arc-1>)
+(ABUTS <arc-1> <circle-1>)

-(NAND-GATE <nand-gate-1>)
+(AND-GATE <and-gate-1>)

-(ABUTS <arc-1> <circle-1>)

-(AND-GATE <nand-gate-1>)
+(NAND-GATE <and-gate-2>)

+(ABUTS <arc-1> <circle-2>)

+(UNINTERPRETED <L1>)

-(ATTACHED <L1> <arc-1>)
-(ATTACHED <L3> <arc-1>)
-(ABUTS <arc-1> <circle-2>)
-(PARALLEL <L1> <L2>)

-(UNINTERPRETED <L1>)
+(NOT-GATE <not-gate-1>)

+(RADIATES <L4> <circle-2>)
+(RADIATES <L5> <circle-2>)
+(CORNER <L4> <L5>)

Fig. 9. Incremental changes to this figure (consisting of visual element additions (+) and deletions (-)) are dynamically reflected in the high-level interpretation of the figure (in bold)

4.4   Handling Default Assumptions

Finally, to allow the HLRD to properly handle incremental LLRD information, we extended the default assumption mechanism in the HLRD's LTMS (Figure 10). While the LTMS already supports simple incremental reasoning, it does not support dynamically changing the high-level interpretation when it depends on default reasoning.

For example, in Figure 9(B), the current visual domain theory supports an interpretation of the figure as both a NAND gate as well as an AND gate. Both interpretations are assumed, and when they are found to be in conflict, the AND gate interpretation is retracted.

This system works well for a diagrammatic reasoner that functions in batch mode, where retracted assumptions do not need to be re-examined. In an incremental reasoner, however, the visual elements and relations that lead to an over-ruled default interpretation may themselves change. When the circle is removed as in Figure 9(C), the standard LTMS assumption mechanism cannot re-assume the AND interpretation because it has already been explicitly retracted.

To handle this problem, the new default assumption mechanism uses two additional node types: a *default assumption node* and an *interpretation-hypothesis node*. The default assumption node is an extension to the existing LTMS node structure, with slots added to store alternative interpretations and to link to an interpretation-hypothesis node. The interpretation-hypothesis node is a standard LTMS node, created as an assumption and justified by the appropriate implicational structure.

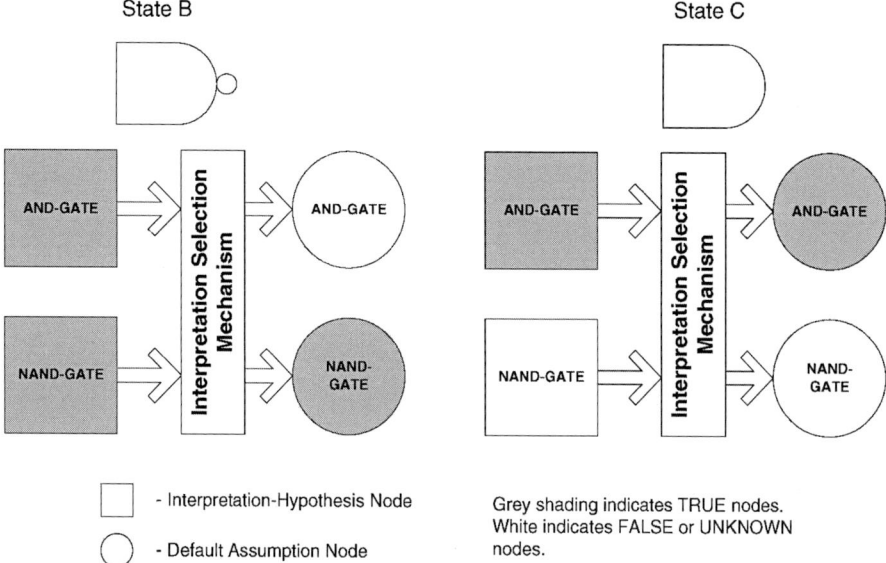

Fig. 10. A default assumption mechanism was added to the truth-maintenance system in the HLRD to allow for clean retraction and re-assumption of default interpretations

These nodes allow all potential interpretations to be available even if some have been previously rejected. When a new composite object is detected, it is *default-assumed*, rather than assumed. This creates both a default assumption node for the new object as well as an interpretation-hypothesis node. Thus, for each potential interpretation, there will be a valid interpretation-hypothesis node. However, at any instance, there only one valid interpretation exists for each set of interpretation hypotheses. The existence of multiple interpretation-hypothesis nodes causes a contradiction within the LTMS, triggering the interpretation selector.

Selection between potential interpretations is handled in a domain-dependent fashion. For example, in the logic circuit domain, the maximally-preferred alternative is the interpretation with the most elements [21]. In the example, the NAND gate would be selected because the AND gate is a subset.

This handles the case where one composite object has two possible interpretations. However, to support dynamically changing interpretations, we must also consider what happens when removing part of an object requires revising our interpretation again. In this case, we may want to revert to a previous interpretation that was discarded because it was not the maximally preferred.

This is handled by activating the selector mechanism when an interpretation hypothesis node is retracted. In the example, this corresponds to removing the circle from the NAND gate, which causes a retraction of the NAND gate interpretation, and the reactivation of the AND gate interpretation. The interpretation selector returns to the next-best alternative interpretation, allowing the AND gate interpretation (and all its high-level consequences) to be reactivated.

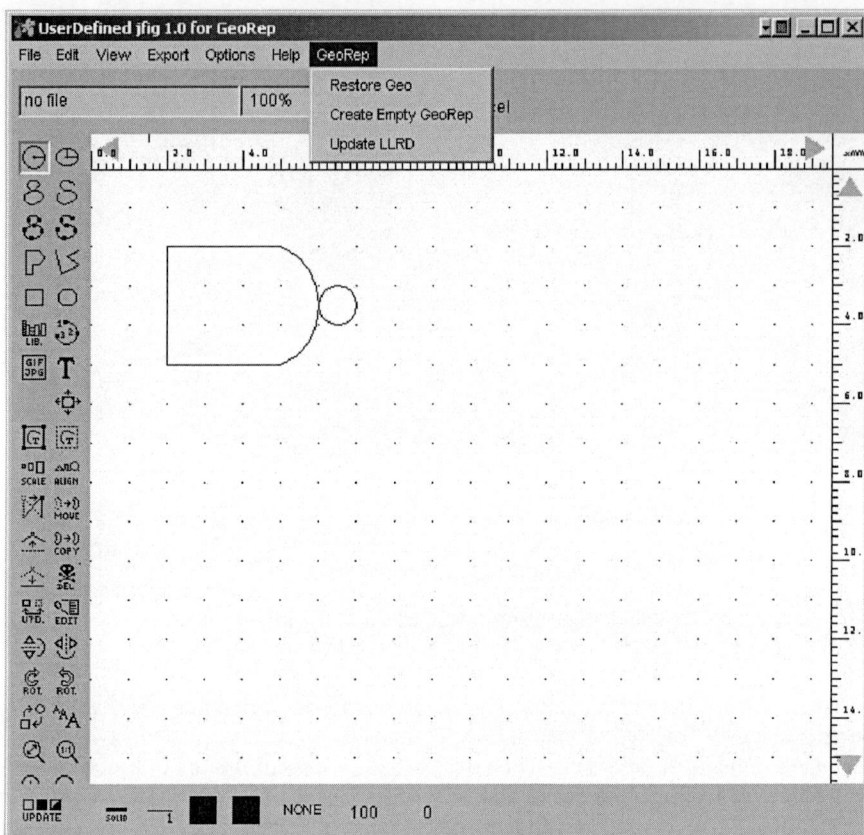

Fig. 11. Drawing a figure in JFIG. An additional menu allows the interface to control the link to the incremental version of GeoRep

5   Integrating Incremental GeoRep into a Drawing Program

We use an existing drawing system, JFig [22], as an interface to GeoRep. JFig is a Java implementation of the well-known XFig program [23], and is freely available on the web. We customized the JFig interface (Figure 11) to notify GeoRep of visual element adds, deletes, and modifies.

Whenever a new element is added or deleted from the diagram in JFig, the corresponding element in GeoRep is added or removed, and the dependency network is updated. The restore command works on the most recently removed object, and makes the corresponding element valid again in GeoRep.

6   Discussion

We have presented an extensible framework for incremental reasoning over a variety of spatial relations. This framework employs jointly exhaustive and pairwise disjoint relations to encapsulate visual reasoning subtasks.

Although the dependency network handles the addition, retraction, and restoration of visual elements, there are many ways in which the dependency network could be used to make more powerful inferences. For example, it could aid in the re-evaluation of modified visual elements. Because each relation node depends on a procedural visual test to choose between its alternative internal relations, once a visual element is modified, the tests must be rerun to determine if the previously chosen internal relation remains valid. If not, a chain of visual tests must be applied to the modified visual element and all proximate elements.

Another potential use of this network is to let the HLRD influence the LLRD by, for example, setting test tolerance values (e.g., the margin for a perpendicular line test). For example, it should be possible to have the HLRD detect quadrilaterals that are almost square, and then modify the *relative-angle* test so that the LLRD recognizes a necessary corner as perpendicular.

Finally, there is the problem of tradeoffs between spatial reasoning and visual processing. Assuming that visual processing is extremely cheap, which spatial relations are really worth caching in this kind of mechanism? In this architecture, we have assumed that even very low-level spatial relations are worth caching in order to test the implications of the architecture. Clearly, however, the utility of caching these relations depends critically on the task and the power of the visual processing system (i.e., the visual operators available to the LLRD).

One other spatial reasoning task that is especially important in geographic reasoning is proper treatment of scaling. Geographic elements or features that are critical at one scale (e.g., sidewalks) may be of little or no interest at larger scales.

Although it would be interesting to discuss how GeoRep might handle scale issues (for example, sets of visual elements could be turned into composite glyphs that would limit processing on its internal elements), GeoRep currently deals with scaling issues through the HLRD's visual domain theory, which means, inevitably, that scale is handled in a domain-dependent fashion.

This provides GeoRep with a lot of flexibility in handling scaling issues. For example, in the COA domain (Figure 5), military units (which are rectangular) have visual extent, but are treated as point objects by the visual domain theory. Other geographic elements, such as "Objective Buford," scale with the picture, since the extent is meaningful geographically.

Finally, we note that this work is only one part of a larger effort to create a next-generation diagrammatic reasoner. This reasoner will combine incremental spatial reasoning with other abilities, such as dynamic reinterpretation of diagrams.

Acknowledgements

We would like to thank Norman Hendrich, the creator of JFig, for very useful advice and suggestions on interfacing to that program. We would also like to thank Ken Forbus and several anonymous reviewers for helpful suggestions.

References

1. Ferguson, R.W., and K.D. Forbus. 2000. GeoRep: A flexible tool for spatial representation of line drawings. *In* Proceedings of the 18th National Conference on Artificial Intelligence. AAAI Press, Austin, Texas. 510-516.

2. Hernández, D. 1993. Maintaining qualitative spatial knowledge. *In* Proceedings of the European Conference on Spatial Information Theory (COSIT), Elba, Italy. 19-22.
3. Freksa, C. 1992. Temporal reasoning based on semi-intervals. *Artificial Intelligence* 54:199-227.
4. Egenhofer, M., and K. Al-Taha. 1992. Reasoning about gradual changes of topological relationships. *In* Theory and Methods of Spatio-Temporal Reasoning in Geographic Space. A. Frank, I. Campari, and U. Formentini, editors. Springer-Verlag, Pisa, Italy. 196-219.
5. Davis, R. 2002. Position statement and overview: Sketch recognition at MIT. *In* AAAI Spring Symposium on Sketch Understanding, Palo Alto, CA.
6. Scher, D. 2000. Lifting the curtain: The evolution of the Geometer's Sketchpad. *The Mathematics Educator* 10:42-48.
7. Ferguson, R.W., R.A.J. Rasch, W. Turmel, and K.D. Forbus. 2000. Qualitative spatial interpretation of Course-of-Action diagrams. *In* Intelligent Systems Demonstrations, 18th National Conference on Artificial Intelligence, Austin, Texas.
8. Ferguson, R.W., and K.D. Forbus. 1998. Telling juxtapositions: Using repetition and alignable difference in diagram understanding. *In* Advances in Analogy Research. K. Holyoak, D. Gentner, and B. Kokinov, editors. New Bulgarian University, Sofia, Bulgaria. 109-117.
9. Ferguson, R.W. 2001. Symmetry: An Analysis of Cognitive and Diagrammatic Characteristics. Ph.D. Dissertation, Department of Computer Science, Northwestern University. Evanston, Illinois.
10. Ferguson, R.W. 2000. Modeling orientation effects in symmetry detection: The role of visual structure. *In* Proceedings of the 22nd Conference of the Cognitive Science Society. Erlbaum, Hillsdale, New Jersey. 143.
11. Ferguson, R.W., A. Aminoff, and D. Gentner. 1996. Modeling qualitative differences in symmetry judgments. *In* Proceedings of the Eighteenth Annual Conference of the Cognitive Science Society. Lawrence Erlbaum Associates, Hillsdale, NJ. 12.
12. Hoffman, D.D., and W.A. Richards. 1984. Parts of recognition. *Cognition* 18:65-96.
13. Rock, I. 1973. Orientation and Form. Academic Press, New York, NY.
14. Allen, J.F. 1983. Maintaining knowledge about temporal intervals. *Communications of the ACM* 26:832-843.
15. Ferguson, R.W. 1994. MAGI: Analogy-based encoding using symmetry and regularity. *In* Proceedings of the Sixteenth Annual Conference of the Cognitive Science Society. A. Ram and K. Eiselt, editors. Lawrence Erlbaum Associates, Atlanta, GA. 283-288.
16. Forbus, K.D., and J. de Kleer. 1993. Building Problem Solvers. The MIT Press, Cambridge, MA.
17. Cohn, A.G. 1997. Qualitative spatial representation and reasoning techniques. *In* Proceedings of KI-97. G. Brewka, C. Habel, and B. Nebel, editors. Springer-Verlag, Freiburg, Germany. 1-30.
18. Cohn, A.G., D.A. Randell, Z. Cui, and B. Benett. 1993. Qualitative spatial reasoning and representation. *In* Proc of the IMACS Workshop on Qualitative Reasoning and Decision Technologies, QUARDET '93. N.P. Carrete and M. Singh., editors. CIMNE, Barcelona. 513-522.
19. Cohn, A.G. 1995. A hierarchical representation of qualitative shape based on connection and convexity. *In* International Conference on Spatial Information Theory COSIT-95, Semmering, Austria.
20. McAllester, D. 1990. Truth maintenance. *In* Proceedings of the Eighth National Conference on Artificial Intelligence. AAAI Press, Boston, MA. 1109-1116.
21. Doyle, J. 1992. Rationality and its roles in reasoning. *Computational Intelligence* 8:376-409.
22. Hendrich, N. 1999. JavaFIG: The Java diagram editor. Computer Science Department, University of Hamburg, Germany.
23. Smith, B.V. 1999. XFig. http://www.xfig.org

MAGS Project:
Multi-agent GeoSimulation and Crowd Simulation[*]

Bernard Moulin, Walid Chaker, Jimmy Perron, Patrick Pelletier,
Jimmy Hogan, and Edouard Gbei

Computer Science Department and Center for Research in Geomatics,
Laval University, Ste Foy, QC G1K 7P4, Canada
{bernard.moulin,walid.chaker,jimmy.perron,
patrick.pelletier,edouard.gbei}@ift.ulaval.ca

Abstract. Geosimulation aims at modeling systems at the scale of individuals and entity-level units of the built environment and provides a new way to simulate how geographic spaces can be used by their future users, particularly in urban environments. In the MAGS Project we are developing a generic software platform for the creation of Multi-Agent Geo-Simulations involving several thousand agents interacting in virtual geographic environments (in 2D and 3D) and endowed with spatial cognitive capabilities (perception, navigation, reasoning). Our approach is currently applied to the simulation of crowd behaviors in urban environments.

1 Introduction

Constrained by the structure of space and communication networks, urban regions form a complex system of interactions which evolves over time. Actors in the system strategically adapt their behaviors according to their own priorities among perceived opportunities. Government agencies need tools that take into account these complex interactions in order to compare planning scenarios on the basis of their global and long-term effects. However, statistical modeling remains unable to simulate these urban processes appropriately. The 'traditional' urban simulation models have been criticized [20] because of their centralized approach, a poor treatment of dynamics, a reduced flexibility and a lack of realism [31]. In addition, an increasing number of researchers think that an adequate forecast of transportation demand should be based on the study of individual mobility behaviors [6]. In such an approach, individual behaviors should be modeled at a spatio-temporal scale which is appropriate for characterizing the nature and importance of the decision processes influencing the transformation of the urban environment. However, applying such an approach involves the development of simulation systems that are able to deal with the simultaneous actions of thousands of actors, integrating their interactions [7], and taking into account the structural constraints of the urban environment (transportation network, localization of infrastructures, distribution of activities and services, etc).

---

[*] The MAGS project is supported by GEOIDE, the Network of Centers of Excellence in Geomatics, Defense Research and Development-Valcartier, the Natural Science and Engineering Council of Canada and le Fonds Québécois de Recherche sur la nature et les technologies

W. Kuhn, M.F. Worboys, and S. Timpf (Eds.): COSIT 2003, LNCS 2825, pp. 151–168, 2003.

Geo-simulation is an approach which aims at modeling systems at the scale of individuals and entity-level units of the built environment. It provides a new way to simulate how geographic spaces can be used by their future users, particularly in urban environments. Torrens [31] indicates that geo-simulation models are in their relative infancy as applied to urban simulation and constitute a new class of simulation models which borrow heavily from developments in geographic information science, artificial intelligence, artificial life, complexity studies and simulation in natural sciences and social sciences outside geography. Applied to urban design, geo-simulation provides the means to study the characteristics of the urban environment by analyzing the interactions of moving agents simulating the behavior of various actors (such as pedestrians and automobiles) in a urban landscape. In distributed artificial intelligence, researchers developed techniques that are used to create multi-agent systems (MASs) composed of agents which are autonomous programs collaborating together to solve problems [18] [32]. MASs are particularly adapted to the simulation of population dynamics in large-scale environments [22] [8] [11]. They are well suited to the exploration of dynamic phenomena in which the interactions of individual entities can be studied at a micro-level and the emergence of behavioral patterns can be observed at a macro-level [29] [27].

Microsimulation has been frequently used in the field of transportation systems analysis during the past decade, especially to model urban travel behaviors in order to predict the spatial and temporal distributions of trips in urban areas. For example, the TRANSIMS system [1], a generic platform for modeling and simulating complex behaviors using actors, provides a series of integrated transportation and air quality analysis models and attempts to simulate the aspects of human behavior that are relevant to transportation planning. Although traffic models often use a multi-actor approach, they typically do not contain models of cognitive aspects of human spatial behavior[1] [8] [31]. Indeed, most traffic models simulate urban phenomena [24] using a cellular automata approach [33] in which space is represented as a uniform lattice of cells. Each cell may be in a finite number of discrete states and can change its state at discrete time steps during the simulation. The cells are subject to a uniform set of rules which drive the overall behavior of the system. Such an approach does not enable actors to move in space autonomously and does not provide mechanisms to simulate basic spatial cognitive capabilities such as perception and memory. Introducing agent's autonomy and cognitive spatial capabilities in geo-simulation models would offer new possibilities for the analysis of phenomena resulting from the decisions and actions of a large number of actors resulting from their interactions with their spatial environment and other actors.

In the MAGS Project we aim at developing a generic software platform for the creation of Multi-Agent Geo-Simulations (MAGS) involving several thousands of agents interacting in virtual geographic environments and endowed with spatial cognitive capabilities. The application domain in which we are currently applying our geo-simulation approach is the simulation of crowd behaviors in urban environments.

Section 2 presents the requirements that we selected for the development of the MAGS platform in the context of recent work on the simulation of crowd behavior and pedestrian flows. As an illustration of the agent's cognitive spatial capabilities,

---

[1] In this paper we cannot review the large body of literature dealing with the cognitive aspects of human spatial behavior. The interested reader can consult an in-depth review on the subject in [21] as well as several papers in the proceedings of the COSIT Conference.

Sections 3 and 4 present the main characteristics of the perception and navigation mechanisms implemented in the MAGS platform. Section 5 briefly outlines the other main agents' characteristics (needs, objectives, etc.) and presents an example of a simulation involving several hundreds of agents moving in a portion of Quebec city. Section 6 presents some performance results and concludes the paper.

2   Requirements for the MAGS Project

Several studies have been carried out on crowd movements in a portion of a urban environment represented on a 2D map and successful simulations have been created on the basis of mathematical models used in physics to simulate flows of particles in constrained environments. At medium and high pedestrian densities, the motion of pedestrian crowds shows similarities with the motion of fluid particles, giving rise to self-organization phenomena [15]. Such approaches have been used to simulate the formation of lanes of pedestrians on busy pavements [14] and to study evacuation strategies [13]. These approaches, which model the interactions between individuals in a quite simplified way (in terms of physical interactions of particles) are successful when simulating the flow of dense crowds in various situations. However, they cannot differentiate between different types of individuals with different goals and behaviors. Jager and his colleagues [17] developed a multi-agent system to simulate clustering and fighting behaviors of two-party crowds. They provided agents with simple rules based on the recognition of own-party agents and other-party agents (as a result of scanning an area 40 by 40 cells around the agent) and the agent's level of aggression motivation. Some simple clustering behaviors emerged from the simulations, which resembled real-world crowd phenomena.

Several systems have been designed to study pedestrian flows and movement at a strategic level. The STREETS System [28] applies an approach similar to the TRANSIMS model in which the simulation of the activities of pedestrians in urban districts follows a two-stage approach. In the first stage, the system exploits socio-economic data sets to predetermine the pedestrians' intended activity schedules which are used to "load" the agents into the simulation component. In the second stage, a simulation generates the movements of a population of agents representing pedestrians [12]. The simulation is influenced by the urban district's spatial configuration, pre-determined activity schedules and the distribution of land-uses. The PEDFLOW System [19] is another multiagent microsimulation system which is used to study conflicting pedestrian flows on a section of sidewalk or in an open or enclosed space with obstacles. Each pedestrian is represented as a single process which makes decisions about the pedestrian's movements. The simulated space is mapped onto a grid. A pedestrian agent occupies a grid element for a length of time required by its walking speed. A shared data structure is used to record the current position of every pedestrian as well as location information about obstacles and the pavement. Displacement behavior is specified in the form of rules which take into account a number of parameters such as preferred gap size, desired walking speed and personal space measure.

The currently available systems for crowd and pedestrian flow simulation rely on very simple agent movements on grids (possibly directed by cellular automata) which do not take into account terrain characteristics and environmental factors that may influence the agents' perception and navigation. In addition, their decision making capabilities are quite basic (simple displacement rules) and group behaviors are non existent or at best very simple. In order to allow agents to simulate cognitive spatial activities in geosimulation systems we agreed upon basic principles: 1) an agent should be able to perceive the spatial environment as well as the objects and other agents surrounding it; 2) the spatial environment and the static objects it contains should be generated from data contained in geographic information systems (GIS) and related databases; 3) agents should be able to efficiently plan their activities based on their internal states and goals as well as the information they perceive in the virtual space.

Taking advantage of our previous experience with the development of *PADI-Simul*, a multi-agent system simulating basic navigation behaviors of hundreds of agents in a 2D sketch of a natural park [5] [23], we selected a set of requirements which involve integrating several technologies into the MAGS platform: GIS, multi-agent systems, 3D real-time animation engine and parallel processing. The set of requirements that we selected is as follows:

We need to create a virtual geographic environment (VGE) in 2D (and possibly 3D) from reliable GIS sources.

We need to create agents of various types, each agent being individualized.

An agent must be able to perceive its environment, to navigate autonomously and to react to changes occurring in the VGE.

The agents' characteristics must reflect various possible states (static, dynamic, possession, etc.) and the agents' behaviors must offer efficient planning capabilities (reactive planning based on objectives).

Agents must be able to display group behaviors and to communicate with other agents.

The system must be optimized and allow simulations involving several thousand agents in relatively large spaces (a portion of a city for example).

An agent needs a memory capability in order to organize the knowledge about the VGE that it obtained from past experience.

Simulation scenarios must be specified easily, including the initialization of agents and the VGE and the introduction of specific events influencing the simulation.

Several teams work on the development of digital representations of urban spaces (see for example [4]) in order to create "virtual cities" that can be explored by designers, urban planners and citizens in order to assess various characteristics of urban projects. People explore these virtual cities thanks to a 3D visualization engine that enables them to control a point that moves in the virtual space as well as virtual cameras that are used to observe the landscape. However, we do not know any system that enables a large number of virtual agents to move in a virtual city, perceiving the landscape around them and acting according to their perceptions. So, we first worked on an approach to generate the VGE as a 3D model in which agents can navigate. To this end, we elaborated a suite of transformation processes based on Geomedia [16] and 3ds Max software [3]. The VGE used in our current simulation is created from topographic data of a portion of Quebec city (scale: 1:20000), a digital elevation model and a data base giving the characteristics of the main buildings. In addition, a module of the MAGS System generates a collection of bitmaps which are used by the agents to obtain knowledge about the space surrounding them (Section 3). At the heart of the MAGS simulation engine there are 5 main modules: 1) a thread manager which coordinates the activities of the other modules; 2) a 3D engine which manages the display of the 2D or 3D VGE and cameras; 3) the agent manager and related modules; 4) the VGE manager and 5) the user interface manager.

As an illustration of the agent's cognitive spatial capabilities, the two following sections present the main characteristics of the perception and navigation mechanisms implemented in the MAGS platform.

3   Agent Perception

Perception is an important agent ability which must be carefully simulated in a 3D VGE if we want that agents exhibit plausible cognitive spatial behaviors [8]. We must bear in mind that simulating visual perception is a resource intensive process which must be optimized if we want to enable thousands of agents to perceive the VGE in real-time. By analogy to human spatial perception, we identified several perception modes for MAGS agents: 1) perception of terrain characteristics (elevation and slopes) in the area surrounding the agent; 2) perception of the landscape surrounding the agent (including buildings and static objects); 3) perception of other mobile agents navigating in the agent's range of perception; 4) perception of dynamic areas with specific properties such as a smoky area or zones having pleasant odors; 5) perception of special events (detonation, explosion, etc.) occurring in the agent's vicinity; 6) perception of messages communicated by other agents. We developed several mechanisms that enable MAGS agents to take advantage of these perception modes.

Encoding Spatial Data. Spatial information is recorded in a raster mode which enables agents to access the information contained in various bitmaps that encode different kinds of information about the terrain characteristics and the objects contained in the VGE. The *HeightMap* is a 2D matrix (or grid) which represents the space of the VGE. It is generated from data contained in a digital elevation model and different layers of the GIS data base defining buildings and all the information that may influence the agents' perception and navigation. Every cell contains a single value indicating the height of the corresponding point relative to the point of lowest elevation in the VGE. Figure 2 presents the plan (x, z) of section A shown in the Height Map of Figure 1. We can observe that building2's height is higher than building1's height. But, the Height Map encodes that the top of building1 is higher than the top of building2 relative to sea level. The agent perception module uses this simple structure to determine the visibility of the matrix cells (using the elevation information). Hence, an agent can perceive the terrain characteristics (slope, elevation) as well as objects that are in its range of perception. The agents' positions are recorded and updated in another data structure called *LocationMap*, a 2D matrix of cells in which every cell contains a pointer to the agent occupying the corresponding position. If an agent has a dimension allowing it to occupy several cells, several pointers refer to the same agent. Hence, it is possible to analyze the visibility of a cell using the *HeightMap* and to determine, using the *LocationMap*, if an agent located at the corresponding position is visible.

Visual Perception. Several researchers have already studied the problem of simulating perception in an environment represented by a height map [9] [2]. The goal of these techniques was to determine the visibility of all the cells of the height map which are in an observer's field of vision. They use lines of sight in order to test the cells' visiblity (labelled as visible or not) from the observer's location. If the observer moves, it is necessary to compute the visibility map corresponding to the new position. This technique is by far too inefficient to simulate the perception of thousands of agents moving in real time. We propose a solution extending Franklin's algorithm in a way that enables agents to perceive the environment as well as other agents in real time. The agent's perception field is represented by an isosceles triangle, the main vertex being at the agent's location, the congruent sides of the triangle limiting the perception field and the bisector of the main angle corresponding to the agent's direction of movement. The length of the bisector corresponds to what we call the perception radius. The angle of perception (Figure 3B) is a parameter that can be adjusted (currently set at 90 degrees).

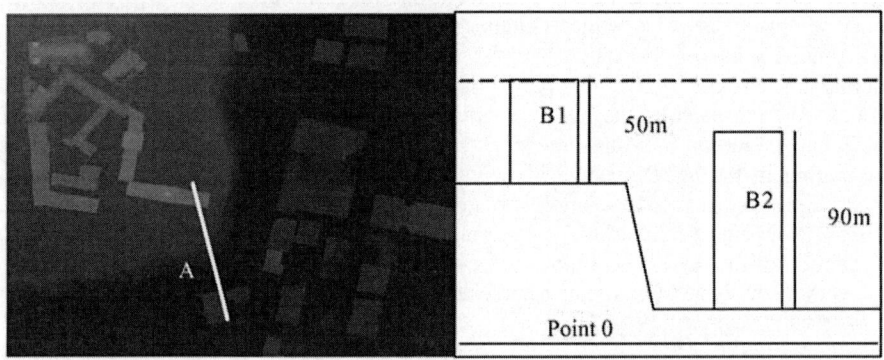

Fig.1. Height Map of a part of Quebec city   **Fig.2.** Plan (x,z) of section A in Fig. 1

We developed an algorithm which computes, for each cell in the agent's field of perception, the line of sight linking the agent's position to the cell in order to determine if the cell is visible or not. This computation takes into account the cell's height and the visibility of the other cells that may block the line of sight. As an illustration, our algorithm browses an environment of 300x300 cells and computes a visibility map in 27 ms on a Pentium 1000Mhz. This algorithm has a complexity $O(n^2)$ in which n represents the perception radius. During a simulation, each agent must compute its own visibility map at every iteration of the simulation engine. It is obviously impossible to allocate 27 ms to each agent in order to browse a height map of 300x300, using a perception radius of 300 cells. The solution is to reduce the length of the perception radius in order to decrease the computing time. Figure 3A shows how the perception computing time increases when we increase the perception radius. In the context of our simulator in which one pixel represents one meter approximately, we can think about limiting the perception of an agent's immediate neighborhood to 20 meters. In these conditions, the necessary time to process an agent's perception is 0,12 ms with a 90 degree angle of vision and a perception radius of 20 cells. If 1000 agents are moving at the same time, the global perception time would be 120 ms. Considering that a cycle in the simulator takes 33ms (to emulate real time), we would allow approximately 250 agents to perceive at every cycle, knowing that in a real-time 3D engine we use 30 cycles per second (hence, 33 ms / cycle).

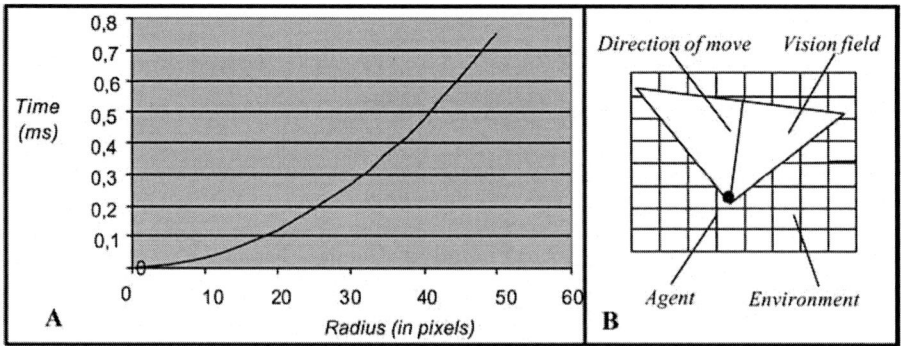

Fig.3A. Calculation time for the vision algorithm   **Fig.3B.** Agent's field of perception

However, limiting the perception to 20 meters does not correspond to people's experience in the real world. Buildings distant from several kilometers can been seen in good visibility conditions. However, it is rather rare that we can distinguish smaller objects moving several kilometers away. Thus, we had the idea to compute a *static visibility map* in a pre-processing phase taking place before the simulations. Each cell of the static visibility map has a list of pointers to the static objects that are visible from this cell. Hence, an agent can directly access the static visibility map in order to determine which static objects (buildings, etc.) can be perceived from its position. This perception mode which is called *static perception* complements the *dynamic perception* mode that allows an agent to perceive in real time all the objects located within its perception radius. It is clear that this perception radius can be increased depending on the performance of the computer and the number of agents that need to perceive during each cycle of the simulation engine. Our approach allows an agent to have a global vision of the VGE (static perception) while keeping a focus on what is happening in front of it (dynamic perception).

Fig.4A. Ariadne Map of Quebec city   **Fig.4B.** Portion of a Zone map

Perception of Roads, Paths and Regions. In order to optimize the agent navigation function, we generate a bitmap called an *Ariadne Map* from the GIS data. In this bitmap streets and paths are colored in red (Figure 4A), and an agent can directly access it to determine which cells around it correspond to a street or path. The Ariadne Map can be generated in several colors in order to differentiate particular categories of ways such as sidewalks, roads, paths, bike paths, etc. Specific areas such as public squares can be perceived by the agents in the VGE. These areas are identified in the GIS and a bitmap called the *Zone Map* is automatically generated (Figure 4B). Each area is associated with a list of properties recorded in an auxiliary file. When an agent enters such a zone, it can access the corresponding information and act accordingly.

Perception of Dynamic Areas (or Volumes). Certain gaseous phenomena such as smoke, dense gases and odors are related to the VGE atmosphere and cannot be modeled using static objects or moving agents. They are associated with areas or volumes whose properties (boundaries, local density, etc.) change dynamically under the influence of external forces like the wind. A good way to simulate such phenomena is to use particle systems [30]. We developed such a module, but we will not go into detail about it in this paper. What is of interest here is the way that agents perceive these phenomena. Our particle system encodes the position of each particle in a *Gas Bitmap* as well as a pointer to a file in which the particle characteristics are encoded (particle type, density). An agent can access this gas bitmap in order to determine if there are some particles at its location. If yes, it can get the information about the particles directly and this information is taken into account by the agent behavior module. Indeed, gaseous phenomena are dynamic and the areas/volumes that they occupy change. Our particle system computes the particles' trajectories and takes snapshots of their positions (encoded in the Gas Bitmap) and sends them to the simulation system every k simulation cycles (k depends on how rapidly the phenomenon changes).

Perception of the Effects of Field-Generating Objects. Agents may react to events occurring in the VGE such as the detonation of a fire cracker and the explosion of a tear gas canister. These events are simulated by what we call *field-generating objects*, which emulate the propagation of sound or light in the atmosphere. The user may specify in the simulation scenario when and where in the VGE such an object will be triggered in order to study crowd behavior characteristics through the reactions of individual agents. These objects may also be created by agents in the VGE such as a policeman throwing a tear gas canister or using an ultra-sound whistle to call his dog. Hence, agents may react in different ways to the effects of field-generating objects. A dog will react to the sound of the ultra-sound whistle, but a person will not. In the MAGS system a field-generating object is implemented as a special type of agent whose behavior reflects the consequences of triggering the object. The field-generating object identifies which agents are susceptible to perceive its effects according to the distance and to the agents' perception capabilities, and sends them a message to warn them about the occurrence of the corresponding event. We chose this approach[2] for efficiency purposes, because it would have been very costly to use a function that enables agents to scan the VGE in order to monitor the activation of field-generating objects. In fact, the messages emitted by field-generating objects simply simulate the transmission of information to the agents perceiving the

---

[2] There is a similarity with the notion of *affordance* [10]. Affordances are defined as what objects or things offer people to do with them. In the same way field-generating objects indicate to the agents how to react to the events that they trigger.

associated phenomenon and mimic the transmission of sound or light in the atmosphere. A field-generating object such as the explosion of a tear gas cannister may also trigger a particle system which generates a gazeous phenomenon (see previous sub-section). Consequently, an agent may first react to the perception of a canister's explosion (after receiving the message from the corresponding field-generating object) and then to the perception of the tear gas emitted by the canister (after accessing the corresponding gas bitmap).

Communication between Agents. In the real world people communicate verbally in different ways, speaking or shouting for example. We simulate verbal communications in a simple way: a MAGS agent can send messages to one or several agents if it wants to communicate with them. A simple computation can limit the maximum distance at which a message broadcasted by an agent can be perceived by other agents depending on the distance separating them from the emitting agent.

4   Agent Navigation

While navigating, agents may either follow paths (we call this navigation mode « *following-a-path-mode* ») or move through open spaces (we call this navigation mode « *obstacle-avoidance-mode* »). When an agent is in the *obstacle-avoidance-mode*, the navigation module accesses the Height Map's portion which is visible by the agent (obtained from the dynamic perception function) in order to evaluate the difficulty of crossing the space separating the agent from its destination. When an agent is in the *following-a-path-mode*, the navigation module accesses the Ariadne Map's portion which is visible by the agent (obtained from the dynamic perception function) in order to compute its next move: this navigation mode requires less computational resources than the obstacle-avoidance-mode. An agent can opportunistically change its navigation mode in order to draw nearer to its destination in the VGE. We also developed a function for collision avoidance which is used to manage the agents' local interactions when they move on the Ariadne Map. As an illustration, Figure 5 shows 12 agents in different situations and aiming at the same destination. Some of them (n° 1 to n° 5) are in the *following-a-path-mode* and the others (n° 6 to n° 12) are in the *obstacle-avoidance-mode*. Let us briefly comment upon the agent navigation behavior using this example. In order to determine its next move, an agent must choose one of the eight cells surrounding its current position. To this end, the agent goes through 6 steps :

Step 1: Direct Beam Computing
Using a ray tracing function, the system computes the direction of the *direct Beam* originating from the agent's current position and aiming at its current destination

Step 2: Path_mode Checking
Whenever it is possible, an agent tries to follow a path. Hence, when the agent is not following a path (it is in the *obstacle-avoidance-mode*) the system calls the perceptual function that tries to return the nearest visible position on the path of the Ariadne map. If such a position is found, the agent switches to the *following-a-path-mode* and takes this position as an intermediate destination (ex: agent n° 9 in Figure 5).

Step 3: Beam Tracing

A beam is traced in the *direct beam* direction until reaching a predetermined distance (for example 20 pixels) called the *Beam_range* which is less than or equal to the perception radius. The *Beam_range* value is much less important when the agent is following a path (ex: agents n° 1 to n° 5 in Figure 5) than when it is moving in an open space (ex: agents n° 6 to n° 12). In the *following-a-path-mode*, the tracing function succeeds when the beam crosses an Ariadne pixel (ex: agents n° 4 to n° 5). In the *obstacle-avoidance-mode* the tracing function succeeds when the beam does not hit an obstacle (ex : agent n° 10).

In both navigation modes, when the beam tracing function fails, the system sends a second beam to the right of the *direct beam*, changing its direction by a *Beam_variation_angle*. If this new beam also fails, the system sends a third beam symmetrical to the second, but this time to the left of the *direct beam*. This process goes on until either the beam tracing succeeds, or the angle between the beam direction and the *direct beam* direction goes beyond a given threshold called the *Beam_maximum_angle*[3]. In Figure 5, agents n° 1, 2 and 3 are following a path, each of them moving in a direction (represented by its beam on the figure) which enables it to stay on the path, since the corresponding *Beam_maximum_angle* is not yet exceeded. Agents n° 6, 7 and 8 are in the *obstacle-avoidance-mode*: the function looks for the direction nearest to the *direct beam* that can be followed without hitting an obstacle.

Step 4: Beam Tracing Exception

If Step 3 failed to trace a beam and the agent is in the *following-a-path-mode*, it goes into the *obstacle-avoidance-mode*. This is the case of agent n° 11 for which the *Beam_maximum_angle* is exceeded: it is now in the *obstacle-avoidance-mode*. If the agent is already in the *obstacle-avoidance-mode*, it is blocked by an obstacle because the *Beam_maximum_angle* is exceeded. In that case a function called *GetOutOf-Blockage* moves the agent away from the obstacle. This is the case of agent n° 12 which is currently blocked. The function *GetOutOfBlockage* will enable it to jump to the neighboring Ariadne path.

Step 5: Oscillation Detection

In order to avoid certain critical situations in which the agent could get lost by following cyclic trajectories, we call a function that detects oscillations. This function records the agent's last position every *k* steps and analyzes the evolution of the distance between this position and the agent's current position. If an oscillation is detected, we call the *GetOutOfBlockage* function.

Step 6: Collision Detection and Agent Displacement

CollisionDetection and *AgentDisplacement* are functions accessing the Location Map. If the location *l* chosen for the agent's displacement is already occupied by another mobile agent, a simple displacement algorithm determines a position next to location *l*.

---

[3] The *Beam_variation_angle* and the *Beam_range* are parameters that are associated to the agent profile and can be adjusted. The *average_beam_range* value is 20 pixels. The average beam _variation_angle is π/32 (or 5.625 degrees). The *Beam_maximum_angle* is another parameter that can be adjusted. The average value is 45 degrees on each side of the direct beam.

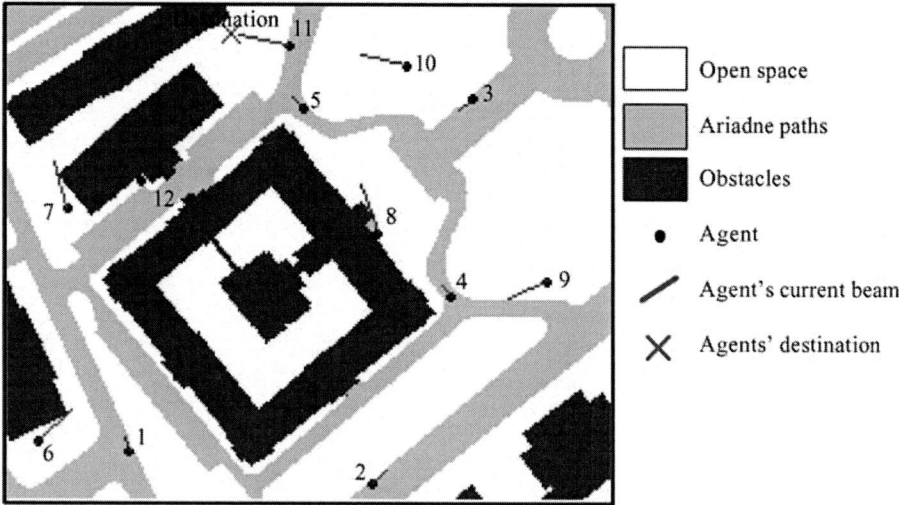

Fig. 5. Different navigation modes

To sum up, the agent navigation module takes advantage of the Height Map, the Ariadne Map, the Location Map and the ray tracing function in order to move the agents to their destinations by opportunistically choosing either the *following a path* or the *obstacle avoidance* modes. This simple mechanism allows the system to perform the simultaneous navigation of a large number of agents (several thousands of agents on a grid of 2048 x 2048 pixels). Besides this navigation mechanisms, we developed a set of basic navigation functions that can be used by the behavior module:

- ChangeDesiredSpeed : change the agent's desired speed
- Flee : flee from another agent's position or from a given location
- MoveInsideDisc : useful to keep an agent moving in a circular area
- MoveAroundDisc : useful to keep an agent moving around a circular area
- Goto : go to an object's location or to a given location on the map
- Follow : follow a mobile agent
- Walk : walk around the VGE by choosing destinations randomly.

5   The Agent's Knowledge

In order to behave autonomously, agents must be able to interact with their environment (VGE, objects and other agents), make decisions with respect to their own states and preferences and act accordingly. We consider five types of agents: mobile, object, field-generating object, cluster and group. *Mobile agents* such as persons and cars are able to move in the VGE. *Object agents* such as buildings and trees are not able to move. *Cluster agents* are dynamically created when several agents gather in a definite area for a certain duration. Moreover, clusters allow more control over member agents at a higher level of abstraction. *Group agents* associate agents having a common characteristic such as policemen. We can also define a group containing other groups. For example, a buildings group contains the different groups of buildings having the same type such as residential, industrial, governmental and commercial buildings. Each agent of any type has its own behavior which depends on its profile.

An agent is characterized by a number of variables whose values describe the agent's state at any given time. We distinguish *stable states* and *dynamic states*. A *stable state* does not change during the simulation and is represented by a variable and its current value. For example, the fact that an agent is respectful of city regulations will not change during the simulation. A *dynamic state* is a state which can possibly change during the simulation. For example, an agent's tiredness can change during the simulation. A dynamic state is represented by a variable associated with a function which is used to compute how this variable changes values during the simulation. The variable is characterized by an initial value, a maximum value, an increase rate, a decrease rate, an upper threshold and a lower threshold which are used by the function. Using these parameters, the system can simulate the evolution of the agents' dynamic states and trigger the relevant behaviors. An agent is also associated with a set of objectives that it tries to reach. The objectives are organized in hierarchies such that elementary objectives are associated with actions that the agent can perform. Each agent owns a set of objectives corresponding to its needs. An objective is associated with rules containing constraints on the activation and completion of the objective. Constraints are dependent on time, on the agent's states, and on the environment's states. The selection of the current agent's behavior relies on the priority of its objectives. Each need is associated with a priority which varies according to the agent's profile. An objective's priority is primarily a function of the corresponding need's priority. It is also subject to modifications brought about by the opportunities that the agent perceives and by the temporal constraints applying on the objective. Each agent of any type has its own behavior depending on its profile. A profile contains a set of roles that an agent can play during a simulation. It is also used to personalize the agent's needs. A role is represented as a tree of objectives. The tree structure allows us to define a behavior at different levels of abstraction. The root nodes are composite objectives and leafs are elementary objectives. The actions are associated with the elementary objectives.

As an illustration we developed a simple scenario in which mobile agents represent people wandering in a VGE of a portion of Quebec city. We also specified a group of buskers who try to find spots in order to entertain the passers-by. Figure 6 presents the simple objective tree of buskers. The main (composite) objective (1.0) is activated if the busker's need for money is greater than a given threshold. Four elementary objectives are associated with *Objective 1.0*. *Objective 1.1* is activated when *Objective 1.0* is active and aims at finding a presentation spot: it triggers an action that makes the busker wander through the city in search of a presentation spot. If the busker perceives a free presentation spot, it reserves this spot, starts to move around it and *Objective 1.1* is completed with success. *Objective 1.2* is activated when *Objective 1.1* is completed with success. The action triggered by *Objective 1.2* is the sending of messages to agents passing by in order to advise them that a show will start soon at this spot. If at least 5 passers-by stay near the spot, *Objective 1.2* is completed with success. If after a certain time there are not enough spectators around the busker's spot, *Objective 1.2* is completed without success. *Objective 1.3* is activated when *Objective 1.2* is completed with success. The action makes the busker agent present its show for a given period of time. Then, *Objective 1.3* is completed with success, *Objective 1.4* is activated and the busker agent sends a message to the spectator agents which are still near the presentation spot in order to ask them for some money.

Elementary objective 1.1 :
To have found a
presentation spot
Activation Rule : There is no
precondition, the objective will be
active when its super-objective
becomes active
Action : Search a zone for a
presentation spot
Completion Rule : If a presentation
spot is perceived and this presentation
spot is free, then make the spot not
free, move around the spot and the
objective 1.1 is completed with success

Composite objective 1.0 :
To have performed a
street play
Activation Rule : If the need for
money is greater than a threshold
Completion Rule : If the
objective 1.4 is completed with
success, then the objective 1.0 is
completed with success

Elementary objective 1.4 :
To have collected money
Activation Rule : If the objective
1.3 is completed with success
Action : Send message to ask
spectators to pay for the show
Completion Rule : If there is no
spectator around, then the need for
money is reduced, make the
presentation spot free and the
objective 1.4 is completed with
success

Elementary objective 1.2 :
To have attracted a
crowd of spectators
Activation Rule : If the objective 1.1 is
completed with success and the objective 1.3
isn't active, ongoing or completed with
success
Action : Send message to gather spectators
Completion Rule 1 : If there are enough
(5) spectators around, then the objective
1.2 is completed with success
Completion Rule 2 : If there are not
enough spectators around, then the
objective 1.2 is completed without success

Elementary objective 1.3 :
To have performed a show
Activation Rule : If the objective 1.2
is completed with success
Action : Perform a show
Completion Rule : If the show was
performed for its predefined duration
time, then the objective 1.3 is
completed with success

Fig. 6. The buskers' tree of objectives

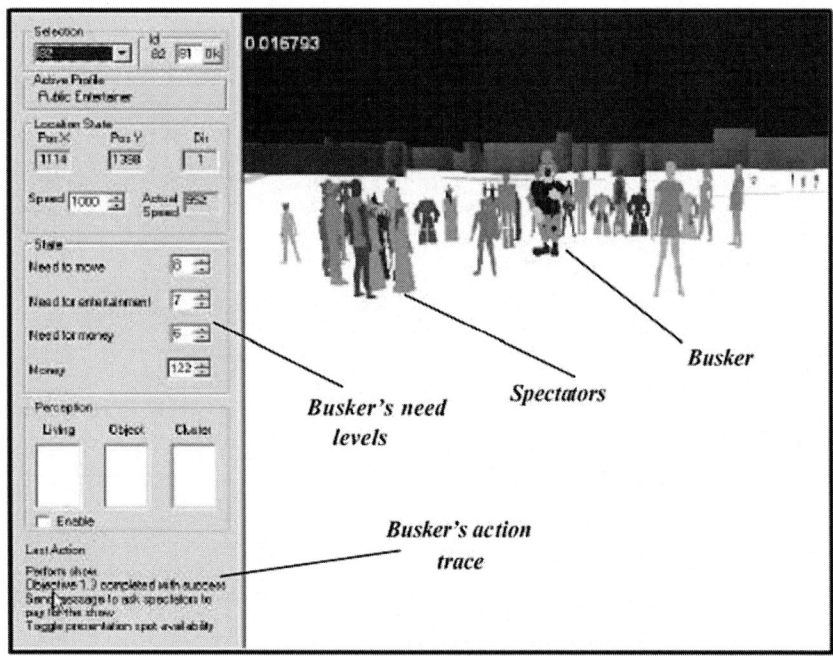

Fig. 7. Busker and audience

After a certain time the busker agent's need for money decreases, it frees the spot and completes *Objective 1.4* with success which leads to the completion with success of *Objective 1.0*. Then, the busker activates another need (which takes the highest priority) that makes the agent wander around in the VGE until its need for money reaches the threshold that triggers *Objective 1.0* again. Passers-by also have an objective tree that emphasizes their need for being entertained by buskers. Space limitations prevent us from presenting this tree in this paper.

Figure 7 presents a snapshot of the simulation which displays the passers-by gathered around a busker in a 3D model of the city. On the left-hand side of the screen we see various characteristics of the selected agent (the busker in this case). At the bottom of this part of the screen, the *Last Action* section shows that the busker has completed *Objective 1.3* and is performing the actions of *Objective 1.4*.

Figure 8 presents a 2D view of people (little dots on the streets) marching in a portion of Quebec city. Figure 9A offers a 3D view that shows people marching and gathering on a square of the city. Figure 9B presents the agent's point of view when moving among other agents. Our system enables the user to manipulate a camera to explore the 3D VGE according to various modes. In Figure 9A we view the scene from above (third person view). In Figure 9B we view the scene from the position of an agent (first person view). The user may view the landscape from any agent's point of view.

The 3D characters associated with the agents are not elaborate since the emphasis of the MAG project was not put on the creation of character animations.

Fig. 8. A peace walk in Quebec city (2D)

Fig. 9A. Observing a march
in Quebec city (from above)

Fig. 9B. Observing the scene from an
agent's point of view

6   Discussion and Conclusions

Since perception and navigation are the two fundamental spatial cognitive capabilities
available to MAGS agents, we wanted to assess the efficiency of our current implementation with respect to the durations of the navigation and perception cycles. Let us
recall that navigation and perception are performed by two separate threads in our
architecture. We carried out tests on the example of agents marching from one location to another in Quebec city (represented on an image of 2048x2048 pixels). The
tests were performed on a Pentium P4, 2.66 GHz with 2 Go of RAM and a Radeon
9700 Pro graphic card. We grouped agents in clusters in order to lessen the load on
the perception thread. In each cluster we have a leader that fully perceives
(*Beam_range* of 30 pixels in the *obstacle-avoidance-mode* and 13 pixels in the
following-a-path-mode). Agents playing the role of followers in the cluster follow the
leader and have a *Beam_range* of 1 pixel. We carried out the tests with various populations of agents (1000, 2000, 3000, 5000, 10 000 agents) and with different cluster
sizes (100, 50, 25 and 15 agents per cluster). In order to make the measurements, we
had to run the system in a debug mode which is less efficient than the normal simulation mode. The navigation and perception cycles do not have the same duration because they require different computing resources. Figure 10 shows the duration of the
navigation cycle for the various populations and cluster sizes. Since an agent moves at
most one pixel (equivalent to about one meter according to the map scale) per cycle,
this means that in the worst case (10 000 agents clustered in groups of 100), the
agents move 6.25 pixels per second. Considering that the average speed of a fast
walking person is 5 km/hour or 1.38 m/s (corresponding to 1.38 pixel/s in the simulation), our simulation shows displacements accelerated by an average factor of 4.5.
This leaves time for the system to devote computing resources to the behavior thread.

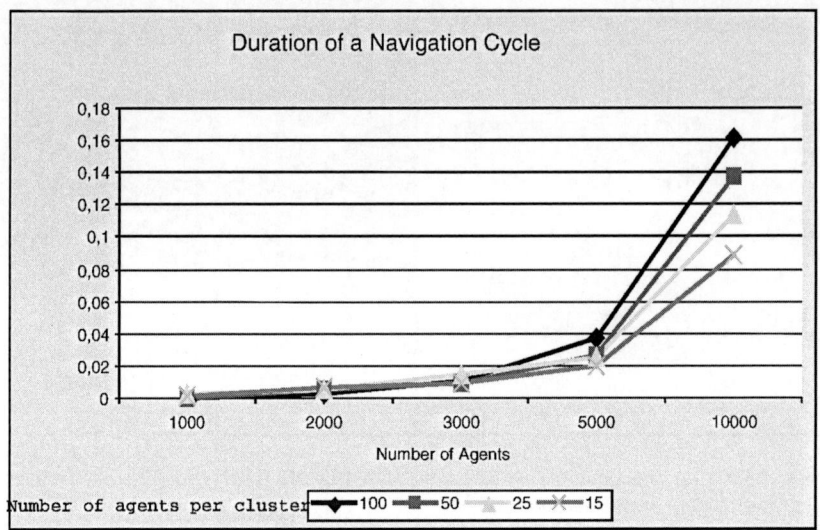

Fig. 10. Duration of a navigation cycle

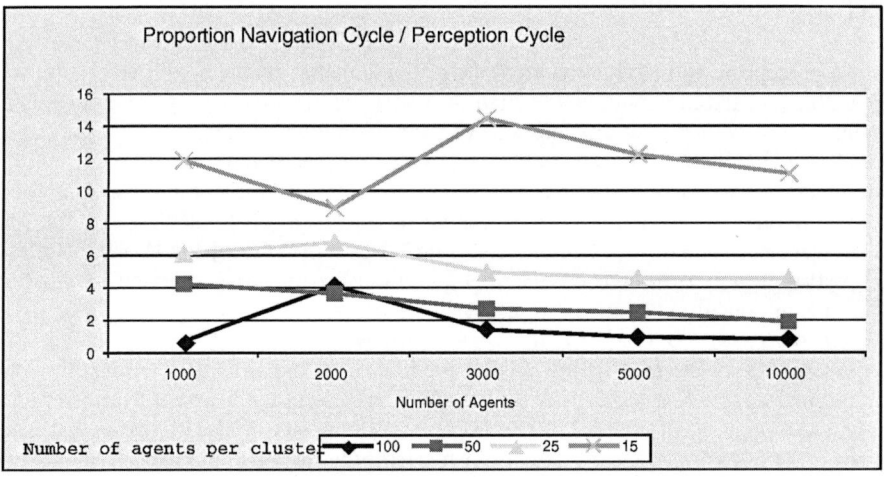

Fig. 11. Ratio of the navigation and perception cycles' durations

Perception takes more time than navigation. Figure 11 shows the ratio of the navigation cycle duration and the perception cycle duration in the different cases of our experiments. Since the navigation and perception threads run in parallel, a ratio of 4 means that an agent will be able to move 4 pixels at most before it gets new information from the perception module. This is not a problem since basic functions such as collision avoidance are part of the navigation cycle. Hence, it is quite acceptable that an agent moves 4 to 6 pixels (equivalent to 4 to 6 meters) before taking into account new information from the dynamic perception function. However, for the worst case of 15 agents per clusters (see Figure 11) the ratio value varies between 9 and 15 times. This is too much. Hence, in our current implementation clusters of at most 25 agents give acceptable results in the debug mode. The performance is better in the simulation mode. Considering the improvement of hardware performance anticipated for the coming years, we expect that the current performance limitations of the MAGS platform will be overcome.

The MAGS system is not fully completed and optimized. For example, we are currently working on the development of the scenario specification module and on the agent's spatial memory capability. We are also exploring various alternatives to distribute the computing load on networked computers. For example, the particle systems used to simulate dynamic areas, such as dense gaz and smoke, require a large amount of computing resources, especially when simulating several phenomena at once (i.e. several tear gaz cannisters exploding in front a group of demonstrators). We developed an initial version of a distributed system in which MAGS and the particle system simulator are on two different computers, the particle system simulator, sending an update of the gas bitmap (see Section 3) at regular time intervals that can be adjusted by the user. However, we need to dedvise a complete distributed solution for MAGS considering the possible distribution strategies available for this kind of applications [25] [26]. However, the current version of the MAGS system and its application to the simulation of crowd behavior show the interest of building agent-based geo-simulation environments as well as the potential of this kind of approach for various kinds of simulations in which agents should exhibit plausible cognitive spatial behaviors [8].

References

1. Beckman, R. J. (ed.): "The Dallas – Fort Worth Study", Los Alamos unclassified report LAUR-97-4502LANL, Los Alamos National Laboratory, Los Alamos NM, available at http://transims.tsasa.lanl.gov/ (1997)
2. De Floriani, L, Magillo, P.: Visibility algorithms on DTMs, *International Journal of Geographic Information Systems*, 8(1), 13–41(1994)
3. Discreet: http://www.discreet.com/products/3dsmax/, (Last visit March 2003)
4. Dodge, M., Doyle, S., Smith, A., Fleetwood, S.: Towards the Virtual City: VR & Internet GIS for Urban Planning, Virtual Reality and Geographical Information Systems Workshop, Birkbeck College (1998)
5. Epstein, S. L., Moulin, B., Chaker, W., Glasgow, J., Gancet, J.: Pragmatism and spatial layout design, in D. Montello (edt.), *Spatial Information Theory: Foundations for Geographic Information Science*, Springer Verlag LNCS 2205,189-205 (2001)
6. Ettema, D., Timmermans, H.: *Activity Based Approaches to Travel Analysis*, Elsevier Science, Amsterdam (1999)
7. Fotheringham, A. S., O'Kelly, M. E.: *Spatial Interaction Models : Formulation and Applications*, Kluwer Academic Publishers, Dordrecht (1989)
8. Frank, A.U., Bittner, S., Raubal, M.: Spatial and cognitive simulation with multi-agent systems, in D. Montello (edt.), *Spatial information Theory: Foundations for Geographic Information Science*, Springer Verlag, LNCS 2205,124-139 (2001)
9. Franklin, W.R.: Applications of analytical cartography, *Cartography and Geographic Information Systems*, 27(3), 225–237 (2000)
10. Gibson, J.: *The Ecological Approach to Visual Perception*, Houghton Mifflin Company, Boston (1979)
11. Gimblett, H. R.: *Integrating Geographic Information Systems and Agent-Based Modeling Techniques for Simulating Social and Ecological Processes*, Oxford University Press (2002)

12. Hacklay, M., O'Sullivan, D., Thurstain-Goldwin, M ., Schelhorn, T.: « So go downtown » : simulating pedestrian movements in town centres, *Environment and Planning B: Planning and Design*, 28(3), 343-359 (2001)
13. Helbing, D., Farkas, I. J., Vicsek, T.: Simulating dynamic features of escape panic, *Nature*, n. 407, 487-490 (2000)
14. Helbing, D., Molnar, P., Schweitzer, F. Computer simulation of pedestrian dynamics, In proceedings of the 3rd International Symposium of SFB 230, *Evolution of Natural Structures*, Sonderforschungsbereich, Stuttgart, Germany, 229-234 (1999)
15. Helbing, D., Molnar, P., Farkas, I. J., Bolay, K.: Self-organizing pedestrian movements, *Environment and Planning B: Planning and Design*, 28(3), 361-383 (2001)
16. Intergraph: Geomedia Professional, http://www.intergraph.com/gis/gmpro (2003)
17. Jager, W., Popping, R., van de Sande, H.: Clustering and fighting in two-party crowds: simulating approach-avoidance conflict, *Journal of Artificial Societies and Social simulation*, vol.4 n.3, http://jasss.soc.surrey.ac.uk/JASSS/4/3/7.html (2001)
18. Jennings, N., O'Hare, G.: *Foundations of Distributed Artificial Intelligence*, Wiley (1996)
19. Kerridge, J., Hine, J., Wigan, M.: Agent–based modelling of pedestrian movements: the questions that need to be asked, *Environment and Planning B: Planning and Design*, 28(3), 327-341 (2001)
20. Lee, D. B.: Retrospective on large-scale urban models, *Journal of the American Planning Association*, 60, 35-40 (1994)
21. Mark, D.M., Freksa, C., Hirtle, S.C., Lloyd, R., Tversky, B.: Cognitive models of geographic space, *International Journal of Geographical Information Science*, vol. 13 no. 8, 747-774 (1999)
22. Moss, S., Davidsson, P.: Multi-Agent-Based Simulation, Proc. of the 2nd Internat. Workshop MASB 2000, Springer Verlag, LNAI, n.1979 (2000)
23. Moulin, B., Chaker, W., Gancet, J. : PADI-Simul, an agent-based software which simulates the behaviors of hundreds of actors in a geographic space, To appear in *Journal on Computers, Environment and Urban Systems* (2003)
24. O'Sullivan, D., Torrens, P.: Cellular models of urban systems, in S. Bandini and T. Worsch (edts.), *Theoretical and Practical Issues on Cellular Automata*, Springer Verlag, also available from Centre for Advanced Spatial Analysis, Working Paper 22, June 2000, www.casa.ucl.ac.uk (2000)
25. Ray, C., Claramunt, C.: Atlas : A distributed system for the simulation of large-scale systems, In Chen, S.-C. et Voisard, A.(edts.), Proceedings of 10th ACM International Symposium On Advances In Geographic Information Systems, 155–162, McLean, VA, ACM Press (2002)
26. Righter, R., Walrand, J. C.:Distributed simulation of discrete event systems. *Proceedings of the IEEE*, 77(1), 99–113 (1989)
27. Sawyer, R.K.: Artificial societies: multiagent systems and the micro-macro link in sociological theory, *Sociological Methods and Research*, vol 31 n3, 325-363 (2003)
28. Schelhorn, T., O'Sullivan, D., Haklay, M., Thustain-Goodwin, M.: STREETS : An agent-based pedestrian model, CASA Working Paper 9, www.casa.ucl.ac.uk (1999)
29. Schillo, M., Fischer, K., Klein, C.T.: The micro-macro link in DAI and sociology, in S. Moss & P. Davidsson (edts.), *Multi-Agent Based Simulation*, Springer Verlag, Lectures Notes in Artificial Intelligence, n. 1979, 133-148 (2000)
30. Stam, J. : Interacting with Smoke and Fire in Real Time, *Communications of the ACM*, vol 43, n 7, 76-83(2000)
31. Torrens, P.M.: Can geocomputation save urban simulation? Throw some agents in the mixture, simmer and wait…, CASA Working Paper 32, www.casa.ucl.ac.uk (2001)
32. Weiss, G. (ed.): Multi-Agent systems, MIT Press (1999)
33. Wolfram, S.: *Cellular Automata and Complexity : Collected papers*, Addison-Wesley (1994)

"Simplest" Paths:
Automated Route Selection for Navigation

Matt Duckham and Lars Kulik

National Center for Geographic Information and Analysis
University of Maine, Orono, ME 04469, USA
{duckham,kulik}@spatial.maine.edu

Abstract. Numerous cognitive studies have indicated that the form and complexity of route instructions may be as important to human navigators as the overall length of route. Most automated navigation systems rely on computing the solution to the shortest path problem, and not the problem of finding the "simplest" path. This paper addresses the issue of finding the "simplest" paths through a network, in terms of the instruction complexity. We propose a "simplest" paths algorithm that has quadratic computation time for a planar graph. An empirical study of the algorithm's performance, based on an established cognitive model of navigation instruction complexity, revealed that the length of a simplest path was on average only 16% longer than the length of the corresponding shortest path. In return for marginally longer routes, the simplest path algorithm seems to offer considerable advantages over shortest paths in terms of their ease of description and execution. The conclusions indicate several areas for future research: in particular cognitive studies are needed to verify these initial computational results. Potentially, the simplest paths algorithm could be used to replace shortest paths algorithms in any automated system for generating human navigation instructions, including in-car navigation systems, Internet driving direction servers, and other location-based services.

Keywords: Navigation, wayfinding, route selection, shortest path, instruction complexity.

1 Introduction

Most people will have had the experience of giving or receiving directions for navigating through an unfamiliar geographic environment. For example, visiting a foreign city for the first time, a tourist might ask a passer-by for directions to a hotel or visitor attraction. In such situations, often what is required is not the *shortest* route to a destination, but the *simplest* route, in terms of how easy it is to explain, understand, memorize, or execute the navigation instructions for the route. Most automated navigation systems rely on computing the solution to the shortest path problem, and not the problem of finding the "simplest" path.

As a motivational example, consider the network in Fig. 1. This network might represent, for instance, the block structure of a road network in a city.

W. Kuhn, M.F. Worboys, and S. Timpf (Eds.): COSIT 2003, LNCS 2825, pp. 169–185, 2003.
© Springer-Verlag Berlin Heidelberg 2003

Assuming the network is embedded in the Euclidean plane, there exist six equivalent optimal shortest routes from intersection i_1 (upper left) to intersection i_9 (lower right)[1]. Without yet defining precisely what is meant by "simplest," intuitively only two of these six routes seem to be optimal simplest routes. Only $(i_1, i_2, i_3, i_6, i_9)$ and $(i_1, i_4, i_7, i_8, i_9)$ avoid the more complex 4-way intersection i_5, and only these two routes require just one "turn." We might describe the route $(i_1, i_2, i_3, i_6, i_9)$ with the instruction sequence "orient yourself, go straight ahead, and turn right at the end of the road." In contrast, the route $(i_1, i_2, i_5, i_8, i_9)$ might require a longer instruction sequence to describe, such as "orient yourself, go straight ahead, turn right at the first intersection, then turn left at the second intersection."

In this paper, we propose an algorithm that can be used to select routes that minimize the complexity of instructions, rather than the distance traveled. The goal is to simplify navigation in an unknown environment when following a route instruction. Following the literature review in Section 2, the algorithm is introduced in Section 3, and its computational properties are reviewed. In Section 4, an empirical comparison of the simplest paths algorithm with the shortest paths algorithm is conducted, using an example road network data set. A discussion of the results, conclusions, and suggestions for further work is contained in Section 5.

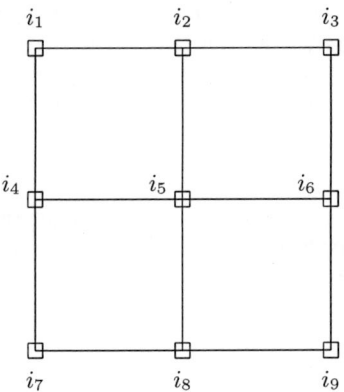

Fig. 1. Example network with six shortest and two "simplest" paths

2   Background

Several cognitive studies have indicated that the form and complexity of route instructions may be as important in human navigation as the overall length of a route. Streeter and co-authors [1,2] looked at the use of verbal instructions in human navigation. They found that human navigators were prepared to select

---

[1] i.e. $(i_1, i_2, i_3, i_6, i_9)$, $(i_1, i_2, i_5, i_6, i_9)$, $(i_1, i_2, i_5, i_8, i_9)$, $(i_1, i_4, i_5, i_6, i_9)$, $(i_1, i_4, i_5, i_8, i_9)$, $(i_1, i_4, i_7, i_8, i_9)$

suboptimal routes, in terms of the total length of a route, in favor of routes that were potentially easier to describe or follow. Golledge ranked ten different criteria used in human route selection, based on experiments using human subjects [3]. While the related criteria of shortest distance and least time were ranked most highly, many other criteria, including number of turns, also ranked highly. Michel Denis and coauthors studied the "informational units" contained within human route descriptions [4]. Amongst other findings, Denis et al. showed that those route descriptions that were considered more likely to prevent users from making errors were preferred by study subjects. Following on from Denis, Tversky and Lee have examined the relationship between pictorial and verbal descriptions of routes [5,6]. Tversky and Lee found both pictorial and verbal route descriptions exhibited considerable redundancy in information, a feature they attribute to the importance of reassurance and error prevention for human navigators. Research by Richter and Klippel [7], into optimal locations for "You-are-here" maps, emphasizes the importance of the overall number of decision points and the number of branches at an intersection as route selection criteria, in addition to the length of a route. Research from the vehicle navigation literature has also confirmed that successful vehicle navigation systems rely as much on clarity of route instructions as length of route [8,9]. Shortest path algorithms only minimize route efficiency, in terms of distance or travel cost, and not route description complexity.

Algorithms for finding an optimal route that is not the shortest path have been proposed by Shapiro et al. [10] as well as Liu [11,12]. The approach of Shapiro et al. can be used to take advantage of the type of roads (major roads versus minor roads), generating a route that prefers major routes. The time to find a route in the graph is considerably reduced, since the algorithm essentially restricts the search to major roads. Liu [11,12] also developed an efficient approach that incorporates road network knowledge to narrow the search for routes, using the fact that major roads partion a road network. However, none of these approaches have focused explicitly on the simplicity of a route description.

The aim of the simplest path algorithm presented in this paper is to minimize the complexity of a route description, based on the amount of information required to negotiate each decision point. Determining how much information is communicated by some route description may itself be a difficult question (see [13,14,15] for a discussion of the information content of geographic information in general and route descriptions in particular). However, the simplest path algorithm presented in this paper does not rely on any particular model of geographic information content, so long as some measure of instruction complexity can be derived for decision points along a route.

2.1   Automated Route Selection for Navigation

In 1986, David Mark [16] published a paper entitled "Automated route selection for navigation" (hence the subtitle for this paper). Mark proposed a modification of the A* shortest path algorithm, which took into account both the total length and the "ease of description" of the route. Based on work of Streeter and co-authors [1,2], Mark [17] classified different intersections according to the complexity of the instructions needed to successfully negotiate that intersection. Using this classification, Mark's algorithm adjusted the weights used in the shortest path computation to preferentially select routes through intersections that could be described using less complex instructions.

The simplest paths algorithm below was motivated in part by the work presented by Mark. However, the algorithm below differs in at least three ways from that presented in [16]. First, the simplest path algorithm does not use distance or any other metric information in its operation. The algorithm computes the simplest paths using only a measure of instruction complexity. Even without any metric information, the results in Section 4 indicate that simplest paths are still comparable in length to shortest paths. Second, the lack of distance information means the simplest path algorithm does not depend on an arbitrary assignment of weights to the relative importance of distance and instruction complexity. Given a weighting function for instruction complexity, the simplest path algorithm proposed here yields a uniquely determined simplest path from start to destination. Third, the algorithm presented below is a single source algorithm, able to efficiently compute the distance from a single source vertex to every other vertex. Recent work by Duckham et al. [18] has indicated that single source route finding algorithms are particularly important in the context of navigation under imprecision, where a user's precise location or destination may be unknown. Other uses of single source route finding might include applications where multiple route options need to be presented to the user, for example finding the routes from a user's current location to a variety of tourist attractions. The computational properties of the algorithm are discussed in more detail in Section 3.1.

3   Formal Model of Simplest Paths

Graphs are a common mechanism for representing networks, such as road networks. A graph G comprises a set of vertices V and edges E connecting those vertices. A weighted graph additionally has a function $w : E \to \mathbb{R}^+$ associating a weight with each edge $e \in E$. Finding the shortest paths, the paths of least cost between vertices in a weighted graph, is a fundamental network analysis function and a classic problem in computation. There have been estimated to be more than 2000 articles published on the topic since the 1950s [19]. As a result, we make no attempt here to summarize the different approaches to computing shortest paths; any introductory textbook on algorithms (e.g. [20]) or artificial intelligence (e.g. [21]) will contain such a summary.

The essential difference between shortest and simplest paths algorithms is that the latter uses a weighting function that associates a weight with each *pair of connected edges* (rather than each edge) in the graph, $w : \mathcal{E} \to \mathbb{R}^+$ where $\mathcal{E} = \{((v_i, v_j), (v_j, v_k)) \in E \times E\}$. The intuition behind these weights is that they should reflect the amount or complexity of information required to negotiate the "decision point" represented by the edge pair (i.e. negotiating the path from

v_i to v_k through intersection v_j). The more information needed, the higher the weight for an edge pair. The later discussion in Section 4.1 contains an example of such a function.

The actual simplest path algorithm is presented in Algorithm 1 below. The graph used in the simplest path algorithm is assumed to be directed (i.e. the direction of the edges is significant), connected (i.e. there exists a path from any vertex to any other vertex), and simple (i.e. there are no edges from a vertex to itself and at most one edge between two different vertices). Algorithm 1 operates by first initializing all edges connected to the starting vertex with zero weight. Thus, the algorithm applies no cost to the initial orientation stage of routing. A more sophisticated algorithm might include weights for initial orientation, perhaps preferentially selecting initial orientations that are easier to explain (for example, orientation towards an easily visible nearby landmark). Next, the algorithm iterates through each edge, minimizing the cumulative instruction complexity. At each iteration, the edge associated with the minimum instruction complexity is selected, and the cumulative instruction complexity from that edge to all connected edges is recalculated. The algorithm interates until all edges have been visited. At no point in the algorithm is any distance information involved in the calculation, only the instruction complexity.

Algorithm 1: Simplest path algorithm

Initial conditions: $G = (V, E)$ is a connected, simple, directed graph; $s \in V$ is the starting vertex; $\mathcal{E}$ is the set of pairs of (directed) edges that share their "middle" vertex, $\mathcal{E} = \{((v_i, v_j), (v_j, v_k)) \in E \times E\}$; $w : \mathcal{E} \to \mathbb{R}^+$ is the graph weighting function; $c_s : E \to \mathbb{R}^+$ stores the weights of the simplest path from s; $S = \{\}$ is a set of visited edges.

Initialize $c_s(e) = \infty$ for all $e \in E$
for *all* $(s, v_i) \in E$ **do**
    set $c_s(s, v_i) = 0$

while $|E \backslash S| > 0$ **do**
    Find $e \in E \backslash S$ such that $c_s(e)$ is minimized
    Add e to S
    for *all* $e' \in E \backslash S$ **do**
        if $(e, e') \in \mathcal{E}$ **then**
            set $c_s(e') = \min(c_s(e'), c_s(e) + w(e, e'))$

The function c_s in Algorithm 1 stores the weights of the simplest paths from s to every destination vertex. To recover the simplest path to a particular destination vertex $d \in V$, we must first find the edge $(v_i, d) \in E$ where $c_s(v_i, d)$ is minimum. Reconstructing the simplest path is then a matter of iterating backwards through the edges, at each iteration choosing the least costly edge (Algorithm 2).

Algorithm 2: Retrieving the simplest path using c_s (see also Algorithm 1)

Initial conditions: Weights c_s from simplest path algorithm (see Algorithm 1), starting vertex $s \in V$ and destination vertex $d \in V$.

Initialize $t = d$, path $p = (t)$
while $t \neq s$ **do**
  Find $(v_1, t) \in E$ such that $c_s(v_1, t)$ is minimized
  Prepend vertex v_1 to path p
  Set $t = v_1$

3.1   Computational Issues

One way of explaining how the simplest path algorithm operates, is by considering the mapping of graph $G = (V, E)$ onto $G' = (E', \mathcal{E})$, which we term the *evaluation mapping* μ. In the evaluation mapping, the set of vertices E' in G' is mapped from the set of edges E in G, where the direction of the edge is ignored (i.e. $(v_i, v_j) = (v_j, v_i)$ in E'). The set of (directed) edges in G' is mapped from the set of pairs of connected (directed) edges in G, introduced above and denoted using $\mathcal{E}$. Note that the evaluation mapping does not form the dual of a graph. Furthermore, the evaluation mapping is not injective and does not necessarily preserve planarity. Fig. 2 provides an example of an evaluation mapping μ, which maps (undirected) edges in G onto vertices in G', and pairs of connected (directed) edges in G onto (directed) edges in G'. Finding the simplest paths from v in G means finding the shortest paths from any vertex in G' which v maps onto (i.e. any edge in E which contains v). This can be accomplished with any single source shortest paths algorithm, like the Dijkstra algorithm (see [20]). Therefore, the simplest paths algorithm in Algorithm 1 can be considered essentially equivalent to first mapping the graph, then finding the shortest path through the mapped graph using a conventional shortest path algorithm.

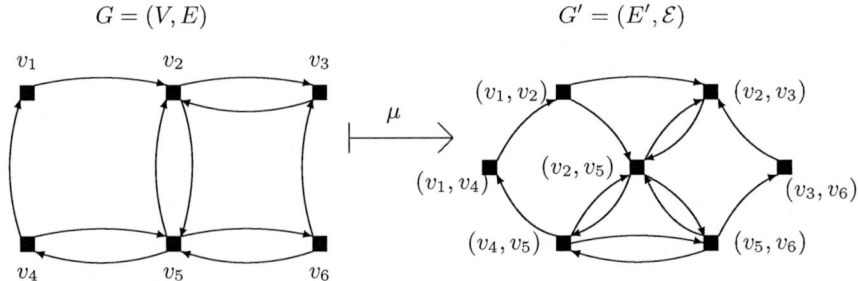

Fig. 2. Evaluation mapping μ between graphs in the simplest path algorithm

It follows that the time complexity of the simplest path algorithm is closely related to the time complexity of the shortest path algorithm. Dijkstra's algorithm has a time complexity of $O(|V|^2)$, where $|V|$ is the number of vertices in the graph G. Therefore, the simplest path algorithm has a time complexity of $O(n^2)$, where n is the number of vertices in G', or equivalently the number of edges in the graph G. In the worst case, the graph G' could have as many as $n = |V|(|V| - 1)$ vertices, leading to a complexity of $O(|V|^4)$ for the simplest path algorithm. However, such a worst case would require a totally connected graph (every pair of vertices is connected by an edge). Most geographic networks can be considered to be planar graphs. Planar graphs have a maximum number of edges $m = 3(n - 2)$. This follows from Euler's Polyhedron Formula: a simple connected planar graph with n vertices, m edges, and f faces satisfies $n - m + f = 2$. As a result, for a simple planar graph the time complexity of the simplest path algorithm is $O(|V|^2)$. For planar graphs with nonnegative edges Henzinger and co-workers [22] recently presented a linear-time algorithm for single-source shortest paths. Since the graph G' might not be planar, the simplest path algorithm can be slower than the shortest path algorithm.

As indicated in Section 2.1, the simplest path algorithm is a single source algorithm. Like the Dijkstra algorithm, it computes the shortest path to every destination in the graph. Mark's original work uses the A* algorithm to compute the distance from a single source to a single destination only [16]. The relationship between A* and Dijkstra is well documented in the literature (e.g. [21]). The key difference between the two algorithms is that the A* algorithm uses a heuristic to preferentially explore more promising paths. For geographic information, the Euclidean distance to the destination forms an ideal heuristic, as it is a consistent underestimate of the network distance to the destination. Unfortunately, there is no natural heuristic for the mapped graph. There is no obvious choice for a non-trivial heuristic function that provides a consistent underestimate of the instruction complexity between any two vertices in the mapped graph. Unless such a heuristic function could be found, A* cannot be used for simplest paths, although this has not yet been fully investigated. In the worst case the computational complexity of the A* algorithm will be $O(n^2)$, the same as for the Dijkstra algorithm. However, the heurisitic means that A* is on average more efficient than Dijkstra where only one route from a single source to a single destination is needed.

4   Comparison of Simplest and Shortest Paths Algorithms

In order to explore the performance and properties of the simplest path algorithm, the algorithm was implemented in Java and tested using an example weighting function and an example road network data set. Section 4.1 below provides details of the weighting function used, with the results following in Section 4.2.

4.1   Choosing a Weighting Function

Any weighting function $w : \mathcal{E} \to \mathbb{R}^+$ can be used in the simplest path algorithm in Algorithm 1. However, as stated in Section 3, the weights should be chosen to reflect the amount or complexity of information required to negotiate the "decision point" represented by a pair of edges $(e_i, e_j) \in \mathcal{E}$. The weights must be ordered so that more complex decision points are associated with higher weights. There are several studies that contain classifications of route instructions that might be used as a basis for a weighting function. For example, Denis et al. provide a model of idealized route instructions [23,4] that might be used. Several other models, including Kuipers' TOUR model [24,25] and the PLAN model of Chown et al. [26], might yield different weighting functions. In implementing the simplest path algorithm in this study, the weighting function described below follows Mark's original work [16], in which the weights were derived from the work of Streeter and coauthors [1,2].

In Mark [16], instructions are classified into *frames*, each frame having several *slots* for different elements of an instruction. A generic turn instruction is modeled as a frame containing a total of 9 slots. Each slot covers information about whether to turn left or right (3 slots), how to recognize when to turn (2 slots), how to recognize if the navigator has gone too far (1 slot), and summary information providing an overview of the turn (3 slots). A turn at a T-junction contains only 6 slots, since it is easy to recognize and not possible to overshoot a T-junction (so information about how to recognize the turn and what to do it the navigator goes too far are unnecessary). A turn in the road that is not at an intersection contains just 4 slots, and so on. The number of slots can be considered as a measure of the information content of the instruction needed to negotiate a decision point, and so was used as the weighting function in this experiment.

Three small deviations from the weights used in Mark [16] seem sensible. First, Mark's weighting contains no cost for going straight on. A navigation system should at least be able to reassure a user that they are on the right track, even if just going straight on, so in our weighting function the cost of straight on was one slot. Second, Mark [16] included a 3-slot weighting for an instruction where the name of a road changed while continuing straight on. The algorithm as implemented does not include this weighting, as this initial work focuses solely on the geometry and topology of the road network, not the attributes. However, such weights could easily be implemented (see discussion in Section 5). Third, Mark's original model does not distinguish between general intersections of different degree. Intuitively, the instruction for turning left at a 3-way intersection requires less information than at a 4-way intersection, which in turn requires less information than for a 5-way intersection. To reflect this wrinkle, the weight for turning left or right at a general intersection vertex v was set to $5 + deg(v)$. This means for a 4-way intersection, the weighting is as in Mark [16], 9 slots, but for a 3-way intersection the weighting is 8 slots, and for a 5-way 10 slots.

The instruction classification and associated weights are illustrated in Fig. 3. Note that this weighting function means that unlike shortest paths, simplest paths are not *symmetric*: the simplest route from A to B may not be the same as the simplest route from B to A. For example, coming from one direction,

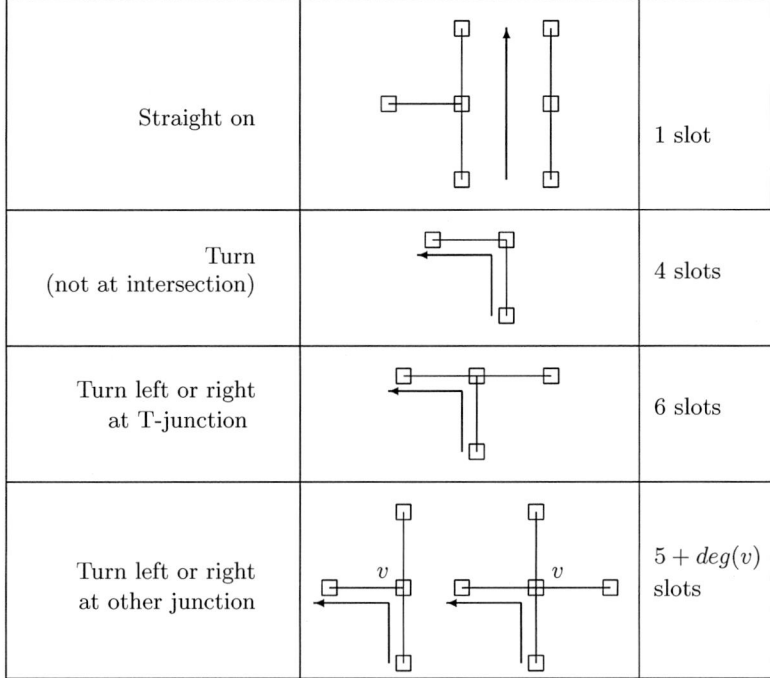

Straight on		1 slot
Turn (not at intersection)		4 slots
Turn left or right at T-junction		6 slots
Turn left or right at other junction		$5 + deg(v)$ slots

Fig. 3. Weighting of different intersection types, based on [16]

turning right at a T-junction is weighted 6 slots, while coming from the other direction, turning left at a 3-way intersection is weighted 8 slots. The weighting function also means that, unlike shortest paths, simplest paths do not satisfy the *triangle inequality*: the simplest path from A to C may be longer than the length of the simplest path from A to B plus the length of the simplest path from B to C. The triangle inequality might be violated because adding the lengths of the simplest path from A to B and from B to C ignores the information required to negotiate the intersection at B, information that is included in the calculation of the simplest path from A to C. Since simplest paths are neither symmetric nor fulfil the triangle inequality, simplest paths cannot be determined by any metric.

4.2   Algorithm Performance

The results of using of the shortest path algorithm on several example road network data sets were encouraging. This paper reports the results of an exhaustive analysis on road network data set for the city of Bloomington, Indiana, USA, which exhibits a wide range of different network configurations, including a dense downtown grid network and sparser suburban networks. Fig. 4 shows a comparison of simplest and shortest paths between two vertices from within a downtown area of Bloomington.

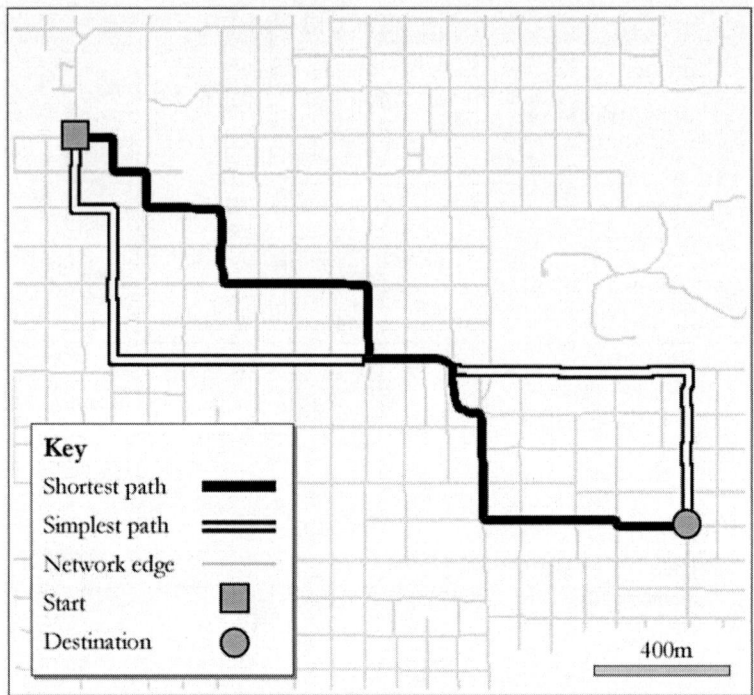

Fig. 4. Example comparison of a shortest and simplest path (approximate scale)

The thick black line in Fig. 4 shows the shortest path. The double black line shows where the simplest path deviates from the shortest path. The simplest path contains only four turns, all made at three-way intersections. The shortest path contains many more turns: a total of 12 turns, made at a range of two-, three- and four-way intersections. The lengths of the two paths are very similar: the simplest path is 2% longer than the shortest path.

The performance of the simplest path algorithm was exhaustively tested for the entire data set. With more than 3000 vertices in the data set, this involved computing the shortest and simplest paths for a total of over 10 million pairs of vertices. A comparison of the lengths of the simplest and shortest paths for one set of 3200 shortest paths from a single source to every other vertex in the data set is shown in Fig. 5. The figure provides a scatter plot of the normalized simplest path length (the ratio of simplest to shortest path lengths), plotted against shortest path length.

In this example, more than 90% of the simplest paths between two vertices are less than 50% longer than the corresponding simplest path. This example is typical: for the entire data set of more than 10 million simplest paths, 93.2% of the shortest paths are less than 50% longer than the corresponding shortest path. Over the entire data set, on average a simplest path is 15.8% longer than the

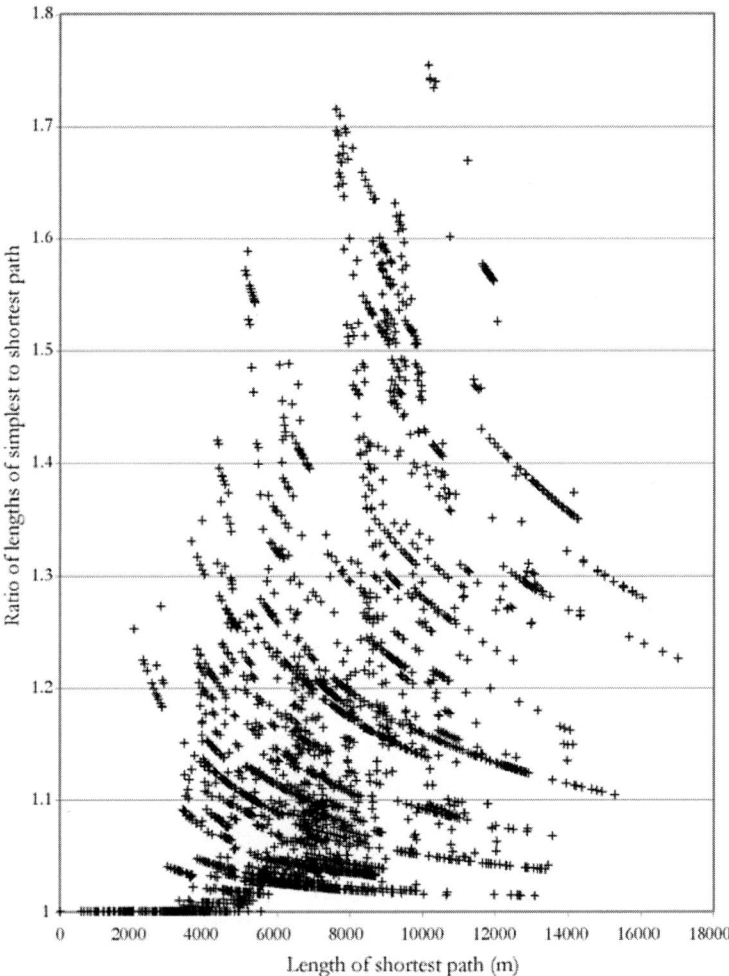

Fig. 5. Scatter plot comparing the simplest and shortest path lengths for a single source

corresponding shortest path. A noticeable but incidental feature of the scatter plot is that it exhibits some strongly correlated "stripes" running from the top left to the bottom right. These occur because the set of all shortest paths from a single starting vertex is strongly correlated, with many of the paths containing similar sequences of edges.

Fig. 6 shows an example of one path where the starting point is some way out of the dense downtown road network. The example in Fig. 6 is drawn from the data set plotted in Fig. 5. The simplest path (shown as double black line) has total length of approximately 10km, 30% longer than the shortest path (thick black line), placing it somewhere near the middle of the scatter plot in Fig. 5.

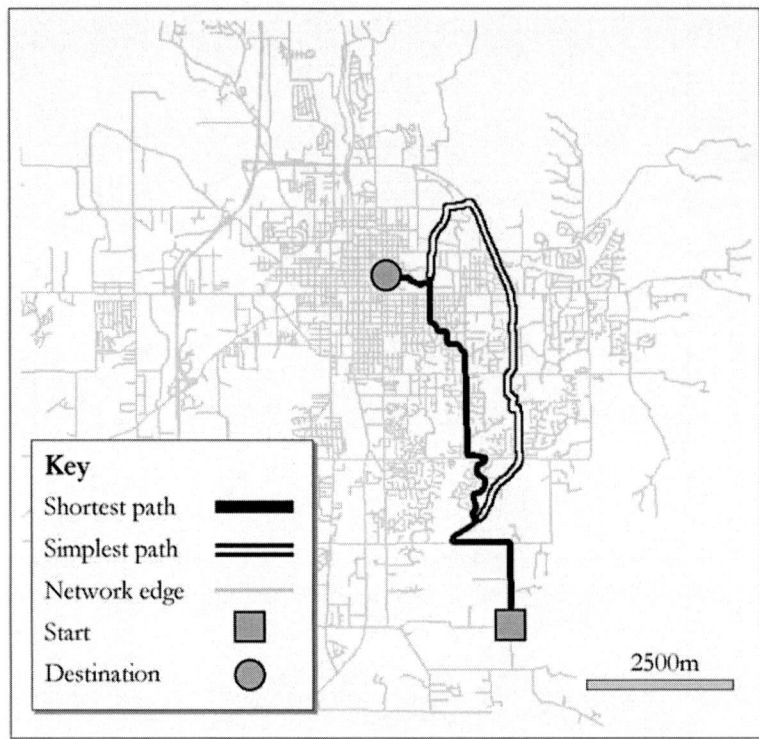

Key

Shortest path

Simplest path

Network edge

Start

Destination

2500m

Fig. 6. Comparison of a typical shortest and simplest path (approximate scale)

Fig. 7 shows the "worst case" outlier simplest path, also drawn from the data set plotted in Fig. 5. This simplest path has a total length of approximately 18km, 75% longer than it's corresponding shortest path, placing it right at the top of Fig. 5. The most noticeable features of the simplest paths in both Fig. 6 and 7 is that they tend to skirt the city center areas, in favor of longer routes through less dense road networks. In the "worst case" example in Fig. 7, the simplest path actually goes south, in the opposite direction from the destination, in order to join a long straight road with relatively few intersections (actually an Interstate highway), and travels north past the destination before switching back south to the destination. Despite being the "worst case," this does not seem an unreasonable route to choose. The route avoids the numerous complicated intersections traversed by the shortest path, at the cost of increased total distance.

Note that Interstate highway has fewer intersections than it would appear from Fig. 7 alone: several minor roads cross the Interstate, but do not intersect it (underpasses or bridges). This is the reason why the simplest path must travel south away from the destination to join the Interstate and north past the destination to leave it. The example highlights that the assumption of a planar graph (see Section 3.1) is an oversimplification for road networks. Note also that the

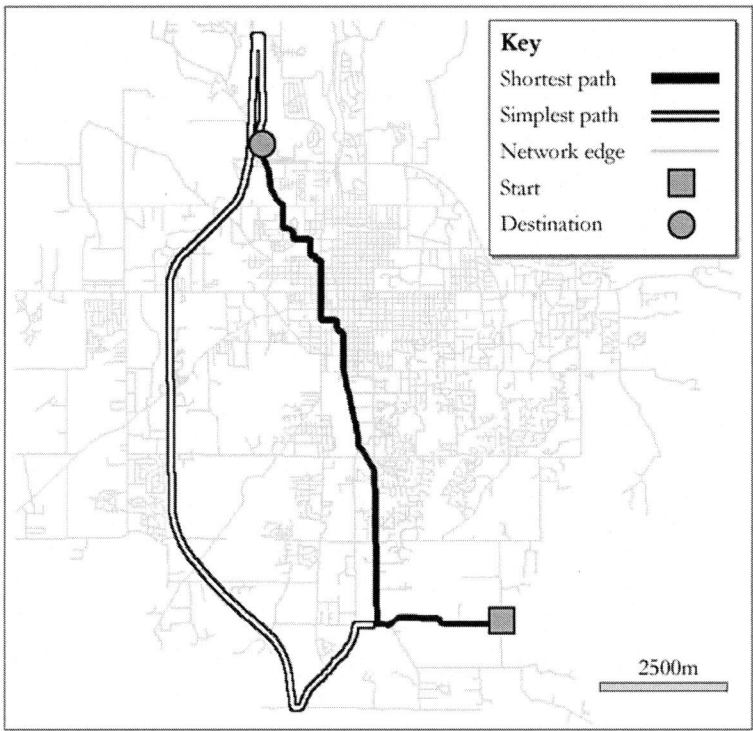

Fig. 7. Comparison of a "worst case" shortest and simplest path (approximate scale)

algorithm has no access to information about what sort of road it is selecting. A more sophisticated algorithm could preferentially select routes that use more important or faster roads. That is not what is happening in Fig. 7: it is simply that the relatively low frequency of intersections and straightness of the highway means it is a low cost route for the simplest path algorithm.

Finally, Fig. 8 summarizes the entire data set of 10 million shortest paths. Fig. 8 shows the spatial distribution of the standard deviation of the normalized lengths of the simplest paths. Each point in Fig. 8 represents the standard deviation of normalized length for all simplest paths starting from that point. The standard deviations have been classified in five quantiles (five classes with equal cardinality).

The figure shows generally low standard deviations (lighter gray data points) in the denser downtown road network regions and in the sparse suburban road network regions (c.f. the road network in Fig. 7). The higher standard deviations (darker gray points) generally occur at the periphery of the dense downtown road networks, at the transition to sparse suburban road networks. This greater variability can be interpreted as a result of the deviations of the simplest path from the shortest path being more pronounced at the periphery of the dense

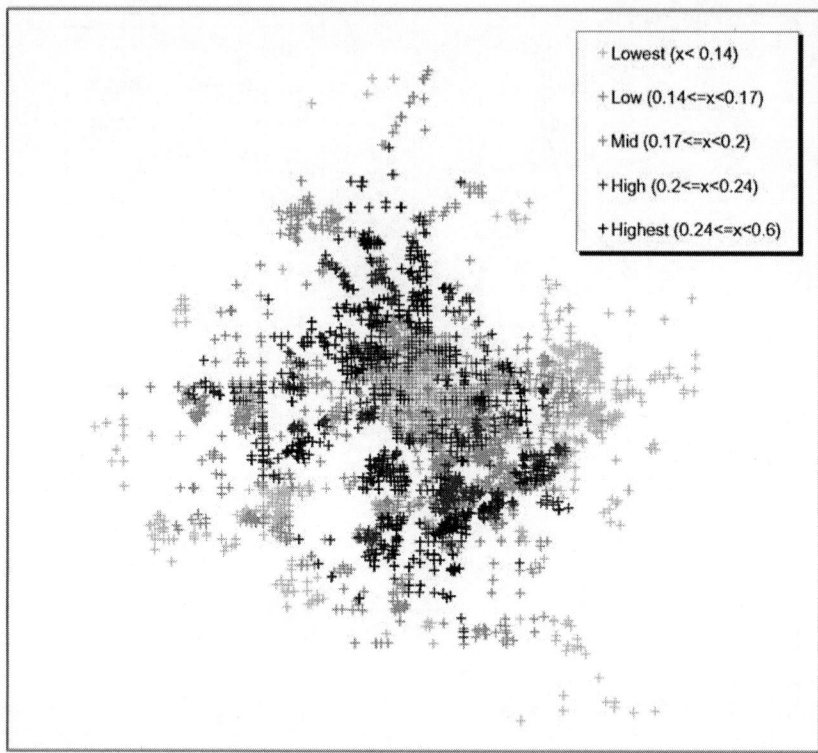

Fig. 8. Spatial distribution of the standard deviation of normalized simplest path length

downtown road networks. Starting locations within the downtown road network have no option but to traverse this denser network. Starting locations well away from the downtown road network require relatively smaller deviations to avoid the denser network if necessary. Those locations at the periphery of the denser downtown road network are more likely to require larger deviations from the shortest path, such as that shown in Fig. 7. A similar diagram showing the spatial distribution of mean normalized simplest path length revealed no appreciable spatial pattern.

5   Discussion and Conclusions

The results of this work are generally encouraging. The simplest paths are on average only 16% longer than the shortest paths, and more than 90% of simplest paths are less than 50% longer than the corresponding shortest paths. In return for paths of slightly longer length, the simplest paths algorithm produces routes through a network that are *cognitively plausible*, in the sense that they are based on a cognitive model of the complexity of instructions needed to complete a route. Further, the simplest paths algorithm requires no distance or other metric information to operate. Diagrams like Fig. 4 provide compelling (if subjective) evidence that the simplest paths are indeed more cognitively plausible that simplest paths. Even the "worst case" examples, like Fig. 7 do not appear to be unreasonable routes. A major component of future work, therefore, should be cognitive studies with human subjects to test the hypothesis that the route instructions based on simplest paths are preferable, less error prone, or easier to explain and use for human navigators, when compared with route instructions based on shortest paths.

The analysis presented in this paper has empirically examined the relationship between the lengths of simplest and shortest paths. While the data set used in the analysis does exhibit a range of different network configurations and densities, further empirical work studying different types of road network might reveal some different properties (for example, the road networks in many European cities, which have developed over longer periods of time). Moreover, the work presented in this paper lacks a theoretical basis for the relationship between the length of simplest and shortest paths. As a result, further theoretical work is needed to explore this relationship, including the investigation, identification, and classification of different types of graph with respect to the properties of simplest and shortest paths through those graphs.

The simplest path algorithm can have a longer computation time than shortest path algorithms. Assuming the graph is planar, the algorithm has at most quadratic computation time. In some cases, graphs may be locally non-planar, as in road networks where roads cross but do not intersect, (e.g. overpasses or bridges). However, road networks still retain relatively low connectivity, so it seems unlikely that these locally non-planar graphs would significantly increase the computational complexity of the simplest paths algorithm. The simplest paths algorithm can efficiently compute the simplest paths from a single source vertex to every other vertex in the graph. This property can be vital for dealing with imprecision in navigation, for example, where the location or destination of a navigation agent may not be precisely known [18].

The simplest path algorithm might easily be adapted to provide more sophisticated behavior. As seen in Fig. 7, the simplest path algorithm sometimes selects major roads simply by virtue of their straighter geometry and less connected topology (i.e. fewer intersections). A modification of the weighting function could easily be used to explicitly prefer certain types of road [10,11,12]. For example, major roads could be preferred in the route selection by making the weights for turning onto a major road smaller, and turning off a major road larger. Conversely, minor roads could be avoided by making the weights for turns onto a minor road larger, and turning off a minor road smaller. Many other possible adaptations of the algorithm could result from other modifications of the weighting function. Another important component of future work, therefore, should be cognitive studies with human subjects to determine what types of weighting functions are most appropriate for human users in a particular contexts.

Acknowledgements

This work was conducted under the SmartMaps project, supported by the National Imagery and Mapping Agency, grant no NMA201-01-1-2003. The authors would like to acknowledge David Mark, for an early conversation which helped initiate many of the ideas for this work, as well as later comments on a first draft which helped shape the paper. In addition, comments and suggestions from Max Egenhofer and Mike Worboys were valuable. This research has also benefited from collaboration with Jörg-Rudiger Sack at Carleton University, Ottawa (NSERC supported Collaborative Research and Development grant). The authors are also grateful to the three anonymous reviewers for their constructive comments.

References

1. Streeter, L., Vitello, D., Wonsiewicz, S.: How to tell people where to go: Comparing navigational aids. International Journal of Man Machine Interaction **22** (1985) 549–562
2. Streeter, L., Vitello, D.: A profile of driver's map-reading abilities. Human Factors **28** (1986) 223–239
3. Golledge, R.: Path selection and route preference in human navigation: A progress report. In Frank, A., Kuhn, W., eds.: Spatial Information Theory: A Theoretical Basis for GIS (COSIT '95). Number 988 in Lecture Notes in Computer Science, Berlin, Springer (1995) 207–222
4. Denis, M., Pazzaglia, F., Cornoldi, C., Bertolo, L.: Spatial discourse and navigation: An analysis of route directions in the city of Venice. Applied Cognitive Psychology **13** (1999) 145–174
5. Tversky, B., Lee, P.: How space structures language. In Freksa, C., Habel, C., Wender, K., eds.: Spatial Cognition. Number 1404 in Lecture Notes in Computer Science, Berlin, Springer (1998) 157–176
6. Tversky, B., Lee, P.: Pictorial and verbal tools for conveying routes. In Freksa, C., Mark, D., eds.: Spatial Information Theory. Cognitive and Computational Foundations of Geographic Information Science (COSIT'99). Number 1661 in Lecture Notes in Computer Science, Berlin, Springer (1999) 51–64
7. Richter, K.F., Klippel, A.: "You-are-here maps": Wayfinding support as location based service. In Moltgen, J., Wytzisk, A., eds.: GI-Technologien für Verkehr und Logistik. IfGI Prints 13, Münster (2002)
8. Burnett, G.: 'Turn right at the traffic lights': The requirement for landmarks in vehicle navigation systems. Journal of Navigation **53** (2000) 499–510
9. May, A., Ross, T., Bayer, S.: Drivers' informational requirements when navigating in an urban environment. Journal of Navigation **56** (2003) 89–100
10. Shapiro, J., J., W., Nir, D.: Level graphs and approximate shortest paths algorithms. Networks **22** (1992) 691–717
11. Liu, B.: Using knowledge to isolate search in route finding. In: Proceedings of the Fourteenth International Joint Conference on Artificial Intelligence, IJCAI 95. Volume 1., Montréal, Québec, Canada, Morgan Kaufmann (1995) 119–125
12. Liu, B.: Intelligent route finding: combining knowledge, cases and an efficient search algorithm. In: 12th European Conference on Artificial Intelligence (ECAI-96), Budapest, Hungary, John Wiley and Sons (1996) 380–384

13. Frank, A.U.: Pragmatic information content—how to measure the information in a route description. In Duckham, M., Goodchild, M.F., Worboys, M.F., eds.: Foundations in Geographic Information Science. Taylor & Francis, London (2003) 47–68
14. Goodchild, M.F.: The nature and value of geographic information. In Duckham, M., Goodchild, M.F., Worboys, M.F., eds.: Foundations in Geographic Information Science. Taylor & Francis, London (2003) 19–31
15. Worboys, M.F.: Communicating geographic information in context. In Duckham, M., Goodchild, M.F., Worboys, M.F., eds.: Foundations in Geographic Information Science. Taylor & Francis, London (2003) 33–45
16. Mark, D.M.: Automated route selection for navigation. IEEE Aerospace and Electronic Systems Magazine **1** (1986) 2–5
17. Mark, D.M.: Finding simple routes: 'ease of description' as an objective function in automated route selection. In: Proceedings, Second Symposium on Artificial Intelligence Applications (IEEE), Miami Beach (1985) 577–581
18. Duckham, M., Kulik, L., Worboys, M.F.: Imprecise navigation. GeoInformatica **7** (2003) 79–94
19. Pallottino, S., Scutellà, M.: Shortest path algorithms in transportation models: Classical and innovative aspects. In Marcotte, P., Nguyen, S., eds.: Equilibrium and Advanced Transportation Modelling. Kluwer, Amsterdam (1998) 245–281
20. Cormen, T., Leiserson, C., Rivest, R., Stein, C.: Introduction to Algorithms. Second edn. McGraw-Hill (2001)
21. Luger, G., Stubblefield, W.: Artificial Intelligence: Structures and strategies for complex problem solving. Third edn. Addison-Wesley, Reading, MA (1998)
22. Henzinger, M.R., Klein, P., Rao, S., Subramanian, S.: Faster shortest-path algorithms for planar graphs. Journal of Computer and System Sciences **55** (1997) 3–23
23. Denis, M.: The description of routes: A cognitive approach to the production of spatial discourse. Cahiers de Psychologie Cognitive **16** (1997) 409–458
24. Kuipers, B.: Representing Knowledge of Large-Scale Space. PhD thesis, Mathematics Department, Massachusetts Institute of Technology (1977) Technical Report 418, M.I.T. Artificial Intelligence Laboratory.
25. Kuipers, B.: Modelling spatial knowledge. Cognitive Science **2** (1978) 129–153
26. Chown, E., Kaplan, S., Kortenkamp, D.: Prototypes, location and associative networks (plan): Towards a unified theory of cognitive mapping. Journal of Cognitive Science **19** (1995)

A Classification Framework for Approaches to Achieving Semantic Interoperability between GI Web Services

Michael Lutz, Catharina Riedemann, and Florian Probst

Institute for Geoinformatics, University of Münster, Robert-Koch-Str. 26-28,
48149 Münster, Germany
{lutzm,riedemann,probst}@ifgi.uni-muenster.de

Abstract. The discovery of services that are appropriate for answering a given question is a crucial task in the open and distributed environment of web services for geographic information. In order to find these services the concepts underlying their implementation have to be matched against the requirements resulting from the question. It is in this matchmaking process where semantic heterogeneity has to be tackled. Whether semantic interoperability can be achieved depends on the quality of the information available to the matchmaker on the semantics of requirements and resources. The explicitness, structuring and formality of this information can differ considerably leading to different types of matchmaking. In this paper a framework is presented for classifying the approaches that are currently employed or proposed for achieving semantic interoperability according to these criteria. The application of the framework is illustrated by analyzing possible solutions to three examples of semantic interoperability problems.

1   Introduction

Geographic information science is currently characterized by a paradigm shift – from providing theories for monolithic systems to theories for open and distributed GIS and their use processes. With this comes a move from standardized data formats to specifications of geographic information (GI) service interfaces [1, 2]. In practice, the number of GI services available on the web is rapidly and continually increasing. Semantic interoperability is a core problem in such an open and distributed environment [3].

In the description of the OpenGIS service architecture [4] the syntactic and semantic aspects of interoperability are defined as follows: "Syntactical interoperability assures that there is a technical connection, i.e., that the data can be transferred between systems. Semantic interoperability assures that the content is understood in the same way in both systems, including by those humans interacting with the systems in a given context."

In the open and distributed environment of GI web services, the components that are to interoperate are not previously known. The starting point is a requester's spe-

W. Kuhn, M.F. Worboys, and S. Timpf (Eds.): COSIT 2003, LNCS 2825, pp. 186–203, 2003.
© Springer-Verlag Berlin Heidelberg 2003

cific question rather than a given system. Discovering the services[1] that are appropriate for answering this question from among a large number of available services is a central task within the GI web services domain [5]. Service discovery will therefore be the focus of this paper.

In order to find an appropriate service the requirements resulting from the requester's question have to be matched against descriptions of the the the available service implementations. It is in this matchmaking process that semantic interoperability is ensured, making it a crucial part of service discovery (Fig. 1).

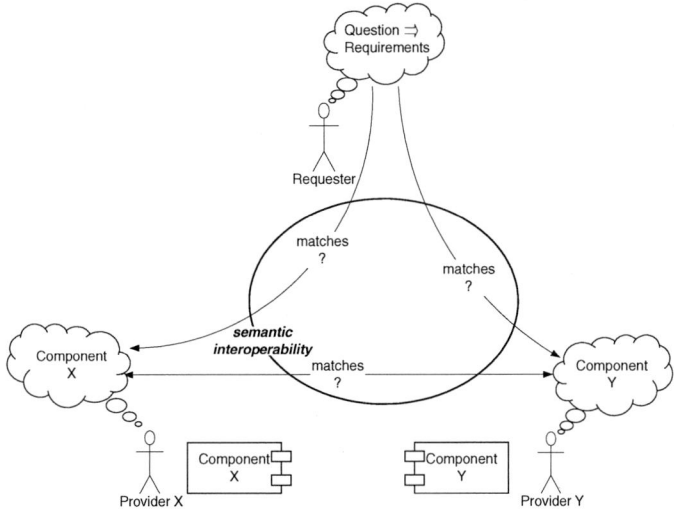

Fig. 1. Semantic interoperability in a GI web services scenario

A large number of languages and technologies have been proposed for service discovery, e.g. for web services in general [6-8], for services and data in the geospatial domain [5, 9, 10], or for software agents and the Semantic Web [11-13][2]. Whether semantic interoperability during service discovery can be achieved in any of these approaches depends on the quality of the information available to the matchmaker on the semantics of requirements and resources. The explicitness, structuring and formality of this information can differ considerably leading to different forms of matchmaking. We are not aware of any framework for classifying the plethora of existing approaches for service discovery with respect to achieving semantic interoperability. Therefore, we propose such a framework in this paper.

The remainder of the paper is structured as follows. In the next section we present several examples of practical problems caused by semantic heterogeneity. The framework for classifying approaches for overcoming semantic heterogeneity is developed in section 0 and applied to the practical problems in section 0. We conclude

---

[1] The notion of *service* in this paper includes both services that can be used to operate on multiple, unspecified datasets (loosely-coupled services) and services that are associated with a specific dataset (tightly-coupled services) [27].

[2] We assume that the reader is familiar with these approaches. Their strengths and weaknesses are outside the scope of this paper and will therefore not be discussed.

the paper by discussing how the results can be applied in the more complex task of (automatic) service composition and by pointing out the next steps for semantic interoperability research along these lines.

2   Examples of Semantic Interoperability Problems

This section presents three examples for problems caused by semantic heterogeneity that we have encountered in our research. They occur in monolithic, at most partially component-based, GIS environments. Nevertheless, they are equally valid for a web service environment.

2.1   Classification of Semantic Heterogeneity

Semantic heterogeneity, the source of semantic interoperability problems, is defined in [14] as the consequence of different conceptualizations and database representations of a real world fact. Two types can be distinguished. *Cognitive heterogeneity* arises when two disciplines have different conceptualizations of real world facts. This becomes a semantic problem when the same names are used for different concepts in both disciplines. Such word pairs are referred to as homonyms. *Naming heterogeneity* refers to different names for identical concepts of real world facts, also called synonyms. The examples subsequently described are classified according to this distinction in order to make sure that both types of heterogeneity are covered.

2.2   Using Topographic Data for Noise Abatement Planning

Situation. To determine which roads could have a considerable noise effect on residential areas those roads touching or crossing residential areas must be identified. German topographic data (Amtliches Topographisch-Kartographisches Informationssystem, ATKIS) contain residential areas and roads as feature classes [15]. Some roads are modeled as lines as shown in Fig. 2 (b).

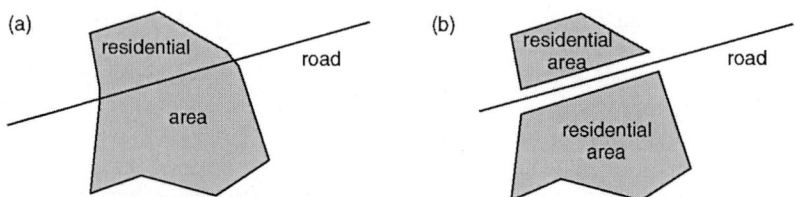

Fig. 2. Different models of roads crossing residential areas

Problem. A user might have the mental concepts of roads and residential areas as depicted in Fig. 2 (a). The system model instead uses representations of roads and residential areas as depicted in Fig. 2 (b). If the user is not aware of the system model (that the terms "residential area" and "road" do not reveal) he might assume that roads overlap residential areas as indicated in Fig. 2 (a) and consequently use the dataset as input for an intersect operation in order to find roads crossing residential areas. However, based on a system model as depicted in Fig. 2 (b) he will not find any roads by doing so, which is correct for the data model of the dataset, but does not meet the user's expectations.

Heterogeneity Type. This example depicts cognitive heterogeneity concerning residential areas; the concepts of user and system regarding the geometric representation are different. The difference is hidden by the homonym "residential area".

2.3   Calculating the Area of Greenland in a Mercator Projection

Situation. In the Mercator map projection features on the reference ellipsoid are projected onto a cylinder touching the equator. This leads to increasing distortion towards the poles and does not preserve areas (see Fig. 3 (b)). For all tasks requiring real world area values, it is not appropriate to calculate the area of features in polar regions, like Greenland, directly from the Mercator projection cylinder (see Fig. 3 (a)).

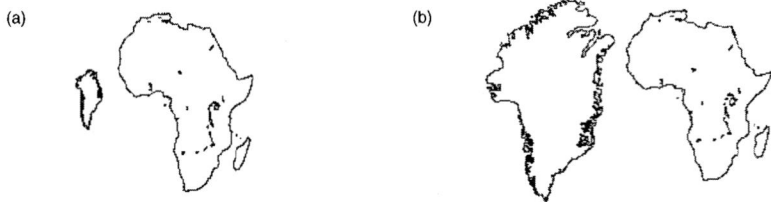

Fig. 3. Greenland and Africa in (a) equal-area Mollweide projection and (b) non-equal-area Mercator projection (images taken from [16])

Problem. Most GIS do not inform users during execution how areas of features are calculated and whether the results reflect the real world area of that feature. The user may expect an area calculation to return the real world area. Such an area calculation would be based on the feature's geometry on the reference ellipsoid. However, if the system's concept of area calculation is based on the feature's geometry on the projection cylinder, the operation will return a completely different result. If the user is not aware of the different concepts of area calculation he will misinterpret the results.

Heterogeneity Type. This example depicts cognitive heterogeneity within the concept of area calculation.

2.4   Topological Operators in GeoMedia and Oracle

This example consists of two parts. First, it describes two operators with the same name and different behavior. Then it describes two operators with different names and equivalent behavior.

Situation. GeoMedia[3] provides a set of topological operators. In addition, it integrates topological operators of the Oracle[4] database system. We look at two GeoMedia operators, called "touch" and "meet", and compare them to an Oracle operator called "touch".

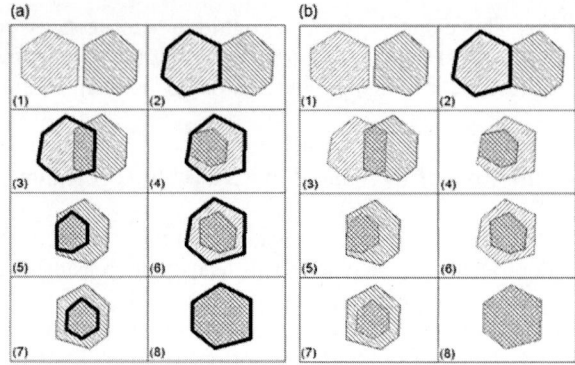

Fig. 4. Regions found (*each marked with a thick line*) by (a) GeoMedia "touch" operator and by (b) Oracle "touch" operator

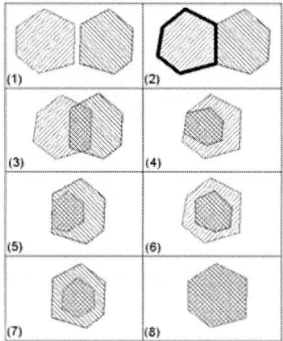

Fig. 5. Region found (*marked with a thick line*) by GeoMedia "meet" operator as well as by Oracle "touch" operator

Problem. We have identified two problems in this example:
1. Although the names are identical, the two "touch" operators of GeoMedia and Oracle return different results (Fig. 4).
2. The GeoMedia "meet" operator and the Oracle "touch" operator, however, find the same regions (Fig. 5 (2)) although they are named differently.

Thus, the names are confusing and misleading, and consequently not useful to the user for deciding if an operation does what he expects.

---

[3] GeoMedia Professional (Intergraph Corp.) V5.0
[4] Oracle 9*i* Release 2 Spatial (Oracle Corp.)

Heterogeneity Type. The first problem is caused by cognitive heterogeneity with the homonym "touch". The second problem demonstrates naming heterogeneity with the synonyms "meet" and "touch".

3   Analysis Framework

There are many approaches to ensuring semantic interoperability in the examples presented in the previous section, e.g. [13, 17, 18]. In this section we present a framework for classifying and analyzing such approaches. We define the term *matchmaking* and present different types of matchmaking. We proceed to differentiate several levels of explicitness, structuring and formality for the information required by the matchmaker.

3.1   Matchmaking

In the literature on agent systems matchmaking is defined as mediating among re-questers and providers of services for some mutually beneficial cooperation [13]. The process of finding an appropriate service for a certain task can be regarded as match-making, too. During matchmaking it is assessed whether (or how well) an available service fits the requirements of the requester.

We distinguish different roles that are played by human actors or system compo-nents that exist in the domain of GI web services:

- the *requester* role, which is (ultimately) always played by a human (end user or web service provider),
- the *provider* role, which is also played by a human (web service provider), and
- the *matchmaker* role, which can be played by either a human (one of the above or an independent *broker*) or a matchmaking service.

Note that the same person can take different roles. For example, the person in the requester or provider role can also be responsible for the matchmaking process. As either human or computer can do the matchmaking, two kinds of matchmaking can be distinguished, which represent two endpoints of a continuum:

- *Purely manual matchmaking.* Manual matchmaking is done by a human actor and occurs in the mind of the matchmaker. The matchmaker decides whether or not some service fits the requester's requirements based on information that is avail-able to him about the service. Manual matchmaking is prone to misunderstandings caused by synonyms and homonyms (section 2.1). In order to mitigate this prob-lem, additional information is collected to reduce ambiguity.
- *Fully automatic matchmaking.* In contrast to manual matchmaking fully automatic matchmaking is always done by a service. This requires *formal* descriptions of re-quirements and service capabilities. These are matched automatically using an al-gorithm such as described in [13, 19].

In cases where some of the required formal descriptions are missing, the existing informal descriptions have to be formalized for automatic matchmaking to be applied. Alternatively, the formal descriptions can be made informal and manual matchmaking can be applied. Informalization becomes necessary because formal descriptions are usually difficult to read for non-experts. It should be noted that automatic matchmaking, too, could lead to results unexpected by the requester. This can either be due to explication or formalization errors (i.e. inappropriate capabilities or requirements descriptions) or inappropriate parameterization of the matchmaking algorithms.

3.2   Levels of Explicitness, Structuring and Formality

The quality of the information (metadata) on requirements and service capabilities that is available to the matchmaker is crucial for the matchmaking task. Which information on requirements and on the service has to be made explicit to the matchmaker depends on who does the matchmaking:

− If the *requester does the matchmaking* the requirements are already available in the matchmaker's mind. Therefore, they do not have to be formalized or even made explicit. However, making the requirements explicit and thus reducing ambiguity can help avoiding misinterpretation.
− If the *provider does the matchmaking* the service capabilities are already available in the matchmaker's mind. Therefore, they do not have to be formalized or even made explicit. However, making the capabilities explicit can help to clarify them and discover inconsistencies.
− If an *independent broker does the matchmaking* both requirements and service capabilities have to be made explicit to the matchmaker. A (possibly standardized) structure and formalization might help the broker to do the matchmaking.

The quality of the metadata can vary along three dimensions:

− *Explicitness of information.* The information can be implicit, i.e. only in someone's mind, or explicit, i.e. written down in some language. It is also important to note how complete the available information is, i.e. whether all the information that is required by the matchmaker is available.
− *Structuring of information.* The structure of the information can be implicit and thus unobservable or explicit or even standardized. We refer to the former as unstructured and to the latter as structured information. There are, of course, different levels of structuring [20].
− *Formality of semantics.* The semantics of the concepts used to describe the service can be expressed in ontologies, which are defined as explicit specifications of conceptualizations [21]. A conceptualization is a set of concepts, their definitions and interrelationships [22]. Ontologies can be expressed both informally and formally, i.e using natural or formal languages. There are also intermediate levels of formality [20].

The classification framework could simply consist of these dimensions. However, they are not independent of each other, e.g. the structuring or formality dimensions do not matter if this information is not explicit. Therefore, we propose five levels of explicitness, structuring and formality.

A. *Completely implicit semantics.* The information exists only in the mind of the provider, requester or matchmaker.
B. *Implicit semantics.* Only names (e.g. „forest data", „web mapping service") but no metadata are made explicit to refer to services or requirements.
C. *Explicit, unstructured, informal semantics.* Metadata are made explicitly available, but in an unstructured form using natural language text.

D. *Explicit, structured, informal semantics.* Metadata are made explicitly available in a structured form, e.g. referring to metadata standards such as ISO 19115 [23] that – usually informally – specify metadata fields and their semantics. However, with the exception of value lists being specified for some fields, the content of the metadata fields is to be given in the form of free natural language text.

E. *Explicit, formal semantics.* The information is made explicitly available referring to formal ontologies.

These categories are somewhat arbitrary as all three dimensions are continuous. However, we think they represent typical examples for approaches to achieve semantic interoperability. This is illustrated by the scenarios presented in the following section.

3.3   Matchmaking Scenarios

In order to illustrate the levels of explicitness, structuring and formality described in the previous section, three scenarios are depicted. In all of them the requester wants to know the location of forest parcels in the German federal state Northrhine-Westfalia (NRW).

The scenarios represent typical approaches to achieving semantic interoperability at three stages of development. The first scenario shows what is possible and widely practiced by users of the World Wide Web today (levels B and C[5]). The second scenario describes the research and industry attempts made in the GI community (level D), most notably in the OpenGIS Consortium (*http://www.opengis.org*) and the ISO Technical Committee 211 (*http://www.isotc211.org*). The last scenario presents ideas that are currently discussed in the Semantic Web and agent systems communities (level E).

Note that the role labeled *requester* in the following figures could either represent an end user who wants an appropriate service to answer his question, or a service provider who wants to find appropriate services to build a complex service that performs a specific task. The actor or component responsible for the matchmaking is highlighted in gray.

Scenario 1 – Manual Matchmaking Based on Names or Unstructured and Informal Information. In the first scenario (Fig. 6) the capabilities of the services are not made explicit by their providers. The only clues for the requester to what the services are doing or which data they provide are their names. One means for finding appropriate services by their name is through a keyword search in an Internet search engine like Google. In such a search the requester can encounter the following problems:

– *No match.* Services that fit the requester's requirements are not found at all, because their names do not match the keywords included in the requester's query. The simplest reason for this are spelling differences or mistakes. Leaving these aside, the problem can be classified as a case of naming heterogeneity (section 2.1): The conceptualizations of requester and provider are sufficiently similar for the task at hand but concepts are given different names (synonyms). This can have

---

[5] Level A is not considered because service discovery becomes extremely difficult or even impossible when the semantics of requirements or capabilities are completely implicit.

several reasons. Either the name of the service or the keywords used in the query are not appropriate, i.e. they do not well reflect the service capabilities or the requester requirements, respectively. Or both keywords and names are appropriate (within their respective domain), but requester and providers belong to different information communities.

- *Unsuitable match.* Services that are found because their name matches the keywords included in the requester's query do not fit the requester's requirements. The conceptualizations of requester and provider are different but are given the same names (homonyms). This case can be classified as cognitive heterogeneity leading to naming conflicts (section 2.1). The possible reasons for this can again be inappropriate names or differing information communities.

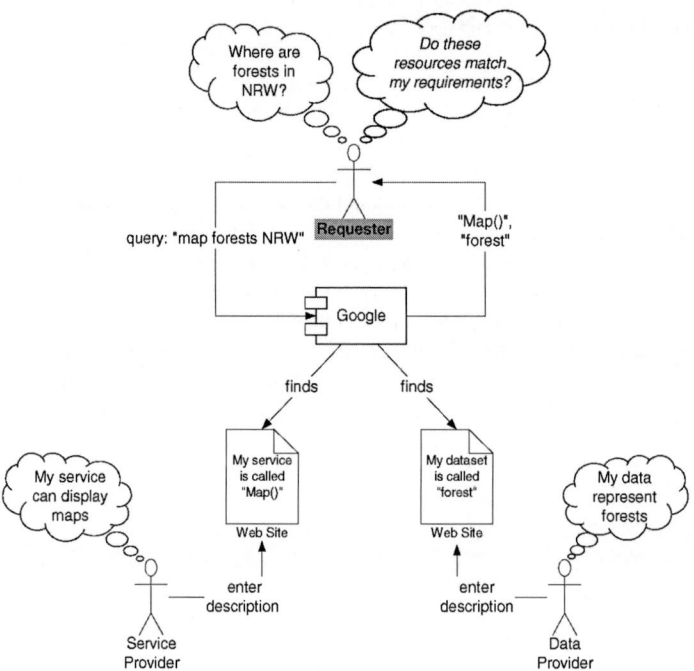

Fig. 6. Manual matchmaking based on names (level B) or unstructured and informal information (level C)

An explicit requirements specification can help the requester to do the matchmaking, because the process of explication often helps to clarify and disambiguate ideas on requirements. Also explicitly describing the service's capabilities rather than only giving a name can improve the matchmaking by reducing guesswork. This is the case of explicit, but unstructured and informal description of semantics. However, misinterpretation is still possible if the descriptions are ambiguous or incomplete. These two cases are currently the most common ones as service descriptions are either informal or missing completely.

Scenario 2 – Manual Matchmaking with Standardized Metadata. In the second scenario (Fig. 7) the providers' conceptualizations are made explicit and are recorded in metadata documents whose structure is well known and which are made available through one (or several) registries. The requester can search a registry using keywords for all of the fields provided by its query interface. He can then use the returned metadata documents to assess whether or not the services fit his requirements. He might need to access other documents that the metadata documents refer to, e.g. a feature type catalog providing definitions for feature classes or ISO standards defining units of measurement.

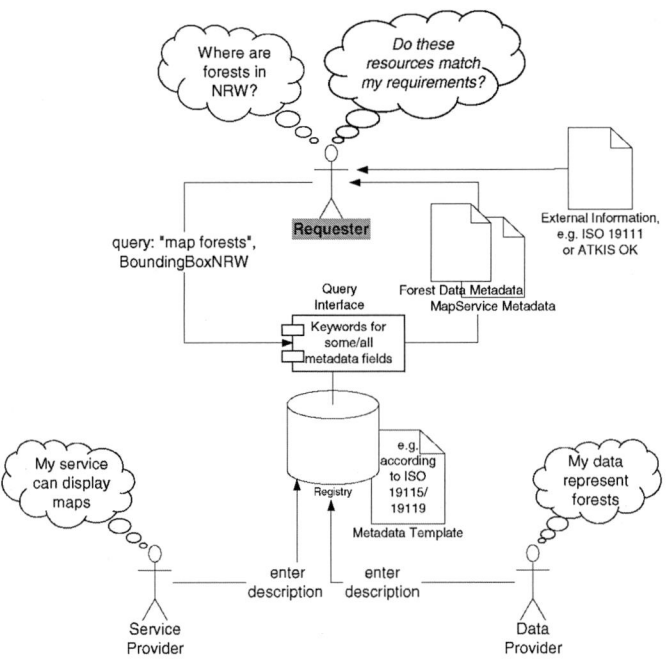

Fig. 7. Manual matchmaking with standardized metadata, e.g. ISO 19115/19119 (level D)

As the matchmaking in this scenario is still based on keywords, the problems described for the first scenario can still occur. There can be ambiguity in either the metadata entries themselves or in the referenced documents (e.g. the feature catalogue). However, this can be considerably reduced by using standardized documents, by providing a controlled vocabulary (e.g. lists of keywords), and by referring to other standardized or at least widely known and agreed-upon documents.

Scenario 3 – Automatic Matchmaking with Formal Metadata. In the last scenario (Fig. 8) the conceptualizations of requester and providers are not only explicit but also formalized. They use concepts from existing *domain ontologies* [24] to formulate their requirements or advertisements, respectively. A service automatically matches the requester's requirements against advertisements stored in its registry using a matchmaking algorithm such as described in [13].

It is assumed that by using formal descriptions of semantics and automatic match-making algorithms problems such as those described in the previous scenarios can be avoided [12, 25]. However, in this scenario, too, problems similar as those identified in the previous scenarios, albeit for different reasons, can occur.

– *No match.* Services that fit the requester's requirements are not found at all because the matchmaking algorithm is too rigorous. In [13] a threshold value has to be specified by the requester indicating which degree of similarity between adver-tisements and requirements is still acceptable.

– *Unsuitable match.* Services that are found do not fit the requester's requirements. This, too, can be caused by the calibration of the matchmaking algorithm. Here, the matchmaking algorithm is too tolerant because the threshold value is too low. An-other possible reason is that the requirements document does not correctly reflect the requester's requirements or the capabilities documents do not correctly reflect the providers' conceptualization of the service. We refer to these kinds of errors as *explication* or *formalization errors*, respectively.

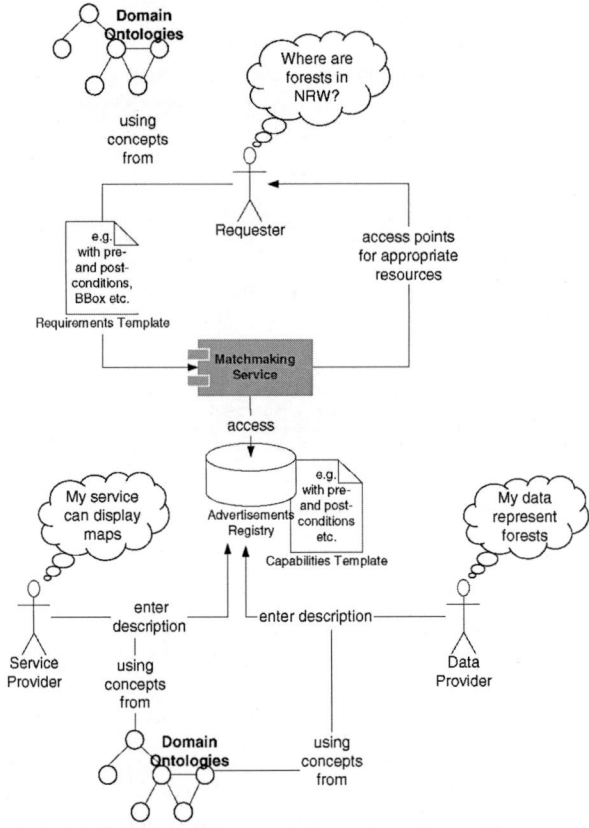

Fig. 8. Automatic matchmaking with formal metadata (level E)

3.4   Likelihood of Misunderstanding

Misunderstandings can occur in all matchmaking scenarios described in the previous sections. Summarizing the arguments from the previous section, Table 1 gives an estimate of the likelihood of misunderstandings for all possible combinations of explicitness, structuring and formality levels described above. It is assumed that the requester does the matchmaking.

Table 1. Likelihood of misunderstandings for different levels of requirements and capabilities descriptions if the requester does the matchmaking. Shading: white – manual matchmaking possible, light gray – manual matchmaking possible but difficult for non-experts, dark gray – automatic matchmaking possible. The scenarios described in the previous section are framed

requirements service capabilities	(completely) implicit	explicit, unstructured, informal semantics	explicit, structured, informal semantics	explicit, formal semantics
completely implicit	matchmaking impossible			
implicit	**I** highly likely	highly likely	highly likely	highly likely
explicit, unstructured, informal semantics	highly likely	likely	likely	likely
explicit, structured, informal semantics	likely	possible	**II** possible / possible (automatic matchmaking limited)	possible / possible (automatic matchmaking limited)
explicit, formal semantics	likely	possible	possible / possible (autom. matchm. limited)	unlikely / **III** unlikely

4   Analysis of Examples

After having presented the framework for classifying matchmaking approaches we show in this section how it can be applied to the examples presented in section 0. In the following tables the first row lists the information required by the matchmaker in order to find resources appropriate for answering the requester's question. The names of the concepts appear in italics. The remaining rows contain an analysis of the availability, quality and source of the information in each of the three scenarios presented in section 0.

4.1   Using Topographic Data for Noise Abatement Planning

This example depicts the requester's attempt to intersect residential areas with roads. This involves a matchmaking process for which information about the requester's conceptualization of *road*, *residential area* and the operators *touch* and *cross* as well as information about the ATKIS geometry model are needed (Table 2).

Table 2. Application of the classification framework to example 1 – Using topographic data for noise abatement planning. (The table is split into two for enhanced readability)

		information required by the matchmaker	
		requester conceptualization of *road* and *residential area*	requester conceptualization of *touch* and *cross*
scenario 1	available	✓	✓
	level	implicit	implicit
	source	requester's mind	requester's mind
scenario 2	available	✓	✓
	level	implicit	implicit
	source	requester's mind	requester's mind
scenario 3	available	✓	✓
	level	explicit, formal	explicit, formal
	source	domain ontology chosen by the requester to describe his task	domain ontology chosen by the requester to describe his task

		information required by the matchmaker		
		ATKIS geometry model for *road* and *residential area*		process model for geoprocessing operations, e.g. *intersect* or *buffer*
scenario 1	available	–	✓	✓
	level	n.a.	implicit	implicit
	source	n.a.	requester's mind (if he is an ATKIS expert) or dataset (accessible via visualization of data, requires GIS expertise)	requester's mind (if he is an expert of the specific GIS) or trial and error (requires GIS expertise)
scenario 2	available	✓		✓
	level	explicit, structured, informal		explicit, structured, informal
	source	The ISO metadata standard supports references to external feature type catalogs like that of ATKIS as well as graphic overviews [23].		The ISO services standard provides a template for describing services [4]. Alternatives are UDDI [7], WSDL [8], Capabilities XML [26]. They focus on operation signatures; descriptions are available only on service level and appear as free text. ISO in addition provides free text descriptions on the operation level. The ISO spatial schema standard provides information for filling such a template [27]. They consist of free text descriptions and formalized operation signatures.
scenario 3	available	✓		✓
	level	explicit, formal		explicit, formal
	source	domain ontology based on ATKIS feature type catalog [15]		(geo)processing domain ontology, e.g. based on ISO spatial schema standard [27]

In scenario 1 the intersection attempt will only be successful if the requester is an expert who is aware of how the ATKIS geometry model will fit his requirements. In scenario 2 the intersection attempt will be successful if the requester is willing to spend the time to access and understand the available metadata and perform the matchmaking manually. In scenario 3 the intersection attempt will be valid even if the requester is no ATKIS expert, because the information needed for the matchmaking is available in formal and explicit form, making automatic matchmaking possible. The result of the matchmaking process may be that the intersection is not possible because the mapping from system to requester concepts would require additional services that are not available. Nevertheless, even in this case the requester is saved from misinterpreting the results of the intersection.

4.2   Calculating the Area of Greenland in a Mercator Projection

This example depicts the requester's attempt to calculate the real world area of Greenland displayed with a GIS using the Mercator projection. This involves a matchmaking process for which information about the requester's conceptualization of area calculation, the system model of area calculation and indirectly information about the attributes of Mercator projections is needed (Table 3).

In scenario 1 the area calculation attempt is likely to lead to misinterpretation as long as the requester is no GI expert. In scenario 2 the area calculation attempt is likely to be canceled. If the requester is willing to spend the time to access and understand the available metadata, he becomes aware of that the calculated area will not meet his requirements of representing the real world area. However, in scenario 2 no further solution is offered. In contrast, in scenario 3 the area calculation attempt may be successful, because all information needed for the matchmaking process is available in formal and explicit form. The requester is made aware of that his requirements differ from the system's abilities. It might be possible to search for a service that is able to calculate the area according to the requester's requirements. In this scenario the requester does not need any knowledge about projections and area calculation operations.

4.3   Topological Operators in GeoMedia and Oracle

This example depicts the requester's attempt to find operations that return geometry features whose boundaries intersect but whose interiors do not. To find the appropriate operations the requester's requirements have to be matched with the systems' capabilities. For the matchmaking process information about the requester's conceptualization of *touch* is needed as well as the process models of the available operations of the systems, in this case GeoMedia and Oracle (Table 4).

In scenario1 the attempt to find the appropriate operation is likely to lead to misinterpretations if the requester is no system expert. In scenario 2 the attempt may be successful if the requester is willing to spend the time to access and understand the available metadata. He then will learn about the meaning of the different operations and will be able to perform a manual matchmaking. In scenario 3 the matchmaking will be successful. Using terms from a domain ontology the requester can specify his requirements formally and explicitly. Based on this formal and explicit specification the appropriate operations can be chosen automatically from among the available operations.

Table 3. Application of the classification framework to example 2 – Calculating the area of Greenland in a Mercator projection

<table>
<tr><td colspan="2" rowspan="2"></td><td colspan="6">information required by the matchmaker</td></tr>
<tr><td colspan="2">requester con-ceptualization of area calculation</td><td colspan="2">system model of area calculation (possibly including attributes of the projection, see next row)</td><td colspan="2">attributes of Mercator projection</td></tr>
<tr><td rowspan="3">scenario 1</td><td>available</td><td>✓</td><td>–</td><td>✓</td><td>–</td><td>✓</td></tr>
<tr><td>level</td><td>implicit</td><td>n.a.</td><td>implicit</td><td>n.a.</td><td>implicit</td></tr>
<tr><td>source</td><td>requester's mind</td><td>n.a.</td><td>requester's mind (if he is a GI expert)</td><td>n.a.</td><td>requester's mind (if he is a GI expert)</td></tr>
<tr><td rowspan="3">scenario 2</td><td>available</td><td>✓</td><td colspan="2">✓</td><td colspan="2">✓</td></tr>
<tr><td>level</td><td>implicit</td><td colspan="2">explicit, structured, informal</td><td colspan="2">explicit, structured, informal</td></tr>
<tr><td>source</td><td>requester's mind</td><td colspan="2">The operation signatures can be described in the same way as for the intersect and buffer operations in Table 2.
The ISO metadata standard [23] provides attributes for operations which can be applied to the dataset. However, the requester has to judge whether the results (e.g. area calculation) fit his expectations. (see same column next row).</td><td colspan="2">The ISO standard for spatial referencing by coordinates provides a free text description indicating for which application a coordinate reference system is valid [28].</td></tr>
<tr><td rowspan="3">scenario 3</td><td>available</td><td>✓</td><td colspan="2">✓</td><td colspan="2">✓</td></tr>
<tr><td>level</td><td>explicit, formal</td><td colspan="2">explicit, formal</td><td colspan="2">explicit, formal</td></tr>
<tr><td>source</td><td>domain ontology chosen by the requester to describe his task</td><td colspan="2">(geo)processing domain ontology, e.g. based on ISO spatial schema standard [27]</td><td colspan="2">domain ontology for projections, e.g. based on ISO standard for spatial referencing by coordinates [28]</td></tr>
</table>

5   Conclusions and Future Work

We have presented a framework for classifying approaches to achieving semantic interoperability in the domain of GI web services. The framework focuses on the process of matchmaking as this is where semantic interoperability is ensured. Therefore approaches to achieving semantic interoperability are classified according to the quality of the information that is available to the matchmaker.

The application of the framework has been illustrated by analyzing existing approaches to solving examples of semantic interoperability problems. In scenario 1 misinterpretations are likely to occur unless the requester is an expert for the components employed. In scenario 2 misinterpretations are less likely if the requester is willing to spend the time to access and understand the available metadata. In scenario 3 misinterpretations are unlikely, even for non-experts, as automatic matchmaking is applied. However, there is still the possibility that the services required for the requester's query are not available.

Table 4. Application of the classification framework to example 3 – Topological operators in GeoMedia and Oracle

		information required by the matchmaker		
		requester conceptualization of *touch*	process models for *touch* operations (GeoMedia and Oracle) and *meet* operation *(GeoMedia)*	
scenario 1	available	✓	–	✓
	level	implicit	n.a.	implicit
	source	requester's mind	n.a.	requester's mind (if he is an expert of the specific GIS) or trial and error (requires GIS expertise)
scenario 2	available	✓	✓	
	level	implicit	explicit, structured, informal	
	source	requester's mind	The ISO services standard provides a template for describing services [4]. The ISO spatial schema standard provides information for filling such a template [27]. See also intersect and buffer operations in Table 2.	
scenario 3	available	✓	✓	
	level	explicit, formal	explicit, formal	
	source	domain ontology chosen by the requester to describe his task	(geo)processing domain ontology, e.g. based on ISO spatial schema standard [27]	

The analysis of practical problems only presents a first application of the framework. We believe the framework to be valuable to the GI research community for structuring the domain of semantic interoperability research, because it supports the following tasks:
– The information required for the matchmaking process can be identified.
– The required information can be classified according to the qualities explicitness, structuring and formality.
– It can be assessed which quality level of the required information is appropriate for the task at hand.
– The different levels of explicitness, structuring and formality can easily be associated to predefined scenarios that indicate possible implementation methods .
– In the combination of the above reasons, researchera can classify their approach and judge whether the applied methods are appropriate for the task at hand.
Future work must look at the role that service discovery plays within the larger task of service composition. It will also be examined whether other sub-tasks play a role in ensuring semantic interoperability in (especially ad-hoc) service composition. For this an abstract model of service composition should be developed. Such a model could be

valuable for the standardization efforts in OGC and ISO TC 211, where the task of service composition has not yet been thoroughly explored.

It also remains an open question whether examples like those presented in this paper represent a specific (i.e. spatial) kind of semantic heterogeneity or whether they can be treated in the same way as other (non-spatial) semantic problems. If the latter turns out to be possible the framework should be adjusted accordingly.

Acknowledgements

Comments from Werner Kuhn to earlier drafts of this paper helped clarify the ideas. The work presented in this paper has been partially supported by the German Ministry for Education and Science as part of the GEOTECHNOLOGIEN program (grant number 03F0369A) and can be referenced as publication no. GEOTECH-23. Furthermore, support from the European Commission through the ACE-GIS (grant number IST-2002-37724) and BRIDGE-IT (grant number IST-2001-34386) projects are gratefully acknowledged.

References

1. Abel, D. J., Gaede, V. J., Taylor, K. L., Zhou, X.: SMART: Towards Spatial Internet Marketplaces. Geoinformatica 3 (1999) 141-164
2. Groot, R., McLaughlin, J.: Geospatial data infrastructure – Concepts, cases, and good practice. Oxford University Press (2000)
3. OGC: OpenGIS Web Services Architecture. OpenGIS Consortium, OpenGIS Discussion Paper OGC 03-025 (2003)
4. ISO/TC-211, OGC: Geogaphic information – Services (ISO/DIS 19119) v4.3. International Organization for Standardization & OpenGIS Consortium (2002)
5. Egenhofer, M.: Toward the Semantic Geospatial Web. In: Proc. The 10th ACM International Symposium on Advances in Geographic Information Systems (ACM-GIS) (2002)
6. OASIS: OASIS/ebXML Registry Services Specification v2.5. OASIS/ebXML Registry Technical Committee (2003)
7. Bellwood, T., Clément, L., Ehnebuske, D., Hately, A., Hondo, M., Husband, Y. L., Januszewski, K., Lee, S., McKee, B., Munter, J., von Riegen, C.: UDDI v 3.0. (2002)
8. Chinnici, R., Gudgin, M., Moreau, J.-J., Weerawarana, S.: Web Services Description Language (WSDL) v1.2. (2002)
9. Reed, C., Nebert, D.: The Importance of Catalogs to the Spatial Web. An OGC White Paper. (2002)
10. OGC: OWS1.2 UDDI Experiment. OpenGIS Consortium, OGC 03-028 (2003)
11. Constantinescu, I., Faltings, B.: Efficient Matchmaking and Directory Services. Swiss Federal Institute of Technology, Techn. Report IC/2002/77 Lausanne, Switzerland (2002)
12. Paolucci, M., Kawamura, T., Payne, T. R., Sycara, K.: Semantic Matching of Web Service Capabilities. In: Proc. 1st International Semantic Web Conference (ISWC2002) (2002) 333-347
13. Sycara, K., Widoff, S., Klusch, M., Lu, J.: Larks: Dynamic Matchmaking Among Heterogeneous Software Agents in Cyberspace. In: Proc. First International Joint Conference on Autonomous Agents and Multi-Agent Systems (2002) 173-203
14. Bishr, Y.: Overcoming the semantic and other barriers to GIS interoperability. International Journal of Geographical Information Science 12 (1998) 299-314

15. AdV-Arbeitsgruppe ATKIS: ATKIS-Objektartenkatalog Basis-DLM. (2002)
16. Furuti, C. A. Useful Map Properties.[Online]. Available: http://www.progonos.com/furuti/MapProj/Normal/CartProp/
17. Visser, U., Stuckenschmidt, H.: Interoperability in GIS - Enabling Technologies. In: Proc. 5th AGILE Conference on Geographic Information Science (2002) 291-297
18. Kuhn, W., Raubal, M.: Implementing Semantic Reference Systems. In: Proc. 6th AGILE Conference on Geographic Information Science (2003)
19. Sycara, K., Klusch, M., Widoff, S., Lu, J.: Dynamic Service Matchmaking Among Agents in Open Information Environments. ACM SIGMOD Record 28 (1999) 47-53
20. Uschold, M.: Knowledge level modelling: concepts and terminology. The Knowledge Engineering Review 13 (1998) 5-29
21. Gruber, T. R.: Toward Principles for the Design of Ontologies Used for Knowledge Sharing. International Journal of Human-Computer Studies 43 (1995) 907-928
22. Uschold, M., Gruninger, M.: Ontologies: Principles, Methods and Applications. The Knowledge Engineering Review 11 (1996) 93-136
23. ISO/TC-211: Geogaphic information – Metadata (ISO/FDIS 19115). International Organization for Standardization (2003)
24. Guarino, N.: Formal Ontology and Information Systems. In: Proc. Formal Ontology in Information Systems (FOIS'98) (1998) 3-15
25. Guarino, N.: Semantic Matching: Formal Ontological Distinctions for Information Organization, Extraction, and Integration. In: Proc. Information Extraction: A Multidisciplinary Approach to an Emerging Information Technology, International Summer School (SCIE-97) (1997) 139-170
26. de La Beaujardière, J.: Web Map Service Implementation Specification. Open GIS Consortium (2002) 82
27. ISO/TC-211: Geographic information – Spatial Schema (ISO/DIS 19107). International Organization for Standardization (2002)
28. ISO/TC-211: Text for FDIS 19111 Geogaphic information - Spatial Referencing by Coordinates. Final Draft Version. International Organization for Standardization, (2002)

Relative Adjacencies in Spatial Pseudo-Partitions

Roderic Béra and Christophe Claramunt

Naval Academy Research Institute, Lanvéoc-Poulmic
BP 600, 29240 Brest Naval, France
{bera,claramunt}@ecole-navale.fr

Abstract. This paper introduces a relative adjacency operator that characterises mutual relationships between regions in a pseudo-partition. The relative adjacency is computerised from the dual graph of a spatial pseudo-partition. It is flexible enough to reflect different degrees and clusters of relative adjacencies by minimising or maximising the effect of neighbouring and remote regions. The properties of the relative adjacency are illustrated by some canonical examples and a case study.

Keywords: Spatial reasoning, relative adjacency, distance, graph analysis

1 Introduction

Qualitative spatial reasoning has been long recognised as a valid support for inferring relationships in geographical spaces. In particular, it has been observed that qualitative inferences are cognitively more expressive than quantitative ones, and that they also support reasoning mechanisms in the absence of complete spatial knowledge [1]. In geographical space, reasoning on spatial entities is supported by representations that mainly involve topological ([2], [3], [4], [5], [6]) and direction relationships ([7], [8], [9], [10], [11]). Those spatial relationships provide useful mechanisms to evaluate the mutual relationships of regions over space.

One of the most effective data structures applied to model geographical spaces is based on a partition of space that forms a categorical coverage derived from a thematic classification ([12], [13]). Categorical coverages have been extensively implemented in GIS systems due to their intuitive character and numerous computational advantages. Spatial partitions have been recently studied formally in [14]. They apply to large-scale spaces, e.g. political coverages where the location of a country can be referenced with respect to its surrounding countries, to smaller-scale spaces such as cadastral systems where several geometrical constraints are verified due to some of the partitions' properties.

Topological relationships in those spatial partitions are mainly based on two operators: adjacency and disjunction. However, and despite their computational efficiency, these operators are not capable of qualifying the relationships between regions distributed in a given spatial partition. Alternatives include spatial statistic operators that estimate the variability of a given property over a spatial distribution but not the

W. Kuhn, M.F. Worboys, and S. Timpf (Eds.): COSIT 2003, LNCS 2825, pp. 204–220, 2003.

manner in which elements of this distribution are interrelated. We believe that in many situations there is an interest in analysing how two given regions relate to one another. This should also help to identify local and global structural patterns in a spatial partition.

The aim of this paper is to take advantage of some of the dual graph properties of spatial partitions, and to introduce a generalisation of the adjacency relationship. We introduce and give a formal definition of a *relative adjacency* operator that evaluates to which degree regions in a given spatial partition are mutually distant in the dual graph derived from adjacency relationships. The operator is flexible enough to evaluate those relative adjacencies at different levels of magnitude, that is, by minimising or maximizing the impact of the outlying regions. Given a reference region in a spatial partition we also show how the relative adjacency operator can support the analysis of the relative distribution of other regions, and how those regions are clustered with respect to that reference region. The remainder of the paper is organised as follows. Section 2 briefly outlines related work. The adjacency operator and its main properties are formalised in section 3. A case study illustrates its potential in section 4. Section 5 investigates how patterns can be derived and computed from the application of the relative adjacency operator. Finally section 6 draws some conclusions.

2   Related Work

The concept of distance, although intuitive, has several meanings depending on its definition and its domain of application. This leads to several forms of distance that range from metric to qualitative and approximate measures. The distance between two regions can be measured quantitatively using basic measures of metric spaces such as the Euclidean, Minkowski or spherical Manhattan distances. Quantitative measures based on distances measured on dual graphs, and derived from spatial partitions, have been applied in many environmental and urban studies. For example, ecological studies evaluate the degree of cohesion using the minimum distance between two regions of the same class or an average distance between the instances of two classes [15]. Refinements of these measures also consider the cumulated influence of perimeters and areas between neighbours in the dual graph of a spatial partition [16]. Statistical analysis has long integrated distance measures to evaluate patterns, clusters, autocorrelations and interrelations in a population of entities distributed in space [17], [18], [19]. The distance factor is particularly important in gravity models or autocorrelation statistical analysis [20].

Some urban studies, regrouped in the emerging field of space syntax, analyse the structural properties of an urban network using integration, connectivity, centrality and clustering measures derived from graph theory principles [21]. Those indices measure the degree of connectivity, structural importance and integration of a node in a network and some clustering indices. Space syntax measures also evaluate connectivity and integration degrees of a graph.

Although distance is a spatial concept *per-se*, it can also be applied by extension to evaluate some semantic relationships. This applies to semantic networks where the nodes of a graph represent the population of interest and edges relationships between those nodes. Those principles have been recently applied to the study of "small-world" properties (i.e. low separation between any two nodes and existence of local clusters) to analyse the structure of social networks [22] and the study of semantic distances between web servers over the Internet [23].

Integration of the context and external factors can provide a qualitative component to a given distance measure. For instance, the Euclidean distance can be modulated by a contextual parameter that takes into account the external environment to derive a proximity value between two given regions [24]. This proximity value reflects the fact, observed in qualitative studies [25], that the distance between a region A and a distant region B should be magnified when the number of regions near A increase, and *vice versa*. The influence of external factors can be generalised to relative size, shape, or event temporal (i.e. travel time) or cultural aspects depending on the context. Qualitative evaluations of distances (e.g. far, near) [26] or the concept of fuzzy distance – usually defined as the *infimum* of the lengths of all paths between them, that is, a form of Hausdorff distance [27] – offer other alternatives to metric distances. Inspired by those recent works and Worboys' approach of a contextual distance [24], the objective of our model is to explore a flexible form of qualitative distance applied to spatial partitions, and based on principles of topological spaces and graph theory.

3   Modelling Principles

The first law of geography states that "Everything is related to everything else, but near things are more related than distant things" [28]. We briefly discuss the differences that one can observe in analysing how near and distant regions are interrelated. First, let us consider the spatial partition derived from American countries. With respect to an adjacency relationship, it appears that Uruguay is more integrated with Brazil than the reverse. This reflects the computational property that the probability of moving randomly one step from Uruguay in the dual graph of the American countries and reaching Brazil is higher than the contrary. Secondly, one should also say that if near things are related, distant things, although less related, are related too and in different ways that reflect their integration versus segregation in the dual graph. This also corresponds to the essential intuition that some of the spatial relationships identified in a spatial partition are not always symmetric [29].

Those different degrees of mutual integration between regions lead us to explore the modelling of an operator that reflects those properties. In order to develop further our modelling approach we introduce some basic properties and definitions of topological spaces. Let $\mathfrak{C}$ be a topological space, and let x be a region of $\mathfrak{C}$, that is a connected subset of $\mathfrak{C}$. x° denotes the interior of x and ∂x its boundary according to the usual notations. The mathematical definition of a partition implies that the elements of the partition don't intersect. However this doesn't fit geographical spaces as adjacent regions share part of their boundaries. Therefore, we propose a slightly relaxed definition of the partition where the union of its elements still gives the set, but where the interiors of any pair of elements don't intersect. The definition of a pseudo-partition is as follows.

Definition 1 – Pseudo-partition.
$X_{pp} = \{x_1, x_2, \ldots, x_n\}$ is a pseudo-partition of the connected subset $X \subset \mathbb{C}$ iff

- $$\bigcup_{i=1}^{n} x_i = X$$

- $\forall x_i, x_j \in X_{pp}, x_i \neq x_j$, then $x^{\circ}_i \cap x^{\circ}_j = \varnothing$

To construct the dual graph derived from adjacency relationships the topological relationship *touch* is required. We take the usual definition as follows.

Definition 2 – Touch.
Let x_1, x_2 be two regions of $\mathbb{C}$, x_1 touches x_2 iff
$$\begin{cases} x^{\circ}_1 \cap x^{\circ}_2 = \varnothing \\ \partial x_1 \cap \partial x_2 \neq \varnothing \end{cases}$$

The operator *touch* is symmetric but not reflexive. We introduce a Boolean operator that manipulates the relationship *touch*, and the definition of an adjacency set.

Definition 3 – Adjacency.
$$Adj(x_1, x_2) = \begin{cases} 1 \text{ iff } x_1 \text{ touches } x_2 \\ 0 \text{ otherwise} \end{cases}$$

Definition 4 – Adjacency set.

The adjacency set of a region x of $\mathbb{C}$ is the union of the regions of $\mathbb{C}$ adjacent to x.
$$Adj(x) = \{x_i \in \mathbb{C} \, / \, Adj(x, x_i) = 1\}$$

Let us denote $CAdj(x)$ the cardinality of $Adj(x)$.

Definition 5 – Dual graph of a pseudo-partition.
The dual graph G of a pseudo-partition X_{pp} is given by the pair (E,N), where N is the set of regions of X_{pp} and E the set of edges subset of the Cartesian product $N{\times}N$, where $(x_i, x_j) \in E$ iff $Adj(x_i, x_j) = 1$.

The dual graph G provides the computational support for deriving the relative adjacency operator whose definition is introduced as follows:

Definition 6 – Relative adjacency.
The relative adjacency $R(x_p, x_r)$ of two regions x_p and x_r of X_{pp} is defined as follows:

$$R(x_p, x_r) = d\, \text{Adj}(x_p, x_r) + (1-d) \sum_{x_i \in \text{Adj}(x_p)} \frac{R(x_i, x_r)}{\text{CAdj}(x_p)} \qquad (1)$$

where $0 < d < 1$

By convention the higher $R(x_p, x_r)$, the higher x_p's integration with respect to x_r (vs. the lower their level of segregation).

The relative adjacency between two regions is defined recursively. It takes into account the importance of the adjacency between the two given regions x_p and x_r (first term) and to which degree the k-neighbourhoods of x_p are relatively adjacent to x_r (second term, where a k-neighbourhood of x_p is defined as a region which is k-step away in the dual graph G). The user defined coefficient d, the so-called *damping coefficient*, balances the relative importance of adjacencies and relative adjacencies. Higher values of d lead to a smaller account of outlying regions, whereas smaller ones minimise the importance of closer regions. The resolution of $R(x_p, x_r)$ for $Card(X_{pp}) = n$ is equivalent in complexity to the resolution of a system of n linear equations with n unknowns. This results from the fact that from a given x_p, all x_r of X_{pp} are recursively used at least once in the expression.

Property 6.1 – Solvability.
The calculation of the relative adjacency $R(x_p, x_r)$ between two given regions x_p, x_r of X_{pp} is always solvable.

Proof: We express the system of linear equation in a matrix form. This matrix is strictly dominant for any value of $d \neq 1$. This ensures existence and uniqueness of the solution. For $d = 1$ the matrix is the identity matrix, also ensuring existence and uniqueness of the relative adjacency (note that in this later case $R(x_p, x_r) = Adj(x_p, x_r)$).

The relative adjacency operator is not symmetric, except in some special cases where symmetry is caused by a symmetrical configuration of the spatial pseudo-partition. Relative adjacency values are non null except for regions disconnected in the dual graph G. The relative adjacency is drawn by the unit interval as it is made of a sum of adjacency values (*i.e.* 0s and 1s) weighed by the cardinality of the adjacency set and the multiplicative coefficient d.

The relative adjacency coefficient considers connectivity values in the dual graph, without accounting for other relations of spatial or even aspatial nature that could enrich this graph (*i.e.*, coefficients allocated to the edges). When considering the dual graph of a given spatial partition, other spatial criteria may interfere, amongst them the length of common borders between two neighbours, the relative importance of regions' surfaces, or geographical centres' mutual distances.

In order to take into account these criteria, we introduce an extension of the relative adjacency operator.

Definition 7 – Extended relative adjacency.

$$R^+(x_p,x_r) = d\varphi(x_p,x_r)\,\text{Adj}(x_p,x_r) + (1-d)\sum_{x_i \in \text{Adj}(x_p)} \frac{R^+(x_i,x_r)}{\text{CAdj}(x_p)} \tag{2}$$

where d is kept as a damping coefficient, as before.

$\varphi(x_p,x_r)$ is a semiotic function that reflects a criterion (or criteria) under study.

This criterion is basically a quantifiable relation between adjacent regions. An even greater number of criteria can be considered in the case of a geographical space, as additional information is often implied (e.g. relative distance between main cities, population or economical centres of gravity).

In the case where the distance between geographical centres is considered, the semiotic function can be defined as follows.

$$\varphi(x_p,x_r) = \frac{\dfrac{1}{D(x_p,x_r)}}{\displaystyle\sum_{x_i \in \text{Adj}(x_p)} \dfrac{1}{D(x_p,x_i)}} \tag{3}$$

The integration of x_p to x_r is considered inversely proportional to the distance between the respective geographical centres. It is normalised by the sum of the inverse distances between the x_i and x_p, ensuring a normalisation drawn by the unit interval as for the relative adjacency.

4   Application of the Relative Adjacency to Some Examples of Spatial Configurations

In order to illustrate the overall behaviour of the relative adjacency operator, we introduce some canonical cases. In the trivial case of a pseudo-partition composed of a unique region x then $R(x,x) = \text{Adj}(x,x) = 0$. The most significant configurations are discussed in the following sub-sections.

4.1   n Regions Mutually Connected

Let us consider the configuration of many regions mutually connected (*i.e.* every region is connected to every other region of the pseudo-partition). This should give an indication of the behaviour of the relative adjacency as connectedness varies. Figure 1 introduces the cases of 2, 3 and 4 regions.

In the case where there are two adjacent regions x and y only (Figure 1 left), relative adjacency values, derived from (1), are given by the two equations below.

$R(y,y) = 0 + (1\text{-}d)\,R(x,y)$
$R(x,y) = d + (1\text{-}d)\,R(y,y)$

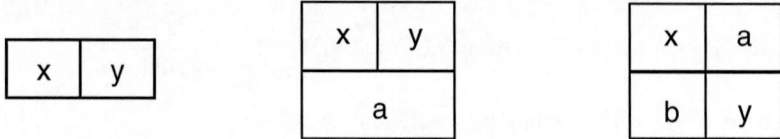

Fig. 1. Total connectedness with 2, 3 and 4 regions.

Those two equations are equivalent to

$$\begin{pmatrix} -1 & 1-d \\ 1-d & -1 \end{pmatrix}\begin{pmatrix} R(x,y) \\ R(y,y) \end{pmatrix} = \begin{pmatrix} -d \\ 0 \end{pmatrix} \Leftrightarrow \begin{pmatrix} R(x,y) \\ R(y,y) \end{pmatrix} = \begin{pmatrix} \dfrac{1}{2-d} \\ \dfrac{1-d}{2-d} \end{pmatrix}$$

For $d = 0.5$ this gives (without loss of generality, and in order to compare the values of the different examples presented hereafter we set $d = 0.5$).

$$\begin{pmatrix} R(x,y) \\ R(y,y) \end{pmatrix} = \begin{pmatrix} 0.66 \\ 0.33 \end{pmatrix}$$

In the case of three adjacent regions (Figure 1 centre), the matrix equation to solve is as follows (derived from (1))

$$\begin{pmatrix} -1 & \dfrac{1-d}{2} & \dfrac{1-d}{2} \\ \dfrac{1-d}{2} & -1 & \dfrac{1-d}{2} \\ \dfrac{1-d}{2} & \dfrac{1-d}{2} & -1 \end{pmatrix}\begin{pmatrix} R(x,y) \\ R(a,y) \\ R(y,y) \end{pmatrix} = \begin{pmatrix} -d \\ -d \\ 0 \end{pmatrix} \Rightarrow \begin{pmatrix} R(x,y) \\ R(a,y) \\ R(y,y) \end{pmatrix} = \begin{pmatrix} 0.8 \\ 0.8 \\ 0.4 \end{pmatrix} \text{ for } d = 0.5.$$

In the case of four adjacent regions (Figure 1 right), relative adjacency values are (derived from (1))

$$\begin{pmatrix} R(x,y) \\ R(a,y) \\ R(b,y) \\ R(y,y) \end{pmatrix} = \begin{pmatrix} 0.85 \\ 0.85 \\ 0.85 \\ 0.42 \end{pmatrix}$$

It is worth noting the increase of the relative adjacency values with the increase of connectedness (e.g. $R(x,y)$ increases from 0.66 to 0.8 to 0.85). This denotes an increase of x's integration with y. Figure 2 illustrates how the relative adjacency R evolves with the damping coefficient d. Figure 2 shows that the $R(x,y)$ values increase with d to reach 1 when $d = 1$ (*i.e.* $Adj(x,y) = 1$), whereas the $R(y,y)$ values (self-relative adjacency) decrease to reach 0 when $d = 1$.

One can also notice the higher initial values of $R(x,y)$ and $R(x,x)$ (when d is close to zero) as connectivity increases (respectively 0.5, 0.66, 0.75). Those configurations also denote the fact that x, y and a are similar in their respective pseudo-partitions.

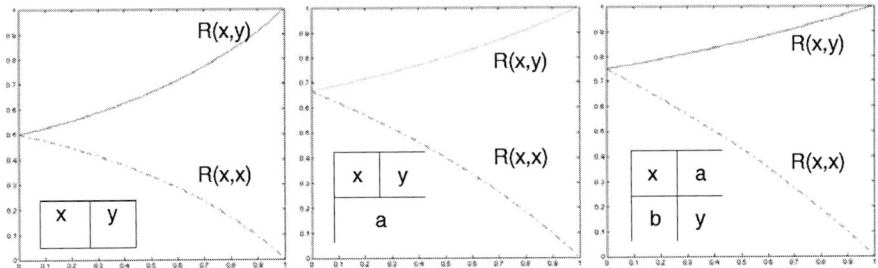

Fig. 2. Relative adjacencies along d for mutually connected regions.

In order to generalise those cases to n regions mutually connected we note that $Adj(x_i, x_j) = 1 \; \forall \; x_i, x_j \in X$ so the dual graph G is complete and the system of equations to resolve can be expressed as follows (derived from (1)). Overall relative adjacency values increase with n.

$$\begin{pmatrix} -1 & \dfrac{1-d}{n} & \cdots & \dfrac{1-d}{n} \\ \dfrac{1-d}{n} & \ddots & \ddots & \vdots \\ \vdots & \ddots & \ddots & \dfrac{1-d}{n} \\ \dfrac{1-d}{n} & \cdots & \dfrac{1-d}{n} & -1 \end{pmatrix} \begin{pmatrix} R(x_1, x_n) \\ \vdots \\ R(x_{n-1}, x_n) \\ R(x_n, x_n) \end{pmatrix} = \begin{pmatrix} -d \\ \vdots \\ -d \\ 0 \end{pmatrix}$$

4.2   n Regions Connected in a Linear Way

We consider the case of n regions connected in a linear way where each region is only adjacent to two regions except for the two located at the extremities which are adjacent to one region only (Figure 3). This configuration should help to evaluate the influence of remoteness on the relative adjacency.

$$\boxed{x} \; \boxed{a} \; ---- \; \boxed{y}$$

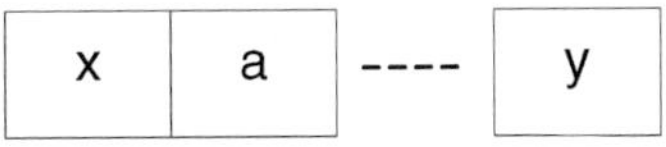

Fig. 3. n regions connected in a linear way.

Let us first consider the case of a chain made of three regions. The relative adjacency equations to solve are

$$R(x, y) = R(y, y) = (1-d) \, R(a,y) \quad \text{and} \quad R(a,y) = d + \frac{1-d}{2}(R(x, y) + R(y, y))$$

$$\text{this gives in a matrix form} \begin{pmatrix} R(x,y) \\ R(a,y) \\ R(y,y) \end{pmatrix} = \begin{pmatrix} (1-d)/(2-d) \\ 1/(2-d) \\ (1-d)/(2-d) \end{pmatrix}$$

$$\text{which implies} \begin{pmatrix} R(x,y) \\ R(a,y) \\ R(y,y) \end{pmatrix} = \begin{pmatrix} 0.33 \\ 0.66 \\ 0.33 \end{pmatrix} \text{ for } d = 0.5$$

$$\text{For five regions, relative adjacency values are} \begin{pmatrix} R(x,y) \\ R(a,y) \\ R(b,y) \\ R(c,y) \\ R(y,y) \end{pmatrix} = \begin{pmatrix} 0.02 \\ 0.04 \\ 0.16 \\ 0.61 \\ 0.30 \end{pmatrix} \text{ for } d = 0.5$$

Figure 4 shows how relative adjacency values change along d. To the left the graph reflects the increase of relative adjacency values along d for adjacent regions and the decrease of relative adjacency values for outlying regions. To the right the graph reflects the increasing importance of relative adjacency values of the adjacent region c when d increases, and the degree of remoteness with respect of y of the other regions.

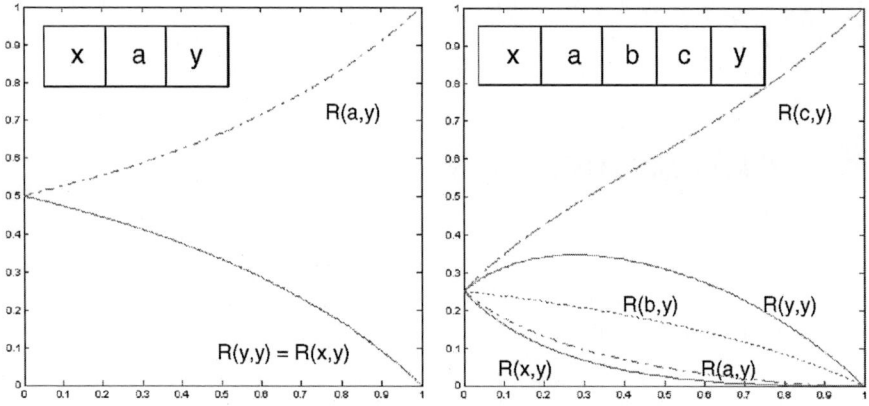

Fig. 4. Relative adjacencies along d for chains of regions.

The decrease of relative adjacencies with the increase of the number of regions in the chain reflects the mutual segregation of outlying regions (e.g. x and y). This reflects the fact that there is only one path that connects x and y in the dual graph. This can be generalised to the case of an n-element chain, where relative adjacency values are derived from the equation below presented in a matrix form. Those figures also show how useful the damping coefficient is in differentiating between different degrees of relative adjacencies.

$$
\begin{pmatrix}
1 & -(1-d) & 0 & \cdots & & \cdots & 0 \\
-\dfrac{1-d}{2} & 1 & -\dfrac{1-d}{2} & 0 & & & \vdots \\
0 & -\dfrac{1-d}{2} & 1 & -\dfrac{1-d}{2} & \ddots & & \\
\vdots & 0 & -\dfrac{1-d}{2} & \ddots & \ddots & 0 & \vdots \\
& & \ddots & \ddots & 1 & -\dfrac{1-d}{2} & 0 \\
\vdots & & & 0 & -\dfrac{1-d}{2} & 1 & -\dfrac{1-d}{2} \\
0 & \cdots & & \cdots & 0 & -(1-d) & 1
\end{pmatrix}
\begin{pmatrix}
R(x,y) \\ R(a_1,y) \\ R(a_2,y) \\ \vdots \\ R(a_{n-3},y) \\ R(a_{n-2},y) \\ R(y,y)
\end{pmatrix}
=
\begin{pmatrix}
0 \\ 0 \\ 0 \\ \vdots \\ 0 \\ d \\ 0
\end{pmatrix}
$$

4.3   Relative Adjacency – Spatial Structure

Let us study how the structure of a given spatial pseudo-partition is reflected by the relative adjacency coefficient. Figure 5 below presents some common spatial structures with the difference being that Figure 5 right has an additional region, *i.e.* a region c adjacent to y but not to x although all regions adjacent to x (alternatively y) are also adjacent to y (alternatively x) in Figure 5 left.

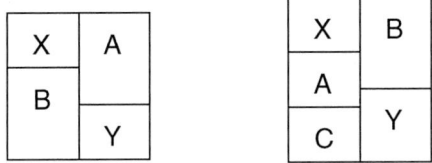

Fig. 5. n regions randomly distributed.

Relative adjacency values for Figure 5 configurations are as follows:

$$
\text{Figure 5 left} \quad
\begin{pmatrix}
R(x,y) \\ R(a,y) \\ R(b,y) \\ R(y,y)
\end{pmatrix}
=
\begin{pmatrix}
0.37 \\ 0.75 \\ 0.75 \\ 0.37
\end{pmatrix}
\text{for } d = 0.5
$$

$$
\text{Figure 5 right} \quad
\begin{pmatrix}
R(x,y) \\ R(a,y) \\ R(b,y) \\ R(c,y) \\ R(y,y)
\end{pmatrix}
=
\begin{pmatrix}
0.38 \\ 0.79 \\ 0.76 \\ 0.79 \\ 0.39
\end{pmatrix}
\text{and}
\begin{pmatrix}
R(x,x) \\ R(a,x) \\ R(b,x) \\ R(c,x) \\ R(y,x)
\end{pmatrix}
=
\begin{pmatrix}
0.35 \\ 0.7 \\ 0.72 \\ 0.24 \\ 0.27
\end{pmatrix}
\text{for } d = 0.5
$$

With respect to the configurations given in Figure 5 we can make several observations. First, x is more integrated with y in Figure 5 left than in Figure 5 right. This results from the fact that the parcels adjacent to x are adjacent to y in Figure 5 left, which is not the case in Figure 5 right (where c is adjacent to y but not to x). Also the adjacency configuration of the regions a and b with respect to x is similar to the one with respect to y. This is not the case in Figure 5 right where one can observe a dissymmetry, that is, relative adjacencies of the form $R(x,y)$ are higher than $R(y,x_i)$. This trend results from the fact that all the parcels adjacent to x are adjacent to y while the reverse is not true. The spatial configurations presented in this section outline several properties of the relative adjacency coefficient. The larger the distance between two given regions in the dual graph G, the generally weaker their relative adjacency. The higher the number of possible paths between two given regions, generally speaking the higher their relative adjacency value. Moreover, the relative adjacency's absence of symmetry reflects the fact that the likelihood of reaching a region b from a region a is generally different from the reverse. Another property is the possibility of balancing the influence of closer neighbours against that of the remote ones. This flexibility is given by the damping parameter d which is user defined.

5   Analysis and Clustering

While the preliminary results given by the relative adjacencies concern the evaluation of the mutual integration of two given regions in a spatial pseudo-partition, global observation of those relative adjacencies should denote some structural properties and patterns. The fact that the relative adjacency is modulated by the damping coefficient d reinforces the panel of measures to observe and analyse. In order to illustrate how some structural patterns can be observed with the relative adjacency operator, we introduce the large-scale pseudo-partition example of the countries of America. The dual graph has a high diversity of node degrees, a highly connected side (Southern America) as well as a poorly connected peripheral side (central and northern part of the continent). Figure 6 shows the dual graph derived from this spatial pseudo-partition.

A first level of analysis is produced by the observation of the quantitative values given by the relative adjacency for a user-defined value of the damping coefficient d. In order to extend the dimension of the relative adjacency values we observe the evolution of relative adjacency values as a function of d. Figures 7 to 9 display these functions for three American countries, one country relatively well connected in the dual graph: Brazil, one country not very much connected (but of interest because it is the only link between two parts of the graph): Panama and one country in between: Chile.

Figure 7 shows the following patterns for the integration of American countries to Brazil. The curve (I) is given by the self-relative adjacency, cluster (II) regroups Brazil's adjacent countries, (III) corresponds to the most remote countries (North and Central America), (VI) gives Panama which is adjacent to Colombia, one of the Brazil's adjacent countries. However, as Panama is not very much integrated with the graph, its curve differs slightly from those obtained for the other 2-neighbours (curve V, Equator and IV, Chile) which are more connected to the 1-neighbours of Brazil.

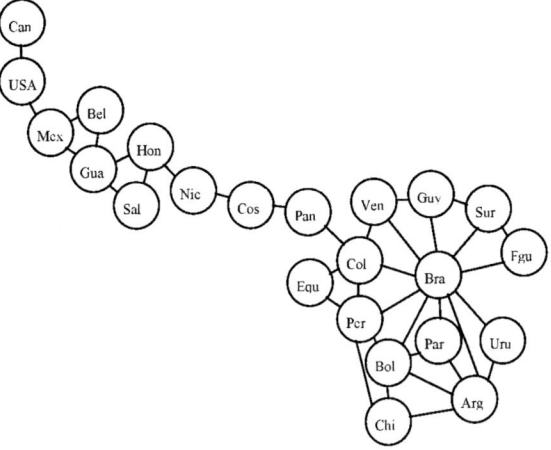

Fig. 6. Dual graph for the pseudo-partition of America's countries.

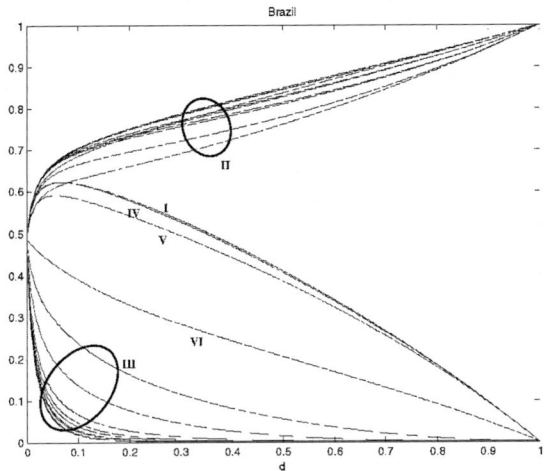

Fig. 7. Brazil's relative adjacencies.

Figure 8 shows that relative adjacency values calculated for Chile are generally smaller than those observed for Brazil; Chile being less integrated than Brazil in the dual graph. Also clusters are much less apparent as they are for the case of Brazil. The curve mainly denotes the degree of remoteness of the countries of America with respect to Chile (e.g. I gives Chile, II Chile's adjacent countries, IV Chile's 2-neighbourhood etc). Additional patterns are even discernible for higher-level neighbourhoods: IIIa denotes the three Guyana countries, IIIb corresponds to Panama while IIIc gives the other countries of Central and North America.

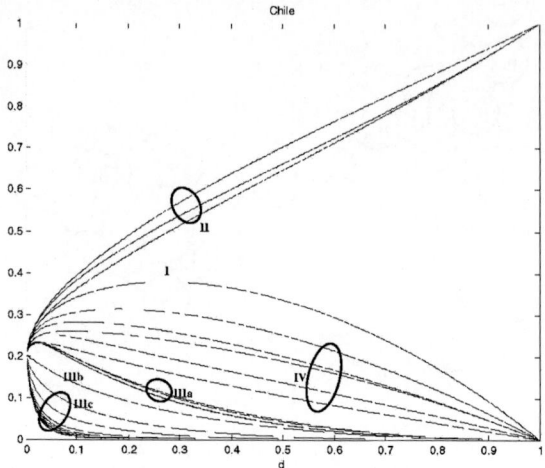

Fig. 8. Chile's relative adjacencies.

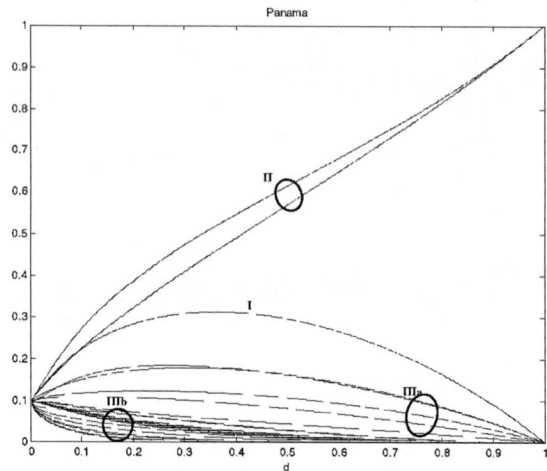

Fig. 9. Panama's relative adjacencies.

Figure 9 presents the case of Panama, a relatively non integrated country although quite central in the graph. This is reflected by lower values of relative adjacencies. As for the previous figures, the curve I is given by the self-relative adjacency, II by adjacent countries (Costa Rica and Colombia), IIIa for 2-neighbourhood, and IIIb for more remote ones. One may notice the intertwining of curves for clusters IIIa and IIIb. This corresponds to the general direction in which the propagation is performed: Nicaragua to the north *vs.* Colombia's direct neighbours to the south for IIIa, and the rest of North America *vs.* the rest of South America for IIIb. This segregated behaviour is explained by fact that the northern branch curves reach their maximum (or minimum) earlier along d because of a smaller connectivity when compared to the southern branch.

In order to illustrate the mutual integration of two given countries, Figure 10 gives the relative adjacencies between Chile and Brazil. This shows that Chile is more integrated with Brazil than the reverse. This is due to the fact that all Chile's 1-neighbourhood are adjacent to Brazil whereas less than a third of Brazil's 1-neighbourhoods are adjacent to Chile. This behaviour is quite similar to the example of Figure 5 right, with Brazil (alternatively Chile) playing a role analogous to y (alternatively x).

Those examples confirm that the configuration of the pseudo-partition as a whole influences relative adjacency values. It is worth noting that when the derivative of the relative adjacency is considered, some general trends can be inferred. For $d = 1$, the derivative of 1-neighbours is strictly positive, whereas those for self-relative adjacency are the most negative. It is also possible to discriminate 2-neighbours as their derivatives are negative for $d = 1$, while other n-neighbours, with $n > 2$, have their derivatives equal to zero. Those figures provide local and global information on mutual relationships in a given spatial pseudo-partition. Local, as degrees of mutual relationships and clusters are qualified for a reference region; global, as relative degrees of integration in the dual graph can be compared across the spatial pseudo-partition.

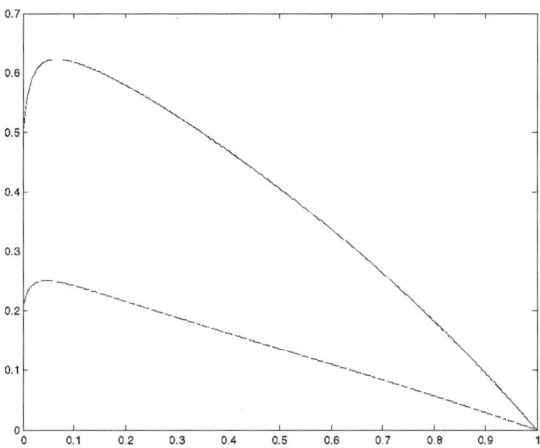

Fig. 10. R(Chile,Brazil) (upper curve) and R(Brazil,Chile) (lower curve).

Let us finally apply the extended relative adjacency to the example of Chile. Figure 11 shows the respective integration of America's countries to Chile when considering the extended relative adjacency. Compared to the relative adjacency values presented in Figure 8, one can remark that Chile's 1-neighbours are more discriminated (i.e. Argentina is more integrated to Chile than Bolivia and Peru). Overall extended relative adjacencies are smaller than relative adjacencies, this reflecting the lower integration of the whole continent to Chile when considering the distance between theirs and Chile's geographical centres.

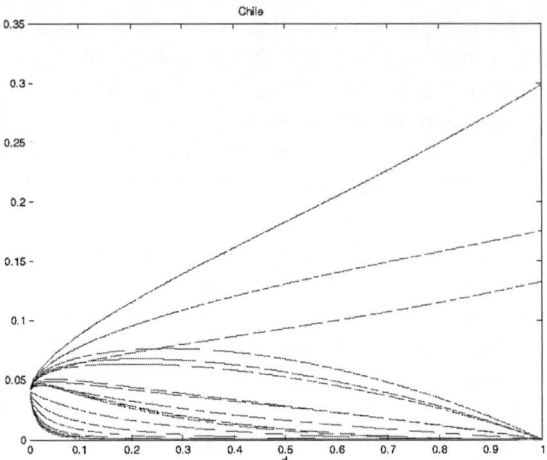

Fig. 11. Chile's extended relative adjacencies.

6   Conclusion

Topological models have proven to be a fundamental support for successful implementation of GIS as they provide useful reasoning primitives. In particular, they help to verify geometrical constraints in spatial partitions. They might even be used in analysing local and global properties and configuration of spatial structures. This paper deals with such an objective and introduces a relative adjacency operator that qualifies degrees of mutual integration in pseudo-partitions. We give a computable expression of the relative adjacency, together with a damping coefficient that outlines either neighbouring or outlying regions.

Such a coefficient and observed trends, derived from its computation, can be modulated in function of the context and properties intrinsic to the underlying spatial pseudo-partition and the phenomenon represented. This dependence on the context has been discussed elsewhere with respect to qualitative distance and qualitative orientation ([30], [31], [32], [29]). Although the relative adjacency measures a mutual integration factor for some given spatial regions at the local level, it can also be considered as a form of qualitative distance, and a means of analysing the structure of a given spatial partition. Relative adjacencies can be also moderated with respect to additional spatial and aspatial properties that will enrich the approach. For example, region sizes and lengths of the boundaries shared by two adjacent regions, distances between centroids are relevant parameters to consider. Further work covers the experimental validations of the approach to evaluate to which degree relative adjacencies correlate cognitive interpretations.

Acknowledgments

The authors thank the referees for their valuable comments and suggestions. They also thank Marius Thériault and Jean-Paul Cheylan for fruitful discussions on early drafts of this paper.

References

1. Cohn, A.G.: Qualitative spatial representation and reasoning techniques. In: Brewka G., Habel C., Nebel, B. (eds.): Proceedings of KI-97. LNAI, Vol. 1303. Springer-Verlag, Berlin (1997) 1-30
2. Pullar, D.V., Egenhofer, M. J.: Towards the defaction and use of topological relations among spatial objects. In: Proceedings of the 3rd International Symposium on Spatial Data Handling. IGU, Colombus (1988) 225-242
3. Egenhofer, M.: Reasoning about binary topological relations. In: Günther, O., Schek, H.J. (eds.): Advances in Spatial Databases. Springer-Verlag, Berlin (1991) 143-160
4. Randell, D.A., Cui, Z., Cohn, A.G.: A spatial logic based on regions and connection. In: Proceedings of the 3rd International Conference on Knowledge Representation and Reasoning. Cambridge, Massachusetts (1992) 165-176
5. Clementini, E., Di Felice, P., Van Oosterom, O.: 1993, A small set of topological relationships suitable for end-user interaction. In: Abel, D.J., Ooi, B.C. (eds.): Advances in Spatial Databases. Springer-Verlag, Singapore (1993) 277-295
6. Cui, Z., Cohn, A.G., Randell, D.A.: Qualitative and topological relationships in spatial databases. In: Abel, D.J., Ooi B.C. (eds.): Advances in Spatial Databases. Springer-Verlag, Singapore (1993) 296-315
7. Freksa, C.: Using orientation information for qualitative spatial reasoning. In Frank, A.U., Campari, I., Formentini, U. (eds.): Theories and methods of spatio-temporal reasoning in geographic space. Lecture Notes in Computer Science, Vol. 639, Springer-Verlag, Berlin (1992) 162-178
8. Frank, A.U.: Qualitative spatial reasoning: cardinal directions as an example. International Journal of Geographical Information Systems, 10(3) (1996) 269-290
9. Sharma, J.: Integrated Spatial Reasoning in GIS: Combining Topology and Direction. PhD Thesis, Department of Spatial Information Science and Engineering, University of Maine, Orono, ME (1996)
10. Papadias, D., Egenhofer, M.J.: Algorithms for hierarchical spatial reasoning. Geoinformatica, 1(3) (1997) 251-273
11. Goyal, R.K., Egenhofer, M.J.: Consistent queries over cardinal directions across different levels of detail. In: Tjoa, A.M., Wagner, R., Al-Zobaidie, A. (eds.): 11th International Workshop on Database and Expert Systems Applications, Greenwich, UK (2000) 876-880
12. Robinson, A. H., Sale, R. D., Morrison, J. L., Muehrcke, P. C.: Elements of Cartography 5th ed., John Wiley & Sons, New York (1984)
13. Frank, A.U., Volta, G.S., Gahegan, M.: Formalisation of families of categorical coverages. International Journal of Geographical Information Systems, Taylor and Francis (1997)
14. Erwig, M., Schneider, M.: Formalisation of advanced map operations, 9th Symp. on spatial data handling, 8a. (2000) 3-17
15. Gustafson, E.J., Parker, G.R.: Relationships between land-cover proportion and indices of landscape spatial pattern. Landscape Ecology, 7 (1992) 101-110

16. Schumaker, N.H.: Using landscape indices to predict habitat connectivity. Ecology, 77 (1996) 1210-1225
17. Haynes, K.E., Fotheringham, A.S.: Gravity and Spatial Interaction Models. Scientific Geography Series. Sage, Newbury Park (1984)
18. Odland, J.: Spatial Autocorrelation. Scientific Geography Series. Sage, Newbury Park (1988)
19. Anselin, L., Getis, A.: Spatial statistical analysis and geographic information systems. The Annals of Regional Science 26 (1992) 19-33
20. Getis, A., Ord, J.K.: The analysis of spatial association by use of distance statistics. Geographical Analysis, 24-3 (1992) 206
21. Hillier, B., Hanson, J.: The Social Logic of Space. Cambridge University Press, Cambridge (1984)
22. Watts, D.J., Strogatz, S.H.: Collective dynamics of small-world networks. Nature, 393 (1998) 440-442
23. Hou, J., Zhang, Y.: A matrix approach for hierarchical web page clustering based on hyperlinks. In: Proceedings of the 3rd Int. Conf. on Web Information Systems Engineering (workshops), Huang, B. et al. (eds.). IEEE Press, Singapore (2002) 207-216
24. Worboys, M.: Metrics and topologies for geographic space. In: Advances in GIS Research II, Kraak, M.J., Molenaar, M. (eds.). Taylor and Francis (1996) 365-375
25. Golledge, R.G., Hubert, L.J.: Some comments on non-Euclidean mental maps. Environment and Planning A, 14 (1982) 107-118
26. Clementini, E., Di Felice, P., Hernandez, D.: Qualitative representation of positional information, Artificial Intelligence, 95(2) (1997) 317-356
27. Klir, G.J., Yuan, B.: Fuzzy Sets and Fuzzy Logic: Theory and Applications. Prentice Hall (1995)
28. Tobler, W.R.: A computer model simulating urban growth in the Detroit region. Economic Geography, 46 (1970) 234-240
29. Duckham, M., Worboys, M.F.: Computational structure in three-valued nearness relations. In: Montello, D. (ed.): Conference in Spatial Information Theory. Lecture Notes in Computer Science, Vol. 2205. Springer-Verlag (2001) 76-91
30. Robinson, V.B.: Interactive machine acquisition of a fuzzy spatial relation. Computers and Geoscience, 16(6) (1990) 857-872
31. Frank, A.U.: Qualitative spatial reasoning about distances and directions in geographic space. Journal of Visual Languages and Computing, 3 (1992) 343-371
32. Gahegan, M.: Proximity operators for qualitative spatial reasoning, Spatial Information Theory: A theoretical Basis for GIS, Frank, A.U., Kuhn, W. (eds.): Lecture Notes in Computer Science, Vol. 988. Springer-Verlag, Berlin (1995) 31-44

A Geometry for Places: Representing Extension and Extended Objects

Hedda R. Schmidtke

Department for Informatics, University of Hamburg
Vogt-Kölln-Str. 30, D-22527 Hamburg, Germany
schmidtke@informatik.uni-hamburg.de

Abstract. The article presents a qualitative region-based approach to the representation of extension. A geometry of incidence and ordering is taken as a basis to characterize the concept of extension founded on the congruence of certain regions (called *places*) which have equal extension into all directions. The notion of extension of regions is derived from the sizes of places—not from the distance between points as in classical geometry—and represented by size intervals. A geometric specification of granular or scale-specific spatial contexts and of the local extension of a region is then derived. Extension relative to a spatial context is used to formally specify conditions under which object regions can be classified e.g. as *punctual, linear,* or *planar* in the context.

Keywords: Representation of extension, extended objects, spatial granularity, spatial context, qualitative spatial reasoning, incomplete spatial knowledge.

1 Introduction

Representing spatial objects as extended is an important topic for research in the foundations of spatial information theory. In the temporal domain time intervals can replace punctual moments, in the spatial domain mereotopological approaches derive points from regions. It should be equally rewarding to represent the size of regions based on regions, not on the distance between points.

The approach taken in this article is to formulate a geometry for sizes that uses only qualitative concepts. The regions thus can be compared for size, but a quantitative description is not necessary. It has been shown that axiomatic geometry can be used to successfully specify geometric concepts so as to represent cognitive concepts: E.g., Eschenbach, Habel, Kulik, and Lessmöllmann [7] show that shape concepts like *corner* can be represented geometrically. Kulik and Klippel [14] present a geometry of cardinal directions based on the grid lines of a map.

The inferential strength of directional calculi (cf. [8,15]) depends on the geometry of points and straight lines. If it can be ensured that objects are punctual in a context, this inferential strength can be retained, for the most part. But given arbitrarily shaped regions, directions between two locations are hard to

W. Kuhn, M.F. Worboys, and S. Timpf (Eds.): COSIT 2003, LNCS 2825, pp. 221–238, 2003.

characterize, and the different possible cases have to be examined carefully (cf. e.g. [11,16]). Topological reasoning, in contrast, is interesting only in the case of extended objects. Therefore, a reasoning system could benefit from a front-end system that determines whether an object is *punctual* or *extended, linear* or *planar* in a context. It is then possible to choose an appropriate representation and reasoning sub-system depending on the extension of the objects in the spatial context. Herskovits [12] underlines the importance of such a classification for cognitive systems.

The extension of a simple object like a building, for instance, can be represented by specifying length and width. Extension in the large-scale space of GIS is a more complex phenomenon. Extended objects, like the river in Fig. 1, differ extremely in extension. This can be exemplified by looking at the means we would choose to cross the river: Close to the mouth a boat or a long bridge is needed, close to the source we can cross the river over a small bridge or even by jumping.

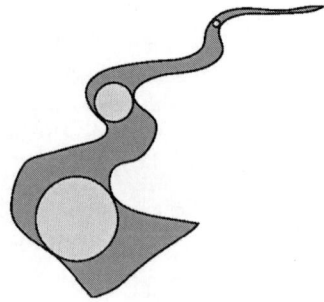

Fig. 1. A river (schematized) varies in width along its course. Depending on the part that is currently in the focus the river is classified as linear (in the middle), or planar (at the mouth).

Goodchild [10] points out that "many geographic attributes are scale-specific": A region that is assigned the attribute *urban* is linked to a specific scale. Zooming in on the region should result in details being displayed. The scale of the current view is the factor that determines whether to represent a city as a punctual location, an urban region, or an aggregation of blocks and streets, etc.

GIS should benefit from a classification of types of extension that includes both *global* and *local* representations for extension. To arrive at such a classification, this article presents a geometry for comparing regions, called *places*, that *have the same extension into all directions*. The name *place* is chosen since regions that have this property can be used well to locate objects within a larger context. A city can be termed a place in a country, but a river cannot, even though there are places *at* or *in* a river. The region of the river stretches too far to be a location, instead it provides several locations along its course.

On the one hand, places are punctual in the sense that extension does not play a decisive role, because it is the same into all directions. On the other hand, places are closely tied to a specific scale, or spatial granularity. A city may be represented as a point only on a large-scale map. Scale depends on size: Two places are of the same size, iff they are both *delimited* by a common *strip*. A strip is a geometric entity consisting of two parallel straight lines and the space in between. The delimited places touch both the straight lines and lie completely inside the strip. This method of comparison employs parallelity as a concept of equidistance.

Places are related to circles of a certain metric. The term *circle* was avoided, since it is used here to denote the circles of the Euclidean metric, which provide one system of places. Another system can be obtained, for instance, using aligned squares, which are important for raster-based conceptualizations of space. Borgo, Guarino, and Masolo [3] and also Bennet [2] have shown that a full characterization of congruence geometry can be axiomatized on a mereotopological basis that defines spheres. In contrast to this approach, the presented geometry characterizes only the congruence of sizes of places.

The rigid framework of straight lines provided by the underlying geometry is used to focus on the extension of the places and abstract from topological notions. The resulting geometry allows a wide spectrum of regions: line segments and curves, unbounded regions, regions with holes, or even certain unconnected regions. The places are restricted only in their extension. The obtained geometry is—except for the places—ontologically neutral wrt. the existence of regions for objects, and can thus be used to link directional and topological calculi[1].

Structure of the Article. The article consists of two parts: An axiomatic specification of a geometry for congruence of size, and a classification of different types of extension built on *local*, i.e. context-dependent, and *global* size intervals of regions. In Sect. 2, the basic geometric notions needed to formulate a concept of the size of regions are presented. The main idea is to obtain a geometric method for comparing the sizes of places. The geometry of places provides a notion of size and extension of regions that is not derived from the distance between points. It is proved to be correct wrt. the concept of distance of classical geometry: the distance between points can be derived from the sizes of places. Additionally, it is shown that the geometry yields a qualitative approach to size, which allows comparison of sizes but not computations on sizes.

In Sect. 3, different types of extension of objects are examined. A size interval derived from the *internal* and *external* extension of an object is used to represent extension. The size interval of an object can be compared to the size interval of another object or a spatial context. A spatial context is defined by the extent of the region of the context and a minimal grain size: Depending on the context region and the grain size, object regions can then be classified as punctual, planar, linear, extended, or local. This classification can be used to choose an appropriate reasoning system (directional or topological) in qualitative

---

[*] A deeper analysis of the ontological consequences of the chosen approach and a detailed specification of the possible link to mereotopological approaches like [5] is beyond the scope of this article.

spatial reasoning, or to formally specify and represent scale-specific geographic attributes more adequately in GIS.

2   A Geometry for Congruence of Places

The basis for the geometric characterization of places is the incidence and ordering geometry of straight lines and points of Eschenbach, Habel, Kulik, and Lessmöllmann [7][2]. To be more exact, axioms (I1)-(I4), which specify the relation of incidence (ι) between straight lines and points, and (β1)-(β7), which characterize the relation of betweenness β between three points, are needed[3].

A first means to compare sizes can be derived from parallelity, if it is conceived as equidistance. Two straight lines are parallel if they do not intersect (D1)[4]:

$$g \parallel g' \overset{\text{def}}{\Leftrightarrow} \neg \exists P : P \iota g \wedge P \iota g' \tag{D1}$$

The axiom of parallelity (A1) and the transitivity of parallelity (A2) can then be formulated:

$$\forall g, P : \neg P \iota g \to \exists g' : P \iota g' \wedge g' \parallel g \tag{A1}$$

$$\forall g_1, g_2, g_3 : g_1 \parallel g_2 \wedge g_2 \parallel g_3 \to g_1 \parallel g_3 \vee g_1 = g_3 \tag{A2}$$

Based on this geometric framework, it is now possible to characterize necessary restrictions for *regions*. The specification is solely based on geometric concepts.

2.1   Projectively Connected Regions

The relation of incidence is widened to cover also the relation between a region and the points that lie in it. Every region A contains at least two points and there is at least one point that is outside the region (A3). Regions differ in the points they contain (A4). This axiom ensures that the regions can be identified uniquely wrt. the points they contain (extensionality).

$$\forall A : \exists P, P', Q : P \neq P' \wedge P \iota A \wedge P' \iota A \wedge \neg(Q \iota A) \tag{A3}$$

$$\forall A, A' : A = A' \leftrightarrow [\forall P : P \iota A \leftrightarrow P \iota A'] \tag{A4}$$

- The geometry is given in a sorted first order logic as the basic format of representation. Variables $g, g', g.$, etc. range over straight lines, $P, Q, R, P', P.$ over points. $A, A', A.$ range over arbitrary regions. Strips and places are special regions, with $t, t', t.$ being variables only for strips and $p, p', p.$ being variables reserved for places.
- Equivalent axiomatizations like the basic axioms for ordering geometry I presented in [16] could also be used. Both are equivalent to the planar version of the axiomatization of incidence and ordering of Hilbert [13]. The axioms are listed in Sect. A.
- For better readability the formulae are abbreviated by saving brackets: The scope of the quantifier is to be read as maximal, i.e. until a bracket closes that was opened before the quantifier. The following order of precedence applies: negation, conjunction, disjunction, implication, biimplication.

Additionally, a geometric equivalent for the topological notion of connection is needed. A very weak characterization that suffices for the purposes of this article and provides a very broad range of regions shall be termed *projective connection*. A region is projectively connected, iff no straight line separates the region without intersecting it (A5). P and Q lie on *different sides* of a straight line g iff neither lies on g and between them is a point R that lies on g (D2).

$$\forall A, g, P, Q : P \iota A \wedge Q \iota A \wedge \text{diffside}(P, Q, g) \rightarrow \exists R : R \iota g \wedge R \iota A \qquad \text{(A5)}$$

$$\text{diffside}(P, Q, g) \overset{\text{def}}{\Leftrightarrow} \neg(P \iota g \vee Q \iota g) \wedge \exists R : \beta(P, R, Q) \wedge R \iota g \qquad \text{(D2)}$$

This concept of connection has the advantage that it allows a very broad notion of regions: Straight lines, line segments, curves and even certain aggregations are regions. The conceptualization of regions is closely related to vision and viability and can thus be very useful for specifying cognitively motivated concepts: An object that occupies a region that is projectively connected cannot be crossed on a straight path through it and it looks solid from outside. Figure 2 shows three projectively connected regions[5].

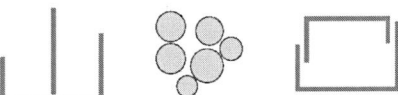

Fig. 2. Examples of projectively connected regions.

Two relations can be defined between arbitrary regions: Given any two regions, A_1 is *part of* A_2 or *lies in* A_2 iff all points lying on A_1 also lie on A_2 (D3). Iff A_1 and A_2 share at least one point, they *intersect* (D4).

$$A_1 \sqsubseteq A_2 \overset{\text{def}}{\Leftrightarrow} \forall P : P \iota A_1 \rightarrow P \iota A_2 \qquad \text{(D3)}$$

$$A_1 \bigcirc A_2 \overset{\text{def}}{\Leftrightarrow} \exists P : P \iota A_1 \wedge P \iota A_2 \qquad \text{(D4)}$$

2.2 Strips

In this section, strips are introduced as special regions that provide a means to compare the sizes of regions. A strip t is fully specified by two parallel straight lines that border it (D5). Since there is no axiom that ensures the existence of arbitrary regions, the existence of strips is ensured separately (A6).

---

[*] The axioms specify some necessary restrictions on regions, but it is left open which regions actually exist. A philosophical analysis of the ontological consequences of different approaches to the concept of object regions has been given by Casati and Varzi [5].

$$\text{border}(t, g_1, g_2) \overset{\text{def}}{\Leftrightarrow} g_1 \parallel g_2 \wedge \forall P : P \iota t \leftrightarrow P \iota g_1 \vee P \iota g_2 \vee \tag{D5}$$
$$\exists Q_1, Q_2 : Q_1 \iota g_1 \wedge Q_2 \iota g_2 \wedge \beta(Q_1, P, Q_2)$$

$$\forall g_1, g_2 : g_1 \parallel g_2 \leftrightarrow \exists t : \text{border}(t, g_1, g_2) \tag{A6}$$

The axioms for parallelity (A1) and (A2) entail that strips exist and that a strip is uniquely characterized by two parallel straight lines.

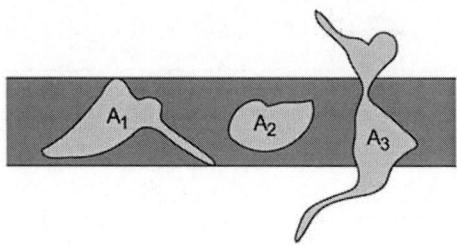

Fig. 3. Relations between a strip t and three regions. $A_{\bullet}$ is *contained* in t, $A_{\bullet}$ *crosses* t. Both relations hold also for $A_{\bullet\bullet}$. Therefore, $A_{\bullet}$ is *delimited* by t. The width of t exactly specifies the width of $A_{\bullet}$ in one direction.

A variant of the relation *cross* $\not\Yup$ of [16] can be defined (D6): A strip t crosses a region A, iff all straight lines that lie in t intersect A. With this relation, strips allow rudimentary comparison of extension into a certain direction: *The strip t delimits $\rightleftharpoons$ the region A*, iff t contains A, and A at the same time crosses t (D7)[6]. Two regions are *co-localized* $\overset{\triangle}{=}$, iff they are delimited by exactly the same strips (D8): A square and the cross spanned by its diagonals, e.g., are co-localized, since they are delimited by the same strips. Figure 3 illustrates the relations of delimitation, containment, and crossing.

$$A \not\Yup t \overset{\text{def}}{\Leftrightarrow} \forall g : g \sqsubseteq t \rightarrow g \bigcirc A \tag{D6}$$

$$A \rightleftharpoons t \overset{\text{def}}{\Leftrightarrow} A \sqsubseteq t \wedge A \not\Yup t \tag{D7}$$

$$A_1 \overset{\triangle}{=} A_2 \overset{\text{def}}{\Leftrightarrow} \forall t : t \rightleftharpoons A_1 \leftrightarrow t \rightleftharpoons A_2 \tag{D8}$$

Since parallelity is not reflexive here, every strip covers some space. With parallelity defined as reflexive, straight lines would also fulfill the requirements for strips. Since strips are the main means of comparison of sizes, reflexive parallelity would imply the existence of a minimal size. Irreflexive parallelity on the other hand implies that only extended regions can have a size.

2.3 Places

Places can now be characterized as special regions that have the same extension in every direction. Using places a notion of size can be derived that supports

---

[•] Note that open regions are not delimited by the closed strips defined here.

comparison of regions. The relation of delimitation ($=$) ensures that arbitrary strips can be compared for their size. There are different models for different systems of places. Circles and aligned squares provide a system of places each (Fig. 5).

Axiom of Coverage. This axiom ensures that the whole space is covered with places: For any point P on a strip t, there is a place p that contains P and is delimited by t.

$$\forall P, t : P \iota t \rightarrow \exists p : P \iota p \wedge p = t \tag{A7}$$

The axiom entails that any point belongs to some place.

Axioms of Delimitation. The next group of axioms ensures the existence of an ordering on places. Figure 4 illustrates the axioms. Two places can be compared wrt. their size by a strip that *delimits* the first and crosses or contains the other (A8). (A9) ensures that places have the same extension in every direction: Assume p_1 lies in t_1 and a place p, which functions as a reference, is delimited by t_1 and also by another strip t_2 that is crossed by p_2. If p_2 is part of any strip t that delimits p_1, then p_2 is delimited by t.

If two places both lie on two strips, then either the strips are identical or the places are co-localized (A10). Finally, it has to be ensured that places are delimited by a strip in every direction (A11): For any place p and straight line g, there are parallels of g that border a strip that delimits p. Without this axiom the places could be open regions or be unbounded in some direction.

$$\forall p_1, p_2 \exists t : p_1 = t \wedge (p_2 \ntrianglelefteq t \vee p_2 \sqsubseteq t) \tag{A8}$$

$$\forall p_1, p_2, t_1, t_2, t : p_1 \sqsubseteq t_1 \wedge p = t_1 \wedge p_2 \ntrianglelefteq t_2 \wedge p = t_2 \wedge \tag{A9}$$
$$p_1 = t \wedge p_2 \sqsubseteq t \rightarrow p_2 = t$$

$$\forall p_1, p_2, t_1, t_2 : p_1 = t_1 \wedge p_2 = t_1 \wedge p_1 = t_2 \wedge p_2 = t_2 \tag{A10}$$
$$\rightarrow p_1 \overset{\triangle}{=} p_2 \vee t_1 = t_2$$

$$\forall p, g : \exists t, g', g'' : g' \parallel g \wedge \mathrm{border}(t, g', g'') \wedge p = t \tag{A11}$$

Relations larger ($\geq$) and smaller ($\leq$) on places can be defined (D9), (D10). A place p_1 is smaller than a place p_2, iff p_1 is delimited by a strip t that is crossed by p_2; correspondingly, p_1 is larger than p_2, iff p_1 is delimited by a strip that contains p_2. The relation of congruence of places $\equiv$ can be derived: Two places that are delimited by a common strip are congruent (D11). $<$ and $>$ are irreflexive variants of $\leq$ and $\geq$, respectively, that exclude congruent places (D12) and (D13).

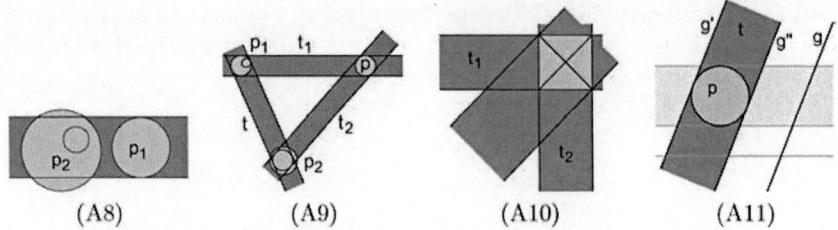

(A8)   (A9)   (A10)   (A11)

Fig. 4. Illustration of the axioms of delimitation.

$$p_1 \leq p_2 \overset{\text{def}}{\Leftrightarrow} \exists t : p_1 \dot= t \land p_2 \not\Subset t \tag{D9}$$

$$p_1 \geq p_2 \overset{\text{def}}{\Leftrightarrow} \exists t : p_1 \dot= t \land p_2 \sqsubseteq t \tag{D10}$$

$$p_1 \equiv p_2 \overset{\text{def}}{\Leftrightarrow} \exists t : p_1 \dot= t \land p_2 \dot= t \tag{D11}$$

$$p_1 < p_2 \overset{\text{def}}{\Leftrightarrow} p_1 \leq p_2 \land \neg p_1 \equiv p_2 \tag{D12}$$

$$p_1 > p_2 \overset{\text{def}}{\Leftrightarrow} p_1 \geq p_2 \land \neg p_1 \equiv p_2 \tag{D13}$$

From the axioms and the definitions, the following theorems can be proved, showing that the relation $<$ is a dense strict linear ordering (transitive (1), asymmetric (2), linear (3), dense (4)), and that $\geq$ and $\leq$ are related as expected (5). Consequently, congruence $\equiv$ is an equivalence relation that yields exactly the possible sizes of places as its equivalence classes (6).

$$\forall p_1, p_2 : p_1 < p_2 \land p_2 < p_3 \rightarrow p_1 < p_3 \tag{1}$$

$$\forall p_1, p_2 : p_1 < p_2 \rightarrow \neg p_2 < p_1 \tag{2}$$

$$\forall p_1, p_2 : p_1 \equiv p_2 \lor p_1 < p_2 \lor p_1 > p_2 \tag{3}$$

$$\forall p_1, p_2, t_1 : p_1 < p_2 \rightarrow \exists p : p_1 < p \land p < p_2 \tag{4}$$

$$\forall p_1, p_2 : p_1 \leq p_2 \leftrightarrow p_2 \geq p_1 \tag{5}$$

$$\forall p_1, p_2 : p_1 \leq p_2 \land p_1 \geq p_2 \rightarrow p_1 \equiv p_2 \tag{6}$$

$$\forall p_1, p_2, t_1 : p_1 \dot= t_1 \land p_2 \not\Subset t_1 \rightarrow \exists t_2 : p_2 \dot= t_2 \land p_1 \sqsubseteq t_2 \tag{7}$$

$$\forall p_1, p_2, t_1 : p_1 \dot= t_1 \land p_2 \sqsubseteq t_1 \rightarrow \exists t_2 : p_2 \dot= t_2 \land p_1 \not\Subset t_2 \tag{8}$$

The properties follow mainly from the axioms of delimitation and the interrelations between *containment* ($\sqsubseteq$) and *crossing* ($\not\Subset$) (7), (8). Density of sizes (4) follows from the density of betweenness.

2.4   Models for the Geometry

The places as described by the axioms are circular in the sense that they stretch over a certain distance in every direction. Figure 5a-c illustrates the axioms of delimitation for three possible types of places. The system of circles is the most intuitive model for places. The equidistance given by the strips that are compared by circles corresponds to Euclidean distance.

Another model is given if we set the squares at a given alignment as the places. The derived concept of equidistance differs from Euclidean distance in the diagonal direction: The diagonal of a square is here of the same length as the sides. Since space may be stretched or compressed into any direction the rectangles of a certain fixed ratio of sides provide also a system of places. This ratio may even vary with the different sizes of places; however, it would be inconsistent with the axioms, if the ratio would vary across space or if the alignment of squares of the same size is not fixed (Fig. 5d). The system of squares can be useful, e.g., to reason about raster-based data. However, the geometry does not characterize a raster, since the cells of the raster cannot be the only places in the geometry; rather, the raster selects certain places of the same size in order to cover space.

Before moving on to the characterization of different types of extension in Sect. 3, a slight digression is taken to compare the resulting axiomatic system to the congruence of line segments in classical geometry, and to prove that indeed a qualitative system of sizes is obtained[7].

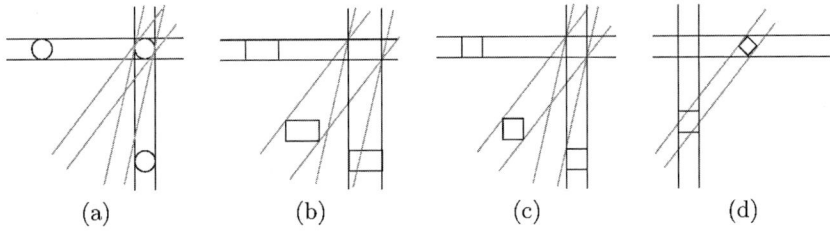

      (a)   (b)   (c)   (d)

Fig. 5. Circles (a), aligned rectangles (b), and aligned squares (c) provide a system of places each. The alignment of squares has to be fixed (d).

2.5 Relation to Congruence of Line Segments

In this section, it is shown how the characterization presented so far can be related to the concept of congruence of line segments in Hilbert's axiomatization of geometry [13]. The definition of congruence of line segments can be derived from the congruence of places, if an additional axiom is introduced that ensures that line segments can be compared. The axiom is thus not necessary for the definitions in Sect. 3. The additional axiom is introduced here to illustrate that the presented concept of congruence can be related to the congruence of line segments of classical geometry in a meaningful way. It will also be shown that the axiomatization of congruence of size as presented here is indeed weaker than the axiomatization of congruence of line segments in [13].

[.] In the terminology of Carnap [4], a system of comparison of values as introduced here would be termed *non-extensive*. He reserves the notion *qualitative* for classificatory systems of values. The term is applied in this article in the broader sense employed in qualitative spatial reasoning.

Axiom of Construction. All previous axioms except (A7) ensured merely comparability of places. (A7) is too weak to support comparison of line segments. For this reason, an additional axiom is needed:

$$\forall t, g, P : t \bigcirc g \wedge \neg g \sqsubseteq t \wedge \neg P \iota g \rightarrow \exists p : p \doteq t \wedge p \bigcirc g \wedge \tag{A12}$$

$$\forall Q : Q \iota p \rightarrow \neg \operatorname{diffside}(g, P, Q)$$

$$\forall t_1, t_2 : t_1 \bigcirc t_2 \rightarrow \exists p : p \doteq t_1 \wedge (p \sqsubseteq t_2 \vee p \not\Subset t_2) \tag{9}$$

(A12) supplements (A7) in that every straight line that intersects a strip can generate on each side a place that is delimited by the strip. A consequence is that all strips that intersect can be compared directly by a place (9).

The length of line segments and thus congruence of line segments can be defined: A line segment is called the *diameter* of a place p, if it connects two points that both incide with the place but also lie on the opposite borderlines of a strip that delimits p (D14). Two line segments are congruent, iff they are diameters of congruent places (D15).

$$\operatorname{dm}(p, P, Q) \overset{\text{def}}{\Leftrightarrow} P \iota p \wedge Q \iota p \wedge \exists t : \overline{PQ} \doteq t \wedge p \doteq t \tag{D14}$$

$$\overline{PQ} \equiv \overline{P'Q'} \overset{\text{def}}{\Leftrightarrow} \exists p, p' : p \equiv p' \wedge \operatorname{dm}(p, P, Q) \wedge \operatorname{dm}(p, P', Q') \tag{D15}$$

The existence of a diameter for every place can be inferred. However, it is not guaranteed that every line segment is a diameter of some place; and systems of places in general do not provide a unique diameter for a given place and strip. Figure 6 shows the system of diagonals of squares, for which both criteria are not fulfilled. Nevertheless, the congruence of line segments directly corresponds

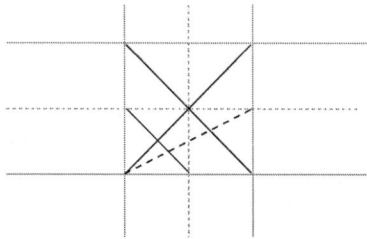

Fig. 6. Places need not to be convex regions. Crosses are sufficient to span the same distances as squares. However, segments stretching into a direction other than vertical, horizontal or diagonal cannot be diameters of such places, and thus are not assigned a length in this case. The diameter in addition is not unique for a given place and strip: Both the diagonal and the vertical line connecting the corners are diameters of the place wrt. the horizontal strip.

to the congruence of places, because it can be proved that if a line segment is a diameter of places p_1 and p_2, then p_1 and p_2 are congruent.

$$\forall P, Q, p, p' : \operatorname{dm}(p, P, Q) \wedge \operatorname{dm}(p', P, Q) \rightarrow p \equiv p' \tag{10}$$

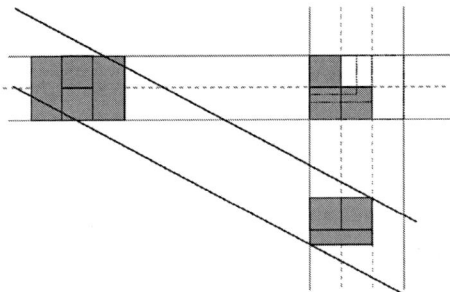

Fig. 7. Model that does not support the addition of lengths: Places grow faster into the horizontal direction than into the vertical direction.

This theorem is important, because it guarantees that the system of congruence of places presented here is compatible with the usual notion of congruence.

Proof of (10): Suppose $p' \leq p$, and t be the strip delimiting both p and $\overline{PQ}$ as required by (D14); t' be the corresponding strip delimiting p' and $\overline{PQ}$. Following from (9) and $p' \leq p$, there are $\hat{t} \sqsubseteq t$ and $\hat{p} = \hat{t}$ and $\hat{p} = t'$. $\overline{PQ} \nparallel t$ and also $\overline{PQ} \nparallel \hat{t}$ holds, because every line segment that crosses t crosses all straight lines of t and therefore also crosses all straight lines of $\hat{t}$. Consequently, $p \nparallel \hat{t}$ but also $p' \nparallel \hat{t}$ holds, because places, like all regions, are projectively connected: Every straight line that incides with a point between two points of the place incides with a point of the place, and P and Q are points of p'. But since $\hat{p} \equiv p'$ holds, $p' \nparallel \hat{t}$ can be valid only, if $p' \sqsubseteq \hat{t}$ is valid, too; and therefore both P and Q lie in $\hat{t}$. Thus, $\hat{t}$ and t have the same delimiting straight lines, i.e. are identical. □

The first two axioms of congruence in [13] can then be derived. In fact, axiom (A12) corresponds directly to the planar version of Hilbert's (III.1). (III.2) ensures that congruence is an equivalence relation, which follows here from the direct correspondence between congruence of places and their diameters. The axiom of addition (III.3) of [13] does not hold. (III.3) says that for two segments $\overline{PQ}$ and $\overline{QR}$ that lie on the same straight line having corresponding congruent segments $\overline{P'Q'}$ and $\overline{Q'R'}$ on another straight line, the segments $\overline{PR}$ and $\overline{P'R'}$ have also to be congruent.

Figure 7 shows a counter example: Assume a system of aligned rectangles similar to that in Fig. 5, but with the difference that the rectangles grow faster in some direction d_1 than in the other d_2. In this case, the ordering constraints still hold between the places of different sizes, but two places p_1 and p_2 joined along the faster growing direction d_1 yield a place of a smaller size than p_1 and p_2 joined along direction d_2.

3   Representing Extension

We can now assign a size to places, but there is still no way to assign a size to regions like the example of the river in Fig. 1. Extended objects have a global maximal primary extension, but differing local secondary extensions. A map depicting the river at its source would need another resolution than a map of the river at its mouth. For extended objects like rivers there is no map depicting all relevant details of the object at every portion. Accordingly, a spatial information system should be sensitive to intended local levels of granularity in user interaction. Relations like proximity but also even directional relations like *north* depend on the extension and thus the local size of the objects [16]. Goyal and Egenhofer [11] provide granularly sensitive directional relations. The question when to depict a river as a line, and when as a two-dimensional region, depends on its local size wrt. the spatial context. The spatial context in addition is influenced by scale. A computer screen for instance has a size and resolution. In GIS, the scale at which the currently selected attributes are valid together with the region that provides the area under inspection are important. To mirror this, a suitable region together with a minimal grain size is called a *scale-specific spatial context*.

In the following, a framework of local size intervals is sketched that can be used to represent the (local) sizes of extended objects in a scale-specific spatial context. The focus in this section lies on the relations between objects and between objects and spatial contexts. The assumed range of regions allowed here is restricted to two-dimensional regions to which a size interval can be assigned.

3.1   Size Intervals

The global primary size of an object is—informally speaking—the smallest place that fully contains the region of the object. Correspondingly, the (local) secondary size is determined by the largest place (in a context place) that is fully contained in the object region. Examples are shown in Fig. 8. The two sizes provide the minimum and maximum of a size interval that describes the local extension of the object. Minimum and maximum are assumed to be given within a range of uncertainty. The inclusion of uncertainty is necessary for two reasons: first, the axiomatic characterization of size does not support arbitrary limiting values as would be needed for ensuring that the maximum/minimum exist; second, geographic information about position is obtained by measurement and thus subject to uncertainty [10]. Generally, GIS as computer systems contain a certain degree of uncertainty. A characterization of uncertainty is beyond the scope of this article. It is simply assumed, that a relation of indifference $\approx$ is given, that holds between two places of *similar* size[8].

Since the equivalence classes of $\equiv$ exactly yield the sizes of places, the following definitions will be defined based on these sizes. Following Galton [9] intervals can be defined from instances. Every pair of places p_1, p_2 that are ordered by

---

[8] The basic modelling can be taken from basic measurement theory (cf. Suppes and Zinnes [17]). A relation of indifference is used in cognitive psychology to model relations of similarity prominent in tasks of classification: If a subject describes colors or temperatures a and b and b and c to be the same, he need not describe a and c to be the same. Uncertainty is expected here to be of this kind.

$p_1 < p_2$ provides a size interval s that has the size of p_1 as its start and the size of p_2 as its final point (A13). The size of a place p lies in a size interval s ($in(p, s)$), iff it lies between the limits or is congruent with one of the limits (D16). Size intervals are fully characterized by the sizes they range over (extensionality (A14)).

$$\forall p_1, p_2 : p_1 < p_2 \leftrightarrow \exists s : start(p_1, s) \wedge end(p_2, s) \tag{A13}$$

$$in(p, s) \overset{\text{def}}{\Leftrightarrow} \exists p_1, p_2 : start(p_1, s) \wedge end(p_2, s) \wedge p_1 \leq p \leq p_2 \tag{D16}$$

$$\forall s_1, s_2 : [\forall p : in(p, s_1) \leftrightarrow in(p, s_2)] \rightarrow s_1 = s_2 \tag{A14}$$

Interval relations can be defined on the intervals (cf. [9] for a thorough review of the theory of interval relations for the domain of time intervals). Reasoning about size intervals does not differ from reasoning about time intervals. For this article, the following relations are especially important: $s_1 \prec s_2$ denotes that all sizes of s_1 are smaller than all sizes of s_2. $s_1 \preccurlyeq s_2$ holds, iff s_1 starts at smaller sizes than s_2, but ends with a size in s_2: s_1 is *smaller but compatible* with s_2. General *compatibility* is given, iff two intervals share a certain size ($s_1 \sim s_2$). A size interval s_1 is *contained in* s_2, iff it shares all sizes with s_1 (s_1 in s_2).

$$s_1 \prec s_2 \overset{\text{def}}{\Leftrightarrow} end(s_1) < start(s_2) \tag{D17}$$

$$s_1 \preccurlyeq s_2 \overset{\text{def}}{\Leftrightarrow} start(s_1) < start(s_2) \leq end(s_1) < end(s_2) \tag{D18}$$

$$s_1 \sim s_2 \overset{\text{def}}{\Leftrightarrow} \exists p : in(p, s_1) \wedge in(p, s_2) \tag{D19}$$

$$s_1 \text{ in } s_2 \overset{\text{def}}{\Leftrightarrow} \forall p : in(p, s_1) \rightarrow in(p, s_2) \tag{D20}$$

3.2   External and Internal Place

The size intervals for regions can now be constructed. It was not necessary for the axiomatization of space to characterize the space a place stretches across. Therefore a square and a cross that spanned it were considered to be equally sufficient for the axiomatization of space. In this section, the places are used as a means to localize and reason about object regions. Therefore it is useful to be able to access the content of places and other regions to ensure that places can be used to localize objects. To achieve this, a simplified variant of the convex hull of a region A is defined: The convex hull $ch(A)$ of a region A contains all points of A and all points between the points of A:

$$P \iota ch(A) \overset{\text{def}}{\Leftrightarrow} P \iota A \vee \exists P_1, P_2 : P_1 \iota A \wedge P_2 \iota A \wedge \beta(P_1, P, P_2) \tag{D21}$$

A place is *external* wrt. a region A, iff it fully contains the region and all places that also contain the region are either larger than p or have approximately the same size (D22). Correspondingly, a place is *internal* wrt. a region A, iff it is fully contained in the region and all places that are also contained in the region are either smaller than p or are approximately of the same size (D23). As demanded in Sect. 1, the internal place should also be defined in a version that mirrors that the secondary size varies with the spatial context. An additional place is used as *context place*: a place p *is internal wrt. a region A and a context p_c*, iff it is fully contained in the region and either p is contained in the context place p_c or p_c is contained in p, and all places that also fulfill these requirements are either smaller than p or are approximately of the same size (D24). Every global internal place is also an internal place local to an external place (11).

$$\text{ext}(p, A) \overset{\text{def}}{\Leftrightarrow} A \sqsubseteq \text{ch}(p) \wedge \tag{D22}$$
$$\forall p' : A \sqsubseteq \text{ch}(p') \to p < p' \vee p \approx p'$$

$$\text{int}(p, A) \overset{\text{def}}{\Leftrightarrow} \text{ch}(p) \sqsubseteq A \wedge \tag{D23}$$
$$\forall p' : \text{ch}(p') \sqsubseteq A \to p > p' \vee p \approx p'$$

$$\text{int}(p_c, p, A) \overset{\text{def}}{\Leftrightarrow} (p \sqsubseteq p_c \vee p_c \sqsubseteq p) \wedge \text{ch}(p) \sqsubseteq A \wedge \tag{D24}$$
$$\forall p' : (p' \sqsubseteq p_c \vee p_c \sqsubseteq p') \wedge \text{ch}(p') \sqsubseteq A \to p > p' \vee p \approx p'$$

$$\forall p, p' : \text{ext}(p, A) \to (\text{int}(p', A) \leftrightarrow \text{int}(p, p', A)) \tag{11}$$

External and internal place are illustrated in Fig. 8. The local extension of appropriate regions can then be described uniquely up to uncertainty (D25).

$$\text{extent}(p_c, s, A) \overset{\text{def}}{\Leftrightarrow} [\exists p_1 : \text{ext}(p_1, A) \wedge \text{end}(p_1, s)] \tag{D25}$$
$$\wedge [\exists p_2 : \text{int}(p_c, p_2, A) \wedge \text{start}(p_2, s)]$$

Given a fixed context place p_c, we can compare—and reason about—the local size intervals of objects. The global extent of an object can be determined by replacing the context-dependent version of the internal place with the context-independent defined in (D23).

Fig. 8. Object regions with internal and external places. For local internal places, the context place is indicated with a dashed line.

3.3   Classification of Scale-Specific Extension

The extension of an object relative to a scale-specific spatial context can be characterized (D26). A *scale-specific spatial context* is determined by a place p

and a size interval s. The extent of p is the maximal size of s. Objects larger than this size are represented only partially in the spatial context. The minimal size of s gives the grain size, below which extension is not represented. An object smaller than this size is punctual, and any part can be localized in the same grain-sized place as the object itself. The *extension s_A of a region A local to a spatial context (p, s)* can be defined as the extension of A in the context of p (D26). And this local extent s_A of A can be compared to the scale s of the spatial context.

$$\text{ssc}(A, s_A, p, s) \overset{\text{def}}{\Leftrightarrow} \text{end}(p, s) \wedge \text{extent}(p, s_A, A) \tag{D26}$$

A classification of local extension can be derived from the relation between s and s_A: An object region A is *punctual, linear,* or *planar, extended* or *local* in a spatial context determined by a place p and a scale given by the size interval s. The relations are illustrated in Fig. 9.

$$\text{punctual}(A, p, s) \overset{\text{def}}{\Leftrightarrow} \exists s_A : \text{ssc}(A, s_A, p, s) \wedge s_A \prec s \tag{D27}$$

$$\text{linear}(A, p, s) \overset{\text{def}}{\Leftrightarrow} \exists s_A : \text{ssc}(A, s_A, p, s) \wedge s \text{ in } s_A \tag{D28}$$

$$\text{planar}(A, p, s) \overset{\text{def}}{\Leftrightarrow} \exists s_A : \text{ssc}(A, s_A, p, s) \wedge s \prec s_A \tag{D29}$$

$$\text{extended}(A, p, s) \overset{\text{def}}{\Leftrightarrow} \exists s_A : \text{ssc}(A, s_A, p, s) \wedge s \preccurlyeq s_A \tag{D30}$$

$$\text{local}(A, p, s) \overset{\text{def}}{\Leftrightarrow} \exists s_A : \text{ssc}(A, s_A, p, s) \wedge (s_A \preccurlyeq s \vee s_A \text{ in } s) \tag{D31}$$

Punctual objects have no relevant extension. They provide simple locations or may not be represented at all, depending on relevance. Reasoning with directional relations between punctual objects is inferentially nearly as powerful as reasoning over points.

Linear objects can be represented by lines, if relevant. Reasoning has to observe extension. Linear objects could border or separate regions in topological reasoning.

Planar objects provide the background of a spatial context, if they are represented at all. Directional reasoning is not possible, topological reasoning is trivial, since any region that is in the context is also part of the planar object. The representation could therefore instead represent parts of the planar object that are local to the spatial context.

Extended objects : If $\preccurlyeq$ holds instead of $\prec$ the object is not planar but extended in the context. The object is represented only partially in the spatial context. Therefore, the reasoning system has to observe extension.

Local objects have no special representations. Approaches that do not represent the scale of a spatial context treat all objects as local objects.

The classification as described above is appropriate only for topologically connected two-dimensional regions. Aggregations, like for instance a forest represented as a tree distribution (cf. Bennett [1]), can also be assigned a size interval, but an aggregation of comparatively small sized atoms would be categorized misleadingly as *linear*, since the minimal size is determined by the size of the

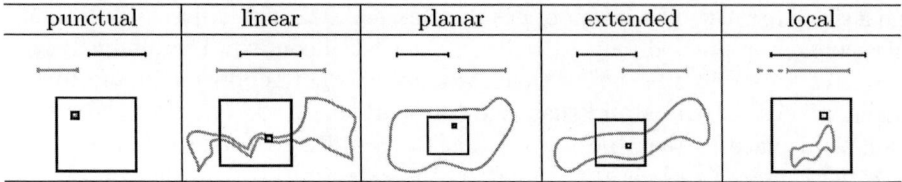

punctual	linear	planar	extended	local

Fig. 9. Types of extension for object regions (grey) in a scale-specific spatial context (black): The extension type is determined by the relation between the size interval of the object and the size interval assigned to the spatial context. The spatial context need not to be a raster, but it has to have a *grain* size, below which extension is not represented.

atoms. At this stage, an object like a forest is best represented using multiple representations: (a) as a forested region on larger scales and (b) as a distribution of trees on smaller scales. The more detailed representation (b) could be chosen, if the simpler representation (a) is not sufficient, e.g., if it is classified as planar. It would need a narrower concept of regions or another characterization of the minimal size to solve this problem.

4   Conclusion & Outlook

The proposed axiomatic system for congruence of places shows that it is possible to define a notion of size without recourse to the distance between points. The resulting geometry is parsimonious wrt. the necessary mathematical concepts on the side of computation: The axioms of delimitation yield just the necessary relations of congruence and ordering of sizes. The system of comparison of sizes that was presented in Sect. 2 provides a means for assigning sizes to spatial regions as demonstrated in Sect. 3. Problems resulting from the definition of size based on the distance of points are moderated.

Size intervals were introduced to represent the extension of appropriate object regions. Reasoning about intervals is a well-investigated branch of qualitative reasoning with a wealth of results that also apply to the introduced size intervals. In addition to the global extension, the local extension of an object region in a scale-specific spatial context was described. Regions can then be classified according to their relative size in a context.

The article presents a Euclidean plane as the geometric basis, but it should be the next step to construct a suitable geometry of strips and places for the spherical globe. The main adjustment needed is to arrive at a concept of parallelity as geometric equidistance that is suitable to compare regions on the globe for their sizes. Another possibility is to adjust the system of strips and places to build upon the grid lines of Kulik and Klippel [14] and the quasi-parallelity of grid-lines. The error is negligible for maps of small regions since the geometry of the sphere comes close to Euclidean geometry in this case. Reasoning about such places would be appropriate for modelling user interaction with maps. Other possible applications include automated map drawing and context dependent qualitative spatial reasoning.

The proposed geometry includes notions of context and uncertainty and a formally precise account of scale-specificity or spatial granularity. This should be useful in all areas of spatial information theory in which context dependency and granularity play a prominent role. The use of axiomatic geometry ensures that the formalization is adequately general to be used for various purposes, but nevertheless, is sufficiently specific to precisely capture properties of spatial granularity and spatial contexts. Future work will include implementations and experimental evaluations of the representational expressiveness gained from the approach.

A   Axioms for Incidence and Ordering Geometry

The axioms for incidence and ordering geometry of [7]. Incidence between straight lines and points is characterized in (I1)-(I4).

$$\forall g : \exists P, Q : \quad P \neq Q \wedge P \iota g \wedge Q \iota g \tag{I1}$$

$$\forall P, Q : \quad \exists g : P \iota g \wedge Q \iota g \tag{I2}$$

$$\forall P, Q, g_1, g_2 : \quad P \neq Q \wedge P \iota g_1 \wedge P \iota g_2 \wedge Q \iota g_1 \wedge Q \iota g_2 \rightarrow g_1 = g_2 \tag{I3}$$

$$\forall g : \exists P : \quad \neg P \iota g \tag{I4}$$

$$\mathrm{col}(P, Q, R) \stackrel{\mathrm{def}}{\Leftrightarrow} P \neq Q \wedge P \neq R \wedge Q \neq R \wedge \exists g : P \iota g \wedge Q \iota g \wedge R \iota g \tag{DC}$$

The axioms of (β1)-(β7) ensure basic ordering constraints in the plane, and that space is planar. Density of points on a straight line follows from (β7).

$$\forall P, Q, R : \quad \beta(P, Q, R) \rightarrow \mathrm{col}(P, Q, R) \tag{β1}$$

$$\forall P, Q, R : \quad \beta(P, Q, R) \rightarrow \beta(R, Q, P) \tag{β2}$$

$$\forall P, Q, R : \quad \beta(P, Q, R) \rightarrow \neg \beta(Q, P, R) \tag{β3}$$

$$\forall P, Q, R : \quad \mathrm{col}(P, Q, R) \rightarrow \beta(P, Q, R) \vee \beta(Q, P, R) \vee \beta(P, R, Q) \tag{β4}$$

$$\forall P, Q, Q', R : \quad \beta(P, Q, R) \wedge \mathrm{col}(Q, Q', P) \rightarrow \beta(P, Q, Q') \vee \beta(Q', Q, R) \tag{β5}$$

$$\forall P, Q : \quad P \neq Q \rightarrow \exists R : \beta(P, Q, R) \tag{β6}$$

$$\forall P_1, P_2, P_3, g : \neg P_1 \iota g \wedge \neg P_2 \iota g \wedge \neg P_3 \iota g \wedge [\exists Q : Q \iota g \wedge \beta(P_1, Q, P_3)] \tag{β7}$$
$$\rightarrow \exists R : R \iota g \wedge (\beta(P_1, R, P_2) \vee \beta(P_2, R, P_3))$$

References

1. B. Bennett. Application of supervaluation semantics to vaguely defined spatial concepts. In D. Montello, editor, *Spatial Information Theory: Foundations of Geographic Information Science*, pages 108–123, Berlin, 2001. Springer.
2. B. Bennett. A categorical axiomatisation of region-based geometry. *Fundamenta Informaticae*, 46:145–158, 2001.

3. S. Borgo, N. Guarino, and C. Masolo. A pointless theory of space based on strong connection and congruence. In L. C. Aiello, J. Doyle, and S. Shapiro, editors, *KR'96: Principles of Knowledge Representation and Reasoning*, pages 220–229. Morgan Kaufmann, San Francisco, California, 1996.

4. R. Carnap. *Philosophical Foundations of Physics*. Basic Books, New York, NY, 1966.

5. R. Casati and A. C. Varzi. *Parts and Places: the Structure of Spatial Representations*. MIT Press, 1999.

6. A. G. Cohn and S. M. Hazarika. Qualitative spatial representation and reasoning: An overview. *Fundamenta Informaticae*, 46(1-2):1–29, 2001.

7. C. Eschenbach, C. Habel, L. Kulik, and A. Leßmöllmann. Shape nouns and shape concepts: A geometry for 'corner'. In C. Freksa, C. Habel, and K. Wender, editors, *Spatial Cognition: An Interdisciplinary Approach to Representing and Processing Spatial Knowledge*, pages 177–201. Springer, Berlin, 1998.

8. A. Frank. Qualitative spatial reasoning about distances and directions in geographic space. *Journal of Visual Languages and Computing*, 3:343–371, 1992.

9. A. Galton. Space, time and movement. In O. Stock, editor, *Spatial and Temporal Reasoning*, pages 321–352. Kluwer, 1997.

10. M. Goodchild. A geographer looks at spatial information theory. In D. Montello, editor, *Spatial Information Theory: Foundations of Geographic Information Science*, pages 1–13, Berlin, 2001. Springer.

11. R. Goyal and M. Egenhofer. Consistent queries over cardinal directions across different levels of detail. In U. Tjoa, R. Wagner, and A. Al-Zobaidie, editors, *11th International Workshop on Database and Expert Systems Applications*, pages 876–880, 2000.

12. A. Herskovits. Language, spatial cognition, and vision. In O. Stock, editor, *Spatial and Temporal Reasoning*, pages 155–202. Kluwer, 1997.

13. D. Hilbert. *Grundlagen der Geometrie*. Teubner, Stuttgart, 1956.

14. L. Kulik and A. Klippel. Reasoning about cardinal directions using grids as qualitative geographic coordinates. In C. Freksa and D. Mark, editors, *Spatial Information Theory: Cognitive and Computational Foundations of Geographic Information Science*, pages 205–220, Berlin, 1999. Springer.

15. G. Ligozat. Reasoning about cardinal directions. *Journal of Visual Languages and Computing*, 9:23–44, 1998.

16. H. R. Schmidtke. The house is north of the river: Relative localization of extended objects. In D. Montello, editor, *Spatial Information Theory: Foundations of Geographic Information Science*, pages 414–430, Berlin, 2001. Springer.

17. P. Suppes and J. Zinnes. Basic measurement theory. In R. Luce, R. Bush, and E. Galanter, editors, *Handbook of Mathematical Psychology*. John Wiley & Sons, New York, 1963.

Intuitive Modelling of Place Name Regions for Spatial Information Retrieval

Thomas Vögele[1], Christoph Schlieder[2], and Ubbo Visser[1]

. Center for Computing Technologies (TZI), Universität Bremen
Postfach 33 04 40, 28334 Bremen, Germany
{vogele,visser}@tzi.de
http://www.tzi.de
. Universität Bamberg
Feldkirchenstraße 21, 96045 Bamberg, Germany
christoph.schlieder@wiai.uni-bamberg.de
http://www.uni-bamberg.de

Abstract. Reasoning about spatial relevance is important for intelligent spatial information retrieval. In heterogeneous and distributed systems like the Semantic Web, spatial reasoning has to be based on light-weight, interoperable and easy-to-use spatial metadata.

In this paper we present an approach to an intuitive and user-friendly creation and application of spatial metadata that are used for spatial relevance reasoning. The metadata are based on discrete approximations of place name regions. Based on knowledge about cognitive aspects of preferred spatial models, our approach allows for the representation and intuitive modelling of indeterminate regions in addition to regions with well-known boundaries.

1 Introduction

Lately, efforts are under way to push for a new quality of information retrieval on the Internet. These efforts tackle the problem that much of the information on the Internet is machine-readable, but not machine-understandable. The goal is to establish a so-called "Semantic Web" in which the semantics of information items can be "understood" and used for "intelligent" information retrieval by machines.

Besides digital maps and other geospatial data that are directly geo-referenced by geographic coordinates, there is a large amount of data that are indirectly geo-referenced through *place names*. Therefore, important aspects of spatial information retrieval on the Semantic Web are the resolution of indi-rectly specified geo-references, and the evaluation of the spatial (in the sense of geographic) relevance of the desired information with respect to spatial queries.

A number of well-established tools like Geographic Information Systems (GIS) and gazetteers support information retrieval based on geographic refer-ence. However, while providing experts with powerful means to process geospatial data, these tools are generally not well equipped to handle common sense spatial knowledge in an user-friendly, intuitive way [4].

W. Kuhn, M.F. Worboys, and S. Timpf (Eds.): COSIT 2003, LNCS 2825, pp. 239–252, 2003.

The Semantic Web is a highly distributed, heterogeneous, and user-centric system. The majority of information seekers and information providers are not experts in the domain of geographic information processing. Intelligent spatial information retrieval on the Semantic Web therefore requires tools that are able to handle intuitive models of common sense spatial knowledge. These tools must provide means to annotate information with machine-understandable spatial metadata, as well as to specify meaningful spatio-thematic queries of the type *concept@location* like: "Find information about *hotels* in the *Schweizer Jura* region". The result set of such a query should be sorted according to thematic and spatial relevance, and it should include near hits as well. For the *concept* part of the query, near hits would mean other types of accommodation (e.g, pensions, bed and breakfast etc.). With respect to *location*, near hits would include all geographic regions that are *spatially relevant* to the *Schweizer Jura*.

In [15] and [14], we outlined an approach to reason about the relative spatial relevance of geographic regions. This approach makes use of an interoperable and light-weight representation scheme that is obtained through a thematic projection of geographic regions onto graph-based abstractions of discrete polygonal tessellations. So far, our representation scheme accounts for regions with well-defined boundaries only. Through a binary mapping, we obtain a discrete representation of a region in form of a *qualitative spatial footprint*.

Obviously, this approach has a number of shortcomings: For one, the approximations of regions obtained through a binary mapping are rather crude. This can be attributed mainly to the limited resolution of the "natural" polygonal reference tessellations (e.g., administrative subdivisions, postal code areas) we use as the basis for our spatial reference model. Secondly, the approach does not yet account for an appropriate representation of the hierarchical partonomic structure of place name regions. And lastly, place name regions with indeterminate boundaries cannot be handled in an intuitive and user-friendly way. For example, a user may want to add a place name region of which he knows only that it is part of another place name region, or that it contains a number of other, smaller regions.

In this paper, we extend our approach to be able to reduce the imprecision of discrete approximations, and to handle regions with incompletely defined boundaries. The paper is structured as follows: In section 2, we will review our basic methods and assumptions for spatial relevance reasoning. In section 3, we will apply an approach for the approximation of regions in discrete space that solves some of the problems associated with the imprecision of spatial footprint approximations. We will also describe a method that uses the improved footprint approximations to infer a partonomic hierarchy of place name regions. In section 4, we will discuss cognitive aspects of *preferred spatial models*, and how they can be used to find a heuristic for the intuitive integration of new place name regions with indeterminate boundaries in an existing place name structure. Finally, in section 5, we will demonstrate the approach using a real-world example, and conclude with a discussion and an outlook.

2   Reasoning about Spatial Relevance

Our approach to spatial information retrieval and reasoning about spatial relevance is focussed on geographic regions that have a well-defined name (i.e., "Kanton Uri, Switzerland"). Such *place names* are frequently used for indirect geo-referencing because they offer a user-friendly and sound method to annotate information with spatial metadata, and to specify spatial queries. Place names are typically organized in place name lists, or *gazetteers*, that use coordinate-based spatial footprints to provide a reference to geographic space [7], [12].

To overcome some of the shortcomings of coordinate-based spatial footprints, and to improve the spatial reasoning capabilities of gazetteers, we introduced *qualitative spatial footprints* [15], [17]. Qualitative spatial footprints are discrete approximations of the regional extent of place names. They are obtained through a binary mapping of such *place name regions* onto a discrete polygonal reference tessellation. This approach is similar to other work in qualitative spatial reasoning (e.g. [6], [2]) in that it extends the representation of regions, and topological relations between regions, to discrete space to provide the basis for a number of practical applications. In our case, discrete approximations of place name regions are the basis for spatial relevance reasoning.

Qualitative spatial footprints encode place name regions in terms of reference units derived from polygonal tessellations, the so-called *standard reference tessellations (SRT)*. Because they provide a set of well-known identifiers (e.g. the names of administrative units), such "natural" polygonal tessellations allow the human user to interact with the system in a more intuitive way than do "artificial" reference tessellations (e.g. regular orthogonal grids, triangulated irregular networks). In addition, many polygonal tessellations provide meaningful decomposition hierarchies that can be used for spatial relevance reasoning. A typical example are tessellations of administrative subdivisions. Here, the semantics of the administrative organization of a geographic region are reflected in the hierarchy of the decomposition tree.

A condensed and interoperable representation of standard reference tessellations is obtained by using graph-based abstractions. With the *connection graph*, we introduced an improved version of neighborhood graphs [9] for this purpose. Connection graphs are planar graphs that encode topologic neighborhood relations between the polygons of a tessellation. That way, polygonal reference tessellations can be reduced to a set of connection graphs (representing neighborhood relations at different levels of granularity) that are interconnected by a decomposition tree (representing the hierarchical partonomy of reference units).

In a first attempt to reason about the relative spatial relevance of objects and regions in geographic space we work with the hypothesis that the relative spatial relevance of two (place name) regions a and b is inversely proportional to both their (spatial) distance $D_H(a, b)$ in the connection graph, and their partonomic distance $D_V(a, b)$ in the decomposition tree. In the simplest case, a linear function can be applied to compute the cumulative distance $D_C(a, b)$ as: $D_C(a, b) = \alpha \times D_H(a, b) + (1 - \alpha) \times D_V(a, b)$. Depending on the context of the query, the weighting factor α is used to bias the computation either towards the evaluation of true spatial distance (i.e. an qualitative approximation of Euclidean distance), or distance within the partonomy (i.e., within a context-dependent hierarchy).

So far, our representation of place name regions uses a binary mapping of regions to discrete space (i.e. the polygonal reference tessellation). The mapping is formalized in such a way that all reference units that intersect the place name region are counted as elements of the region's qualitative spatial footprint. Depending on the resolution of the reference tessellation, the resulting approximation of a region can be very coarse. In general, the approximation overestimates the extent of the place name region because even a marginal intersection of the region with a reference polygon is enough to add this polygon to the spatial footprint. This problem could be solved by using a reference tessellation with a higher resolution. Unfortunately, this is not always possible for two reasons: Firstly, "natural" polygonal standard reference tessellations are often available only on specific levels of resolution. And secondly, using fine grained reference tessellations would be counter-productive to our goal of providing an intuitive and user-friendly method for the (manual) creation of spatial models because qualitative spatial footprints that consist of large sets of reference units cannot be handled without the help of GIS tools.

In the following section, we will describe how we can improve the discrete representation of place name regions by using an *upper* and a *lower* approximation. We will show how this improved representation can be used to infer the hierarchical *partonomy* of place name regions.

3   Discretizing Place Name Regions

To provide better discrete approximations for qualitative spatial footprints we use an approach adopted by Worboys [18]. He applies rough set theory to define the *upper* and the *lower approximation* of discretized regions. Translated to our application, the *upper approximation* $\bar{S}_p$ of a place name region p defines the maximum extent of the region in terms of the units of a polygonal reference tessellation. It is consistent with the approximation derived by a simple binary mapping of the region onto the reference tessellation. To describe this mapping in terms of topologic relations between a region and the units of a reference tessellation, we use RCC-5 [1], [3].

The RCC-5 calculus is based on the region connection calculus known as RCC-8 (e.g., [10]). For two regions a and b, it distinguishes five jointly exhaustive and pairwise disjoint (JEPD) topologic relations (figure 1). In terms of RCC-5, $\bar{S}_p$ can be expressed as

$$\bar{S}_p = \{r | EQ(r,p) \lor PP(r,p) \lor PO(r,p)\} \qquad (1)$$

The *lower approximation* $\underline{S}_p$ of p denotes the minimum extent of the place name region in discrete reference space, i.e. all reference units that *definitely* belong to the spatial footprint. Or, expressed in terms of RCC-5

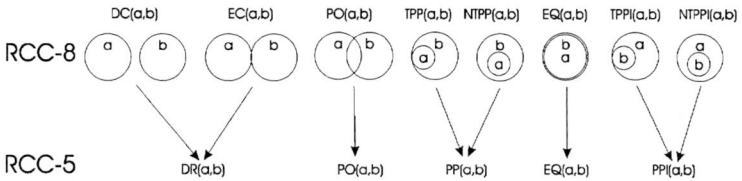

Fig. 1. RCC-8 and RCC-8, after [3]

$$\underline{S}_p = \{r | EQ(r,p) \vee PP(r,p)\} \tag{2}$$

Depending on the underlying assumptions, the complexity of topological relations that can be described between such imprecise regions varies considerably. [3] have shown 46 possible relations between regions of the "egg-yolk" type. A much simpler approach is taken by Worboys who, for two vague regions a and b represented as rough sets, developed a three-valued logic where a is *definitely part of* b if all of the four relations hold: $\bar{S}_a \subseteq \bar{S}_b$, $\bar{S}_a \subseteq \underline{S}_b$, $\underline{S}_a \subseteq \bar{S}_b$, and $\underline{S}_a \subseteq \underline{S}_b$. The region a is *definitely not part of* b if all four relations are false. For all other possible combinations, a is said to be *maybe part of* b.

For the case where the *boundary region* $B_a = \bar{S}_a \cap \underline{S}_a$ of a discretized region a is not more than one reference unit thick (see section 5), we can make the assumption that B_a is small compared to $\bar{S}_a$. Under this assumption, we can simplify the approach by Worboys even further by saying that only if a is *definitely not part of* b, the two regions are disjunct. This allows us to develop a practical solution to infer the partonomy of place name regions on the basis of their discretized approximations: We can say that $PP(a,b)$ holds as long as the lower approximation of a is a subset of the upper approximation of b. Or, expressed in terms of the reference units of a polygonal tessellations:

$$PP(a,b) \rightarrow \{r | r \in \underline{S}_a \wedge r \in \bar{S}_b\} \tag{3}$$

Using this approach, we find that for the five place name regions a, b, c, d and t depicted in Figure 2, we can infer the following partonomic relations: $PP(a,t)$, $PP(b,t)$, $PP(c,t)$, $PP(d,t)$, and $PP(a,b)$. To obtain a meaningful hierarchical partonomy that can be expressed in an *directed acyclic graph (DAG)*, these partonomic relations have to be ordered, and redundant arcs have to be removed. The result is a hierarchical place name structure like the one shown in figure 3.

4   Defining New Place Names

So far, our approach addressed only discrete representation of place name regions for which the spatial extend is more or less well-known. In some cases, even polygonal representations may be available, allowing for an automated creation of the respective qualitative spatial footprints.

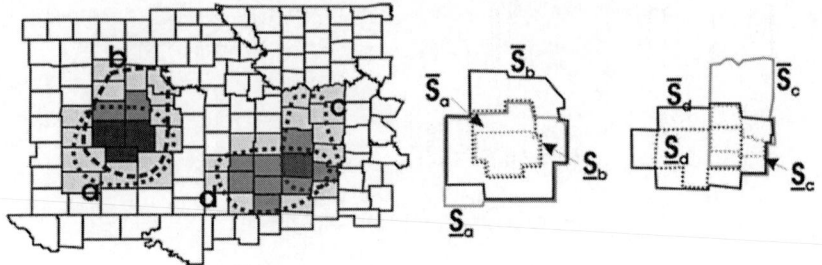

Fig. 2. Configurations of extensionally-defined place name regions

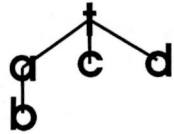

Fig. 3. Inferred hierarchy of place name regions

However, there may be situations where the boundaries of a place name regions are not known, or not well-defined. For example, the manager of the tourist office of the *Frankenwald* region in Germany may want to add a new place name denoted *Frankenwald mining region* in an existing place name structure of the area (see section 5). This *mining region* may not have well-defined boundaries, and the only information available is that it is part of a certain super-region (the *Frankenwald*), and that it contains a number of sub-regions (e.g. the *Münchberger Hochfläche*). In the following we will describe a method to add such *incompletely specified* regions to a place name structure through the intensional definition of spatial relations. This method is based on assumptions drawn from the application of research on the cognitive aspects of *preferred spatial models*.

4.1   Preferred Spatial Models

Indeterminacy arises from incomplete spatial information. Typically, the user specifies a new place name not by giving an extensional definition for it in terms of regions of the tessellation, but rather by stating a few relations to known place names such as: the new place is part of such and such old place. Incomplete descriptions of configurations of spatial regions are generally compatible with a large number of geometrical realizations. Consider the RCC-5 formulas $PP(a,d)$, $PP(b,d)$, and $PP(c,d)$ stating proper part relations between regions a, b, c, and d. Each configuration of regions C_1, C_2, and C_3 shown in figure 4 constitutes a geometrical model of the three formulas. Most computational approaches (e.g. constraint satisfaction systems or logical theorem provers) avoid to construct such models explicitly. A rare exception is the work of [5] who describes the advantages of a preference-based approach to reasoning with cardinal

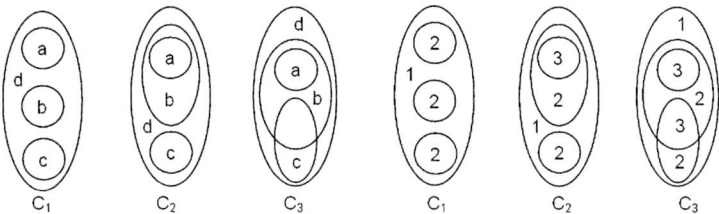

Fig. 4. Spatial configurations and their boundary overlap complexity

directions. The avoidance of models in spatial reasoning is in sharp contrast with the way in which human problem-solvers proceed when solving spatial relational inference tasks. They tend to build a mental model in visuo-spatial working memory representing a specific configuration compatible with the information given to them [8], [16]. This first mental model acts as the preferred model in the reasoning process: relations valid in the preferred model tend to be considered as being valid generally, i.e. valid in all other models too. Mental model-based reasoning therefore shows characteristic reasoning errors (preference biases).

The cognitive preference of certain models over others is not only a shortcoming. In verbal communication about space, preferences can be exploited to reduce the amount of what needs to be said. If, for instance, a speaker states that three regions a, b, and c are part of a region d without adding any further information about the relationship between a, b, and c this triggers a pragmatic reasoning process in the listener which attempts to reconstruct the configuration intended by the speaker. Probably, the process will come up with a configuration similar to C_1 rather than to C_3. To describe C_2, the speaker would have to make explicit that a is a part of b. To describe C_3, even further information would be necessary. In other words, if a simple configuration is intended, it can be described with a few statements, whereas complex configurations need more details.

What determines the simplicity of a spatial configuration? Or, putting it into terms amenable to psychological experiments: Which are the preferred mental models in spatial relational inference tasks? The issue has been studied most comprehensively for the spatial version of the one-dimensional interval calculus [11]. However, this line of research has not yet addressed the RCC calculus, which is the most interesting for spatial relevance computation. We will briefly summarize the main results and derive from this review a hypothesis about preferences for RCC-5.

The best explanation for preferences currently available is that they are the result of spatial chunking in working memory which significantly reduces memory load for the preferred mental models, whereas other models require more memory resources. Details of the spatial chunking process have been described by [13]. Two of its characteristic features are: (1) A configuration with singularities, i.e. touching objects, never acts as preferred model. (2) There is a preference to avoid overlaps where possible and to keep objects disjoint.

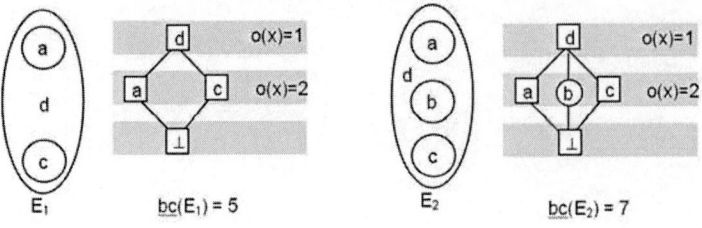

Fig. 5. a: Starting configuration; b: Result of processing PP(b,d)

Note that these principles do not suffice to account for all preferences empirically found. They fail, for instance, to explain how the process of mental model construction depends on the order in which the information is processed. An explanation accounting for almost all preferences is given in Schlieder [13]. Nevertheless, the principles stated above allow us to formulate an educated guess about preferences among RCC relations. Because of (1) we restrict attention to RCC-5, the region-based topological calculus that does not treat configurations with singularities as special cases (TPP, $TPPI$, and EC in RCC-8). In order to deal with (2), we propose the following measure for boundary overlap complexity in a con-figuration of connected regions without holes in the Euclidean plane.

Definition: *$T(R)$ is the tessellation of the plane induced by the boundaries of the regions $R = \{r1,...,rn\}$. Each cell c of $T(R)$ is assigned an overlap number $o(c)$ which is the number of regions from R of which c is a part. The boundary overlap complexity $bc(R)$ is the sum of $o(c)$ for all c from $T(R)$.*

For an example refer to figure 4 where the overlap numbers are shown for the tessellations arising from configurations C_1, C_2, and C_3. The following boundary overlap complexities are found: $bc(C_1) = 7$, $bc(C_2) = 8$, $bc(C_3) = 11$. Sorting according to increasing complexity (diminishing simplicity) results in the ordering $C1$, $C2$, $C3$ which matches well with the intuitive notion of complexity of an arrangement of regions. With the measure for boundary overlap complexity, we can now state our hypothesis about preferred models for RCC-5 in precise terms: In a set of models, the models with least boundary overlap complexity are going to be preferred.

This notion of preferred models helps to solve the problem of processing a specification of a new place name that is input into the system by a user. Consider the situation depicted in figure 5. Three place names are already defined in the place name structure with extensions a, c, and d all of which are aggregates of regions from the reference tessellation. To the right of the geometrical model, the partonomic structure is depicted. It is a directed acyclic graph representation of the proper-part relations holding between the cells of the tessellation induced by the boundaries of the regions a, c, and d. The user starts to specify a new place name d and the system incrementally processes the user's input. As first input a RCC-5 formula - or its natural language equivalent - stating $PP(b, d)$ is processed. The new region b is integrated into the partonomic structure in

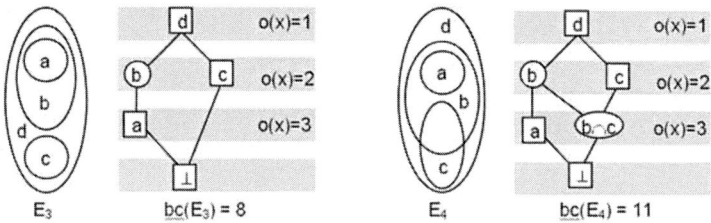

Fig. 6. c: Result of processing PPI(b,a); d: Result of processing PO(b,c)

such a way that the boundary complexity increases least (figure 5b). Similarily, the input $PPI(b, a)$ is processed. It requires making a proper part of b, which involves a local rearrangement of the partonomic structure (figure 6c). Finally, after processing $PO(b, c)$, the geometrical model shown in (figure 6d) is obtained.

Note that the algorithm proceeds incrementally following a greedy strategy which always minimizes boundary overlap complexity. It may well be that at a certain stage, an input from the user cannot be integrated into the model so far constructed, although there exists a model satisfying the constraints represented by the place name structure as well as those stated by the user. This failure of not always finding a solution when a solution exists does not come as an surprise given the complexity of the underlying constraint satisfaction problem which is NP-complete. However, the algorithms works - just as human model-based reasoning - as a very fast approximation to the instantiation problem. With the kind of simple place name specifications that users typically provide, we do not consider the incompleteness of the algorithm a real problem for application.

4.2   Indeterminate Place Name Regions

As we have seen above, we can add a place name region with indeterminate boundaries to an existing place name structure by *intensionally* defining its mereologic relations with respect to the other place name regions. However, because qualitative spatial footprints play a key role in our approach to infer the relative spatial relevance of place name regions, the need arises to determine the discrete approximation of such an intensionally-defined place name region. This can be done with the help of a set of simple rules based on the spatial configurations derived from the preferred spatial models described above.

In Figure 7, we give an example of a very simple place name structure consisting of 2 extensionally-defined place name regions a and b. We can add a new place name region X by specifying two binary relations, namely $PP(X, a)$ and $PP^{-1}(X, b)$. We can then say about the spatial footprint S_X of X that S_X must be at least as large as the lower approximation $\underline{S}_a$ of b, and that S_X cannot be larger than the upper approximation $\bar{S}_a$ of a. Consequently, we can say about the upper and the lower approximation of X that $\bar{S}_X \subseteq \bar{S}_a$, and $\underline{S}_X \supseteq \underline{S}_b$.

Of course, the approximation of X is very coarse and may be too far from the "true" extent of the actual place name region to be useful for spatial relevance

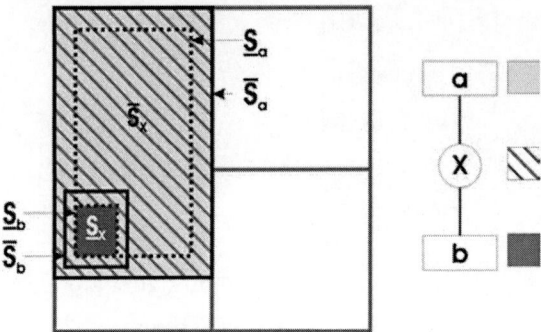

Fig. 7. Upper and lower approximation of the intentionally-defined region X

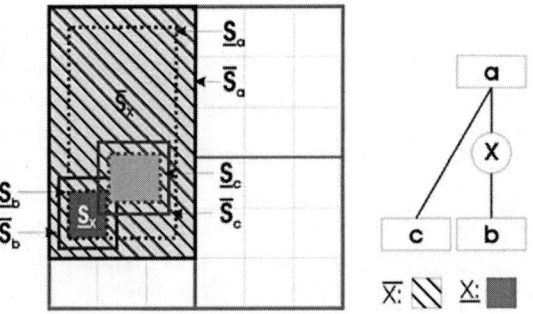

Fig. 8. Refining the approximation of X by adding c

reasoning. However, the quality of the approximation can be improved through the specification of additional spatial relations to other place names, or by adding new extensionally-defined place name regions to the structure.

In figure 8, for example, a new extensionally-defined region c, which is a child of a, is added to the structure. While the lower approximation of X remains the same, we can say that its upper approximation is reduced to the upper approximation of aA, *minus* the lower approximation of c.

$$\bar{S}_X = \bar{S}_a \cap \underline{S}_c \qquad (4)$$

In general we can say that the upper approximation $\bar{S}_X$ of an intensionally-defined place name X can be estimated to be equal to the upper approximation of its parent *parent*, minus the union of the lower approximations of all children of *parent*, except the children of X.

$$\bar{S}_X = \bar{S}_{parent} - \bigcup \underline{S}_p\{p|PP(p, parent)\}\backslash \underline{S}_p\{p|PP(p, X)\} \qquad (5)$$

The lower approximation $\underline{S}_X$ of X can be estimated to be equal to the union of the lower approximations of all place names p_i that are *part-of* X.

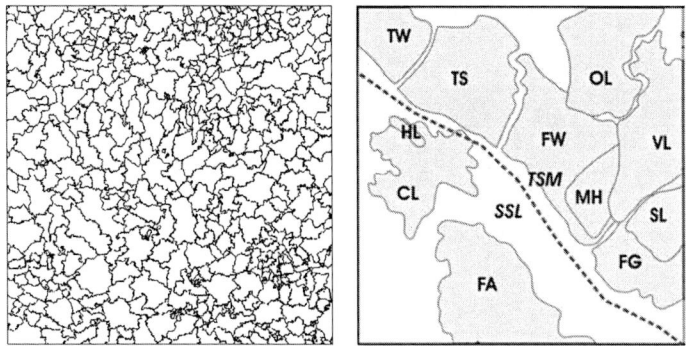

Fig. 9. (a) Polygonal tessellation of counties; (b) Natural regions

$$\underline{S}_X = \bigcup \underline{S}_p \{p | PP(p, X)\} \qquad (6)$$

For regions that do not have children the lower approximation is undefined. In this case, we say that the lower approximation is equal to the upper approximation. Likewise, for regions that do not have parents (or that are children of the virtual top node of a hierarchy), the rules described above have to be modified. We say that the upper approximation of such regions is equal to the union of the upper approximations of all their children.

By applying simple set arithmetics and the rules established in equations 5 and 6, we can compute an approximated spatial footprint for any intensionally-defined place name X that was added to an existing place name structure and thereby transform X into a "regular", extensionally-defined region. The result is a new consistent structure of extensionally-defined place name regions.

5 Example

In a real world example we can demonstrate the discrete approximation of place name regions and the computation of a partonomic hierarchy for such regions. For this purpose, we have chosen an area in the north-eastern part of Bavaria in Germany. Our spatial reference system is a polygonal tessellation of administrative units (counties). We use this reference system to create qualitative spatial footprints for a number of *natural regions*. These place name regions have well-defined boundaries (in fact, they are available as polygonal digital data) (figure 9).

To discretize the place name regions, we project the polygons onto the reference tessellation and obtain a spatial footprint with an upper and a lower approximation for each region. Figure 10 shows the upper and the lower approximation of the *Frankenwald (FW)* region. Figure 11 depicts the upper and lower approximation of the *Münchberger Hochfläche (MH)* region. As we can see in figure 11c , the lower approximation of region MH is a subset of the upper

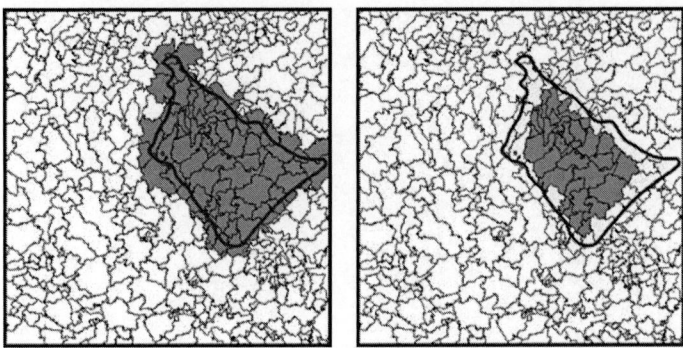

Fig. 10. (a) Upper and (b) lower approximation for the place name region *Frankenwald*

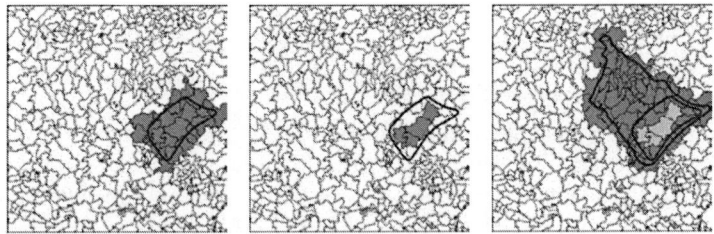

Fig. 11. (a) Upper and (b) lower approximation for the place name region *Münchberger Hochfläche*, (c) *MH* is proper part of *FW*

approximation of region FW. Applying the rules described in 3 we can conclude that the relation $PP(MH, FW)$ holds. Using this procedure, we can compute the partonomic structure for all regions represented through well-defined polygons (solid line) in figure 9.

In a next step, we may want to add two super-regions, TSM and SSL to the place name structure. In figure 9b, these regions are outlined by a dotted line to express the fact that their exact extend is not known. We can add these regions by specifying the respective partonomic relations with respect to the other regions in the place name structure. Following the rules established in 4, we can estimate the lower and upper approximations of these super-units based on the lower and upper approximations of their children.

Finally, we would like to add the *Frankenwald mining region (MR)* to the place name structure. Of this region we only know that it is part of the north-western part of the *Frankenwald (FW)*. We simply specify the relation $PP(MR, FW)$, and the region is added to the structure as a child of FW (figure 12). Note that MR is modelled to be a sibling of MH because our algorithm tries to minimize overlap complexity (see 4.1). If MH were to be a sub-region of MR, we would have to explicitly state this fact. Following the rules set up in section 4.2, the upper approximation of MR is estimated to be equal to the upper approximation of FW, minus the lower approximation MH. Because MR

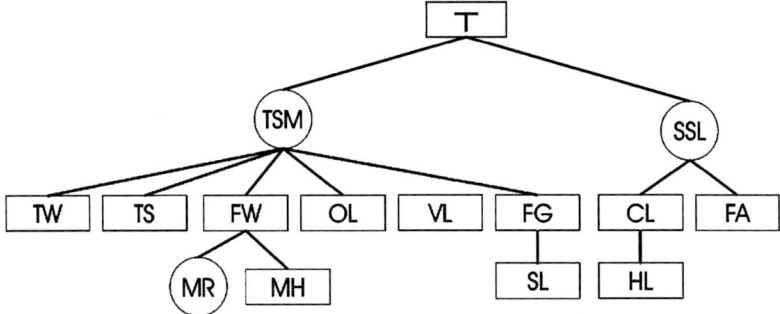

Fig. 12. Place name structure of natural regions

has no children, its lower approximation is undefined and defaults to equal the upper approximation.

The final result is the place name structure shown in figure 12. It represents the hierarchical structure of the place name regions based on both the evaluation of topo-mereologic relations between extensionally-defined regions, and the inclusion of intensionally-defined regions. For each place name region in this structure, a (explicitly stated or computed) discretized approximation is available.

6   Discussion and Outlook

In this paper we used spatial metadata based on discrete approximations of regions for spatial relevance reasoning in information retrieval tasks that are typical for distributed systems like the semantic web. We extended our work on the representation of place name regions to allow for the representation of vague regions. Our focus was the development of methods that can be used for an intuitive and user-friendly modelling of common-sense spatial knowledge. These include the evaluation of preferred spatial models as the basis for heuristics that are applied to guide the interactive design of place name structures.

References

1. Brandon Bennett. Spatial reasoning with propositional logics. In J. Doyle, E. Sandewall, and P. Torasso, editors, *4th International Conference on the Principles of Knowledge Representation and Reasoning (KR94)*, San Francisco, 1994. Morgan Kaufmann.
2. Thomas Bittner and John G. Stell. Vagueness and rough location. *GeoInformatica*, 6(2):99–121, 2002.
3. Anthony G. Cohn and N.M. Gotts. The 'egg-yolk' representation of regions with indeterminate boundaries. In P.A. Burrough and Andrew Frank, editors, *Geographic Objects with Indeterminate Boundaries*, pages pp. 45–55. Taylor and Francis, London, 1996.

4. Max J. Egenhofer and David M. Mark. Naive geography. In *Spatial Information Theory, A Theoretical Basis for GIS*. Springer, 1995.
5. Andrew Frank. Qualitative spatial reasoning: Cardinal directions as an example. *International Journal of Geographical Information Systems*, 10(3):269–290, 1996.
6. Antony Galton. The mereotopology of discrete space. In Christian Freksa and D.M. Mark, editors, *Proceedings of the International Conference on Spatial Information Theory COSIT'99*, Spatial Information Theory: Cognitive and Computational Foundations of Geographic Science, pages 251–266, Stade, Germany, 1999. Springer, Berlin.
7. Linda L. Hill. Core elements of digital gazetteers: placenames, categories, and footprints. In J. Borbinha and T. Baker, editors, *ECDL 2000*, Research and Advanced Technology for Digital Libraries, pages 280–290, Lisbon, Portugal, 2000.
8. M. Knauff, R. Rauh, C. Schlieder, and G. Strube. Mental models in spatial reasoning. In C. Habel & K. Wender C. Freksa, editor, *Spatial Cognition*, pages 267–291. Springer, Berlin, 1998.
9. Marien Molenaar. *An Introduction to the Theory of Spatial Object Modelling*. Research Monographs in Geographical Information Systems. Taylor and Francis, London, Bristol, 1998.
10. D.A. Randell, Z. Cui, and Anthony G. Cohn. A spatial logic based on regions and connection. In *3rd International Conference on Knowledge Representation and Reasoning*, pages 165–176, San Francisco, 1992. Morgan Kaufmann.
11. R. Rauh, C. Hagen, C. Schlieder, G. Strube, and M. Knauff. Searching for alternatives in spatial reasoning: Local transformations and beyond. In *Proceedings of the TwentySecond Annual Conference of the Cognitive Science Society*, pages 871–876, Mahwah, NJ, 2000. Lawrence Erlbaum Associates.
12. W.-F. Riekert. Erschließung von fachinformationen im internet mit hilfe von thesauri und gazetteers. In C. Dade and B. Schulz, editors, *Management von Umweltinformationen in vernetzten Umgebungen, 2nd workshop HMI*, Nürnberg, 1999.
13. C. Schlieder. The construction of preferred mental models in reasoning with allen's relations. In C. Habel and G. Rickheit, editors, *Mental models in discourse processing and reasoning*. Elsevier Science, Oxford, 1999.
14. Christoph Schlieder and Thomas Vögele. Indexing and browsing digital maps with intelligent thumbnails. In *In : Proceedings of the International Symposium on Spatial Data Handling (SDH) 2002*, Ottawa, Canada, 2002. Springer.
15. Christoph Schlieder, Thomas Vögele, and Ubbo Visser. Qualitative spatial reasoning for information retrieval by gazetteers. In Dan Mortello, editor, *Conference on Spatial Information Theory (COSIT) 2001*, Morro Bay, California, 2001.
16. A. Vandierendonck and G. De Vooght. Mental models and working memory in temporal reasoning and spatial reasoning. In V. De Keyser, G. d'Ydewalle, and A. Vandierendonck, editors, *Time and the dynamic control of behavior*, pages 383–402. Hogrefe und Huber, Göttingen, 1998.
17. Thomas Vögele and Christoph Schlieder. Spatially-aware information retrieval with place names. In *to appear in: Proceedings of the 16th international FLAIRS conference*, St. Augustine, FL, 2003. AAAI Press.
18. M. F. Worboys. Imprecision in finite resolution spatial data. *GeoInformatica*, 2:257–279, 1998.

Convexity in Discrete Space

Anthony J. Roy and John G. Stell

School of Computing
University of Leeds
Leeds, LS2 9JT, UK
{anthonyr,jgs}@comp.leeds.ac.uk
http://www.comp.leeds.ac.uk/qsr/dtg/

Abstract. This paper looks at Coppel's axioms for convexity, and shows how they can be applied to discrete spaces. Two structures for a discrete geometry are considered: oriented matroids, and cell complexes. Oriented matroids are shown to have a structure which naturally satisfies the axioms for being a convex geometry. Cell complexes are shown to give rise to various different notions of convexity, one of which satisfies the convexity axioms, but the others also provide valid notions of convexity in particular contexts. Finally, algorithms are investigated to validate the sets of a matroid, and to compute the convex hull of a subset of an oriented matroid.

Keywords: Convexity axioms, alignment spaces, affine spaces, convex spaces, convex hull, discrete geometry, oriented matroids, cell complexes, matroid algorithms.

1 Introduction

The concept of a convex region plays an important part in many practical computations in GIS. It is also a fundamental component in the applications of spatial information theory in other areas. As Boissonnat and Yvinec note [BY98, p125]: "Convexity is a fundamental notion for computational geometry, at the core of many computer engineering applications, for instance in robotics, computer graphics, or optimization." Convexity has been studied both in point-based quantitative theories of space, based on numerical coordinates, and also in the context of qualitative spatial reasoning. As an example of the latter, we note that the region-connection calculus was extended [CB+97, p287] by adding a one place function which returned the convex hull of a region.

There has recently been much interest within the spatial information theory community in discrete notions of space, e.g. [Win95,Gal99,MV99,Ste00,RS02]. If discrete space is to provide an adequate foundation for the representation of geographic information, it will need to be equipped with a notion of convexity. In this paper we examine how convexity may be defined in discrete space, and how it is possible to provide algorithms to compute the convex hull of a discrete region.

The traditional definitions of convexity (in terms of line segments) in Euclidean space are not applicable to discrete space, so it is necessary to investigate

W. Kuhn, M.F. Worboys, and S. Timpf (Eds.): COSIT 2003, LNCS 2825, pp. 253–269, 2003.

a more abstract axiomatic formulation, in terms of alignment axioms. This we do in Section 2 where we present Coppel's axiomatization of alignment, convexity and affine spaces[Cop98].

We then look at different structures for discrete geometry, (oriented) matroids and cell complexes, showing how each can represent different spatial properties. We give a concrete example using these structures to give a discrete model of convex space, based on $\mathbb{Z}^2$ and derived from a discrete affine space on $\mathbb{Z}^2$.

Cell complexes are becoming an ever more popular model for discrete space due to their topological properties; we investigate how different notions of convexity are possible in this setting. Several definitions of convexity are given, including one derived from a matroid structure on the cells of the complex. We investigate these notions to see which axioms for convexity are satisfied by each, and how each may have its uses even if not strictly a convex geometry.

It is important that convexity is not just studied in the abstract: it is necessary to show how computations can be carried out. We provide some algorithms for the computation of convex hulls of oriented matroids in Section 5. A brief introduction to the necessary oriented matroid concepts is provided earlier in Section 3.1.

2   Axiomatizing Convexity

This section gives an axiomatization of convex and affine geometries in a very general setting. The concepts themselves come from the corresponding concepts of convex and affine geometries defined on a vector space over the real numbers, derived from linear combinations of points in a vector space. Given a set of vectors $v_0, \ldots, v_k$, and a set of scalars, $\lambda_0, \ldots, \lambda_k$, a linear combination of these vectors given the set of scalars is a vector given by:

$$\sum_{i=0}^{k} \lambda_i v_i$$

A *subspace* of $\mathbb{R}^n$ is a subset $S \subset \mathbb{R}^n$ which contains all linear combinations of points in S (i.e. closed under linear combinations). Examples include $\mathbb{R}^n$, the origin $\mathbf{0}$, and any line passing through the origin. Note that the origin will always be in a subspace, since all scalars can be set to zero.

Affine combinations are restricted linear combinations; the scalars λ_i must add up to exactly 1. This gives rise to the notion of affine sets, the sets $S \subset \mathbb{R}^n$ closed under affine combinations. The affine sets in $\mathbb{R}^n$ include all translations of the set of subspaces. An affine set is often characterized as containing all of the lines through each pair of its points.

Convex combinations are restricted affine combinations; not only must the scalars add up to 1, but they must also lie in the range $[0, 1]$. Convex sets are sets closed under convex combinations and are often characterized as sets containing all line segments between each pair of its points.

It is clear from the above definitions that linear, affine and convex sets are closely related; in fact all linear sets are affine and all affine sets are convex.

The notion of aligned spaces given below is an abstraction of convex and affine geometries.

2.1   Alignment Spaces

The concept of alignment spaces was introduced by Coppel[Cop98] as a general axiomatization of convexity. The main focus of his work is on continuous settings, but we show here how the axioms for alignment spaces can be successfully applied to discrete spaces.

In Euclidean space a set, S, is said to be convex if for every pair of points in S the straight line between them is also in S. We can show that this definition does not directly extend to discrete space. If we define a straight line between two points, a and b, in $\mathbb{Z}^n$ as being the set of all points collinear with those points, and between them. That is, $line(a,b) = \{x \in \mathbb{Z}^n | x = \lambda a + \gamma b, \lambda + \gamma = 1, 0 \leq \lambda, \gamma\}$. Figure 1 shows that such a definition of convexity does not fit with an intuitive notion of convex set. In this example, the three points alone are by the naive definition a convex set.

The following definition begins to capture some of the properties we would expect a convex set to possess.

Definition 1 (Alignment Space). *An* alignment space *is a pair* $A = (X, \mathcal{C})$ *where* X *is a set and* $\mathcal{C}$ *is a set of subsets of* X *such that the following axioms hold:*

(A0) $\varnothing \in \mathcal{C}$ *(Empty set)*
(A1) $X \in \mathcal{C}$ *(Top set)*
(A2) $\forall Z \subseteq \mathcal{C},\ Z \neq \varnothing \Rightarrow \bigcap Z \in \mathcal{C}$ *(Intersection)*
(A3) $\forall Z \subseteq \mathcal{C},\ Z \neq \varnothing$ *if* Z *is totally ordered by inclusion, then* $\bigcup Z \in \mathcal{C}$
(Qualified union)

We call the sets in $\mathcal{C}$ *aligned sets. The intersection of all aligned sets containing a set* $S \subseteq X$ *is called the* alignment hull *of* S, *denoted* $[S]$. *This intersection always exists due to (A2) above.*

Proposition 1. *Note that the following properties hold of any alignment hull:*

H0 $[\varnothing] = \varnothing$
H1 $S \subseteq [S]$
H2 $S \subseteq T \Rightarrow [S] \subseteq [T]$
H3 $[[S]] = [S]$
H4 $[S] = \bigcup\{[T] : T \subseteq S \text{ and } T \text{ is finite}\}$

The proof for this proposition can be found in [Cop98].

In brief the axioms state that the entire set and the empty set are aligned, arbitrary intersections of aligned sets are aligned, and the union of totally ordered sets of aligned sets are also aligned. These are properties we would expect of convex sets. These axioms alone are not enough however; Figure 1 shows a set of

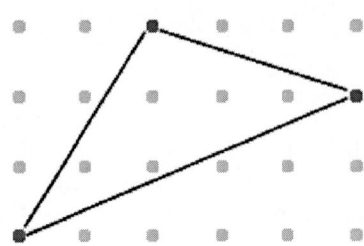

Fig. 1. A 'convex' set according to
the naive definition.

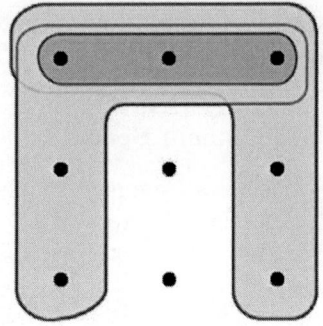

Fig. 2. A subset of 2^X that satisfies the alignment axioms.

points (just the vertices of the triangle) that is aligned but would not consider
convex.

Figure 2 gives a second example showing two sets of points. Let $\mathcal{C}$ be a set
containing these two sets along with their intersection, the entire space X and the
empty set. Then, $\mathcal{C}$ satisfies all four axioms, and is hence an aligned space. This
is an aligned set, but is not convex in our usual understanding of the concept.

2.2   Convex Geometries

To overcome the problem just mentioned, a further axiom, known as the anti-
exchange property, is added.

Definition 2 (Anti-exchange Property). *The anti-exchange property is*
characterized by the following axiom:

(AE) $\forall x, y \in X$, $\forall S \subseteq X$, *if* $x \neq y$ *and* $y \in [S \cup x]$ *and* $y \notin [S]$ *then* $x \notin [S \cup y]$
     (Anti-exchange)

Clearly the example in Figure 2 fails to satisfy this additional axiom. Take
for example S to be one of the two depicted subsets, and x and y as distinct
points not in S then $S \subseteq X$, $x \neq y$ as required, $y \in [S \cup x]$ since $[S \cup x] = X$
(X is the largest aligned set containing $S \cup x$), $y \notin [S]$ since S is the smallest
aligned set containing itself. However, $x \in [S \cup y]$ since X is again the smallest
aligned set containing the union of S and y.

We shall refer to aligned spaces for which the anti-exchange property holds
as *convex geometries*, and the alignment hulls in this special case *convex hulls*.
Note that the term usually refers to geometries defined with points and segments
as primitives (see [Cop98]), but in this paper we have already seen that such an
axiomatization would not be applicable in the case of discrete space.

Figure 3 shows an example of a space that satisfies the five axioms A0-A3
and AE. Note that though all axioms are satisfied, there are subsets missing that
we would normally consider to be convex.

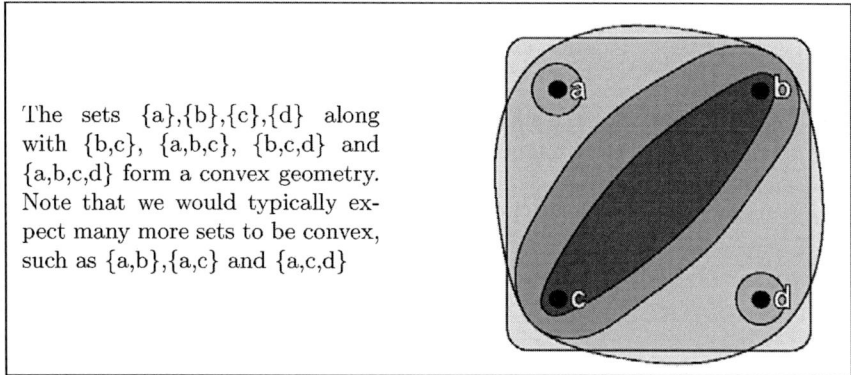

The sets {a},{b},{c},{d} along with {b,c}, {a,b,c}, {b,c,d} and {a,b,c,d} form a convex geometry. Note that we would typically expect many more sets to be convex, such as {a,b},{a,c} and {a,c,d}

Fig. 3. A convex geometry.

2.3 Affine Geometries

The exchange property essentially says that if an element, x, is in the hull of a set $S \cup y$, then y is in the hull of the set $S \cup x$. i.e. x and y are interchangeable for purposes of generating hulls. Due to the difference between this axiom and the convexity (anti-exchange) axiom, we denote the hull in this case by angle brackets, e.g. $\langle S \rangle$.

Definition 3 (Exchange Property). *The exchange property is characterized by the following property:*

(E) $\forall x, y \in X, \ \forall S \subseteq X, \ $ *if* $x \neq y$ *and* $y \in \langle S \cup x \rangle$ *and* $y \notin \langle S \rangle$ *then* $x \in \langle S \cup y \rangle$ *(Exchange)*

Note that since $y \notin \langle S \rangle$ *and* $y \in \langle S \cup x \rangle$ *then* $x \notin \langle S \rangle$.

Together with (A0)–(A3), this axiom gives us the notion of an affine geometry, and the alignment hulls in this case shall be referred to as *affine hulls*. The intuition behind the notion of affine hull of a set S is that it is the smallest affine set of the given space containing S. For example, given two points in Euclidean space, the affine hull of those points is the straight line passing through them.

3 Structures for Discrete Geometry

We consider here two structures for discrete geometry; (oriented) matroids and cell complexes.

Matroids (and oriented matroids) are abstract structures consisiting of a finite set of objects (that is the set is in one to one correspondence with a finite subset of the integers) and a set of subsets of that set which satisfy certain properties - i.e. a combinatorial structure. The set of axioms obeyed by the subsets of the matroid are given below, and as we shall see the resulting structure is rich enough to enable us to model many important geometric constructs such as linear independence, bases, subspaces etc. Due to the discrete nature of the structure, matroids also allow us to do computations in a robust manner - no rounding errors occur since we simply manipulating finite sets to get our results.

Cell complexes can be thought of as a partition of space into objects of different dimension. In two dimensions for example they comprise of 0-dimensional points, 1-dimensional line segments and 2-dimensional regions. A simple cell complex structure, commonly called a cartesian complex, comprises of a set of grid points taking values from $\mathbb{Z}^n$ as the 0(dimensional)-cells, the vertical and horizontal (open) line segments between those cells as the 1-cells, and the square region between 4 0-cells and 4 1-cells as the 2-cells. Other examples include simplicial complexes, and more general partitions such as the triangulation in Figure 4.

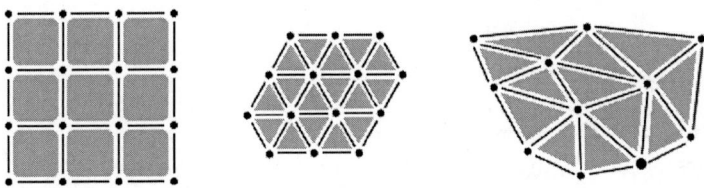

Fig. 4. Cell complexes: Cartesian Complex; Simplicial Complex; Triangulation.

The following sections show that the two structures can be effectively used together to provide robust calculation with a useful topology.

3.1   Capturing the Geometry of Discrete Space with Matroids

Though matroids are essentially an abstract combinatorial structure, they have roots in linear algebra, and consequently capture a good amount of geometric structure. In fact (H1)-(H3) and (E) are the axioms of a closure operator on a matroid. In this section we give a brief introduction to independence axioms on matroids, and circuit axioms on oriented matroids (matroids with some additional structure).

Definition 4 (Matroid: Independence axioms).
   A Matroid is a pair $M = (E, \mathcal{I})$ where E is a finite set, and $\mathcal{I}$ is a collection of subsets of E (called independent subsets*) satisfying the following axioms:*

   (I1) $\varnothing \in \mathcal{I}$
   (I2) *If $I \in \mathcal{I}$ and $I' \subseteq I$ then $I' \in \mathcal{I}$*
   (I3) *If $I_1, I_2 \in \mathcal{I}$ and $|I_1| < |I_2|$, then $\exists e \in I_2 - I_1$, $I_1 \cup \{e\} \in \mathcal{I}$*

   $\mathcal{I}$ *are the* independent sets *of the matroid M; E is called the* ground set.

In the following sections we describe how discrete spaces can be modelled using matroids, and in particular how it is possible to construct an oriented matroid structure with which to develop a convex geometry. Some of the combinatorial structure previously mentioned will be used in these later sections, and are defined as follows:

Bases: Let $M = (E, \mathcal{I})$ be a matroid defined in terms of its independent sets. The maximal independent sets are called *bases* of the matroid.

Circuits: A set $D \subseteq E$ is called *dependent* if $D \notin \mathcal{I}$. The dependent sets which are minimal in E are called the *circuits* of the matroid.

Subspaces: The *rank* of a subset $S \subseteq E$ is the size of the largest independent set it contains; $\rho(S) = max(|X| : X \subseteq S, X \in \mathcal{I})$.

A *subspace* or *flat* of a matroid is a subset $F \subseteq E$ where for every $x \in E - F$, $\rho(F \cup \{x\}) = \rho(F) + 1$; that is a flat is maximal with respect to its rank. The rank of the matroid M, ρM, is the rank of E. It should be noted that flats can be equivalently characterized as the fixed points of a closure operator.

A *hyperplane* is a flat whose rank is one less than the rank of E.

Co-circuits and Duality: The set theoretic complement of a hyperplane, h, in M is called a *cocircuit* of M. This is the co-circuit *associated with* h. A property of the cocircuits, $\mathcal{I}^*$, of a matroid, M, is they form the set of circuits in the dual matroid $M^* = (E, \mathcal{I}^*)$. This is a vital result which we use in constructing an oriented matroid later in this paper.

So far, the matroid structure we have is powerful enough to model certain geometric constructs, such as independent sets, flats and so on. However to enable us to talk about convexity we need a little more structure; we introduce a partition (orientation) on the circuits to get an *oriented matroid*. First we shall define the concept of *signed sets*.

Definition 5 (Signed Set). *A* signed set X *is a set* $\underline{X}$ *along with a partition* (X^+, X^-) *of that set where* $X^+ \cup X^- = \underline{X}$ *and* $X^+ \cap X^- = \varnothing$.

$\underline{X}$ *is called the* support *of* X, X^+ *is the set of* positive elements *and* X^- *is the set of* negative elements.

The opposite *of a signed set* $X = (X^+, X^-)$, *is* $-X = (X^-, X^+)$.

An oriented matroid has signed sets in place of ordinary sets as its circuits, and hence we must alter the axioms accordingly.

Definition 6 (Oriented Matroid: Circuit Axioms).

An oriented matroid *is a pair* $O = (E, \mathcal{C})$, *where* $\mathcal{C}$ *is a collection of signed subsets of a finite set* E *satisfying the following axioms:*

(OC1) $\varnothing \notin \mathcal{C}$

(OC2) $\mathcal{C} = -\mathcal{C}$ *(symmetry)*

(OC3) $\forall X, Y \in \mathcal{C}$, *if* $\underline{X} \subseteq \underline{Y}$ *then* $X = Y$ *or* $X = -Y$ *(incomparability)*

(OC4)

$$\forall X, Y \in \mathcal{C}, \ X \neq -Y, \ and \ e \in X^+ \cap Y^-$$
$$\exists Z \in \mathcal{C} \ such \ that \ Z^+ \subseteq (X^+ \cup Y^+) \setminus \{e\} \ and$$
$$Z^- \subseteq (X^- \cup Y^-) \setminus \{e\}$$

(weak elimination)

The oriented matroid structure gives us access to various useful constructs, such as the notion of halfspaces. The power of this additional structure will be seen in Section 4.1.

3.2   Modelling Discrete Space with Cell Complexes

Cell complexes are becoming ever more popular as an alternative model of discrete space to the standard pixel/point approaches, primarily for their topological properties. In approaches based in $\mathbb{Z}^n$, sets of points have a trivial topology, that is the open and closed sets in the topology are the same. Attempts at adding a notion of boundary to sets of points do not properly address this: problems crop up such as regions and their complements do not share a boundary, boundaries have non-zero area, and other such anomalies (see [Kov89]).

Cell complexes are an alternative to the $\mathbb{Z}^n$ based approaches, adding higher dimensional entities, to 'fill in the gaps' between points. The advantage of doing so is that cell complexes give a far richer topological structure; this allows, for example, notions of open and closed sets, boundary and connectedness.

Formally, a cell complex is a triple (C, B, dim). C is a set, B is an irreflexive, anti-symmetric and transitive bounding relation (read $(a, b) \in B$ as 'a bounds b'), and dim is a function taking elements of the C to an integer value representing the dimension of the element. The only other constraint is that if a bounds b then $dim(a) < dim(b)$.

There are many ways of modelling a discrete space with cell complexes. One example which we look at further in Section 4.2 take the points of $\mathbb{Z}^2$ as the elements of 0 dimension (0-cells), horizontal and vertical pairs of adjacent points as the 1-cells, and 'squares' of four adjacent points as the 2-cells. The bounding relation is defined in the obvious way. This structure is known as a *(2 dimensional) cartesian* cell complex (see [Web01] for further details).

4   Digital Geometries on $\mathbb{Z}^2$

Section 3 looked at two structures with which we may represent discrete space. In this section we show how convexity can be represented in these structures. In each case we give convexity operators which satisfy the axioms for a convex geometry given in Section 2.1. Interestingly, we also show that there are other convexity operators definable on cell complexes which would be appropriate in different contexts, but do not necessarily satisfy the axioms of a convex geometry.

4.1   Matroids and Convexity on $\mathbb{Z}^2$

We have already established various results about convexity; now we look at how oriented matroids can be used as a model of convexity, and at the same time abstract certain properties of digital space. We begin by looking at a matroid, M, whose ground set will be a finite subset of $\mathbb{Z}^2$, and whose independent sets are the affine independent subsets of the space: the sets S satisfying $\forall x, y \in S$, $x \neq y$, $\nexists z \in S$ $\lambda x + \mu y = z$ for $\lambda, \mu \neq 0, \lambda + \mu = 1$. In addition, M must be of rank 3 (the largest independent sets being triples of non-collinear points), i.e. the ground set should not be equivalent to a lower dimensional subspace embedded in $\mathbb{Z}^2$.

Now, the affine sets of the space, i.e. the sets, S, for which the condition $\forall x, y, z \in S \exists \lambda, \mu \in \mathbb{Q}$, $\lambda + \mu = 1$, $x = \lambda y + \mu z$ holds, are easily shown to be the flats of M: they are maximal sets of collinear points (hence subspaces), and the largest independent set is of rank 2 (any pair of points). Hence the set of all flats forms an affine geometry - see Section 3.1. The maximal (proper) flats are the hyperplanes of the matroid, and hence the cocircuits are the set theoretic complements of these. The hyperplanes of M correspond to sets of collinear points.

An example follows to help visualize the different sets:

Example: The illustration in Figure 5 is a rank 3 subspace of $\mathbb{Z}^2$. The following lists give the elements of this subspace which are independent, dependent, circuits etc.

The cocircuits of M form the circuits of a dual matroid, M', and we partition the circuits into positive and negative components depending on which side of the defining hyperplane they fall. (Note that this step is not generally feasible, as it assumes the underlying geometry of the discrete space.) This partitioning of the circuits of M' gives us the oriented matroid OM.

Example: The following pairs give the signed circuits of an oriented matroid on the points in Figure 5: ({c},{d,e}), ({b,d,e},∅), ({c},{a,d}), ({a,b,d},∅), ({b,c},∅), ({a},{e}) and their opposite pairs.

We define halfspaces to be the complement of the positive elements of a circuit. This gives us the means to generate a convex geometry on the oriented matroid. We define a *convex subset* of an oriented matroid as being any subset of its ground set equal to the intersection of halfspaces of the matroid. This set of convex subsets indeed satisfies the axioms for a convex geometry:

Theorem 1. *Let $X \subseteq \mathbb{Z}^2$ be a finite set, and $M = (X, (C^+, C^-))$ be an oriented matroid defined in the manner described above. Then, the set of convex subsets of M, $\mathcal{C}$, is a convex geometry on X.*

Proof.

A0 Take any maximal independent set. Then, take the set of hyperplanes containing the maximal proper subsets of this set. Now take the halfspaces such

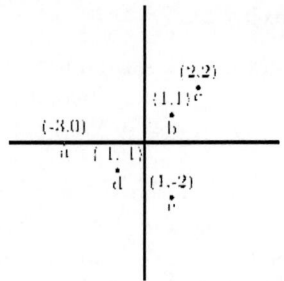

Independent Sets:	∅, {a}, {b}, {c}, {d}, {e}, {a,b}, {a,c}, {a,d}, {a,e}, {b,c}, {b,d}, {b,e}, {c,d}, {c,e}, {d,e}, {a,b,c}, {a,b,d}, {a,b,e}, {a,c,d}, {a,c,e}, {b,c,e}, {c,d,e}
Bases:	{a,b,c}, {a,b,d}, {a,b,e}, {a,c,d}, {a,c,e}, {b,c,e}, {c,d,e}
Dependents:	{a,d,e}, {b,c,d} and all sets of 4 or more points.
Circuits:	{a,d,e}, {b,c,d}, {a,b,c,e}
Subspaces:	(rank 0) ∅ , (rank 1) all singleton sets, (rank 2) {a,b}, {a,c}, {b,e}, {c,e}, {a,d,e}, {b,c,d}
Hyperplanes:	{a,b}, {a,c}, {b,e}, {c,e}, {a,d,e}, {b,c,d}
Co-circuits:	{c,d,e}, {b,d,e}, {a,c,d}, {a,b,d}, {b,c}, {a,e}

Fig. 5. A subset of $\mathbb{Z}^{\bullet}$.

    that the signed circuit part of the halfspace does not contain any elements of the independent set. Then the intersection of these halfspaces is empty.

A1 Since the space is finite, it is therefore bounded. Take any hyperplane such that the remainder of the space is entirely to one side, then we have a halfspace which is in fact the entire space.

A2 Since all convex sets are defined as an intersection of halfspaces, the intersection of sets of convex sets are by definition intersections of sets of halfspaces, and hence are convex sets themselves.

A3 Given a set of convex sets totally ordered by inclusion, the largest of these will be their union, and is of course a convex set.

AE (Idea of proof) We need to show that for two points $x, y \notin [S]$ that if $y \in [S \cup \{x\}]$ then $x \notin [S \cup \{y\}]$. We can show that x is an extreme point ('outside vertex') of $S \cup \{x\}$. Since $y \in [S \cup \{x\}]$, then $[S \cup \{x\}] = [S \cup \{x\} \cup \{y\}]$ and so x is an extreme point of $S \cup \{x\} \cup \{y\}$. We can show that a hyperplane exists separating an extreme point from the rest of the set, and hence a halfspace exists containing $S \cup \{y\}$ but not x. Hence $x \notin [S \cup \{y\}]$.

    This shows that the oriented matroid structure is rich enough to give us both convex and affine geometries, and demonstrates how these concepts are closely related. Note that in fact due to the way in which the proof works, the theorem holds more generally for finite subsets of $\mathbb{R}^2$.

4.2   Convexity Operators on Cell Complexes in $\mathbb{Z}^2$

We model a finite subset C of $\mathbb{Z}^2$ here with a 2 dimensional cartesian cell complex. This complex takes singleton sets of points of C as its 0-cells $(dim(\{(n,m)\}) = 0)$, sets of points of the form $\{(n,m), (n, m+1)\}$ or $\{(n,m), (n+1,m)\}$ as 1-cells, and sets $\{(n,m), (n+1,m), (n, m+1), (n+1, m+1)\}$ as 2-cells. The bounding relation bnd is simply the subset relation between cells - a bounds b iff $a \subset b$. The complex will be denoted $\mathbb{C} = (C, bnd, dim)$.

We define an oriented matroid on the cell complex in the way described in Section 4.1, where the ground set is the set of 0-cells. Given a halfspace H of the matroid, we call the set of cells $\{c \in C : c \subseteq \bigcup H\}$ a *cc-halfspace*. It corresponds to all of the cells either on, or to one side of the hyperplane defining the halfspace. Note that a cc-complex will always be a *closed subcomplex*, in the sense that if a cell is in the subcomplex, then so will all of its bounding cells. Dually, if a cell is in an *open subcomplex* then all of the cells that it bounds will be in the subcomplex. Correspondingly, we define a *oc-halfspace* as the smallest open subcomplex containing a cc-halfspace.

This definition of cc- and oc-halfspaces leads us to two distinct notions of convex hull, closed and open convex hulls:

Closed Convex Hull: The closed convex hull of a set S is the intersection of all cc-halfspaces containing S.

Open Convex Hull: The open convex hull of a set S is the intersection of all oc-halfspaces containing S.

These definitions both apply the notion of convexity as intersecting halfplanes directly to cell complexes. These definitions would be particularly applicable if we wish to restrict the subcomplexes we are interested in to the closed (open) subcomplexes, the choice between the two may depend on whether we take a point or pixel view of the world.

We should note here that though these convexity operators do indeed satisfy the axioms A1-A3 (and the closed convexity operator also satisfies A0), neither satisfies the anti-exchange axiom AE. The first diagram in Figure 6 shows a set S (and its closed convex hull), and labelled cells x and y. The second diagram shows the hull of $S \cup x$ - note that this is the same as the hull of $S \cup y$. This shows that y is in the hull of $S \cup x$, but also x is in the hull of $S \cup y$, and so the anti-exchange axiom fails.

Centred Convex Hull: The centred convex hull of a set S is the convex hull of S when the cell complex is considered as an oriented matroid whose ground set is the set of cells, treating cells as their centre points for the purposes of generating the matroid.

This notion simply treats the cell complex as a grid of points twice the size of the set of 0-cells, and as such is clearly a convex geometry. Note that this notion of convexity treats cells of different dimension as if they were all 0-dimension.

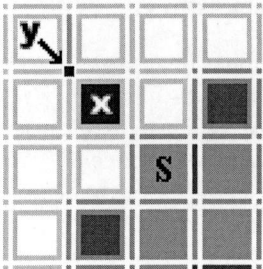 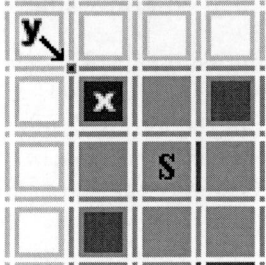

Fig. 6. Closed convex hulls do not satisfy the anti-exchange axiom.

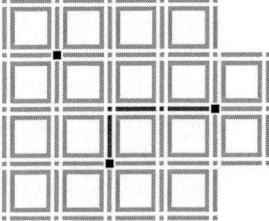

Fig. 7. Closed Convex Hull.

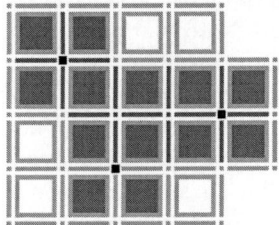

Fig. 8. Open Convex Hull.

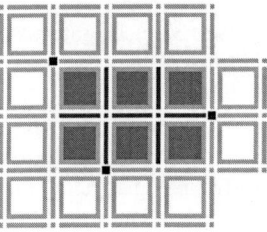

Fig. 9. Embedded Convex Hull.

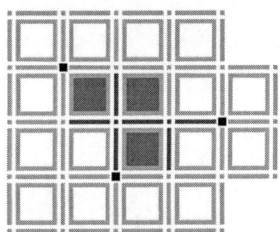

Fig. 10. Centred Convex Hull.

Embedded Convex Hull: The embedded convex hull of a set S is the image in the cell complex of the convex hull of S when considered as embedded in Euclidean space.

This is a convenient way of defining a convexity operation on a cell complex, as it applies well known operations in Euclidean space to the cells of a cell complex. However, it is a matter of interpretation as to how the image of a set in Euclidean space in a cell complex is defined. If we define it as the set of cells which overlap the (Euclidean) hull (see [Web01] for example), then we have a hull which is not idempotent.

Since we are working in a discrete space, we would expect that some aspects of a more traditional continuous notion of convexity will be lost. Indeed the notions of convexity we have investigated can all produce results that may seem at odds with convexity in a Euclidean context, even if they do form a convex geometry. For example there is no guarantee that set which is convex under the closed or centred definitions are connected, embedded convexity is not idempotent, and closed and open convexity do not satisfy the anti-exchange axiom.

Each operator has its uses however, the closed (open) hull is particularly useful in a context where the only sets of interest are closed (open) (see for example [RS02]). The centred convexity operator would be suited to any situation where a robust method of finding the convex hull is desired, and where the hull must satisfy the axioms of convexity. The embedded convexity operator would be most suited to applications where convexity operations are already defined for Euclidean space, and a simple mapping from this to cell complexes is desired.

5   Convexity Algorithms for Oriented Matroids

This section looks at data structures and algorithms using oriented matroids.

5.1   Matroid Data Structure

The basic structure of a Matroid is a set and a set of subsets of that set. The set data structure can be efficiently implemented as a hashtable which is restricted to ignore duplicated items (see for example the Java HashSet class). Such a structure will have constant complexity for checking the *member* relation between an element and a set, and linear complexity for checking the *subset* relation between sets. The structure should be immutable, as there should be no reason to alter the sets once they are initially stored.

The validateIndependents algorithm validates a Matroid under the independence axioms. Axiom I2 states that every subset of an independent set is also independent. We need only check the maximal proper subsets of each independent set however, since due to the transitivity of the subset relation and due to the fact that if a maximal proper subset is independent we will also be checking *its* maximal proper subsets. The following algorithm returns the set of maximal proper subsets of a set (complexity $O(n)$):

Algorithm maximalSubsets(s)
Input: A set s.
Output: A set *out* of the maximal subsets of s.

```
1. create new set out;
2. for (each e of s)
3.    add s-{e} to out
4. end for;
5. return out;
```

We now give an algorithm which determines whether or not the exchange axiom holds for a pair of sets s_1, s_2 (complexity $O(mn)$ where m and n are the cardinality of each input set):

Algorithm exchange(s1,s2)
Input: Two sets s1 and s2
Output: true if the exchange axiom holds for these sets,
false otherwise.

```
1. exchanged := false;
2. diff := s2 - s1;
3. for (each e of diff)
4.    temp := s1 ∪ {e};
5.    exchanged := exchanged OR temp is in independents;
6.    if (exchanged = true ) return true ;
7. end for
8. return false ;
```

The following algorithm validates the independent sets of a matroid (complexity $O(n^2m)$ where n is the number of independent sets, and m is the cardinality of the ground set):

Algorithm validateIndependents(M)
Input: A matroid M=(groundset, indeps)
Output: true if the validation is successful, false otherwise.

```
 1. allSubsets := true ; exchange := true ;
 2. for (each i in indeps)
 3.    if (i is not a subset of groundset) return false ;
 4.    allSubsets := allSubsets AND maximalSubsets(i) ⊆ indeps;
 5.    if (allSubsets = false ) return false ;
 6.    for (each i2 in indeps)
 7.       if (|i| < |i2|) exchange := exchange AND exchange(i, i2);
 8.       if (exchange = false ) return false ;
 9.    end for;
10. end for;
```

Note that though the complexity is quite high on these operations, we only need to perform validation once, when the matroid is first created. In fact, if we have mathematical proof that a particular set and set of subsets satisfies the axioms, validation will be unnecessary.

5.2   Operations on Oriented Matroids

Since the axioms for an oriented matroid are different from (and not equivalent to) the axioms of matroids, the validation algorithm is different. We denote the opposite of a signed circuit $c = (c^+, c^-)$ by $-c = (c^-, c^+)$. The complexity is $O(m^3n^2)$ for a oriented matroid with ground set of cardinality n and m circuits. Again, this is a one-time check, and unnecessary in the presence of a proof that a structure satisfies the axioms.

Algorithm validateCircuits(M)
Input: An oriented matroid M=*(groundset, circuits)*
Output: true if the validation is successful, false otherwise.

```
 1. for (each c in circuits)
 2.    if (c is not a subset of groundset) return false ;
 3.    if (c has no elements) return false ;
 4.    if (c ≠ -c) return false;
 5.    for (each c2 in circuits)
 6.       if (c2 ⊆ c AND c2 ≠ c) return false ;
 7.       if (c and c2 are equal) skip to next iteration of loop;
 8.       setInter := intersection(c⁺, c2⁻);
 9.       setUnionPos := union(c⁺, c2⁺);
10.       setUnionNeg := union(c⁻, c2⁻);
11.       inter : for (each a in setInter)
12.          testPos := setUnionPos - a;
13.          testNeg := setUnionNeg - a;
14.          for (each c3 in circuits)
15.           hasCirc := c3⁺ ⊆ testPos AND c3⁻ ⊆ testNeg;
16.            if (hasCirc = true) skip to next iteration of inter;
17.           return false;
18.          end for;
19.       end for;
20.    end for;
```

The following algorithm generates the set of halfspaces of an oriented matroid. The terminology used is based on the notion that the circuits of the matroid we have in mind are generated by the cocircuits of a matroid whose independent sets are linearly independent in the Euclidean sense. Of course the algorithm will work on any oriented matroid.

Algorithm halfspaces(M)
Input: An Oriented matroid M=*(groundset, circuits)*
Output: A set of subsets of the ground set of M, the *halfspaces*.

```
1. create new set halfspaces
2. for (each c in circuits )
3.    create new set subset
4.    subset = groundset - getPositiveCircuit(c )
5.    add subset to halfspaces
6. return halfspaces
```

The function *getPositiveCircuit(c)* simply returns the positive elements of the circuit. The algorithm then simply has the effect of collecting together the set theoretic complements of these half-circuits. The convex hull algorithm works directly with the set of halfspaces of an oriented matroid:

Algorithm convexHull(M,points)

Input: An Oriented matroid, M=(groundset, signed_circuits), and a subset of points of groundset.

Output: A subset of groundset corresponding to the convex hull of points in M.

```
1. out := groundset;
2. hspaces := halfspaces(M);
3. for (each h in hspaces)
4.    if points ⊆ h then out := intersection(out,h);
5. end for ;
```

Clearly the convex hull operator works with the Oriented Matroid structure to give a simple and robust algorithm for convex hulls. Once the matroid is instantiated, the algorithm for generating the convex hull is quite efficient, involving just one loop. If the number of elements in the ground set is n, and the number of circuits is m, then the complexity of the hull operator will be $O(nm)$.

6   Conclusions and Further Work

We have shown that it is possible to apply Coppel's axiomatization for convexity in a discrete setting. The nature of the axioms make oriented matroids a natural structure for capturing much of the geometry required in such a setting, and we have given a construction of oriented matroids from subspaces of $\mathbb{Z}^n$.

The oriented matroid structure can be superimposed on cell complexes, and convexity operations deriving from this underlying matroid ensure that the axioms of convexity are satisfied by particular convexity operations. We have also briefly investigated the properties of different convexity operations, and considered the sort of setting that each might be appropriate in.

Algorithms for obtaining a convex hull through the oriented matroid structure are shown, along with algorithms for determining that a set of subsets are indeed the independent sets, or signed circuits of a (oriented) matroid.

There is plenty of scope for further development to this work. The algorithms given are not very efficient at present. However we are working on improving these, using techniques similar to those used in convexity algorithms in Euclidean geometry by considering notions of extreme points. We also aim to provide algorithms that generate halfspaces 'on the fly', rather than the memory and time intensive method we currently have in generating all of the halfspaces. We also intend to look further at convexity on cell complexes, with the intent of applying convexity to qualitative spatial reasoning in discrete space.

References

BLS⁺ 99.   A. Björner, M. Las Vergnas, B. Sturmfels, N. White, and G. Ziegler. *Oriented Matroids*, volume 46 of *Encyclopedia of Mathematics and its Applications*. CUP, 1999.

BY98.   J-D. Boissonnat and M. Yvinec. *Algorithmic Geometry.* Cambridge University Press, 1998.

CB* 97.   A. G. Cohn, B. Bennett, et al. Qualitative spatial representation and reasoning with the region connection calculus. *GeoInformatica*, 1:275–316, 1997.

Cop98.   W. A. Coppel. *Foundations of Convex Geometry.* Cambridge University Press, 1998.

dvOS99. M. de Berg, M. van Krevald, M. Overmars, and O. Schwarzkopf. *Computational Geometry Algorithms and Applications.* Springer, 1999.

Eck01.   Ulrich Eckhardt. Digital lines and digital convexity. *Lecture Notes in Computer Science*, 2243:209–228, 2001.

Gal99.   A. Galton. The mereotopology of discrete space. In C. Freksa and D. Mark, editors, *Spatial Information Theory. Cognitive and Computational Foundations of Geographic Information Science. International Conference COSIT'99*, volume 1661 of *Lecture Notes in Computer Science*, pages 251–266. Springer-Verlag, 1999.

Knu92.   Donald E. Knuth. *Axioms and Hulls.* Lecture Notes in Computer Science. Springer-Verlag, 1992.

Kov89.   V. A. Kovalevsky. Finite topology as applied to image analysis. *Computer Vision, Graphics and Image Processing*, 46:141–161, 1989.

Kov92.   V. A. Kovalevsky. Finite topology and image analysis. *Advances in Electronics and Electron Physics*, 84:197–259, 1992.

MV99.   C. Masolo and L. Vieu. Atomicity vs infinite divisibility of space. In C. Freksa and D. Mark, editors, *Spatial Information Theory. Cognitive and Computational Foundations of Geographic Information Science. International Conference COSIT'99*, volume 1661 of *Lecture Notes in Computer Science*, pages 235–250. Springer-Verlag, 1999.

Oxl92.   James G. Oxley. *Matroid Theory.* Oxford Graduate Texts in Mathematics. Oxford University Press, 1992.

Ros79.   Azriel Rosenfeld. Digital topology. *American Mathematical Monthly*, 86(8):621–630, 1979.

RS02.   A. J. O. Roy and J. G. Stell. A qualitative account of discrete space. In *Proceedings of GIScience 2002*, volume 2478 of *Lecture Notes in Computer Science*, pages 276–290, 2002.

Ste00.   J. G. Stell. The representation of discrete multi-resolution spatial knowledge. In A. G. Cohn, F. Giunchiglia, and B. Selman, editors, *Principles of Knowledge Representation and Reasoning: Proceedings of KR2000*, pages 38–49. Morgan Kaufmann, 2000.

von75.   Rabe von Randow. *Introduction to the Theory of Matroids.* Lecture Notes in Economics and Mathematical Systems. Springer-Verlag, 1975.

Web94.   Roger Webster. *Convexity.* OUP, 1994.

Web01.   Julian Webster. Cell complexes and digital convexity. *Lecture Notes in Computer Science*, 2243:272–284, 2001.

Webar.   Julian Webster. Cell complexes, oriented matroids and digital geometry. *Theoretical Computer Science*, to appear.

Wel76.   D. J. A. Welsh. *Matroid Theory.* Academic Press, 1976.

WF00.   Stephan Winter and Andrew U. Frank. Topology in raster and vector representation. *Geoinformatica*, 4(1):35–65, 2000.

Win95.   S. Winter. Topological relations between discrete regions. *Lecture Notes in Computer Science*, 951:310–327, 1995.

Stratified Rough Sets and Vagueness

Thomas Bittner and John G. Stell

˙ IFOMIS - University of Leipzig
thomas.bittner@ifomis.uni-leipzig.de
˙ School of Computing, University of Leeds
jgs@comp.leeds.ac.uk

Abstract. The relationship between less detailed and more detailed versions of data is one of the major issues in processing geographic information. Fundamental to much work in model-oriented generalization, also called semantic generalization, is the notion of an equivalence relation. Given an equivalence relation on a set, the techniques of rough set theory can be applied to give generalized descriptions of subsets of the original set. The notion of equivalence relation, or partition, has recently been significantly extended by the introduction of the notion of a granular partition. A granular partition provides what may be thought of as a hierarchical family of partial equivalence relations. In this paper we show how the mechanisms for making rough descriptions with respect to an equivalence relation can be extended to give rough descriptions with respect to a granular partition. In order to do this, we also show how some of the theory of granular partitions can be reformulated; this clarifies the connections between equivalence relations and granular partitions. With the help of this correspondence we then can show how the notion of hierarchical systems of partial equivalence classes relates to partitions of partial sets, i.e., partitions of sets in which not all members are known. This gives us new insight into the relationships between roughness and vagueness.

1 Introduction

In processing geographic information, handling multiple levels of detail is of considerable practical importance. This is true both of cartographic generalization [MLW95], where the geometric presentation of the data is a major factor, and also of 'model-oriented generalization' in the sense of [M⁺95]. In model-oriented generalization, the relevant attributes of the data are not geometric, but might for example be thematic classifications. In such a case the generalization might replace several distinct specific classifications with one more general one. As, say, in the process of ignoring the distinction between different kinds of road (motorways, major roads, minor roads, etc) and reducing to the single concept 'road'. A conceptually similar kind of generalization can be performed on raster data when deliberately reducing the resolution. In this case a number of pixels, which might be given a number of different colours could be replaced by a single pixel bearing just one colour.

An alternative terminology is used in Jones [Jon97, p271] where *semantic generalization* is described as being "...concerned with the meaning and function of a map and it depends on being able to identify hierarchical structure in the geographical information." This hierarchical structure has been used in making formal theories of the process

W. Kuhn, M.F. Worboys, and S. Timpf (Eds.): COSIT 2003, LNCS 2825, pp. 270–286, 2003.

of semantic generalization. The most obvious is a thematic classification given as a tree, but a richer notion of hierarchical structure is found, for example, in the studies of a lattice of resolution by Worboys [Wor98a,Wor98b].

In investigating the theory of semantic generalization we find the notion of equivalence relation, or partition, is a fundamental ingredient. In collapsing multiple kinds of road to a single one, we are imposing an equivalence relation on the available themes and putting the various kinds of road into the same *equivalence class*. To this equivalence class we give the label 'road'. In the example of raster data, the equivalence relation groups together the pixels at the more detailed level which become a single pixel at the coarser level of detail. This example may also exhibit a second equivalence relation which taking the labels of the more detailed pixels amalgamates them to a single equivalence class which is used to label the single pixel at the coarser level.

The basic way in which an equivalence relation is used may be summarized as follows. An equivalence relation groups together entities which are in some sense similar. Each collection of 'similar' entities forms new a single entity, called an equivalence class. A subset of the original set of entities can be given a rough description by specifying the extent to which each of the equivalence classes lies within the subset. In the most basic approach, this extent can be one of the three: wholly, partly, and not at all. Within geographic information, the use of equivalence relations has been explored in the context of rough sets [BS01], and the extension of equivalence relations on sets to the analogous structure on graphs has also been considered [Ste99]. A formal theory of partitions of space was provided by Erwig and Schneider [ES97].

An equivalence relation allows us to model the passage from one level of detail to another, but does not, on its own, model the considerably more than two levels of detail which are needed in practice. To deal with several levels of detail, a new concept has been proposed: the granular partitions of Bittner and Smith [BS03a]. A granular partition can be seen as an extension of the concept of equivalence relation, and it is the purpose of this paper to examine how the rough descriptions of the theory of rough sets can be extended from ordinary equivalence relations to the multi-level world of granular partitions.

The paper is structured as follows. To generalize the use of partitions in the study of roughness to granular partitions it is useful to present the theory of granular partitions in a new way (section 3 below), and to prepare for that we review the key notions of roughness (section 2 below). In section 4 we introduce systems of hierarchically ordered stratified rough sets. The ordering hereby corresponds to the degree of roughness of the underlying equivalence classes. In section 5 we generalize the notion of stratified rough sets by considering partial equivalence classes or equivalence classes in partial sets [MMO90]. In section 6 the notion of rough set is generalized in order to take into account vagueness. Conclusions are presented in section 7.

In places the paper is rather technical. This apparent complexity seems unavoidable and arises from the interaction between the granular partitions and the rough set concepts. This interaction produces a more intricate theory than is found in either of the two ingredients separately. Despite the technicality, the topic is, as explained above, one of considerable importance and we have provided examples in the paper which are designed to illustrate the main concepts.

2 Labelled Partitions and Rough Sets

In this section we introduce the notions of K-labelled partitions and rough sets. We show that maps are an important class of K-labelled partitions and that rough sets can be used in order to approximate objects with indeterminate boundaries.

2.1 Labelled Partitions

A partition here is understood in the standard mathematical sense: the subdivision of a set into jointly exhaustive and pairwise disjoint subsets via a corresponding equivalence relation. Partitions of a set, X, are often identified with functions of the form $f : X \to K$ which are surjective (that is where for every $k \in K$, there is some $x \in X$ for which $fx = k$).

Given such a function $f : X \to K$, we obtain a partition of X into subsets of the form $[x]_k = \{x \in X \mid fx = k\}$ where $k \in K$. The same partition however can arise from different functions. Consider, for example the subdivision of a part of the plane into subsets indicated by the 12 squares in Figure 1(i). In Figure 1(ii) and (iii) we have two different labelled versions of the same partition: $f_1 : X \to \{1, 2, \ldots, 11, 12\}$, and $f_2 : X \to \{a, b, \ldots, k, l\}$. Two functions $f : X \to K$ and $f' : X \to K'$ give rise to the same labelled partition if and only if there is a bijection $\varphi : K \to K'$ such that $\varphi f = f'$.

A surjective function $f : X \to K$ thus corresponds to something more than a partition of X: it is a partition of X together with a labelling (by the elements of K) of the blocks of the partition. It is useful to use the terms *blocks* and *cells* so that blocks are subsets of the partitioned set X, whereas cells are labels for these blocks. It may be helpful to imagine that the cells are labelled boxes or locations which are used to house the elements of X. The distinction between cells and blocks is then the distinction between a location and the contents of that location. To emphasize the importance of the labelling, we make the following definition.

Definition 1. *Let K and X be sets. Then a K-**labelled partition of** X is a surjective function from X to K.*

An important class of K-labelled partitions are maps (in the cartographic rather than the mathematical sense). Consider the left part of figure 2 which shows a part of the United States. The labelling function f here maps every point of the United States to names of federal states (Montana, Wyoming, Idaho, etc.).

2.2 Rough Sets

Given a labelled partition of X (i.e. a surjective function $f : X \to K$ for some K) we obtain rough descriptions of the subsets of X in terms of the extent to which the cells are occupied by the subset. So, for any $Y \subseteq X$ we obtain a function $Y \boxplus f : K \to \{\mathsf{T}, \mathsf{B}, \mathsf{F}\}$. This function, read as 'Y is coarsened by f' is defined as follows.

$$(Y \boxplus f)\, k = \begin{cases} \mathsf{T} & \text{if } \forall x \in f^{-1}k \; x \in Y \\ \mathsf{B} & \text{if } \exists x, x' \in f^{-1}k \; x \in Y \text{ and } x' \notin Y \\ \mathsf{F} & \text{if } \forall x \in f^{-1}k \; x \notin Y \end{cases}$$

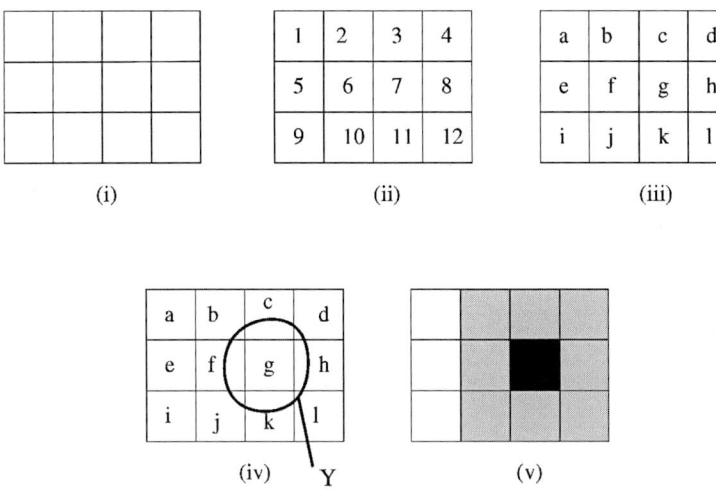

Fig. 1. A partition of a subset of a plane (i) with two different labellings (ii) and (iii) and a subset Y (iv) and its egg-yolk representation (v).

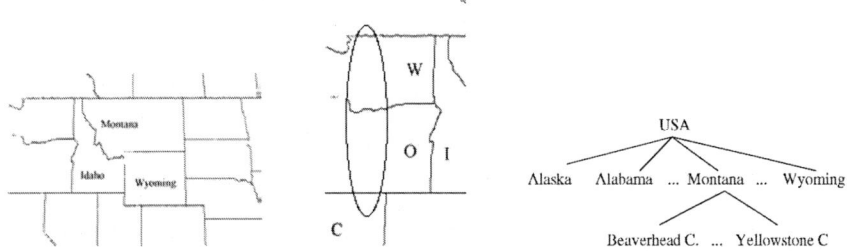

Fig. 2. A k-labelled partition (left); A rough approximation wrt. a k-labelled partition (middle); A stratified labelled partition (right).

where f^{-1} is the inverse image of f, so that $f^{-1}k$ means $\{x \in X \mid fx = k\}$. The notation T, B, F is chosen as these three values are the concepts *True*, *Both*, and *False*. This is because if $(Y \boxplus f)\,k = \mathsf{T}$ then k is definitely in Y; if the value is F, then k is definitely not in Y; if the value is B, then k is both in and not in Y. The structure resulting from a coarsening operation is a *rough set* as defined in [Paw82].

The intuition is that the value of $(Y \boxplus f)\,k$ is T, B, or F according as the cell k is completely, partially or not at all occupied by elements of the subset Y. Consider 1 (iv). Again, let X be the set of points of the part of the plane and let $Y \subset X$ a subset. The rough set approximation of Y with respect to the labelled partition f_2 is given below.

$k \in K$	a	b	c	d	e	f	g	h	i	j	k	l
$(Y \boxplus f_2)\,k$	F	B	B	B	F	B	T	B	F	B	B	B

The rough subset $Y \boxplus f$ can be represented by a pair of ordinary subsets of K: $\langle (Y\boxplus f)^{-1}\{\mathsf{T}\},\ (Y\boxplus f)^{-1}\{\mathsf{T},\mathsf{B}\}\rangle$, leading to the usual 'egg-yolk' pictures (Figure 1 (v)).

Here $(Y \boxplus f)^{-1}\{\mathsf{T}\}$ is the set $\{x \in (f^{-1}k) \mid (Y \boxplus f)\, k = \mathsf{T}\}$ and marked by the black square. Correspondingly $(Y \boxplus f)^{-1}\{\mathsf{T},\mathsf{B}\}$ is the set $\{x \in (f^{-1}k) \mid (Y \boxplus f)\, k \in \{\mathsf{T},\mathsf{B}\}\}$ and corresponds to the union of the black and grey squares in the figure.

In the remainder we will use the phrases 'the rough set $Y \boxplus f$' and 'the (rough) approximation of Y with respect to the labelled partition f' synonymously.

Rough set approximations play an important role for the representation of objects with indeterminate boundaries [BS02], [BS03b]. Consider figure 2. In the middle we have a K labelled partition of the northwestern US and we have the Cascade mountains (CM), indicated by the ellipse, which cover parts of the states Washington (W), Oregon (O), and California (C). The rough set representation of the cascade mountains is $(CM \boxplus f_{USA})\, W = \mathsf{B}$, $(CM \boxplus f_{USA})\, O = \mathsf{B}$, $(CM \boxplus f_{USA})\, C = \mathsf{B}$, $(CM \boxplus f_{USA})\, I = \mathsf{F}$, etc. Rough set representations do not force us to draw crisp boundaries where no crisp boundaries exist.

3   Granular Partitions

Maps are often organized hierarchically. Consider the political subdivision of the US. Here we have counties which form states, which themselves form the US as a whole. This structure is visualized in the right part of figure 2. In this section we introduce the notion of K labelled stratified partition in order to take this hierarchical structure into account.

3.1   Cell Granulations

Above we considered only unstructured sets. Now we consider sets of cells upon which a *tree structure* has been defined.

Definition 2. *A **cell tree** is a finite, partially ordered set of cells, $(K, \leq)$, which forms a tree. The partial order, $\leq$, is called the sub-cell relation, and the maximum element in this order will be the root of the tree. If a cell tree additionally satisfies the constraint that no node have just a single descendant then it is said to be **branching**.*

Consider figure 3 which shows a cell tree K with elements a, b, c, d, e, f, g, and h. Here the cell a is the root of the tree and we have $x \leq y$ if and only if the nodes x and y are connected by a line going upwards, or by a sequence of such lines.

The tree structure gives rise to a lattice (middle of figure 3), the elements of which are the cuts of the tree, defined as follows [RS95]:

Definition 3. *For any element k of a cell tree K, let $\mathsf{d}(k)$ denote the set of immediate descendants of k. A **cut** in K is a subset of K defined inductively as follows:*

1. *$\{r\}$ is a cut, where r is the root of the tree,*
2. *let C be a cut and $k \in C$ where $\mathsf{d}(k) \neq \varnothing$, then $(C - \{k\}) \cup \mathsf{d}(k)$ is a cut.*

It follows that the sets $\{a\}$, $\{b,c\}$, $\{d,e,f,c\}$, $\{b,g,h\}$, and $\{d,e,f,g,h\}$ are cuts in the tree K in the left part of figure 3.

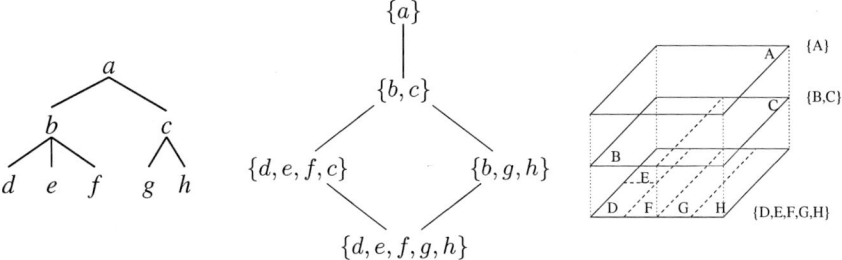

Fig. 3. A cell granulation (left), the corresponding cut lattice (middle), and the corresponding hierarchical subdivision of the point-set A (right).

Let C and D be cuts in the cell tree K. The cuts of a tree form a lattice ordered by $C \sqsubseteq D$ if for each $c \in C$ there is some $d \in D$ with $c \le d$. This lattice will be referred to as the *cut lattice* of the cell tree K. The cut lattice of our example cell tree is shown in the middle of figure 3.

The cut lattice carries additional structure, which we will discuss now. Given cuts $C \sqsubseteq D$ there is a function $\ell : C \to D$ where $c \le \ell c$. The facts (i) that C and D are cuts in a tree K and that (ii) $C \sqsubseteq D$ ensure that ℓ is a function. For example, in the cut lattice seen in the middle of figure 3, with cuts $C = \{d, e, f, c\}$ and $D = \{b, c\}$, the function ℓ is the following mapping: $d \mapsto b, e \mapsto b, f \mapsto b, c \mapsto c$. Note that when $C = D$ the function ℓ will be the identity.

We have seen that a set of cells structured as a tree gives rise to a lattice, the elements of which are sets of cells, and that these sets of sets are related by functions. All this structure can be derived from the tree, but it is often more convenient to deal with it directly than to always be thinking of it as generated from the tree. Thus we will refer to the lattice and associated structure as a **cell granulation**; it consists of:

1. A lattice, $(\mathcal{L}, \sqsubseteq)$, of levels of detail
2. for each level of detail $i \in \mathcal{L}$ there is a set of cells $\mathcal{L}_i$, and
3. for each pair i, j of levels of detail, where $i \sqsubseteq j$, there is a function $\ell_{ij} : \mathcal{L}_i \to \mathcal{L}_j$.

It should be noted that not every structure of the above form will be a cell granulation, as only lattices of a certain form can arise as lattices of cuts of trees. The cell granulation is derived from the cell tree, and will be denoted simply as $\mathcal{L}$ when there is no danger of confusing this with the underlying lattice. If it is necessary to emphasize the dependence on the tree K we can write $\mathcal{L}(K)$ rather than just $\mathcal{L}$. Cuts or *levels of detail* or *levels of granularity* will be referred to by their index i or by the corresponding set $\mathcal{L}_i$.

3.2 Stratified Labelled Partitions

Having described the granulation structure on the set of cells, we now see how these are used to construct stratified labelled partitions. Recall that in the ordinary case a partition of a set X labelled by a set of cells K is a surjective function from X to K. In the granular case, the role of K is taken by the cell granulation $\mathcal{L}(K)$ introduced in section 3.1 above, so it remains to explain what plays the role of the surjective function in the ordinary case.

Definition 4. *Let $\mathcal{L}$ be the cell granulation derived from a cell tree K, and let X be a set. Then a K-labelled stratified partition consists of for each $i \in \mathcal{L}$ a partial and surjective function $f_i : X \to \mathcal{L}_i$ such that whenever $i \sqsubseteq j$*

$$(\forall x \in X) \, (\ell_{ij} \, f_i \, x = f_j \, x)$$

whenever the left hand side of the equation is defined.

The introduction of partial functions here is significant, and is motivated by the theory of granular partitions. At a particular level of detail, we allow that the collection of cells, or labels, at our disposal may not cover all the entities to be classified. It should remembered that the definition of a partial function allows for the function to be undefined for some elements of its domain, but it does not exclude the possibility that the function is total. Thus, partiality corresponds to the *potential* for having unclassified entities, it does not mean that there have to be some things which are unclassified.

Consider the right part of figure 3 which shows the subdivision of the point set A in subsets which form partitions of A at different levels of detail. (In this example we use capital letters to denote sets and corresponding non-capital letters for their labels.) At the top level we have the set A as a whole. At the intermediate level we have a partition of A formed by the subsets B and C. At the finest level we have a partition of A formed by the subsets D, E, F, G, and H. Also, the subsets D, E and F form a partition of the set B and the subsets G and H form a partition of C. For every partition of the set A into subsets there is now a corresponding labelled partition:

$$
\begin{aligned}
&f_1 : A \to \{a\}, && f_2 : A \to \{b, c\}, \\
&f_3 : A \to \{d, e, f, c\}, && f_4 : A \to \{b, g, h\}, \\
&f_5 : A \to \{d, e, f, g, h\}.
\end{aligned}
\tag{1}
$$

One can see that every co-domain of the labelled partitions $f_1 \ldots f_5$ corresponds to a cut in the cell granulation K formed by the cells $\{a, b, c, d, e, f, g, h\}$ depicted in the left part of figure 3. Now consider the labelled partitions f_4 and f_2 and assume $x \in G$. It then follows that $f_4 \, x = g$. Since g is a subcell of c we have $\ell_{42} \, g = c$. On the other hand, since $x \in G$ and $G \subset C$ we also have $x \in C$. Consequently we have $f_2 \, x = c$ and hence $f_2 \, x = (\ell_{42} \, f_4) \, x$.

Definition 4 can be neatly summarized by a diagram in the ordered category of sets and partial functions:

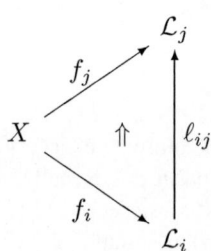

3.3   Granular Partitions

We shall now establish the correspondence between the notion of a granular partition introduced by [BS03a] and the notion of stratified labelled partitions introduced above. Basic components of a granular partition are a cell tree K, a corresponding set X, and mappings between them. However a granular partition does not have multiple surjective functions from X to cuts in K but rather a single order-preserving mapping, π, from K into the powerset of X. This notion of granular partition is very general. In this subsection we will establish the equivalence of labelled stratified partitions and a class of specific, particularly well-formed granular partitions:

Definition 5. *Let $(K, \leq)$ be a cell tree, X be a set, $\mathcal{P}_+ X$ denote the set of non-empty subsets of X, and let $\pi : K \to \mathcal{P}_+ X$ be a function such that for all $k_1, k_2 \in K$,*

$$(i)\ \ k_1 \leq k_2 \Leftrightarrow (\pi\, k_1) \subseteq (\pi\, k_2),$$
$$(ii)\ \pi k_1 \cap \pi k_2 \neq \varnothing \Rightarrow (k_1 \leq k_2 \ or \ k_2 \leq k_1).$$

The triple $\Pi = ((K, \leq), X, \pi)$ is then called a strict mereological monotonic granular partition. Condition (i) expresses the constraint that π be an order-isomorphism.

This particular class of granular partitions is such that the mapping π preserves the tree-structure of K, which is equivalent to saying that the subsets of X singled out by π have a tree structure (with respect to the subset relation) which is isomorphic to that of the cell tree K.

Consider the left and right part of figure 3. A granular partition then is a triple $\Pi = ((K, \leq), A, \pi)$ such that $(K, \leq)$ is as depicted in the left part of the figure and π is defined as follows: $\pi\, a = A, \pi\, b = B, \ldots, \pi\, h = H$, where capital letters refer to sets in the right part of the figure.

Given a cell granulation, we can define $\pi : K \to \mathcal{P}_+ X$ by $\pi k = \{x \in X \mid \lambda_i x = k$ for some $i\}$. The following result shows that this construction provides a strict mereologically monotonic granular partition provided that the cell tree is branching (no node having just a single descendant).

Theorem 1. *If the cell granulation $(\mathcal{L}(K), X, \lambda_1, \ldots \lambda_n)$ with $\lambda_i : X \to \mathcal{L}_i$ is a K branching labelled stratified partition then $\Pi = ((K, \leq), X, \pi)$ is a strict mereologically monotonic granular partition.*

Proof   First we show that if $\pi k_1 \cap \pi k_2$ is non-empty then either $k_1 \leq k_2$ or $k_2 \leq k_2$. If k_1 and k_2 are unrelated in the order then some cut, say q, must contain both of them. But then $\lambda_q x = k_1$ and $\lambda_q x = k_2$ contradicting the unrelatedness of k_1 and k_2.

Next we tackle one half of the first condition for a strict mereologically monotonic granular partition. Suppose that $k_1 \leq k_2$ and let $\lambda_i x = k_1$. Then there must be $j \geq i$ with $k_2 \in \mathcal{L}_j$, and $\ell_{ij} k_1 = k_2$. Hence $\lambda_j x = k_2$, and so $x \in \pi k_2$.

Finally, we have to show that if $\pi k_1 \subseteq \pi k_2$ then $k_1 \leq k_2$. As $\pi k_1 \cap \pi k_2$ is non-empty then either $k_1 \leq k_2$ or $k_2 \leq k_1$. If $k_2 \leq k_1$ then we have $\pi k_1 = \pi k_2$. The possibility that $k_2 < k_1$ can be excluded. For k_1 must have another descendant besides k_2, say k', at level i where $k_2, k' \in \mathcal{L}_i$. Now, as λ_i is surjective, there are distinct x, x' where

$\lambda_i x = k_2$ and $\lambda_i x' = k'$. But $\lambda_j x' = k_1$ for some j, as k' is a descendant of k_1, and so $\pi k_1 \neq \pi k_2$. Hence, having ruled out $k_2 < k_1$, we conclude $k_1 \leq k_2$.

$\square$

We note that if the original cell tree is not necessarily branching, then we can only prove that π is an order homomorphism (i.e. $k_1 \leq k_2 \Rightarrow \pi k_1 \subseteq \pi k_2$).

In the opposite direction, we can start with a strict mereologically monotonic granular partition and construct a K-labelled stratified partition. For each cut i, $\lambda_i x$ is defined if there is $k \in \mathcal{L}_i$ with $x \in \pi k$. In this case, $\lambda_i x = k$. That this construction has the appropriate properties is established in the following result.

Theorem 2. *If* $\Pi = ((K, \leq), X, \pi)$ *is a strict mereologically monotonic granular partition then the cell granulation* $(\mathcal{L}(K), X, \lambda_1, \dots \lambda_n)$ *with* $\lambda_i : X \to \mathcal{L}_i$ *is a K labelled stratified partition.*

Proof   The λ_i are well defined, for if $x \in \pi k_1 \cap \pi k_2$ we have $k_1 = k_2$ as $k_1 < k_2$ is impossible for distinct elements of the same cut. The λ_i are clearly surjective. It remains to show that if $i \leq j$ and $\lambda_i x$ is defined, then $\lambda_j x$ is defined and $\ell_{ij} \lambda_i x = \lambda_j x$. If $\lambda_i x = k \in \mathcal{L}_i$ then we can find $k' \in \mathcal{L}_j$ with $k \leq k'$, thus $\pi k \subseteq \pi k'$ and $x \in \pi k'$. As $k \leq k'$ we get $\ell_{ij} k = k'$ and so $\ell_{ij} \lambda_i x = \lambda_j x$.

$\square$

It follows that the notions *strict mereologically monotonic granular partition* and $\mathcal{K}$ *labelled stratified partition* are equivalent. In the remainder we focus onto the latter.

4   Stratified Rough Sets

As mentioned in section 2 above, an ordinary labelled partition $f : X \to K$ provides for each $Y \subseteq X$ a rough set $Y \boxplus f$. What happens to this process when we have a stratified labelled partition? In order to answer this question we now extend the notion of stratified rough set introduced by [Yao99].

Let $((K, \leq), (\mathcal{L}, \sqsubseteq), \ell, \lambda_1, \dots \lambda_n)$ a stratified labelled granular partition with a *total* surjective function of the form $\lambda_i : X \to \mathcal{L}_i$ for each level of detail $\mathcal{L}_1, \dots, \mathcal{L}_n$ in $(\mathcal{L}, \sqsubseteq)$. [Yao99] then defines a stratified rough set as a sequence of rough sets $(Y \boxplus \lambda_1), \dots (Y \boxplus \lambda_n)$ as follows. Let $(Y \boxplus \lambda_i)^{-1}\{T\}$ be the 'egg', and $(Y \boxplus \lambda_i)^{-1}\{T,F\}$ be the union of 'egg' and 'yolk' in the corresponding egg-yolk representation of Y at the level of detail formed by $\mathcal{L}_i$ (remember Figure 1 (v)). Then whenever $i \sqsubseteq j$ the 'egg' at level i is a subset of the 'egg' of level j which itself is a subset of Y, which is in turn a subset of the union of 'egg' and 'yolk' at level i and so on:

$$(i) \ (Y \boxplus \lambda_i)^{-1}\{T\} \subseteq (Y \boxplus \lambda_j)^{-1}\{T\} \subseteq Y,$$
$$(ii) \ Y \subseteq (Y \boxplus \lambda_i)^{-1}\{T,B\} \subseteq (Y \boxplus \lambda_j)^{-1}\{T,B\}$$

Let $((K, \leq), (\mathcal{L}, \sqsubseteq), \ell, \lambda_1, \dots \lambda_n)$ as defined above and let $\omega_{ij}^p = \{(Y \boxplus \lambda_i) \, k \mid k \in (\ell_{ij}^{-1} \, p)\}$ be the set of approximation values under $(T \boxplus \lambda_i)$ with respect to the subcells $(\ell_{ij}^{-1} \, p) \subseteq L_i$ of the cell $p \in \mathcal{L}_j$. We then define a stratified rough set as follows:

Definition 6. *A stratified rough set is a family of rough sets* $(Y \boxplus \lambda_1), \dots (Y \boxplus \lambda_n)$, *such that whenever* $i \sqsubseteq j$ *then there exists a mapping* $\alpha_{ij} : \{T, B, F\} \to \{T, B, F\}$ *such that the following holds:*

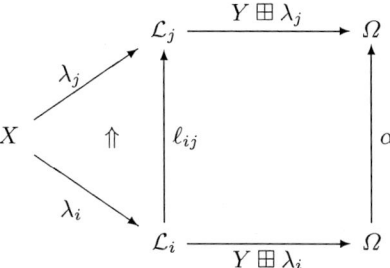

Fig. 4. Rough sets at different levels of granularity in a *cumulative* granular stratified partition.

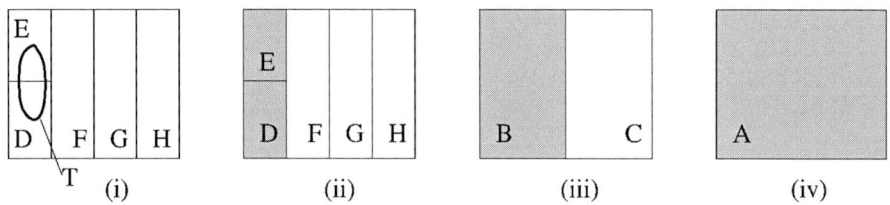

Fig. 5. A stratified rough set representations of Y at different levels of detail.

$$(\alpha_{ij} (Y \boxplus \lambda_i)) k = ((Y \boxplus \lambda_j) \ell_{ij}) k$$

with

$$(\alpha_{ij} (Y \boxplus \lambda_i)) k = \begin{cases} T & \textit{iff } \omega_{ij}^{(\ell_{ij} \ k)} = \{T\} \\ F & \textit{iff } \omega_{ij}^{(\ell_{ij} \ k)} = \{F\} \\ B & \textit{otherwise} \end{cases}$$

Correspondingly we can draw the commutative diagram in figure 4.

Consider figure 5 which corresponds to the labelled stratified partition shown in figure 3 with λ_i corresponding to f_i in equation (1). In figure 5 (i) we have six subsets of the set A five of which form a partition and one (T) which lies skew to this partition. Figures 5 (ii – iv) show stratified rough sets representations of T at different levels of detail. Here the gray color of the set E in figure 5 (ii) indicates that $(T \boxplus \lambda_5) e = B$. Similarly the gray color of the set B in figure 5 (iii) indicates that $(T \boxplus \lambda_2) b = B$. The white color of the set G in figure 5 (ii) indicates that $(T \boxplus \lambda_5) g = F$.

Let ω_{ij}^p be as defined above. In figure 5 (ii) we have $\omega_{54}^b = \{B, F\}$ and $\omega_{21}^a = \{B, F\}$, and hence $(\alpha_{54} (T \boxplus \lambda_4)) b = B$ and $(\alpha_{21} (T \boxplus \lambda_4)) a = B$.

5   Rough Sets in Non-cumulative Labelled Stratified Partitions

An important assumption in the previous section was that in the stratified labelled partition $((K, \leq), (\mathcal{L}, \sqsubseteq), \ell, \lambda_1, \dots \lambda_n)$ the λ_i are *total* surjective function. Consider the K labelled partition depicted as a map of the United States in the left part of figure 2. That the labelling function is total here means that there is no 'white space' or no undiscovered land in the space covered by this map. In this case we also say that the underlying granular partition is *cumulative*.

Definition 7. *Let $((K, \leq), (\mathcal{L}, \sqsubseteq), \ell, \lambda_1, \ldots \lambda_n)$ be a labelled stratified partition. The level of granularity $\mathcal{L}_i$ is cumulative if and only if the function λ_i is total. The partition as a whole is cumulative if each level of granularity is cumulative.*

However there *are* maps with 'white space', unexplored territories, or not well understood domains. In order to take this into account we now generalize the notion of stratified rough sets by giving up this constraint of cumulativeness and allow the λ_i to be partial surjective functions. What results corresponds to what Mislove calls murky sets [MMO90] in the theory of partial sets and to what Bittner and Smith call non-cumulative granular partitions [BS03a]. In Mislove's terminology we now consider stratified rough sets in labelled partitions of murky sets. Roughly, murky sets are sets which are such that we do not know all of their members. In the terminology of Bittner and Smith we consider rough approximations with respect to non-cumulative granular partitions [BS03b].

If the underlying labelling functions λ_i are total surjective functions, then the rough set representations at a coarser levels of detail can be derived from a rough set represented at finer level of detail. Consider levels of detail $i \sqsubseteq j$. Given a rough set $(Y \boxplus \lambda_i)$ we can determine the rough set $(Y \boxplus \lambda_j)$ in the way described in definition 6. In general, however, we cannot assume that the underlying labelling functions are total because this assumes complete knowledge about the underlying set which may not be available.

Under circumstances where the labelling functions λ_i are not total it will be impossible to define a unique generalization mapping α_{ij} in the way shown in figure 4. Moreover a multitude of generalization mappings, each yielding one possible generalization of the rough set at hand will be needed. The example shown in figures 6 and 7 will help to explain this.

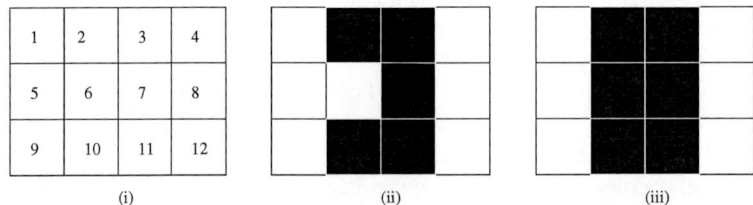

(i)   (ii)   (iii)

Fig. 6. A set Q, with 12 elements (i), a 5 element subset $R \subset Q$ (ii), and a 6 element subset $S \subset Q$ (iii).

In figure 6 (i) we see the set Q with 12 elements, each of them labelled by a natural number. Five of these form the subset $R \subset Q$ and six of them form the subset $S \subset Q$ (figure 6 (ii) and (iii)). The set Q can be given a stratified labelled partition, using the cell tree and the granularity lattice shown in the left and middle of figure 3, and the mappings λ_i given in table 1. The table is read as follows: (row 1) the mapping λ_1 maps

Table 1. The mappings $\lambda_\bullet, \lambda_\bullet, \lambda_\bullet$ with $\mathcal{L}_\bullet = \{a\}$, $\mathcal{L}_\bullet = \{b, c\}$, and $\mathcal{L}_\bullet = \{b, g, h\}$.

$(\lambda_i \, x) =$	values of λ_i and x
a	$(\lambda_\bullet \, 1) \ldots (\lambda_\bullet \, 12)$
b	$(\lambda_\bullet \, 1), (\lambda_\bullet \, 5), (\lambda_\bullet \, 9)$
	$(\lambda_\bullet \, 1), (\lambda_\bullet \, 5), (\lambda_\bullet \, 9)$
c	$(\lambda_\bullet \, 2), (\lambda_\bullet \, 3), (\lambda_\bullet \, 6), (\lambda_\bullet \, 7), (\lambda_\bullet \, 10), (\lambda_\bullet \, 11)$
g	$(\lambda_\bullet \, 2), (\lambda_\bullet \, 3)$
h	$(\lambda_\bullet \, 10), (\lambda_\bullet \, 11)$

all elements of Q onto the label a; (row 2) λ_2 maps the elements 1, 5, and 9 onto b, and so does λ_4. The other rows follow the same pattern. The mappings targeting the granularity levels $\{d, e, f, c\}$ and $\{d, e, f, g, h\}$ are omitted here.

Table 1 tells us that the mapping λ_1 is surjective and total. The other mappings are surjective but partial. No λ_i with $i > 1$ maps the elements $4, 8, 12 \in Q$ to any cell in their target domain $\mathcal{L}_i$. Moreover λ_4 in addition also fails to map the elements $6, 7 \in Q$.

Consider figure 7: (i) depicts the rough set representation of R *and* S for the level of granularity $\mathcal{L}_4 = \{b, g, h\}$; (ii) depicts the rough set representation of R for the level of granularity $\mathcal{L}_2 = \{b, c\}$, and (iii) depicts the rough set representation of S for $\mathcal{L}_2$. We have $\mathcal{L}_4 \sqsubseteq \mathcal{L}_2$. The color grey indicates the approximation value B as in $(R \boxplus \lambda_2) \, c = $ B, black indicates the approximation value T as in $(R \boxplus \lambda_4) \, g = $ T, and the diagonal line pattern represents the approximation value F as in $(R \boxplus \lambda_4) \, b = $ F. The white spaces in the figures 7 (i–iv) indicates the partial character of the mappings λ_2 and λ_4.

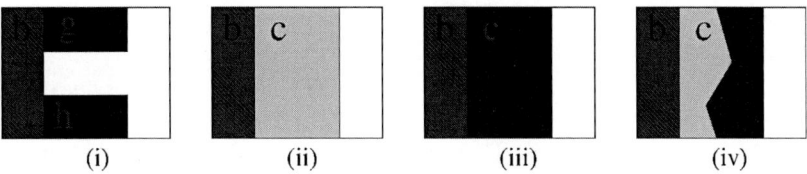

Fig. 7. Rough set representations of R and S at the levels of granularity $\lambda_\bullet$ and $\lambda_\bullet$.

Two significant features appear in this example:

1. At the level of granularity $\mathcal{L}_4$ we cannot distinguish between the sets R and S – both are are represented by the rough set depicted in figure 7 (i).
2. The rough approximation of R with respect to $\mathcal{L}_2$ cannot be derived from that at the finer level of detail $\mathcal{L}_4$ using a generalization mapping α_{24} as defined in definition 6 – applying the generalization mapping defined in 6 to the rough set $(R \boxplus \lambda_4)$ yields the rough set depicted in figure 7 (iii) and not the one depicted in (ii) as one would want.

This is due to the fact that the function λ_4 is to a larger degree partial than λ_2: g and h do not make up the whole of c. From the rough set representation of R and S at the finer level of detail $\mathcal{L}_4$ alone we are unable to determine whether the part of Q labelled

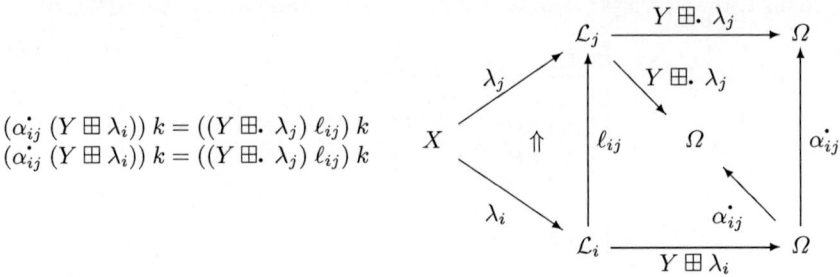

$$(\alpha_{ij}^{\cdot} \ (Y \boxplus \lambda_i)) \ k = ((Y \boxplus. \ \lambda_j) \ \ell_{ij}) \ k$$
$$(\alpha_{ij}^{\cdot} \ (Y \boxplus \lambda_i)) \ k = ((Y \boxplus. \ \lambda_j) \ \ell_{ij}) \ k$$

Fig. 8. The multiplicity of possible generalizations in non-cumulative labelled stratified partitions.

c is wholly or only partly covered by R and by S respectively. Consequently, given tha partial character of the mapping λ_4 and the rough set depicted in figure 7 (i) the two rough sets depicted in the figures 7 (ii) and (iii) equally good candidates for being the result of performing a generalization on (i). This is indicated in figure 7 (iv).

It follows that we need to extend the notion of generalization mapping α_{ij} which was set out in definition 6 in order to take into account the non-cumulative character of the underlying labelled stratified partition. Let $Y \boxplus \lambda_i$ be a rough set based on a non-cumulative granularity-level $\mathcal{L}_i$, let p be a cell belonging to granularity level $\mathcal{L}_j$ and let $i \sqsubseteq j$. We then need to distinguish three cases:

1. If we have $\{(Y \boxplus \lambda_i) \ k \mid k \in (\ell_{ij}^{-1} \ p)\} = \{\mathsf{T}\}$ then there might be elements of the underlying set X which are not labelled at granularity-level $\mathcal{L}_i$ which may or may not belong Y. Therefore we need to have two generalization mappings α_{ij}^{l} and α_{ij}^{l+1} such that $(\alpha_{ij}^{l}(Y \boxplus \lambda_i)) \ p = \mathsf{T}$ and $(\alpha_{ij}^{l+1}(Y \boxplus \lambda_i)) \ p = \mathsf{B}$.
2. If we have $\{(Y \boxplus \lambda_i) \ k \mid k \in (\ell_{ij}^{-1} \ p)\} = \{\mathsf{F}\}$ then, again, there might be elements of X which are not labelled at granularity-level $\mathcal{L}_i$ which may or may not belong Y. Therefore we need two generalization mappings α_{ij}^{l} and α_{ij}^{l+1} such that $(\alpha_{ij}^{l}(Y \boxplus \lambda_i)) \ p = \mathsf{F}$ and $(\alpha_{ij}^{l+1}(Y \boxplus \lambda_i)) \ p = \mathsf{B}$.
3. If we have $\mathsf{B} \in \{(Y \boxplus \lambda_i) \ k \mid k \in (\ell_{ij}^{-1} \ p)\}$ then we can apply definition 6.

Now compare the generalization from a cumulative level of granularity with generalization from from a non-cumulative level of granularity. In a cumulative level of granularity there is a unique generalization function doing the transformation job. When we generalize from a non-cumulative level of granularity $\mathcal{L}_i$ to a level of granularity $\mathcal{L}_j$ with a single cell then there may be *two* generalization functions: α_{ij}^{1} and α_{ij}^{2}. This case is represented in figure 8: The generalization mappings α_{ij}^{1} and α_{ij}^{2} satisfy the equations in the left of the figure. A corresponding diagram representation is given in the right part of the figure.

The more cells the target level of granularity $\mathcal{L}_j$ has two cells the more generalization functions need to be added. This reflects the phenomenon of *vagueness* which is caused by the non-cumulativeness of the underlying stratified partition.

6   Rough Sets and Vagueness

In the previous section we dealt with the problem of vagueness by adding more and more generalization transformations – each yielding one possible rough set at the targeted level of granularity. An alternative way of dealing with the problem of vagueness is to introduce the notion of *vague rough set* and to provide an unique generalization transformation between vague rough sets. The idea hereby is to considerer sets of approximation values rather than sets of possible approximations.

6.1   Vague Rough Sets

Let $((K, \leq), (\mathcal{L}, \sqsubseteq), \ell, \lambda_1, \ldots \lambda_n)$ be a labelled non-cumulative granular partition with $\lambda_i : X \to \mathcal{L}_i$. In order to represent vagueness we consider the following subsets:

$$\tilde{\Omega} = \{\{\mathsf{F}\}, \{\mathsf{B}\}, \{\mathsf{F}\}, \{\mathsf{T}, \mathsf{B}\}, \{\mathsf{B}, \mathsf{F}\}, \{\mathsf{T}, \mathsf{B}, \mathsf{F}\}\}$$

The ordering of $\tilde{\Omega}$ corresponding to the subset relation is given in the diagram in figure 9.

Given a subset $Y \subseteq X$ we define a *vague* rough set as a mapping of signature $(Y \boxtimes \lambda_i) : K \to \tilde{\Omega}$ (notice the difference between $\boxplus$ and $\boxtimes$). The value of $(Y \boxtimes \lambda_i) \, k$ is interpreted as a disjunction of possible relations between the subsets Y and $(\lambda_i^{-1} \, k)$. For example, the value of $(Y \boxtimes \lambda_i) \, k$ is $\{\mathsf{B}, \mathsf{F}\}$ if either Y contains some but not all of elements of $(\lambda_i^{-1} \, k)$ or if there is no overlap between Y and $(\lambda_i^{-1} \, k)$ at all. Under this interpretation the ordering in the diagram in figure 9 represents an increasing degree of vagueness.

Let $\mathcal{L}_i$ be a non-cumulative level of granularity. The rough set $X \boxplus \lambda_i$ is a *crisping* of the vague rough set $Y \boxtimes \lambda_i$ if and only if for every cell k the label $(X \boxplus \lambda_i) \, k$ is one of the disjuncts in $(Y \boxtimes \lambda_i) \, k$ [1]:

$$\mathsf{CR} \, (X \boxplus \lambda_i)(Y \boxtimes \lambda_i) \equiv \forall k \in \mathcal{L}_i : (X \boxplus \lambda_i) \, k \in (Y \boxtimes \lambda_i) \, k$$

Consider figure 7(iv). Let $N \subset Q$ be a set of which we know only the vague rough set representation corresponding to the figure: $(N \boxtimes \lambda_2) \, b = \{\mathsf{F}\}$ and $(N \boxtimes \lambda_2) \, c = \{\mathsf{T}, \mathsf{B}\}$. Crispings of $N \boxtimes \lambda_2$ then are $R \boxplus \lambda_2$ and $S \boxplus \lambda_2$ as depicted in 7(ii) and 7(iii).

6.2   Generalization of Vague Rough Sets

We now discuss generalization transformations of vague rough sets of the form $(Y \boxtimes \lambda_i)$ from granularity level $\mathcal{L}_i$ to $\mathcal{L}_j$ with $i \sqsubseteq j$. Let $\omega_{ij}^p = \{(Y \boxtimes \lambda_i) \, k \mid k \in \ell_{ij}^{-1} \, p\} \subset \mathcal{P} \, \tilde{\Omega}$ be the set containing the sets of approximation values under $(Y \boxtimes \lambda_i)$ with respect to the cells $(\ell_{ij}^{-1} \, p) \subseteq \mathcal{L}_i$.

Consider table 2 and assume sets $O, P, U \in Q$ of which we only know their vague rough set representation with respect to the granularity level $\mathcal{L}_4$ as given in columns δb, δg, and δh of the table. In column ω_{42}^c we have the subset of $\mathcal{P} \, \tilde{\Omega}$ corresponding vague rough set in column δ with respect to the cells $g, h \in (\ell_{42}^{-1} \, c)$.

---

[1] In cumulative granular partitions crisping is more complicated. See [Bit03] for details.

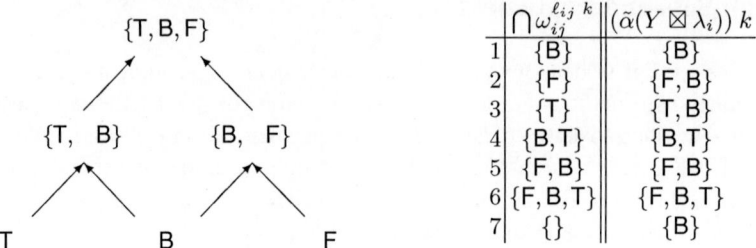

Fig. 9. Representing vagueness as sets of labels.

Table 2. Examples for the generalization of vague rough sets from granularity level $\mathcal{L}$. to $\mathcal{L}.$.

δ	$\delta\, b$	$\delta\, g$	$\delta\, h$	$\omega^c.$	$\bigcap \omega^c.$	$\tilde{\alpha}(\delta\, c)$
$O \boxtimes \lambda.$	{B}	{T,B}	{F,B}	{{T,B},{F,B}}	{B}	{B}
$P \boxtimes \lambda.$	{B}	{T,B,F}	{F,B}	{{T,B,F},{F,B}}	{B,F}	{B,F}
$U \boxtimes \lambda.$	{B}	{T}	{F}	{{T},{F}}	{}	{B}

We define the generalization mapping $\tilde{\alpha}_{ij} : \mathcal{P}\,\tilde{\Omega} \to \mathcal{P}\,\tilde{\Omega}$ which transforms vague rough sets from granularity level $\mathcal{L}_i$ to granularity level $\mathcal{L}_j$ with $i \sqsubseteq j$ by reading the table in figure 9 row-wise as follows (using ω be a shorthand for $\omega_{ij}^{\ell_{ij}\ k}$):

Row1: if $\bigcap \omega = \{\mathsf{B}\}$ then $(\tilde{\alpha}\,(Y \boxtimes \lambda_i))\,k = \{\mathsf{B}\}$

...

Row7: if $\bigcap \omega = \{\}$ then $(\tilde{\alpha}(Y \boxtimes \lambda_i))\,k = \{\mathsf{B}\}$

Consider table 2. In the last two columns we see the values $\bigcap \omega_{42}^c$ and $\tilde{\alpha}(\delta\, c)$ according to the table in figure 9 for the corresponding rough sets in column δ.

Definition 8. *A stratified vague rough set is a family of vague rough sets* $(Y \boxtimes \lambda_1), \ldots$ $(Y \boxtimes \lambda_n)$, *such that whenever* $i \sqsubseteq j$ *then there exists a mapping* $\tilde{\alpha}_{ij} : \mathcal{P}\,\{T, B, F\} \to$ $\mathcal{P}\,\{T, B, F\}$ *as defined in table 9 such that the following holds:*

$$(\tilde{\alpha}_{ij}\,(Y \boxtimes \lambda_i))\,k = ((Y \boxtimes \lambda_j)\,\ell_{ij})\,k.$$

One then can verify that if $((K, \le), (\mathcal{L}, \sqsubseteq), \ell, \lambda_1, \ldots \lambda_n)$ is a labelled non-cumulative granular partition and $(Y \boxtimes \lambda_i)$ is a vague rough set then $\tilde{\alpha}(Y \boxtimes \lambda_i)$ is the result of applying the generalization mapping $\tilde{\alpha}_{ij}$ to $(Y \boxtimes \lambda_i)$ if and only if every crisping of $\tilde{\alpha}(Y \boxtimes \lambda_i)$ is the result of a crisp generalization α, of a crisping of $(Y \boxtimes \lambda_i)$.

7   Conclusions and Further Work

In this paper we have shown how the technique of making rough descriptions of a subset with respect to an equivalence relation can be extended to descriptions with respect to a granular partition. The work has also revealed a new way of looking at a granular partition as a generalization of an equivalence relation. In this generalization, a set of names of equivalence classes is replaced by a tree structure and certain subsets of the tree are extracted to form labels for equivalence classes. In this way we obtain a hierarchy of equivalence classes. This is relevant to Spatial Information Theory because (a) most spatial representations, in particular maps, are granular partitions, (b) those representations are often hierarchical [PM97]; and (c) because approximations with respect to sets of equivalence classes are important in order to deal with vagueness and indeterminacy inherent in many geographic phenomena.

This identification of the way in which the equivalence classes at the various levels of detail relate to each other is an important contribution. It enables us to understand the relationship of granular partitions to the stratified map spaces of Stell and Worboys [SW98]. The stratified map space concept is applicable to problems involving level of detail in temporal data, as for example in the work of Hornsby and Egenhofer [HE99]. The extension to rough descriptions using granular partitions for temporal data is one area for further work which we intend to pursue.

Another area for further work is to extend the results of this paper to richer structures than sets. In particular, graphs represent a significant challenge, and have clear connections with practical issues in spatial information theory. To carry out the extension to graphs would entail replacing the set which is subjected to the family of equivalence relations in a granular partition, by a graph. This would require identification of the appropriate generalization of equivalence relations for the richer context. A number of possibilities for such a generalization have been discussed in the literature [Ste99], and it is possible that more than one could be made to work with granular partitions. If the work were extended in this way, we would expect it to yield new techniques for the rough description of networks, such as those of roads, railways etc.

Acknowledgements

Support for the first author from the the Wolfgang Paul Program of the Alexander von Humboldt Foundation and the National Science Foundation Research Grant BCS-9975557: Geographic Categories: An Ontological Investigation, is gratefully acknowledged. The second author acknowledges support from EPSRC under the project Digital Geometry and Topology: An Axiomatic Approach with Applications to GIS and Spatial Reasoning.

References

Bit03.   T. Bittner. Indeterminacy and rough approximation. In *Proceedings of FLAIRS 2003*. AAAI Press, 2003.

BS01.   T. Bittner and J. Stell. Rough sets in approximate spatial reasoning. In W. Ziarko and Y. Yao, editors, *Proceedings of the Second International Conference on Rough Sets and Current Trends in Computing (RSCTC'2000)*, volume 2005 of *Lecture Notes in Computer Science (LNCS)*, pages 445–453. Springer-Verlag, 2001.

BS02.   T. Bittner and J.G. Stell. Vagueness and rough location. *GeoInformatica*, 6:99–121, 2002.

BS03a.   T. Bittner and B. Smith. A theory of granular partitions. In M. Duckham, M. F. Good-child, and M. F. Worboys, editors, *Foundations of Geographic Information Science*, pages 117–151. London: Taylor & Francis, 2003.

BS03b.   T. Bittner and B. Smith. Vague reference and approximating judgments. *Spatial Cognition and Computation*, 3(2), 2003.

ES97.   M. Erwig and M. Schneider. Partition and conquer. In S. C. Hirtle and A. U. Frank, editors, *Spatial Information Theory, International Conference COSIT'97, Proceedings*, volume 1329 of *Lecture Notes in Computer Science*, pages 389–407. Springer-Verlag, 1997.

HE99.   K. Hornsby and M. Egenhofer. Shifts in detail through temporal zooming. In A. Camelli, A. M. Tjoa, and R. R. Wagner, editors, *Tenth International Workshop on Database and Expert Systems Applications. DEXA99*, pages 487–491. IEEE Computer Society, 1999.

Jon97.   C. B. Jones. *Geographical Information Systems and Computer Cartography*. Longman, 1997.

M⁺95.   J. C. Müller et al. Generalization - state of the art and issues. In J. C. Müller, J. P. Lagrange, and R. Weibel, editors, *GIS and Generalisation: Methodology and Practice*, pages 3–17. Taylor and Francis, London, 1995.

MLW95.   J. C. Müller, J. P. Lagrange, and R. Weibel, editors. *GIS and Generalisation: Methodology and Practice*. Taylor and Francis, London, 1995.

MMO90.   M. Mislove, L. Moss, and F. Oles. Partial sets. In R. Cooper, K. Mukai, and J. Perry, editors, *Situation Theory and Its Applications I*, number 22 in CSLI Lecture Notes, pages 117–131. Center for the Study of Language and Information, Stanford, CA., 1990.

Paw82.   Z. Pawlak. Rough sets. *Internat. J. Comput. Inform*, 11:341–356, 1982.

PM97.   D. Papadias and Egenhofer M. Algorithms for hierarchical spatial reasoning. *Geoinformatica*, 1(3), 1997.

RS95.   P. Rigaux and M. Scholl. Multi-scale partitions: Application to spatial and statistical databases. In M. Egenhofer and J. Herrings, editors, *Advances in Spatial Databases (SSD'95)*, number 951 in Lecture Notes in Computer Science. Springer-Verlag, Berlin, 1995.

Ste99.   J. G. Stell. Granulation for graphs. In C. Freksa and D. Mark, editors, *Spatial Information Theory. Cognitive and Computational Foundations of Geographic Information Science. International Conference COSIT'99*, volume 1661 of *Lecture Notes in Computer Science*, pages 417–432. Springer-Verlag, 1999.

SW98.   J. G. Stell and M. F. Worboys. Stratified map spaces: A formal basis for multi-resolution spatial databases. In T. K. Poiker and N. Chrisman, editors, *SDH'98 Proceedings 8th International Symposium on Spatial Data Handling*, pages 180–189. International Geographical Union, 1998.

Wor98a.   M. F. Worboys. Computation with imprecise geospatial data. *Computers, Environment and Urban Systems*, 22:85–106, 1998.

Wor98b.   M. F. Worboys. Imprecision in finite resolution spatial data. *GeoInformatica*, 2:257–279, 1998.

Yao99.   Y.Y. Yao. Stratified rough sets and granular computing. In R.N. Dave and Sudkamp. T., editors, *Proceedings of the 18th International Conference of the North American Fuzzy Information Processing Society*, pages 800–804. IEEE Press, 1999.

Communicating Vague Spatial Concepts in Human-GIS Interactions: A Collaborative Dialogue Approach

Guoray Cai[1,3], Hongmei Wang[1,3], and Alan M. MacEachren[2,3]

[1] School of Information Sciences and Technology, Penn State University
University Park, PA 16802
{cai,hwang}@ist.psu.edu
[2] Department of Geography, Penn State University, University Park, PA 16802
maceachren@psu.edu
[3] GeoVISTA Center, Penn State University, University Park, PA 16802

Abstract. Natural language requests involving vague spatial concepts are not easily communicated to a GIS because the meaning of spatial concepts depends largely on the contexts (such as *task*, *spatial contexts*, and *user's personal background*) that may or may not be available or specified in the system. To address such problems, we developed a collaborative dialogue approach that enables the system and the user to construct shared knowledge about relevant contexts. The system is able to anticipate what contextual knowledge must be shared, and to form a plan to exchange contextual information based on the system's belief on who knows what. To account those user contexts that are not easily communicated by language, direct feedback approach is used to refine the system's belief so that the intended meaning is properly grounded. The approach is implemented as a dialogue agent, *GeoDialogue*, and is illustrated through an example dialogue involving the communication of the vague spatial concept *near*.

1   Introduction

Current geographical information systems (GIS) do not support human work effectively because users must interact with geographical data through formally defined textual or graphical query languages using keyboards and mouse. Recent research has paid more attention to human communication modalities as potential alternative modes of human-computer interactions, because it is relatively effortless for people to express their information needs in this way. One component in this line of research has been the use of natural language for submitting request to GIS[1-5]. These works have emphasized the scale-dependent and imprecise nature of natural language in describing spatial relations. However, actual implementations of natural language interfaces for GIS have been very limited (for examples see [4, 6] [7] [8] [9]. To overcome the difficulties of natural language for expressing metric details of spatial relations[10], many authors have explored the possibility of using gestures as more intuitive styles for interacting with spatial data. For example, pen-based gestures were used in *spatial query-by-sketch* [11] and *QuickSet[12]*. Free hand gestures were featured in *GIS Wallboard[13]*, *iMap[14-16]*, and *DAVE_G[17]*.

W. Kuhn, M.F. Worboys, and S. Timpf (Eds.): COSIT 2003, LNCS 2825, pp. 287–300, 2003.

The main challenge for using human modalities (speech and gesture in particular) to interact with computers is that human communication are highly contextualized by the shared knowledge of the participants. Such contextual knowledge are often not available or properly captured in computer systems, which is the reason that spoken language requests to computer systems often appear to be incomplete, ambiguous, or vague. In this paper, we address the challenge of communicating spatial information requests that involve vague spatial concepts during human-GIS interaction. In particular, we focus on the concept of "near", which is as a classic example of such concepts[3, 18].

Communicating spatial concepts to a computer system in natural language is especially challenging for a number of reasons. First, geographic concepts do not have unique and precise mapping to quantitative representation of space[10, 19-21]. Second, human uses of spatial language are qualitative and ambiguous, allowing meaning to vary with the contexts of use. In contrast, computer representations of spatial concepts are quantitative and precise[10], leaving no room to adapt to user's conception of spatial concepts. Currently, formal computational models exist that allow GIS to form a quantitative representation of spatial concepts communicated through natural language. Examples of such computational models include those for distance relations[18, 22], directional relations[23, 24], and topological relations[10, 25-28]. These models, which mostly follow Rosch's prototypical categorization theory of human cognition[29, 30], have serious limitations when used in human-GIS interaction. Human-GIS interactions have to deal with both the inherent uncertainties and context-varying nature of spatial concepts. Formal models of spatial concepts are useful in capturing the inherent uncertainty of spatial objects and spatial relations, but do not have the properties of human-human communication with contextualized interpretation and negotiation. Even formal models of common sense geography [31] in natural language do not have the necessary formalisms and parameters to incorporate contextual information that are crucial for communicating spatial concepts in human-GIS interaction.

Human-GIS interactions are usually situated in a complex set of contexts including the geographic site and situation, the task or problem being addressed, and the background (knowledge, experiences, and preferences) of the individual users. The spatial concepts communicated need to be interpreted within such relevant contexts[3, 19, 21, 32-37]. However, computer systems usually ignore the contexts when translating linguistic reference of spatial concepts into formal representations of meanings. This constitutes a reason for the system to misinterpret the user's request and to cause breakdowns in human-GIS interaction[19].

The *collaborative dialogue approach*, as proposed in this paper, follows the principles of human-human communication in dealing with vague spatial concepts. Human-human conversation is grounded in a significant amount of knowledge (or context) shared among participants, which allows the hearer to disambiguate vague concepts. However, adequate sharing of relevant context may not always be established before a vague concept is communicated, causing difficulties for the hearer to construct an unambiguous mental model of that concept that matches the meaning intended by the speaker. In such situations, it is common that human will engage in dialogues that to acquire the meaning of the concept involved[22]. Dialogues as a general human communication strategies serve a diverse functions in communicating and sharing knowledge as well as coordinating actions. For the particular situations of communicating vague spatial concepts, dialogues make two kinds of collaboration possible: (1) soliciting and sharing additional contextual information, and (2) seeking feedback and confirmation. Applying such principles to human-GIS interactions, the *collaborative dialogue approach* addresses a number of computational issues, such as how to represent the belief states of context sharing between the user and the system, how to relate newly communicated knowledge with what has been shared so far, and how to build system intelligence for generating dialogues consistent with the goals of advancing context sharing. It is also about dialogue strategies for grounding linguistic message in a shared meaning reliably through confirmation and visual feedback. The user and the system is brought into collaboration[38], and follows the metal state view of human-human communication and collaborative planning[39, 40]. No other existing natural language based GIS [4, 17, 41, 42] have the abilities to initiate this kind of dialogue.

In order to test the feasibility of our collaborative dialogue approach, we implemented a prototype system, *GeoDialogue*, which is a conversational agent that is able to engage in collaborative dialogue with the user. When a vague spatial concept is detected in a dialogue, GeoDialogue is able to invoke proper schemas (stored in its knowledge-base) about what contextual information needs to be shared. Such schemas are central organizing structures that link contextual knowledge to meanings of spatial concepts and guide the planning of communicative actions in exchanging contextual knowledge. As the dialogue proceeds, the user and the system will mutually adapt to each other and reconstruct meaning representation that reflects the shared contexts about tasks, spatial and geographical environment, and personal backgrounds. A common understanding of a spatial concept is collaboratively constructed and visually shared, resulting in a meaning that is neither what the user initially perceives nor what the system initially estimates. Currently, the collaborative dialogue approach and the *GeoDialogue* agent are already integrated as part of the core reasoning engine of the Dialogue-Assisted Visual Environment for Geoinformation (DAVE_G) [43]which is a generic environment in which the user can interact with the GIS through multimodal interactions.

The rest of the paper is organized as follow. In section 2, we discuss major factors affecting human-computer communication of spatial concepts. In section 3 we will describe our methodology, which incorporates a collaborative dialogue approach that captures, represents and uses the contextual information for understanding to a vague spatial concept. The section 4 will introduce implementation of this approach in the prototype system, the GeoDialogue agent. In the section 5, a sample dialogue is presented to illustrate the collaborative dialogue approach in GeoDialogue. Finally, we conclude with discussion on the implications of this development to the future natural language interfaces to GIS.

2   Factors Affecting the Communication of Spatial Concepts

Natural language terms may carry spatial concepts that refer to spatial objects or spatial relationships. When human communicate a spatial concept to a GIS through linguistic terms, there are two types of factors that present problems for a successful communication. First, the spatial concept itself may be inherently vague[44, 45]. Second, the meaning of a spatial concept may be dependent on a large number of contexts within which the concept is used. For human-GIS interactions, major parts of

such contexts are spatial contexts (including geographical space [46] and environmental space[32, 36]), task contexts[10], and personal contexts of the individual users.

2.1   Inherent Uncertainties of Spatial Objects and Spatial Relations

There are two types of inherent uncertainties in the processing of spatial concepts[44,
45]. First, a spatial concept may refer to the spatial objects that have indeterminate
boundaries. For example, mountainous areas are not easily determined because the
boundary between a "mountainous area" and the surrounding areas is inherently un-
certain. The second type of uncertainties is the fuzzy nature of spatial relations. For
example, the concept *'near'* has long been considered to be an ill-defined[47]. It exhibits classic characteristics of a radial category, and it includes a range from exem-
plars to borderline cases[21, 36]. As an example, consider a request for a map showing 'hotels near *Miami'*. A user may consider hotels within 20 miles to *Miami* to be
near, and those beyond 60 miles from *Miami* as *far*. However, a hotel that is 40 miles
away from *Miami* may be considered neither near nor far from *Miami*.

2.2   Contextual Factors for Interpreting Spatial Concepts

Contextual factors about a particular use of a spatial concept refer to the knowledge
that human uses to constrain the meaning of communication. To reach a common
understanding of a vague concept, e. g. *near*, the system and the user need to share
knowledge about the relevant contexts that affect the understanding of the vague concept. Among many potential contextual factors that may affect how people understand
spatial concepts, we focus here on three of them: task, spatial contexts, and background of the user.

   When a user requests a map to be displayed by a GIS, it is often because that the
user is trying to perform a domain task that has some information needs. The task
becomes an important part of the use context for spatial concepts (similar view was
expressed in[10]). For example, the same request "show me a map *near* Miami" may
be made by a person-*A* who is in a task situation of selecting a grocery store, and by a
person-*B* who is planning a vacation for the weekend. However, person-*B* is likely to
expect a map showing a larger geographical area comparing with person-*A*.

   There are evidences that the meaning of spatial concepts, such as *"near"*, is also
dependent on the spatial context[18, 32, 36]. As detailed by Montello[48], human
cognition supports multiple and qualitatively different conceptualizations of space.
Two of these, which he labels environmental space and geographical space, are relevant here. Environmental space is the behavioral-scale space of physical entities, such
as cities and buildings, in the real world. Worboy's [36, 37] and Fisher [32] provided
evidences that environmental space affects people's interpretation of *near*. Geographical space is usually learned via symbolic representations, such as a map. It can
also affect the understanding of spatial concepts, such as the distance relationship of
near[18, 22, 46]. Since a map conveys information about geographical space in human-GIS interactions, properties of the map, such as the map scale[18, 22], the contents, and the distribution of objects in the map, affect people's perception on the
relationship of *near*.

Individual users may differ in cultural and education background, personal experiences and preferences that affect how they form mental models geographical phenomena and assign meanings to vague concepts, such that they may perceive the same concept differently[36, 37]. In his experiments on the individual differences on the meaning of spatial relationships, Robinson [18] demonstrated that there are significant semantic variations among individuals even for relatively simple geographical datasets. Even in the same spatial context and task context, different individuals are often found to be different in determining whether specific locations are near or not.

3   Methodology

In this section, a collaborative dialogue approach is proposed as a more effective way to facilitate communication of vague spatial concepts in human-GIS interactions. In a collaborative dialogue system, the system and the user can construct and negotiate (perhaps through multiple exchanges) a shared understanding of the meaning of a vague concept by exchanging contextual information. The vague spatial concept *near* is used here to illustrate our approach.

In our approach, the process of communicating the meaning of vague spatial concepts between the system and the user is modeled as a collaborative discourse, following the theory of the collaborative discourse developed by Grosz and Sidner[49]. The process of communicating vague spatial concepts has all the central elements of a typical collaborative discourse: (1) it involves a shared goal among participants (which is to achieve common understanding of a vague spatial concept); (2) it requires a shared pool of knowledge that is initially distributed among participants; and (3) it uses communicative actions as the way to achieve collaboration. According to Grosz and Sidner[49], a collaborative discourse has three interrelated structures: a linguistic structure, an intentional structure and an attentional state. The theory believes that there is always an intention behind any communicated messages (utterances), and they functionally contribute to the achievement of larger intentions of the dialogues within which messages are organized. The key to understand collaborative dialogue is to have a proper model for the intentional structure of the collaboration. A computational model, the recipe graph or *Rgraph*, was developed by Lochbaum [40, 50] to represent the intentional structure of a collaborative discourse. It is based on the *SharedPlan* theory[39, 51], which was developed to model mental attitudes that a group of agents must hold during a collaborative planning process.

The collaborative discourse theory provides the foundation for modeling any extended dialogues in which the participants of the dialogue have a common goal to achieve but each of the participants takes on different roles in their collaboration. In the collaboration process of a dialogue involving a vague spatial concept, participants exchange information about relevant contexts so that part of individual belief and knowledge become shared and integrated into a consistent set of knowledge held by all the participants. As a dialogue proceeds, the system uses an Rgraph [40, 50] to keep track of the dynamically changing set of knowledge and beliefs communicated and shared during a dialogue process. The *Rgraph* consists of actions and the belief of the agents being modeled. Each node in the *Rgraph* is a complex data structure which records an intended action together with a set of beliefs that the system and the user have towards that action. An action can be either basic (an action that can be executed directly) or complex (an action that requires further communication and coordination with the user before it can be executed). For each complex action, the system chooses a recipe from the recipe library maintained in its knowledge-base. A recipe specifies a way of achieving an action, and it usually includes knowledge pre-conditions (represented as *parameters*), subactions, and other constraints necessary for executing the recipe. The belief status of an agent on an action (or a parameter) represents the agent's belief on the status of collaboration on the action (or the parameter). An example of an Rgraph is shown in Figure 1. The belief status on actions and parameters and their corresponding meanings are described in Table 1.

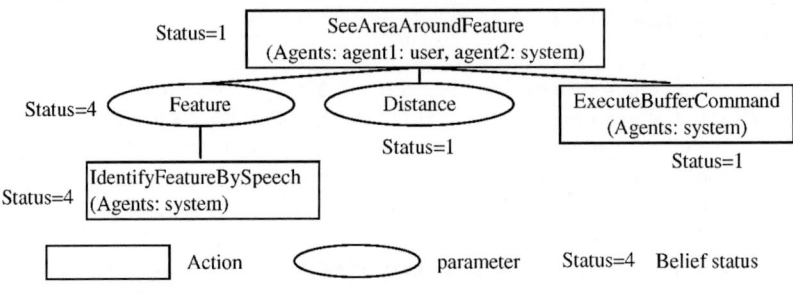

Fig. 1. Structure of Rgraph

With the representation of the dialogue status as an Rgraph, the effect of dialogue interactions will be the evolution of the *Rgraph* from the one that was only partially instantiated towards the one that is fully elaborated. The evolution begins with adding an action for achieving the common goal to the *Rgraph*. This action is the root action of the Rgraph. Then the system will understand and respond to the subsequent messaged based on the knowledge encoded in the Rgraph. Four steps of reasoning are conducted for each new input from the user: recognizing the intention of the utterance, explaining the intention by associating it with the Rgraph, elaborating the modified Rgraph by volunteering additional knowledge or performing executable parts of the Rgraph, and inferring the goal of generating questions or responses. If necessary, the system will call a GIS component to execute basic actions related to GIS commands and generate map responses, such as actions for doing buffering, displaying layers, and doing selections. By maintaining an Rgraph representation of a collaborative discourse, the system decides the process flow of its collaborative behavior (interpretation, elaboration, execution, response, and request,) by reasoning based on the Rgraph.

Table 1. Belief status numbers and their meanings

Status	Meaning on an action	Meaning on a parameter
1	The agent being modeled believes that the action has not been done.	The agent being modeled believes that a value for the parameter can be identified from the current discourse context
2	The agent being modeled believes that the result of the action is needed to be negotiated with the user.	The agent being modeled believes that the value of the parameter needs to be negotiated with the user.
3	The agent being modeled believes that the action fails.	The agent being modeled believes that instantiation of the parameter fails.
4	The agent being modeled believes that the action was successfully executed.	The agent being modeled believes that instantiation of the parameter succeeds.

Now we describe how the collaborative dialogue approach works for human-GIS communication on the meaning of spatial concepts. At the start of the process for communicating a vague spatial concept, the user and the system share the common goal that the system needs to show a map result matching the user's request, that is, the system needs to have a common understanding to the vague spatial concept with the user. However, there may be different initial sets of contextual knowledge and individual beliefs about the spatial concept that were available to the system and the user. The user is usually more knowledgeable about specifying the task and has a set of personal background towards the perception of spatial concepts that may not be available to the system. On the other hand, the system generally has more detailed knowledge on the environmental and geographical context due to its storage of spatial data, which may not be available in the memory of the human user. As the dialogue proceeds, both sides seek to maximize the shared knowledge by keeping track of what knowledge are already shared and what are to be further communicated. As each side shares more contextual knowledge with the other, they modify their beliefs on the correct understanding to the spatial concept, until finally they come to a common understanding to it.

As discussed in section 2, a successful communication of a spatial concept between a system and a user requires the sharing of three categories of contextual information. In our collaborative dialogue approach, each property of the contextual information is represented as an optional parameter of a recipe that represents a context-sensitive model of the spatial concept. For example, when the system receives a request (from a user) for a map *near* a place, it immediately realizes that an action *SeeAreaNear-Feature* is to be planned. Suppose that the system adopts a "buffer zone" model of *near,* then a recipe as in figure 4 will be activated. Since the system understands that identifying the parameter *Distance* is a pre-condition for executing the action *SeeAreaNearFeature*, the system will search for a recipe to identify this parameter. Suppose that the system knows two strategies to acquire this parameter: (1) by making an arbitrary guess and ask the user to correct it though visual interaction (this is labeled as the '*direct approach*'), or (2) by pursuing sharing of contextual knowledge (this is labeled as the '*indirect approach*'). The system may choose to pursue either approach at any given time of the interaction, depending on the system's belief on which one is more effective. Suppose that the second (indirect approach) is applied initially. A recipe *IndentifyDistanceFromContext* (see the lower part of Figure 5) is instantiated, which includes parameters representing contextual knowledge such as map scale, the goal of the task, the vehicle (if any) that the user will use in the task, and the user's personal preferences on the *near* relationship. These parameters are optional to execute *IndentifyDistanceFromNear*. When a contextual parameter, such as task, is missing, the system realizes that there are opportunities to improve the system's understanding of near if such contextual information can be provided by the user. Hence, the system will initiate a subdialogue to request it from the user. If the user responded with the requested information, the system will re-estimate a buffer distance and generate a map showing a revised understanding of *near*. Since not all the contextual factors have equal influence on the model of *near*, our algorithm uses task knowledge as the primary determinant, and others factors (such as spatial contexts and personal preferences) as subsequent modifiers.

4   Implementation of the Approach

We implemented the collaborative dialogue approach as part of a research prototype, *GeoDialogue*, a natural language human-GIS dialogue manager. The architecture of GeoDialogue is shown in Fig. 2. It includes modules for Syntactic Parsing, Semantic Interpretation, Dialogue Control, Query formation and Map generation, and Response control. The Dialogue control module is the central intelligence module that maintains dynamic knowledge and dialogue context, we well as performing automated reasoning on collaborative plans. All modules have access to a Static Knowledge Base (Static KB) which provides knowledge about language structure (grammar and terms), task advancement, and information content of GIS databases.

The dialogue usually begins with a request from the user to the system. The user's natural language request is first analyzed by the Syntactic Parsing module so that words are grouped into meaningful phrases. The parsing result is sent to the Semantic Interpretation module to extract intended actions and associated parameters and constraints. If sufficient information has been accumulated for the system to issue a well-formed GIS request, then the job is continued by the Query Formation & Map Generation module, which translate active request into a GIS query and coordinate the execution of this query through a standard GIS query interface. In our current implementation, an ArcIMS service is used as the GIS component.

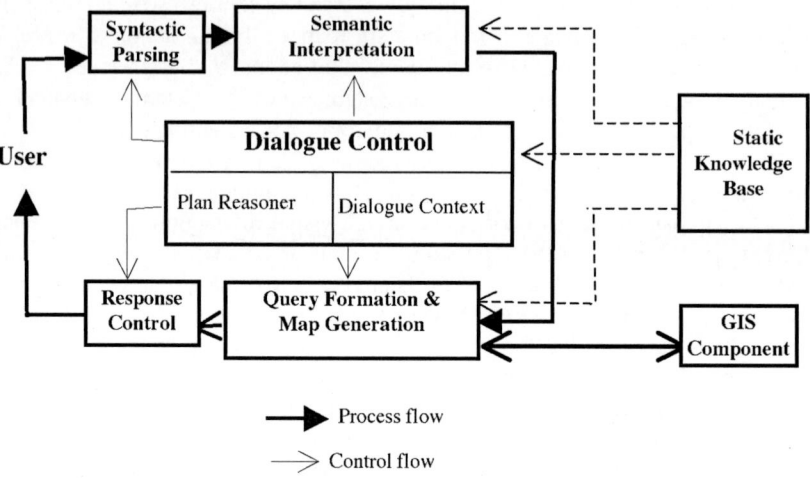

Fig. 2. Architecture of GeoDialogue

The Response Control module assembles response messages and coordinates the presentation of response to the user. A response message may include several components, such as a map generated by a GIS, summary status, reports of problems, and requests for further information, or clarification questions. To make the response as natural and effective as human communications, dialogue controller implements dialogue planning and response strategies as part of its *plan reasoner* functions.

The knowledge about understanding natural language words/phrases, actions for achieving goals and relevant datasets needed in the collaboration are stored in the Static Knowledge Base module. In GeoDialogue, the Static KB is implemented as a relational database. The Dynamic Knowledge module stores all information involving in the dynamic discourse context and a dynamic map context. The discourse context is updated dynamically as the dialogue proceeds and the map context is updated as a new map response is generated by the system. As described in the third section, the collaborative discourse context is modeled as an Rgraph.

5   Example User-System Dialogue Involving *Near*

In this section, a sample dialogue is given to illustrate how the collaborative dialogue approach is used in human-GIS communications involving a vague concept *near*. Within this interaction scenario (Fig. 3), the system knows extensive information about the geographical space, which can be shared with the user through an interactive map. The user has detailed information about the task, which can be communicated to the system through natural language input. Personal background and preferences are hard to be articulated in language, and these factors are taken into account using the direct approach (discussed in section 3). The combination of these dialogue strategies ensures that a shared understanding of the vague spatial concept 'near' is grounded in the shared contextual knowledge, and the process of such communication is more efficient than using only one of these strategies. For the purpose of convenience for description, we assume the user to be female.

The example dialogue begins with the user's request U1, which involves a vague concept 'near'. The system interprets the user's intention of U1 as "to see a map covering an area near some features", and forms a plan of 'show map' action using *SeeAreaAroundFeature* as the recipe. So, an *Rgraph* is initiated with the action *SeeAreaAroundFeaure* as its root action after the interpretation of U1 and the current focus in the Rgraph is the root action. The system elaborates this action with its recipe, *SeeAreaAroundFeature*, and updates the *Rgraph* (Fig. 4)

Based on the understanding of input U1, the system starts to reason about what actions should be taken next. By inspecting the Rgraph, the system found that there are two knowledge pre-conditions, 'Feature' and 'Dis-

Note: The dialogue begins with an empty map shown on the screen.
  (U- the user; G-the system, GeoDialogue)

(**U1**): I would like to see area near *Miami*.
(**G1**): (show a map) What are you going to do with this map?
(**U2**): Oh, I am just planning a trip to *Miami* and will have fun there by car.
(**G2**): (show a new map) Here it is. Is this what you want?
(**U3**): Yes, but I want to see a wider area near *Miami*.
(**G3**): (shows a new map) Is this what you want?
(**U4**): wider.
(**G4**): (shows a new map) Is this what you want?
(**U5**): The previous one is fine.
(**G5**): (Show the previous map) Here it is.

Fig. 3. An example dialogue involving near between the system and the user

tance', that need to be satisfied before the 'buffer zone' model' of the near concept can be calculated. Since *Miami* is a feature mentioned in the user's input, it is used to instantiate the parameter *Feature* (through a basic action *IdentifyFeatureBySpeech*). However, there is no clue from the user on how to determine the second parameter

'Distance'. Assuming that the system has a number of stored model of 'near' that could potentially be used to determine the buffer distance, but it need help from the user to select an appropriate model based on the use contexts. For this reason, the system chooses to use the 'indirect approach' by instantiating a recipe, *IdentifyDistanceFromContext*, pursue the goal of context sharing. The recipe *IdentifyDistanceFromContext* is added to the *Rgraph* by attaching it to the parameter *Distance* node. The action *IdentifyDistanceFromContext* contains four (4) optional parameters, *Map_Scale*, *Task_Goal*, *Task_Vehicle* and *User_Preference* (For simplicity in describing the approach, we implement only four parameters for the three categories of contextual information here).   While reasoning about a way to get knowledge on the 'task_goal' parameter, the system believes that the user has the knowledge about the intended task, and subsequently generates a question (G1 in Figure 3). In the same time, the system will generate a distance estimate (by executing the basic action *ExecuteEstimationOnNear*) using generic model (such as the Fuzzy Logic model). Although such an estimate has little chance to match with the user's intended meaning, it at least allows the system to generate a visual response. The Rgraph at this moment is shown in Fig. 5.

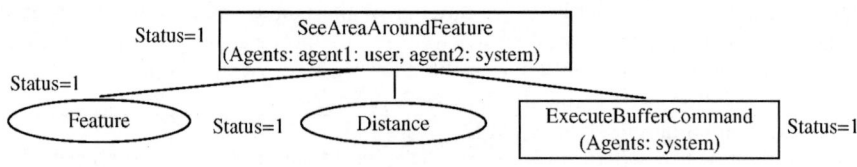

Fig. 4. Initial Rgraph

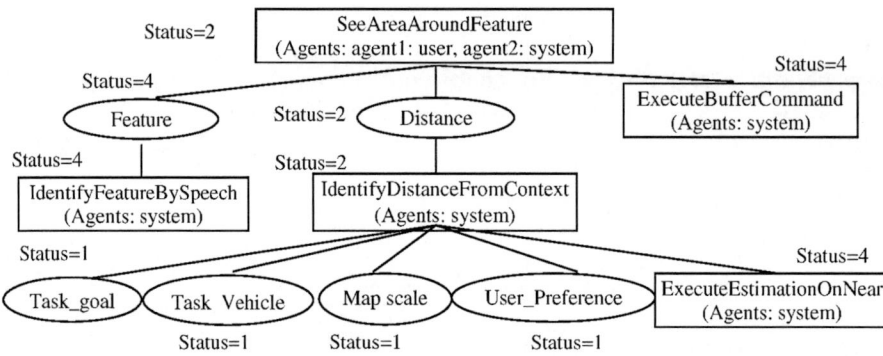

Fig. 5. A snapshot of the Rgraph for the example dialogue (at G1).

The map response in the (G1) of Figure 3 serves two purposes. First, the map provides a shared geographical context (e.g, the distribution and sizes of features on the map) for the user to reflect on his understanding of 'near'. Second, it serves as a representation of how the system currently understands 'near'. In the same time, the system also realizes (by consulting the Rgraph in Figure 5) that the first parameter *Task_Goal* information is missing, which means that task information (that may be known to the user) has not been shared with the system. So it generates a question to the user in G1 in order to elicit the task information from the user.

After the user receives the (G1) response, she replies U2, which provides information not only about the goal of the task but also the information about the vehicle that will be used in the task. U2 provides information to instantiate two parameters: *Task_Goal* and *Task_Vehicle*. Additionally, the system can get the map scale information from the current map context. After acquiring more contextual information such as the task ("having fun by car") and the current map context, the system modifies its belief on the buffer distance (such as '2 hours of driving distance', or '100 miles') as its new understanding of *near*.

Although important contextual information about the tasks and spatial contexts are already communicated and shared at this time, the system and the user may still differ in their understanding due to the user's specific characteristics. The user provides her own preferences in U3 to answer C2: "Yes, but I want to see a wider area near *Miami*." From the system's Static KB, the system knows that "wider" is a vague concept and it, in the current discourse context, represents longer distance. The system modifies the *Distance* value to a larger value based on the value of *Distance* estimated in the last time through a formal method and generates a new map response and a question response G3 to the user.

After the dialogue exchange at G3, the user still wants a wider area. Her reply in U4 led the system to generate a question in G4 and a new map with a larger value of the parameter *distance* than the previous one (with G3 together). With sharing the map context shown by the system, the user finally changes his previous understanding (in U4) of *near* and thinks the map sent with G3 is fine. Based on his reply in G5, the system generates a map by using the same distance used in the previous map (sent with G3). Finally, the system and the user come to the common understanding of *near* based on their shared knowledge of the relevant context through a collaborative dialogue process.

6   Discussions and Conclusions

In this paper, we have developed a collaborative dialogue approach to facilitate communication of vague spatial concepts in Human-GIS interaction. This approach allows contextual information to be captured, represented and shared through mixed-initiative dialogues. The dialogue between the system and the user is modeled according to the collaborative discourse theory developed by Grosz and Sidner [49]. For the vague spatial concept *near*, the context factors that we consider in this paper include the task information, the geographical information, and the user's individual background and preferences. The system contributes more knowledge on generating a shared map context, and the user contributes more on providing the task information and the individual information.

The distinctive feature of our approach is that it is based on a well-received theory, the *SharedPlan* theory of multi-agent collaboration [51]. The theory models the mental attitudes for an agent to participate in a collaborative discourse. By actively modeling the intentional structure of the dialogue as it progresses, the system is able to adapt to users with different level of contextual knowledge. The system participates in the dialogue by interleaving the actions of understanding the user's request, finding a plan to achieve it (which may include collaboration with the user), executing the plan, and generating responses to the user.

Supporting communications of vague spatial concepts between a user and the system is the key to enable natural language based GIS. Towards this goal, the collaborative dialogue approach, as proposed in this paper, is not an alternative to the existing formal models of spatial concepts. Instead, the two approaches are complementary to each other. In particular, our approach emphasize the role of shared contextual knowledge in communicating a vague concept, while formal models of spatial concepts captures the inherent uncertainties of spatial objects and relationships themselves. Ideally, the system should be able to adapt its behavior in a situated interaction that may range from no contextual information at all to the sharing of full knowledge about the contexts.

Our experience with developing collaborative dialogue interfaces [17] for GIS indicates that there are unexplored potentials to incorporate the principles of human-human communication into interfaces for geographical information systems. As exemplified in this paper, dialogues and collaborations are fundamental strategies for human to communicate vague (spatial) concepts. When people's communication involves a vague spatial concept, people usually exchange relevant knowledge about the context to come to a common understanding of the concept. Further research is needed to understand how these multiple contextual factors interact in affecting the meaning of vague spatial concepts, and to develop computational models that are more cognitively plausible.

Acknowledgement

We would like to acknowledge the funding and support from the National Science Foundation under Grant No.BCS-0113030. This work has benefited from discussion with Sven Fuhrmann and Isaac Brewer and has received programming assistance from Ping Xia and Levent Bolelli.

References

1. Frank, A. U. and Mark, D. M.: Language issues for GIS. In: Macguire, D., Goodchild, M. F. and Rhind, D.(eds): *Geographical Information Systems: Principles and Applications*, Wiley, New York (1991) 147-163
2. Mark, D. M. and Frank, A. U.: User interfaces for Geographic Information Systems: report on the specialist meeting. National Center for Geographic Information and Analysis (1992)
3. Mark, D. M., Svorou, S. and Zubin, D.: Spatial terms and spatial concepts: Geographic, cognitive, and linguistic perspectives. *Proceedingsof International Geographic Information Systems (IGIS) Symposium: The Research Adgenda:*, Arlington, VA. (1987) 101-112
4. Shapiro, S. C., Chalupski, H. and Chou, H. C.: Linking ARC/INFO with SNACTor. Santa Barbara, California.: National Center for geographic Information and Analysis (1991)
5. Neal, J. G. and Shapiro, S. C.: Intelligent Multi-Media Interface Technology. In: Sullivan, J. W. and Tyler, S. W.(eds): *Architectures for Intelligent Interfaces: Elements and Prototypes*, Addison-Wesley, Massachusetts (1991) 11-44
6. Zue, V., Glass, J., Goddeau, D., Goodine, D., Leung, H., McCandless, M., Phillips, M., Polfroni, J., Seneff, S. and Whitney, D.: Recent progress on the MIT voyager spoken language system. *Proceedings of the ICSLP*, (1990) 1317-1320

7. Lokuge, I. and Ishizaki, S.: Geospace: An interactive visualization system for exploring complex information spaces. *CHI '95 Conference Proceedings*, New York, 1995 (1995) 409-414
8. Wang, F.: Towards a natural language user interface: An approach of fuzzy query. *International Journal of Geographical Information Systems* 2 (1994) 143-162
9. Wang, F.: Handling Grammatical Errors, Ambiguity and Impreciseness in GIS Natural Language Queries. *Transactions in GIS* 1 (2003) 103-121
10. Egenhofer, M. J. and Shariff, A. R.: Metric details for natural-language spatial relations. *ACM Transactions on Information Systems* 4 (1998) 295 - 321
11. Egenhofer, M. J.: Query Processing in Spatial-Query-by-Sketch. *Journal of Visual Languages and Computing* 2 (1997) 403-424
12. Cohen, P. R., Johnston, M., McGee, D., Oviatt, S., Pittman, J., Smith, I., Chen, L. and Clow, J.: QuickSet: multimodal interaction for distributed applications. *Proceedings of the fifth ACM international conference on Multimedia*, Seattle, Washington (1997) 31 - 40
13. Florence, J., Hornsby, K. and Egenhofer, M. J.: The GIS wallboard: interactions with spatial information on large-scale displays. In: Molenaar, M.(eds): *International Symposium on Spatial Data Handling*, 7. Taylor and Francis, Delft, The Netherlands (1996) 449-463
14. Sharma, R., Pavlovic, V. I. and Huang, T. S.: Toward a multimodal human computer interface. *Proceedings of the IEEE* 5 (1998) 853-69
15. Sharma, R., Poddar, I., Ozyildiz, E., Kettebekov, S., Kim, H. and T.S. Huang: Toward Interpretation of Natural Speech/Gesture: Spatial Planning on a Virtual Map. *Proceedings of ARL Advanced Displays Annual Symposium*, Adelphi, MD (1999) 35-39
16. Kettebekov, S., Krahnstöver, N., Leas, M., Polat, E., Raju, H., Schapira, E. and Sharma, R.: i2Map: Crisis Management using a Multimodal Interface. *ARL Federate Laboratory 4th Annual Symposium*, College Park, MD (2000)
17. Rauschert, I., Agrawal, P., Fuhrmann, S., Brewer, I., Wang, H., Sharma, R., Cai, G. and MacEachren, A.: Designing a User-Centered, Multimodal GIS Interface to Support Emergency Management. *ACM International Symposium on Advances in Geographical Information Systems*, (2002)
18. Robinson, V. B.: Individual and multipersonal fuzzy spatial relations acquired using human-machine interaction. *Fuzzy Sets and Systems* 1 (2000) 133 - 145
19. Duckham, M., Mason, K., Stell, J. and Worboys, M. F.: A formal approach to imperfection in geographic information. *Computers, Environments and Urban Systems* (2001) 89-103
20. Goodchild, M. F.: Sharing imperfect data. In: Onsrud, H. J. and Rushton, G.(eds): *Sharing geographic information*, Rutgers, New Jersey (1995) 413-425
21. Fisher, P.: Sorites paradox and vague geographies. *Fuzzy Sets and Systems* (2000) 7–18
22. Robinson, V. B.: Interactive Machine Acquisition of a Fuzzy Spatial Relation. *Computers and Geosciences* 6 (1990) 857-872
23. Peuquet, D. J. and Zhan, C.-X.: An algorithm to determine the directional relationship between arbitrarily-shaped polygons in the plane. *Pattern Recognition* (1987) 65-74
24. Matsakis, P., Keller, J. M., Wendling, L., Marjamaa, J. and Sjahputera, O.: Linguistic Description of Relative Positions in Images. *IEEE TRANSACTIONS ON SYSTEMS, MAN, AND CYBERNETICS—PART B: CYBERNETICS* 4 (2001) 573-588
25. Egenhofer, M. J. and Franzosa, R.: Point-set topological spatial relations. *International Journal of Geographical Information Systems* 2 (1991) 161-174
26. Zhan, F. B. and Lin, H.: Overlay of Two Simple Polygons with Indeterminate Boundaries. *Transactions in GIS* 1 (2003) 67-81
27. Hazelton, N. W., Bennett, L. and Masel, J.: Topological structures for 4-dimensional geographic information systems. *Computers, Environment, and Urban Systems* 3 (1992) 227-237
28. Zhan, F. B.: Topological Relations Between Fuzzy Regions. *Proceedings of the 1997 ACM Symposium on Applied Computing*, San Jose, CA (1997) 192-6
29. Rosch, E.: Principles of Categorization. In: Rosch, E. and Lloyd, B.(eds): *Cognition and Categorization*, Erlbaum, Hillsdale, NJ (1978)

30. Rosch, E.: On the Internal Structure of Perceptual and Semantic Categories. In: Moore, T.(eds): *Cognitive Development and the Acquisition of Language*, Academic Press, New York, NY (1973) 111–144
31. Egenhofer, M. J.: User Interfaces. In: Nyerges, T., Mark, D. M., Laurini, R. and Egenhofer, M. J.(eds): *Cognitive Aspects of Human-Computer Interaction for Geographic Information Systems*, Kluwer Academic Publishers, Dordrecht, The Netherlands (1995) 143-145
32. Fisher, P. and Orf, T.: An investigation of the meaning of near and close on a university campus. *Computers, Environment and Urban Systems* (1991) 23–25
33. Mark, D. M.: Spatial Representation: A Cognitive View. In: Maguire, D. J., Goodchild, M. F., Rhind, D. W. and Longley, P.(eds): *Geographical Information Systems: Principles and Applications*, 1. John Wiley & Sons, New York (1999) 81-89
34. Mark, D. M. and Frank, A. U.: NCGIA Initiative 2, "Languages of Spatial Relations," Closing Report., Santa Barbara, CA: National Center for Geographic Information and Analysis (1992)
35. Landau, B. and Jackendoff, R.: "What" and "where" in spatial language and spatial congnition. *Behavioral and brain sciences* (1993) 217-265
36. Worboys, M. F.: Nearness relations in environmental space. *International Journal of Geographical Information Science* 7 (2001) 633–651
37. Worboys, M. F.: Communicating geographic information in context. *Meeting on Fundamental Questions in GIScience*, Manchester (2001)
38. Terveen, L. G.: Overview of human-computer collaboration. *Knowledge-Based Systems* 2-3 (1995) 67-81
39. Grosz, B. J. and Kraus, S.: Collaborative plans for complex group action. *Artificial Intelligence* 2 (1996) 269-357
40. Lochbaum, K. E.: A collaborative planning model of intentional structure. *Computational Linguistics* 4 (1998) 525-572
41. Oviatt, S. L.: Multimodal interfaces for dynamic interactive maps. *Proceedings of the Conference on Human Factors in Computing Systems (CHI'96)*, (1996) 95-102
42. Oviatt, S. L. and Cohen, P.: Multimodal interfaces that process what comes naturally. *Communications of the ACM* 3 (2000) 45-53
43. Rauschert, I., Sharma, R., Fuhrmann, S., Brewer, I. and MacEachren, A.: Approaching a New Multimodal GIS-Interface. *Proceeding of the 2nd International Conference on GIS (GIScience)*, CO, USA (2002)
44. Zhan, F. B.: A Fuzzy Set Model of Approximate Linguistic Terms in Descriptions of Binary Topological Relations between Simple Regions. In: Matsakis, P. and Sztandera, L. M.(eds): *Applying Soft Computing in Defining Spatial Relations*, Physica-Verlag, Heidelberg, Germany (2002) 179-202
45. Robinson, V. B. and Frank, A.: About different kinds of uncertainty in collections of spatial data. *Proceedings of 7th International Symposium on Computer-Assisted Cartography*, (1985)
46. Lloyd, R. and Heivly, C.: Systematic distortions in urban cognitive maps. *Annals of the Association of American Geographers* 2 (1987) 191-207
47. Denofsky, M.: How near is near? , Cambridge, MA: MIT AI Lab (1976)
48. Montello, D.: Scale and multiple psychologies of space. *Spatial Information T heory: A Theoretical Basis for GIS (COSIT'93)*, Marciana Marina, Elba Island, Italy (1993)
49. Grosz, B. J. and Sidner, C. L.: Attention, intentions, and the structure of discourse. *Computational Linguistics* 3 (1986) 175--204
50. Lochbaum, K. E.: Using Collaborative Plans to Model the Intentional Structure of Discourse. *Devision of Applied Sceince*, Cambridge: Harvard Univsersity (1994) 158
51. Grosz, B. J. and Kraus, S.: Collaborative plans for group activities. *Proceedings IJCAI 93*, Chambery (1993) 367-373

Wayfinding Choremes

Alexander Klippel

Universität Bremen, SFB/TR 8 Spatial Cognition
Bibliotheksstraße 1, 28359 Bremen, Germany
klippel@informatik.uni-bremen.de

Abstract. How can we represent spatial information in maps in a cognitively adequate way? The present article outlines a *cognitive conceptual* approach that proposes primitive conceptual elements from which maps can be constructed. Based on work in geography that starts with abstract models of geographic phenomena, namely *modelisation chorematique* by R. Brunet (1980, 1987), we coin primitive conceptual elements of route directions *wayfinding choremes*. Sketch map drawings were analyzed as they obey the same medial constraints as maps but are constructed in a way that provides insights into human conceptions. A distinction between *structural* and *functional* aspects of wayfinding presents a useful method to gain further knowledge about human conceptualizations and leads to a practicable cognitive conceptual approach to map construction.

Keywords: Conceptual structuring processes, route directions, wayfinding, cognitive conceptual approach to map making, spatial primitives.

1 Schematic Conceptual Representations

All maps are schematic maps. However, there is a category of maps that intentionally violates cartographic design constraints, such as sketch maps that distort, for example, distance information. For the purpose of this paper, we will refer to these maps as "schematic". Schematic maps reveal characteristics of human spatial information processing since certain aspects of space are simplified, omitted, and/or distorted with the assumption that users do not need the altered information. The most useful information is selected and structured to reflect human conceptualizations of the spatial domain (Freksa, Barkowsky, and Klippel, 1999; Tversky, 2000). Schematization approaches usually employ bottom-up—data-driven—methods similar to cartographic generalization. Stepwise they reduce and focus the information content of rich knowledge sources, for example, by shape simplification algorithms and other related research (e.g., Barkowsky, Latecki, and Richter, 2000; Cabello et al., 2001).

Despite findings from the cognitive science community on the positive effects of information reduction by schematization (e.g., Clark, 1989) cartographers continue to speak of 'maps that have to lie' (e.g., Monmonier, 1996) as they cannot represent the world one to one. In our opinion, schematic maps are not the 'bigger liars' but are representations that focus on relevant information and, thereby, accomplish valuable support for information processing. Tversky (2003) identified benefits of schematization; we have regrouped them and added further aspects:

W. Kuhn, M.F. Worboys, and S. Timpf (Eds.): COSIT 2003, LNCS 2825, pp. 301–315, 2003.
© Springer-Verlag Berlin Heidelberg 2003

- Cognitive and conceptual considerations – Schematization is a prerequisite to deal with the abundance of information available as, for example, the capacity of working memory is restricted. It also fosters the integration of information and thereby speeds decision making.
- Perceptual considerations – Visual complexity has been early on noted as one of the major drawbacks on information processing (e.g., Dobson, 1980; Phillips & Noyes, 1982).
- Informatics and AI perspective – Schematic information qualifies for compact data formats and computational efficiency. Less rich geometries—for example, ordering information and topology—constitute major research fields in qualitative spatial reasoning (e.g., Vieu, 1997).
- Technical aspects – New means for information presentation, for example, small displays, do not allow for great detail.

With the dawn of experimental cartography (Eckert, 1921/1925; Robinson, 1952) and especially since cognitive questions play a greater role in cartography (e.g., Petchenik, 1975; Medyckyj-Scott & Board, 1991; MacEachren, 1995; for an overview see Montello, 2002) map design improved by incorporating a cognitive perspective. However, most cognitively motivated studies focus on the thematic content or the overall design of maps and not on spatial components.

We illustrate this within the theoretical framework of graphic semiotics by Bertin (1974). Bertin grounds his work on visual variables in analyzing information depictable by the properties of the plane. For diagrams these are the horizontal and the vertical axis, which can be used to illustrate all kinds of information, for example, representing the changing amounts of precipitation during a year for a certain place on the earth's surface. While this conveniently visualizes the relation of two kinds of information, maps encounter the problem (or the advantage, see, e.g., Palmer, 1978; Freksa, 1999) that the two dimensions of the plane are reserved for representing *locational spatial information*—the *geographic component* in Bertins terminology. This affords the use of the *third dimension*—Bertin's visual variables, for example, color or hue—for depicting any other information. Whereas cognitive cartographic research focuses often on the third dimension, spatial cognition research is primarily concerned with locational spatial information.

Brunet (1980, 1987) proposed a singular approach that emphasizes conceptual spatial information. Even though he might not have characterized his own work thus, his approach could be viewed as cognitively oriented. Coming from analytical geography (e.g., Wirth, 1979), he established simple models that characterize—according to his theory—every possible spatial situation. Examples for these models are 'meshes' (French *maillage*), indicating the partitioning of a region, or 'contact', characterizing processes at boundaries. In correspondence to his theoretical models Brunet proposed graphical counterparts constituting the basic components of the maps he advocated. Following Bertin (1974) he subdivided every model according to three cartographic primitives in *point-like*, *linear*, and *areal* models. As a fourth synthetic but important primitive he added the *net* (French *reseau*). He chose the term *modelisation chorematique* or chorematic modeling for his theory composing *choreme* of the Greek word for space (*chorus*) and the suffix *–eme*. The basic set of choremes is combinable like the letters of the alphabet, hence, as we communicate 'all' things by combining letters, by combining choremes we cartographically communicate 'all' geographic phenomena. Fig. 1 depicts a part of his choreme table showing three basic models—meshes, attraction, and contact—for the cartographic primitives—point, line, area, and net.

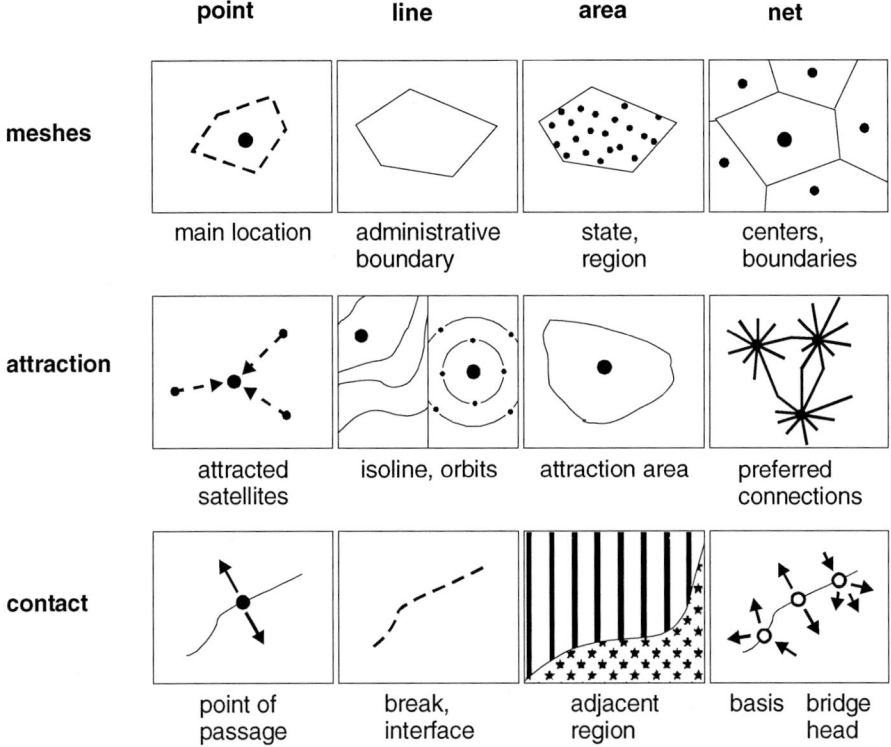

Fig. 1. The table of *choremes* by Brunet (extract, Brunet, 1987, p.191)

Although Brunet motivates his theory not explicitly by cognitive science research, he claims that by applying his graphic models to map making maps speak for themselves "Le langage de la carte est dans la forme, l'arrangement et la signification des distribution qu'elle montre. Les formes élémentaires sont les sémes de ce langage ; la syntaxe est dans leurs relations ; le message entier dans la configuration de la distribution" (Brunet, 1987, p. 190). And indeed, we find correspondences comparing his models with how humans structure environmental information, like, for example, image schemata (Johnson, 1987; Lakoff, 1987), hierarchies (e.g., Hirtle & Heidorn, 1993), or perceptual primitives (e.g., Biederman, 1987).

This—in our terminology—conceptual approach to map making constitutes the starting point for the ideas presented in this paper. Whereas Brunet built a theory for geography and geographic knowledge in general we focus our work on the domain of wayfinding and route directions for three major reasons:

- The present work defines a cognitive adequate way for representing spatial information essential for wayfinding; a domain for which Brunet's choremes are not applicable.

- We aim at automating map making appropriate for a given wayfinding situation. Therefore, the basic conceptual elements should not change from map to map but rather be the invariants. Brunet's choremes provide general means for structuring spatial knowledge but change greatly if represented in maps.
- People's conceptualizations change depending on the given domain and the events that take place in this domain. Hence, to elicit human conceptualizations we have to be as specific as necessary.

Analogous to Brunet's work we coin primitive conceptual elements of wayfinding and route directions *wayfinding choremes*. Combined, wayfinding choremes represent (nearly) all necessary route information in street networks. The conceptual models will be associated with pictorial representations. Hence, the term *wayfinding choremes* is systematically ambiguous as it denotes human conceptual as well as graphical entities.

2   Structure and Function

What is the difference between thinking of an intersection per se and thinking of an intersection at which one has to perform a specific action? Ample research explains how humans schematize spatial information at different scales (see Montello, 1993) from general organizational–structural aspects (e.g., Stevens & Coupe, 1978) to small scale characteristics of it (e.g., Evans, 1980; Moar & Bower, 1983). Schematization can be applied to objects (e.g., an intersection), as well as to actions (e.g., *turn right at the next intersection*). Schematizations of actions were inferred from object schematizations and resemble recent discussions of *events* (cf. e.g., Zacks & Tversky, 2001; Casati & Varzi, 1996). Following Quine (1996), Zacks and Tversky (2001) convincingly argue that events can be treated analogous to objects. Although they differentiate between actions and events—blowing out a candle is categorized as an action whereas a candle blown out on a windy day is an event—the distinction is not pertinent to the present work and the terms will be used interchangeably.

Wayfinding actions take place in environmental spatial structures which consist of objects and relations between objects. Hence, we differentiate conceptualizations of objects and conceptualizations of actions. Moreover, we are concerned with the representation of adequate spatial information in a spatio-analogical medium, therefore, our main focus are spatial aspects of map-like representations. The following terminology will stress these distinctions. With *structure* we refer to the object level, i.e. the spatial structure as physically present in the environment, with *function* we indicate the event/action level, or, to be more precise, the structural aspects demarcated by an action. This—in our opinion important distinction—is also reflected in the differentiation between *path* and *route* (Montello, in press) denoting a behavioral pattern as route, whereas the term path is reserved for a physical structure.

Again, the terms action / function / route build one pillar indicating a behavioral pattern, and object / structure / path group together to form a second pillar denoting the level of the physical reality. Fig. 2 clarifies this distinction for the domain of route directions and wayfinding.

The left side of Fig. 2 depicts a part of a city street network. From a structural perspective this network can be partitioned, for example, according to human spatial concepts, into intersections, i.e. branching points, and path segments. Additionally, the right side of Fig. 2 shows some behavioral patterns within the same network. The actions performed assign a different meaning to the objects identified on the structural level, for example, an intersection becomes a decision point. The route, which—in its character—is isomorphic to a path, demarcates the functionally relevant parts of the intersection within the structural framework.

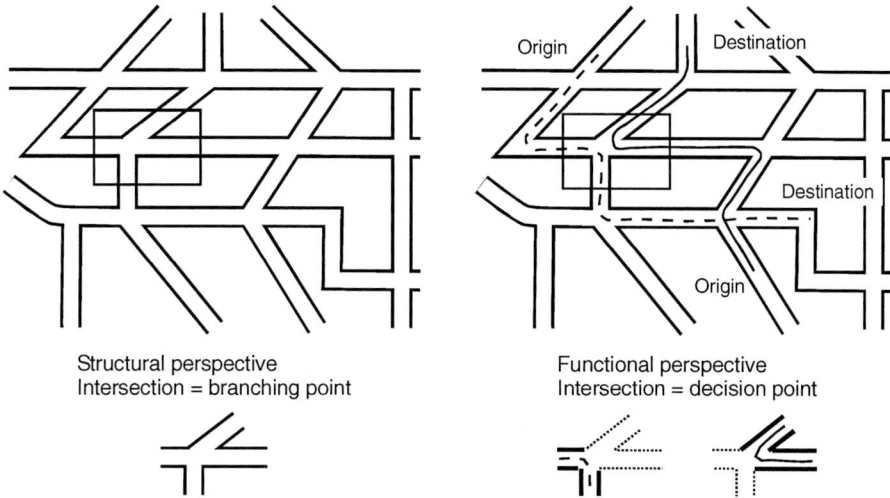

Fig. 2. Distinguishing paths (structural perspective) from routes (functional perspective)

3   Conceptualizing Route Direction Elements

Tversky and her coworkers (1998, 1999, and 2000) proposed that common conceptual structures underlie both, verbal and pictorial route directions. They advocate two toolkits for route directions containing primitive elements that establish a basic set for each of the two forms of external representation. They emphasize that the semantic correspondence of these toolkits can be used to translate between them. On the other hand, the basic elements of route directions in their pictorial toolkit stress structural aspects of intersections—additional arrows complement functional information— while the verbal direction toolkit relies primarily on functional aspects, i.e. most elements in their direction toolkit use verbs of motion. In the following, we focus on decision points as they constitute the most important aspects in route directions (Denis, 1997; Allen, 1997).

In the previous section we have discussed the distinction between structural aspects of a wayfinding situation as opposed to functional aspects of the behavioral pattern of following or planning a route. Rethinking the approach of conceptual elements— especially for pictorial route directions—from this perspective poses the question if the structure, i.e. the configurational information of a branching point, is the primitive element, or if not functional aspects, i.e. the demarcated parts of a decision point activated by a specific action, require more attention.

The structure of *the* prototypical intersection seems to be evident, i.e. two paths meeting at a right angle. However, the concept is invalidated when an uneven number of branches like the case of 5-way intersections occurs, when more branches have to be arranged (6-way intersection), or when the provided concept is underspecific, like in the case of star-shaped intersections.

Thinking about the complications to conceptualize these spatial structures and having in mind the proposed importance of a functional perspective leads to the following questions: First, is more attention necessary on functional aspects? And, second, if the representation of route directions—inherently on the action side—should not seek to identify rather functional than structural prototypes.

In order to gain evidence for the differentiation between structure and function and the corresponding differences in advocating conceptual primitives we examined human conceptualizations of route direction elements in an experimental setting.

Methods

Participants
19 participants volunteered for the study, 8 female and 11 male. They were native German speakers between 20 and 33, most of them holding an academic degree.

Design
Each participant constructed 42 drawings of a list of spatial expressions. Either these spatial expressions were general spatial concepts important for route directions, like 'intersection' (*Kreuzung*) or 'turn right', or they were actually parts of route directions like 'at the star shaped intersection you turn right' (*an der Sternkreuzung biegst Du rechts ab*). The verbal spatial expressions were systematically varied according to different types of intersections and to prototypical directions covering most of the actions required at decision points found in route traveling in outdoor networks on an average level of abstraction. These functional aspects, i.e. the actions to be taken at intersections, were chosen from models of qualitative spatial reasoning (e.g., Frank, 1992; Hernandez, 1994; Raubal, 2001). These are expressions necessary to give directions according to an 8-direction model. The resulting seven—the 'going back' direction is not examined here—functional expressions are: sharp right (scharf rechts), right (rechts), half right (schräg rechts), straight (geradeaus), half left (schräg links), left (links), sharp left (scharf links). 'straight', 'right', and 'left' are referred to as *basic (turning) concepts*, if modified by 'sharp' or 'half' as *specific turning concepts*. The directions were pretested to see if they were understood by participants.

The six concepts that actually require a direction change were tested for every intersection, i.e. 3-way (3-er Kreuzung), 4-way (only referred to as intersection (Kreuzung)), 5-way (5-er Kreuzung), 6-way (6-er Kreuzung), and star shaped (Sternkreuzung). The 'straight' (geradeaus) concept was only tested for 'intersection' (Kreuzung) and '6-way intersection' (6-er Kreuzung). Additionally, the participants obtained written concepts for, for example, intersection (Kreuzung), turn right (rechts abbiegen), straight (geradeaus), or turn right at the 3rd intersection (an der dritten Kreuzung rechts).

Material

The participants were provided with 44 single pages. The first page carried general instructions. Each of the following pages had a spatial expression printed on the top margin of the page leaving the rest of the page as drawing space. The last page contained a questionnaire on general participant information.

Procedure

Participants provided a graphical representation, i.e. a drawing, for each spatial expression within the space supplied. They could pursue the task in a self-paced manner and were unrestricted regarding orientation and scale.

Results

The main results we will focus on provide evidence for the importance of distinguishing functional and structural aspects of route direction elements and of people's conceptions of functional route features, i.e. the question what is the basis for wayfinding choremes. Additionally, some general results are reported on as they add further insight into human conceptualizations of route direction elements. First, we will discuss some of the participants' drawings and then we analyze conceptualizations of turning concepts at decision points.

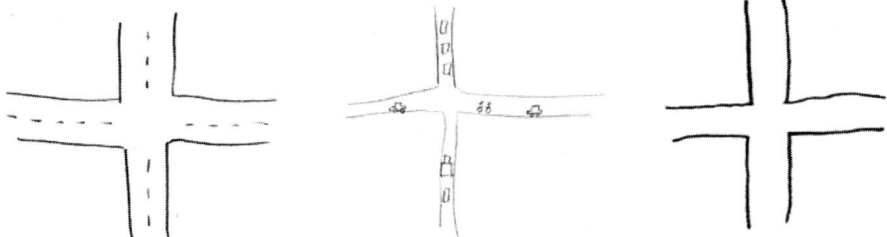

Fig. 3. Drawing of the concept of an *intersection* as a *structural concept*

The prototypical drawing of an intersection as a *structural concept*—the participants received the expression *intersection*—meets the expectations entirely, i.e. a 4-way intersection where the branches meet at right angle (see Fig. 3). This prototypical concept is also adhered to when the spatial expression included one of the following three *basic (turning) concepts* at a 4-way intersection: *turn right*, *go straight*, and *turn left*. 84.2% of the participants followed this scheme for the turn-right-concepts, 84.2% for the turn-left-concept, and 100% for the go-straight-concept[1].

---

[1] Some exceptions in more detail: Participant 2 used the 'prototypical' concept of an intersection throughout his (4-way) intersection drawings without varying it, no matter what kind of action was required at the intersection. Participant 4 ignored the difference between the basic concept, i.e. *left* and *right*, and the *sharp* modification resulting in 'identical' depictions. Participant 8 and 9 used 3-way intersections for the basic concepts *left* and *right*. One participant drew only the functional concept.

When the action required at a 4-way intersection became more specific, i.e. the basic turning concepts were modified either by *sharp* or *half*, for example, *turn half left at the intersection*, the prototype of the intersection concept disappeared (see Fig. 4). This resulted in a changing number of branches ranging from three to five and a differing orientation of the branches that were not functionally involved. These differences did not only occur between subjects but also within subjects—then, for different turning concepts.

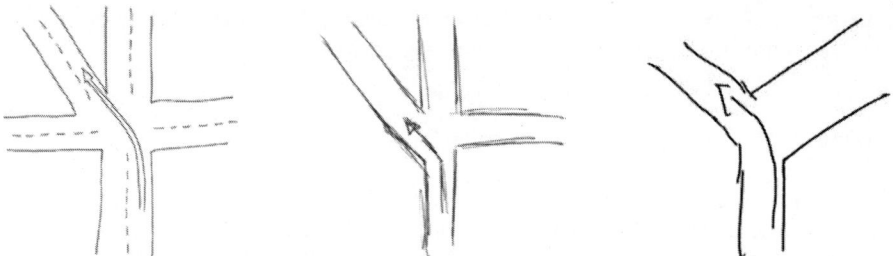

Fig. 4. The *intersection* concept with a superimposed *specific action*

Hence, the prototypical concept of an intersection holds as long as the action necessary corresponds to one of the three basic turning concepts but disappears if more specific actions are compulsory, i.e. there is no prototypical 4-way intersection if one has to turn half left or half right at this intersection.

Likewise, the missing intersection prototype became apparent when the participants were required to draw intersections that do not match a 90° increment scheme, like the 3-, 5-, or 6-way intersection, or the underspecific star-shaped one (see Fig. 5).

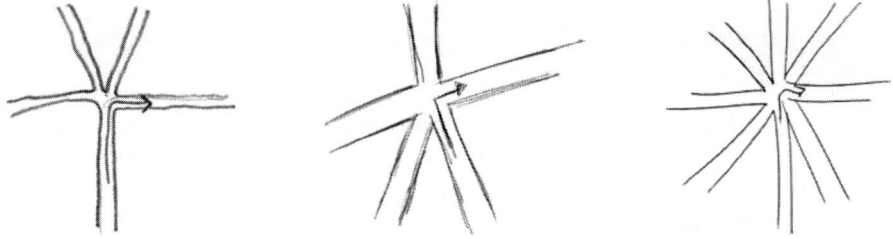

Fig. 5. Drawings for the turning concept *right* at a star-shaped intersection

On the other hand, the examples show that whereas the structure of the intersections changes, the turning concepts, i.e. the functional aspects, seem to be a constant factor of the participants' conceptualization (their drawings).

Hence, we now turn to the analysis of the conceptualization of turning concepts and put forth the following hypotheses: 1) Prototypical turning concepts, i.e. the functional aspect, are a stable factor in people's conceptualization of route direction elements independent of the type of intersection at which they are required. 2) This holds equally for all 6 turning concepts specified according to an 8-direction model.

Table 1 displays the results of the drawings with respect to the intersections' functional aspects, i.e. the question if prototypical conceptual turning concepts exist for the 6 specified direction changes. A participant's drawing was counted as a prototypical turning concept if it matched with a 45° increment according to an 8-direction model.

The data shows that participants greatly agree on the prototypicality of turning concepts, hence, the functional aspects of intersections in route directions seem to be the constant factor. This holds for each of the 5 types of intersections and for the 'pure' turning concepts, i.e. the one not specifying an intersections. And this also holds for each of the 6 turning concepts. The values range from 63.16% for the *sharp right* turning concepts at a 3-way intersection to various 100% agreements, for example, *left* at a 4-way intersection or *half left* at a 6-way intersection. The mean agreement to the six prototypical turning concepts ranges from 14.3 for the *sharp right* turning concept to 18.3 for the *half left* turning concept out of 19 (from 75.44% to 96.49%).

Table 1. Results (N=19; absolute frequencies and percent values (in brackets)) for the prototypicality of turning concepts (sr (sharp right), r (right), hr (half right), hl (half left), l (left), sl (sharp left)). ‚Pure' in the intersection column denotes turning concepts without the specification of a type of intersection, for example, *turn right*

(N = 19) abs & (%)	sr	r	hr	hl	l	sl
pure	14 (73.68)	19 (100.00)	19 (100.00)	18 (94.74)	19 (100.00)	15 (78.95)
3-way	12 (63.16)	14 (73.68)	18 (94.74)	19 (100.00)	14 (73.68)	16 (84.21)
4-way	15 (78.95)	18 (94.74)	18 (94.74)	18 (94.74)	19 (100.00)	16 (84.21)
5-way	17 (89.47)	17 (89.47)	18 (89.47)	19 (100.00)	16 (84.21)	14 (73.68)
6-way	13 (68.42)	16 (84.21)	16 (84.21)	19 (100.00)	14 (73.68)	17 (89.47)
star	15 (78.95)	13 (68.42)	18 (94.74)	17 (89.47)	14 (73.68)	16 (84.21)
mean	14,33 (75.44)	16,17 (85.09)	17,83 (93.86)	18,33 (96.49)	16 (84.21)	15,67 (82.46)

Discussion

The study evidence a distinction between structural and functional aspects in the conceptualization of basic route direction elements, i.e. turning concepts at decision points. This difference is extremely relevant, especially for complex route elements that can be found in many European downtown areas, for example, Trier. From a structural perspective not every intersection can be prototypicalized in the same way, i.e. *the* prototypical intersection as externalized by the participants (see Fig. 3, 4, and 5). Beyond this aspect the data analysis reveals that the required action is of uttermost importance. Functionally relevant aspects play a major role in the conceptualization and prototypicalization of route directions elements. These aspects seem to be important, especially in situations in which a prototypical representation cannot be expected—i.e. intersections with a number of branches that do not allow for a regular 90° division of space—or if the turning concept affords a specific action, like *turn*

half left. The reported results show a common ground for a functional characterization of turning concepts at decision points according to an 8-direction model, rather
than relying strictly on structural prototypes of intersections. This offers a new perspective on schematizing spatial information in maps, i.e. applying prototypical route
direction elements obtained from externalizations of conceptualized actions. Hence, if
the domain comprises actions, the schematization has to consider them, as they are the
focus of the wayfinder while structural information play a secondary role. Prototypical functional elements in route directions are termed *wayfinding choremes*. Their
pictorial counterparts are obtained by externalizing conceptualizations of primitive
wayfinding actions into a spatio-analogical medium (see Fig. 6).

Furthermore, the data shows some differences within the conformity with prototypical turning concepts which will be looked at in greater detail as they reveal some
peculiarities about intersections in interaction with turning concepts. The *half left* and
half right directions at the 3-way intersection were the most consistently represented.
Compared with this result, the basic turning concepts at this type of intersection were
rather weak. Even though not significant, this effect can be explained as some participants equated *take the right part of the fork* with *turn right*. Consequently, the intersections were depicted in fork shape which does not allow for a prototypical *turn
right* concept, i.e. a 90° angle. The comparatively low values for the basic turning
concepts at 6-way and star-shaped intersections can be partially explained by the fact
that some participants drew the intersection before they drew the turning concept. As
they used the same shape for these intersections during the entire experiment the correct representation of basic turning concepts were not possible.

Fig. 6. Prototypical turning concepts (drawings) and identified *wayfinding choremes* (bold
lines) within an 8-direction model

4   Applying Wayfinding Choremes to Map Construction

We now sketch our ideas on applying wayfinding choremes to map construction. We will only outline the main ideas, as this is work in progress. General and commonly known aspects of cognitive spatial information processing relevant for the adequate depiction of route information are only briefly mentioned.

The basic procedure consists of three steps. Assume a scenario where we either provide mobile wayfinding assistance or, like in some internet applications, information on parts of the route. The first step *focuses* on the relevant decision point. We have two classes of decision points: those that require a direction change [DP+] and those that do not [DP-]. As shown, for example, in Klippel, Tappe, and Habel (2003) higher order route elements, i.e. superordinate directions like *turn right at the third intersection*, are a strong organizational aspect of route directions. Hence, we apply simple combination rules on wayfinding choremes. Basically, wayfinding choremes of the type *straight*, i.e. [DP-], can be grouped and combined with wayfinding choremes of type [DP+], corresponding to conceptualization of, for example, *turn right at the third intersection*. The focus lies on these combined wayfinding choremes, i.e. higher order route direction elements.

The second step *replaces* functionally relevant parts of intersections with one of the six basic wayfinding choremes for a directional change if the subsequent decision point is of the type [DP+], or it replaces the functional aspects of a higher order structure, i.e. a combination of wayfinding choremes, respectively. The functionally involved branches of an intersection are categorized according to an 8-direction model. Only the functionally relevant parts are replaced as these are the important aspects of a given intersection and they are the ones for which we can advocate a minimal number of primitive prototypical realizations, i.e. the wayfinding choremes.

The last step *aligns* the resulting wayfinding choreme map with the direction of traveling, according to proven benefits of alignment on information processing (e.g., Adeyemi, 1982; Levine, 1982; Warren & Scott, 1993).

Problems of underspecificity are solved by the depiction of the corresponding spatial structure and in the spatio-analogical medium. Whereas verbal route direction elements like *turn right at the intersection* can lead to confusion if the direction matches two alternatives, this problem is solved in a map-like representation by the perseverance of ordering information[2].

5   Conclusions and Outlook

The cognitively adequate representation of spatial information is still an open question. We approached this problem for the domain of wayfinding by setting of from work on graphical primitives, namely the choremes of Brunet and the direction toolkit by Tversky and Lee. Distinguishing functional and structural aspects of the given domain provided valuable insights into human conceptualizations. The study on sketch map drawings led us to advocate a set of six prototypical turning concepts

---

[2] The only problem occurs if the categorization according to the 8-direction model results in a violation of this information; this may be solved by further refining the model.

corresponding to a functional conceptualization of route direction elements. For these primitives (plus the concept for *straight*) we dub the term *wayfinding choremes*. The approach can be termed cognitive conceptual or top-down as it does not schematize a rich set of spatial information but starts from abstract spatial conceptualizations.

A combination of conceptual elements and veridical information is possible and desirable as it bears resemblance to human information processing. Some information is conceptualized while other information is used for identification purposes or is simply taken from the environment (e.g., Raubal & Worboys, 1999). Thereby, we stress the conceptualization and identification of the relevant information.

Tversky and Lee (1999) have argued that their two toolkits are transduceable into each other; wayfinding choremes realize the next step as they, as well as the corresponding verbal route direction elements, correspond to human conceptualizations of wayfinding actions, hence, both are functionally determined. Based on wayfinding choremes a grammar for route directions can be specified. Together with rules of their combination, we obtain a simple, yet powerful framework for characterizing routes and route directions. By no means have we claimed that the work is finished yet as it is applicable only to 'normal' spatial situations in street networks. Variant spatial structure, like, for example, rotaries, are not specified. On the other hand, unusual spatial structures complicate conceptualization. As wayfinding choremes are based on human conceptualizations they further corroborate, for example, research on route complexity (e.g., Richter & Klippel, 2002).

We plan to extend our wayfinding choreme approach to other forms of assistance and to the characterization of further spatial situations, for example, pedestrians using public transportation facilities. Other domains comprise different concepts (see for example, work by Raubal (2002) on wayfinding in built environments; Kuhn, 2001; Timpf, 2002). The current set of wayfinding choremes needs to be complemented, for example, by conceptualizations for *through the hall* or *up the stairs*.

The work at hand offers also a new perspective on the characterization of events, or, to be more specific, actions. As the conceptualization study shows, it is possible to represent actions in a static spatio-analogical medium without requiring any animation. We suppose that highlighting the relevant information and providing a cognitive conceptualization, i.e. a wayfinding choreme, makes additional information, like arrows, unnecessary. This leads to a further reduction of visual clutter, which has been noted early as a major drawback on information processing (Phillips & Noyes, 1982).

Acknowledgements

Funding by the Deutsche Forschungsgemeinschaft (DFG) is gratefully acknowledge (projects: Aspectmaps (Spatial Priority Program on Spatial Cognition) and MapSpace (SFB/TR-8 Spatial Cognition)). I am in particularly indebted to Heike Tappe for productive discussion. I thank Dan Montello, Paul Lee, Kai-Florian Richter and three anonymous reviewers for comments that have increased the paper's clarity. Special thanks to Christian Freksa, Markus Knauff, and Lars Kulik.

References

Adeyemi, E.O. (1982). The effect of map orientation on human spatial orientation performance. *The Cartographic Journal*, 19(1), 28-33.

Allen, G.L. (1997). From knowledge to words to wayfinding: Issues in the production and comprehension of route directions. In S.C. Hirtle & A.U. Frank (Eds.), *Spatial information theory: A theoretical basis for GIS* (pp. 363-372). Berlin: Springer.

Barkowsky, T., Latecki, L.J., and Richter, K.-F. (2000). Schematizing maps: Simplification of geographic shape by discrete curve evolution. In C. Freksa, W. Brauer, C. Habel, and K.F. Wender (Eds.), *Spatial Cognition II. Integrating abstract theories, empirical studies, formal methods, and practical applications* (pp. 41-53). Berlin: Springer.

Bertin, J. (1974). *Graphische Semiologie. Diagramme, Netze, Karten*. Berlin: de Gruyter.

Biederman, I. (1987). Recognition by components: A theory of human image understanding. *Psychological Review*, 94(2), 115-147.

Brunet, R. (1980). La composition des modéles dan l'analyse spatiale. In *L'Espace Géographique*, 4: 253-265. (engl. tranlsation (1993). Building models for spatial analysis.)

Brunet, R. (1987). *La carte, mode d'emploi*. Paris: Fayard-Reclus.

Cabello, S., de Berg, M., van Dijk, S., van Kreveld, M., & Strijk., T. (2001). Schematization of road networks. In *Proceedings of the 17th ACM Symposium on Computational Geometry* (pp. 33-39). Boston.

Clark, A. (1989). *Microcognition: Philosophy, cognitive science, and parallel distributed processing*. Cambridge, MA: MIT Press.

Eckert, M. (1921/1925). *Die Kartenwissenschaft. Forschungen und Grundlagen zu einer Kartographie als Wissenschaft. 2 Vol*. Berlin: de Gruyter.

Casati, R. & Varzi, A.C. (1996). *Events*. Aldershot, England; Brookfield, Vt.: Dartmouth.

Denis, M. (1997). The description of routes: A cognitive approach to the production of spatial discourse. *Cahiers de Psychologie Cognitive*, 16, 409-458.

Dobson, M.W. (1980). The influence of the amount of graphic information on visual matching. *The Cartographic Journal*, 17(1), 26-32.

Evans, G.W. (1980). Environmental cognition. *Psychological Bulletin*, 88, 259-287.

Frank, A.U. (1992). Qualitative spatial reasoning about distances and direction in geographic space. *Journal of Visual Languages and Computing*, 3, 343-371.

Freksa, C. (1999). Spatial aspects of task-specific wayfinding maps. A representation-theoretic perspective. In J.S. Gero & B. Tversky (Eds.), *Visual and spatial reasoning in design* (p. 15-32). Sydney: Key Centre of Design Computing and Cognition.

Freksa, C., Barkowsky, T., and Klippel, A. (1999). Spatial symbol systems and spatial cognition: A computer science perspective on perception-based symbol processing. *Behavioral and Brain Sciences*, 9(4), 616-617.

Hernández, D. (1994). *Qualitative representation of spatial knowledge*. Berlin: Springer.

Hirtle, S.C. & Heidorn, P.B. (1993). The structures of cognitive maps: Representation and Processes. In T. Gärling & R.G. Golledge (Eds.), *Behavior and environment: Psychological and geographical approaches* (pp. 170-192). Amsterdam: Elsevier.

Johnson, M. (1987). *The Body in the Mind*. Chicago: University of Chicago Press.

Klippel, A., Tappe, H. and Habel, C. (2003). Pictorial representations of routes: Chunking route segments during comprehension. In C. Freksa, W. Brauer, C. Habel & K. Wender (Eds.), *Spatial Cognition III. Routes and Navigation, Human Memory and Learning, Spatial representation and Spatial Reasoning (pp. 11-33)*. Berlin: Springer.

Kuhn, W. (2001). Ontologies in support of activities in geographical space. *International Journal of Geographical Information Science*, 15(7), 613-631.

Lakoff, G. (1987). *Woman, fire, and dangerous things. What categories reveal about the mind*. Chicago: University of Chicago Press.

Levine, M. (1982). You-are-here maps - Psychological considerations. *Environment and Behavior*, 14(2), 221-237.

MacEachren, A.M. (1995). *How maps work. Representation, visualization, and design.* New York: The Guilford Press.

Medyckyj-Scott, D. & Board, C. (1991). Cognitive cartography: A new heart for a lost soul. In J.C. Müller (Ed.), *Advances in cartography* (pp. 201-230). London: Elsevier.

Moar, I. & Bower, G.H. (1983). Inconsistency in spatial knowledge. *Memory and Cognition*, 11(2), 107-113.

Monmonier, M. (1996). *How to lie with maps* (2nd ed.). Chicago: University of Chicago Press.

Montello, D.R. (1993). Scale and multiple psychologies of space. In A.U. Frank & I. Campari (Eds.), *Spatial information theory: A theoretical basis for GIS* (pp. 312-321). Berlin: Springer.

Montello, D.R. (2002). Cognitive map-design research in the twentieth century: Theoretical and empirical approaches. *Cartography and Geographic Information Science*, 29(3): 283-304.

Montello, D.R. (in press). Navigation. In P. Shah & A. Miyake (Eds.), *Handbook of visuospatial cognition*. Cambridge: Cambridge University Press.

Palmer, S. (1978). Fundamental aspects of cognitive representation. In E. Rosch & B.B. Lloyd (Eds.), *Cognition and categorization* (pp. 259-303). Hillsdale: Lawrence Erlbaum.

Petchenik, B.B. (1975). Cognition in cartography. In *Proceedings of the International Symposium on Computer-Assisted Cartography (Auto-Carto II), September 21-25* (pp. 183-193).

Phillips, R.J. & Noyes, L. (1982). An investigation of visual clutter in the topographic base of a geological map. *Cartographic Journal*, 19(2), 122-132.

Quine, W.V. (1996). Events and reification. In R. Casati & A.C. Varzi (Eds.), *Events* (pp. 107-116). Aldershot, England; Brookfield, Vt.: Dartmouth.

Raubal, M. (2001). *Agent-based simulation of human wayfinding: A perceptual model for unfamiliar buildings.* Ph.D. thesis, Technical University of Vienna.

Raubal, M. (2002). Wayfinding in built environments: The case of airports. Solingen: Natur & Wissenschaft.

Raubal, M. & Worboys, M. (1999). A formal model of the process of wayfinding in built environments. In C. Freksa & D.M. Mark (Eds.), *Spatial information theory. Cognitive and computational foundations of geographic information science* (pp. 381-399). Berlin: Springer.

Richter, K.-F. & Klippel, A. (2002). You-are-here-maps: Wayfinding support as location based service. In J. Möltgen & A. Wytzisk (Eds.), *GI-Technologien für Verkehr und Logistik. Beiträge zu den Münsteraner GI-Tagen 20./21. Juni 2002* (pp. 357-364). Münster: IfGIprints, 13.

Robinson, A.H. (1952). *The look of maps*. Madison: University of Wisconsin Press.

Stevens, A. & Coupe, P. (1978). Distortions in judged spatial relations. *Cognitive Psychology*, 10, 422-437.

Timpf, S. (2002). Ontologies of wayfinding: A traveler's perspective. *Networks and spatial economics*, 2, 9-33.

Tversky, B. (2000). What maps reveal about spatial thinking. *Developmental Science*, 3(3), 281-282.

Tversky, B. (2003). Navigating by mind and by body. In C. Freksa, W. Brauer, C. Habel & K.F. Wender (Eds.), *Spatial Cognition III. Routes and Navigation, Human Memory and Learning, Spatial representation and Spatial Reasoning*. Berlin: Springer.

Tversky, B. & Lee, P. (1998). How space structures language. In C. Freksa, C. Habel, K.F. Wender (Eds.), *Spatial Cognition. An interdisciplinary approach to representing and processing spatial knowledge* (p. 157-175). Berlin: Springer.

Tversky, B. & Lee, P. (1999). Pictorial and verbal tools for conveying routes. In C. Freksa & D.M. Mark (Eds.), *Spatial information theory. Cognitive and computational foundations of geographic information science* (51-64). Berlin: Springer.

Tversky, B., Zacks, J.; Lee, P.U., Heiser, J. (2000). Lines, blobs, crosses and arrows: Diagrammatic communication with schematic figures. In M. Anderson, P. Cheng, and V. Haarslev (Eds.), *Theory and Application of Diagrams - First International Conference, Diagrams 2000, Edinburgh, Scotland, UK, September 1-3, 2000 Proceedings* (pp. 221-230). Berlin: Springer.

Vieu, L. (1997). Spatial representation and reasoning in AI. In O. Stock (Ed.), *Spatial and temporal reasoning* (pp. 3-41). Dodrecht: Kluwer.

Warren, W.H. & Scott, T.E. (1993). Map alignment in traveling multisegment routes. *Environment and Behavior*, 25(5), 643-666.

Wirth, E. (1979). *Theoretische Geographie. Grundzüge einer theoretischen Kulturgeographie.* Stuttgart: Teubner.

Zacks, J.M. & Tversky, B. (2001). Event structure in perception and conception. *Psychological Bulletin*, 127, 3-21.

Testing the First Law of Cognitive Geography on Point-Display Spatializations

Daniel R. Montello, Sara Irina Fabrikant,
Marco Ruocco, and Richard S. Middleton

Department of Geography, University of California at Santa Barbara
Santa Barbara, CA 93106 USA
{montello,sara,ruocco,richard}@geog.ucsb.edu

Abstract. Spatializations are computer visualizations in which nonspatial information is depicted spatially. Spatializations of large databases commonly use distance as a metaphor to depict semantic (nonspatial) similarities among data items. By analogy to the "first law of geography", which states that closer things tend to be more similar, we propose a "first law of cognitive geography," which states that people *believe* closer things are more similar. In this paper, we present two experiments that investigate the validity of the first law of cognitive geography as applied to the interpretation of "point-display spatializations." Point displays depict documents (or other information-bearing entities) as 2- or 3-dimensional collections of points. Our results largely support the first law of cognitive geography and enrich it by identifying different types of distance that may be metaphorically related to similarity. We also identify characteristics of point displays other than distance relationships that influence similarity judgments.

1   Introduction

Information visualization may be defined as "the art, science, and technology of making visual representations for knowledge discovery" (by analogy to a definition of cartography offered by [1]). The design and implementation of information visualizations has attracted a great deal of attention from various disciplines working on human-computer interaction [e.g., 2, 3]. However, most of this development has proceeded relatively uninformed by relevant behavioral and cognitive science on human perception and cognition, and with little empirical evidence to support claims that interactive visualization tools indeed amplify people's cognition (good examples of the application of perceptual and cognitive science to cartography and information visualization may be found in [4, 5, 6]). We believe basic-science cognitive research is necessary to establish solid theoretical foundations for the design of effective information visualizations. Here we report research that contributes to this objective.

We are especially interested in a class of information visualizations known as "spatializations" [7, 8]. Spatializations are information visualizations in which nonspatial information is depicted as a spatially extended entity, such as a natural or urban landscape. The use of such spatial metaphors is very common in information visualization. Spatialization involves a two-step process of transformation [9, 10, 11]. First, a geometric and semantic generalization procedure transforms large complex data do-

W. Kuhn, M.F. Worboys, and S. Timpf (Eds.): COSIT 2003, LNCS 2825, pp. 316–331, 2003.
© Springer-Verlag Berlin Heidelberg 2003

mains into their basic information components. Mathematical transformations create a semantically defined coordinate system that rearranges a set of data items based on their content and functional interrelationships. Secondly, the spatialized data abstractions are graphically (sometimes *carto*graphically) rendered for visual examination and exploration.

In the experiments we report below, we focus on perhaps the most basic principle for rendering information spatializations—the distance-similarity metaphor. This metaphor states that elements closer together on information displays will be understood by users to be more similar. The *distance-similarity metaphor* reflects a basic principle we call *the first law of cognitive geography,* as expressed in the context of information displays. The first law of cognitive geography states that people believe closer things to be more similar than distant things. We identify this law by analogy to the geographical principle that closer places on the earth's surface actually tend to be more similar, whether in rainfall amounts, landforms, linguistic dialects, or foods consumed. This occurs because closer places tend to interact more with each other (movement of matter, energy, information). As a descriptive generalization and predictive heuristic (even occasionally an explanatory mechanism), this principle is so fundamental to analytic geographic thinking that Tobler [12] dubbed it *the first law of geography.* The first law may extend to information spatializations insofar as spatializations afford experiencing information spaces as if they were geographic spaces (or spaces more generally). Our research may thus be understood as an empirical investigation of the first law of cognitive geography in the context of information displays—are closer features on information displays perceived or conceived to be more similar? If so, what aspect of "closeness" corresponds to similarity, and what is the function relating the two?

Although the distance-similarity metaphor is very appealing intuitively, empirical evidence of its validity in information displays is scarce [see 13]. The few usability studies that do exist are not placed within the context of a coherent theoretical spatial framework. Commonly, semantic similarity is mapped metaphorically onto distance (relative location) in the graphic information space, usually metric distance of the straight-line or direct Euclidean type. The widespread and uncritical application of this metaphor may be problematic when information spaces resemble geographic spaces, such as landscapes [14], natural terrain [15], or urban spaces [16]. Geometry in geographic space is not just Euclidean, and in fact, it is not just metric. In other words, similarity can be graphically suggested in terms of several types of "distance," especially when distance is understood broadly to include a variety of expressions of separation (temporal, topological, etc.). Information spaces will be more usable if they are based on a sound theoretical and empirical framework, including those concerning cognitive aspects of space and place.

Our work ties research on information visualization to research on the psychology of distance perception and cognition [e.g., 17, 18, 19]. Distance research has characterized the psychophysical functions describing the relation of physical to psychological distance, identified factors that influence the extent of psychological distance, modeled processes used to estimate distances during perception or from memory, and more. Below we discuss some distance research that bears on the way graphical appearance may affect perceived distance in information displays. Our work also connects information visualization to a large body of research on the psychology of similarity [e.g., 20, 21, 22]. This research has described various perceptual and conceptual factors that influence judgments of similarity, modeled how different factors are combined or given differential attention as a function of context or training, modeled the role of similarity relationships in the categorical organization of stimuli, and more. Although this research has much to say about similarity judgments made from information displays, it does not specifically address in detail how people interpret different forms of distance in terms of representing similarities.

Alternatively, one may consider information displays from the perspective of graphic perception. The displays may be meant to suggest or represent landscapes, but they are small graphic displays, either static or dynamic. Phenomena that influence the perception of distance or similarity among graphic elements in pictorial displays might be expected to influence judgments about the entities represented metaphorically by the graphic elements. One such phenomenon that has been studied systematically for well over a century is the *filled-space illusion* [23, 24]. Intervals separating graphical elements that are empty (blank) appear shorter than do intervals containing intervening elements. If distance is interpreted by users to signify similarity in an information display, we expect filled intervals to appear longer and lead to judgments of lower similarity. However, though the effect is fairly reliable, it is modest in strength and would not be expected to affect similarity much in situations where the interval distances being compared are fairly different. Another relevant perceptual phenomenon is the *vertical illusion,* also called the *horizontal-vertical illusion* [25, 26]. This refers to the fact that vertically-oriented lines in the visual field appear longer than horizontal lines of the same length. Again, if distance is interpreted to signify similarity, we expect the vertical illusion to bias judgments of similarity accordingly.

Perceptual phenomena may influence judged similarity not just through an influence on perceived distance but through an influence on perceptual grouping. As discussed by the Gestalt school of perceptual psychologists, including Wertheimer and Koffka [reviewed in 27, 28], the arrangement of features in a picture or graphical image will influence the perceived thematic or group membership relations of elements. The *laws of perceptual organization* describe stimulus conditions under which separate elements in graphical displays will be grouped into coherent higher-order features. Under certain conditions, for example, a concentration of points will be seen as a cluster or object. Those conditions include various aspects of graphical similarity, proximity, continuation, familiarity, and so on. If graphical elements are perceived to be part of a higher-order feature, users may make a categorical or typological inference that the separate elements are more similar (they may or may not perceive them to be closer). We refer to this as the *emergent-feature effect.*

In this paper, we present results from two experiments on perhaps the simplest of spatialization metaphors—an area or cloud of points, each point meant to signify an information-bearing entity such as a book or news story. We call this a *point-display spatialization.* In both experiments, we examined the way people interpret information displays consisting of points. We wanted to find out how people would interpret the graphical layout of the displays in order to arrive at judgments of semantic similarity among the documents. On different trials, participants viewed and made similarity judgments while viewing different point displays; the different displays varied the spatial relationship between two pairs of comparison points (documents) and varied the context provided by other points in the display. In addition to the point-display trials, both experiments also tested other spatialization metaphors, including networks, regions, and network-region hybrids. But in order to give the present paper a clear focus and manageable size, we focus exclusively on the point displays here. Results for the additional display types will be reported in a forthcoming paper.

2   Experiment 1

In our first experiment, we investigated how users interpret point-display spatializations. In addition to an initial exploration of how users interpret such displays, we wanted to test our data-collection methods, including our similarity comparison request and rating scale. We showed research participants computer displays of simple point displays, explaining that the points represented documents. Three of the document points were labeled 'A,' '1,' and '2;' participants were asked to compare the similarity of A and 1 to the similarity of A and 2. A 9-point scale was provided for them to express their judgments of similarity on an interval scale. Thus, the main task of the experiment involves comparing the relative similarity of two pairs of comparison document points. We varied the point displays across trials with respect to the direct distances between the comparison points and the possible emergence of a cluster feature.

2.1   Methods

2.1.1 Participants. Forty-four students (25 males and 19 females) from an undergraduate regional geography class took part in the experiment, with a mean age of 21.0. They received a small amount of course credit in return for their participation.

2.1.2 Materials. Participants viewed computer displays composed of black points, created using ESRI ArcMap®. Each point was intended to represent a single document in a digital database. In each display, three points to be compared for similarity were labeled with red text as 'A,' '1,' and '2' (see Figure 1). Participants were prompted to "compare the similarity between document A and document 1 with the similarity between document A and document 2." They rated similarity on a 9-point scale ranging left to right from '5' to '1' and then back up to '5' (5-1-5 was transformed to 1-9 for analysis). On the left, '5' was labeled "Documents A and 1 are much more similar to each other." In the middle, '1' was labeled "1 and 2 are equally similar to A." On the right, '5' was labeled "Documents A and 2 are much more similar to each other." In this paper, we refer to the pair of documents A and 1 as 'A:1,' and that of A and 2 as 'A:2.'

Participants viewed 10 different point-display trials in a block (as mentioned above, they also viewed 30 additional trials involving other display metaphors). The point displays were varied to allow comparisons of the effects of different visual aspects of the point displays on judgments of similarity, specifically (1) straight-line distance, (2) vertical illusion, and (3) emergent cluster of points (the potential phenomenon that an aggregation of neighboring points would be perceived as a cluster of related documents). Graphical elements that were not thought to affect similarity judgments between the comparison points (such as the absolute location of the point on the screen) were varied non-systematically.

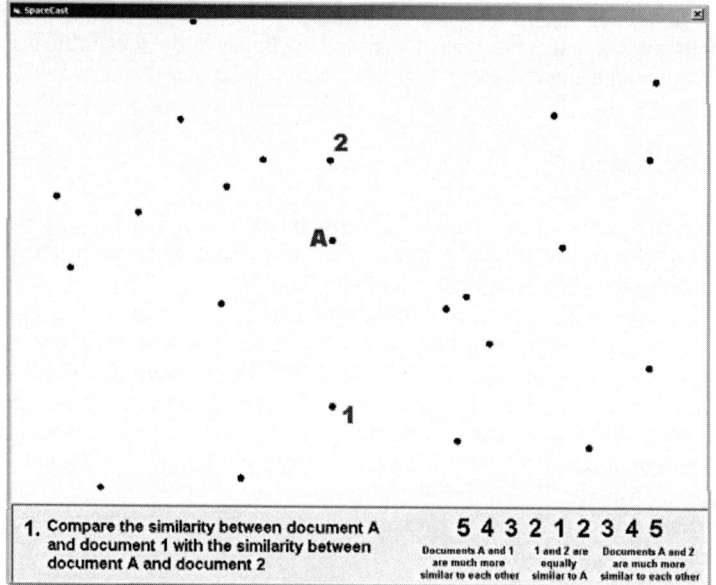

Fig. 1. Sample screenshot from an Experiment 1 trial, showing display, similarity question, and rating scale as they appeared to participants. The 5-1-5 scale shown to participants was transformed into a standard 1-9 scale for data analysis and description

Participants were introduced to the concept of similarity, the style of the trials, and the format of the response scale through three practice trials at the beginning of the test. To avoid priming any particular equivalence between distance and similarity, the practice trials prompted judgments of non-distance similarity (e.g., by asking for a comparison of the similarity of images of a pet dog, a domestic cat, and a tiger). Participants also responded to 11 pre-test questions asking about their personal background, including questions on age, sex, the presence of visual impairments (including specifically color blindness), as well as their formal experience in particular areas such as cartography and GIS. After the main test questions, participants responded to 28 post-test questions, such as how useful they thought each display type was for rating similarity and how easy it was to judge similarities for each display type. Participants also indicated how they had judged similarity and whether the displays reminded them of anything.

The experiment was administered using a Windows 2000 Pentium III personal computer. The interface was programmed with Microsoft Visual Basic 6.0. Images were projected onto a back-projection screen using an RGB color projector, generating an image size of 1.8 meters wide and 1.4 meters high, at 0.6 meters above the floor. Participants sat at a viewing table 2.7 meters in front of the screen, resulting in a horizontal viewing angle of approximately 37°. A standard mouse and keyboard were used to answer questions. Answers were recorded automatically and stored digitally, including the time required to make similarity judgments. Response time was measured as the elapsed time in milliseconds between when the trial display appeared on the screen and the moment the participant proceeded to the next trial.

2.1.3 Procedure. Prior to each test, participants were briefed that they would be presented with a series of trials about "diagrams that show an information collection from our computer database. The database contains documents such as news stories, books, and journal articles." Participants were told that each document would be shown as a single point. No information was provided on how to judge similarity. Participants were assured that there were no right or wrong answers and were asked not to waste time, as their answers would be timed. Participants then answered the pre-test questions and performed the practice trials. Following that, participants responded to the main test trials organized into four separate blocks (the block of point displays plus three blocks for the other metaphor types mentioned above), so that participants rated all trials of one display type before turning to another type. The block of point trials was presented in a counterbalanced order with the network and region blocks, so that an equal number of participants responded to point trials first, second, or third. All participants responded to network-region trials last. Trials within each block were presented in a different randomized order for each participant. After completing the main test trials, participants answered the post-test questions, were marked down for credit, and thanked for their participation.

2.2   Results and Discussion

Similarity ratings were treated as 9-point interval scales, by scoring a response of '5' to the far left ("A and 1 much more similar") as a '1,' a response of '5' to the far right ("A and 2 much more similar") as a '9,' and a response of '1' in the middle ("1 and 2 equally similar to A") as a '5.' Thus a mean rating less than 5 indicates that participants saw A:1 as more similar, while a mean rating greater than 5 indicates they saw A:2 as more similar. Participants apparently did equate direct distance with (dis)similarity, rating closer document pairs as more similar. Figure 2a shows an example where the actual direct distance between A:1 equals that between A:2. Participants gave this a mean rating of 5.1, which is not significantly different from 5.0, a rating of exactly equal similarity between A:1 and A:2. In contrast, Figure 2b shows an example where the actual direct distance between A:1 is much less than between A:2. On this trial, participants rated the relative similarity as 4.0, which indicates that A:1 is seen as significantly more similar than A:2. In fact, on all four of the point-display trials in which the direct distances between A:1 and A:2 differed, participants rated the closer pair as significantly more similar.

On three of the six point-display trials in which the direct distances between A:1 and A:2 were equal, participants rated the two pairs as not significantly different in similarity. On the other three, therefore, participants rated the two pairs that were actually equally distant as being significantly different in similarity. These trials contradict the basic principle of direct distance equaling similarity. Two perceptual phenomena explain these trials. First is the vertical illusion, in which vertical separations appear longer than horizontal separations. Figure 2c shows an example where the actual direct distance between A:1 equals that between A:2, but A:2 is vertically displayed while A:1 is horizontally displayed. As we anticipated, participants rated the relative similarity as 4.4, which indicates that A:1 is seen to be significantly more similar than A:2. The second relevant perceptual phenomenon is the emergent-feature effect. Figure 2d shows an example where the direct distance between A:1 again equals that between A:2, but A and 2 are members of an emergent cluster of points. In this case, participants rated the relative similarity of the two pairs as 6.7, which indicates that A:2 is seen to be significantly more similar than A:1.

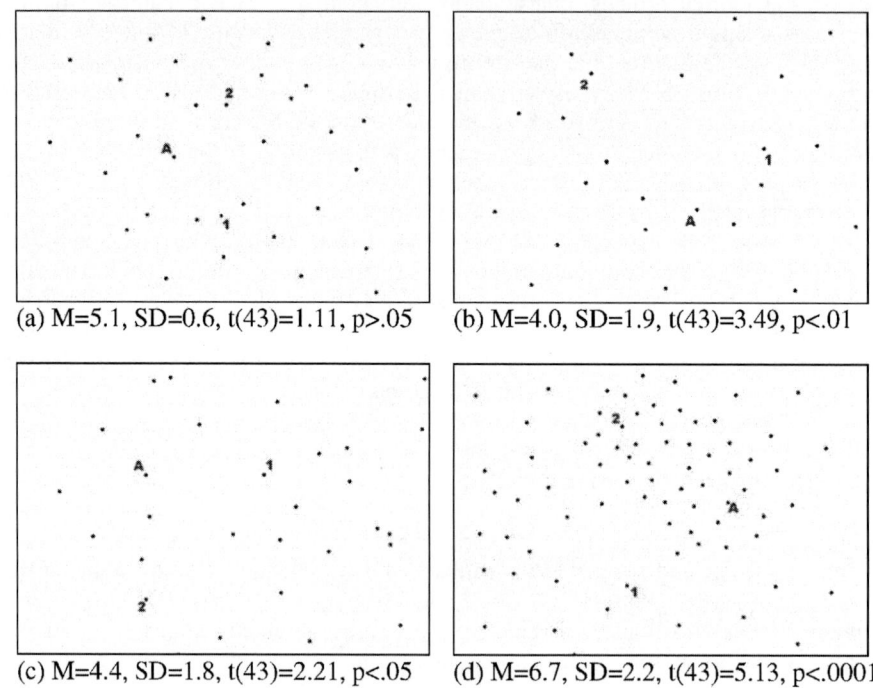

(a) M=5.1, SD=0.6, t(43)=1.11, p>.05   (b) M=4.0, SD=1.9, t(43)=3.49, p<.01

(c) M=4.4, SD=1.8, t(43)=2.21, p<.05   (d) M=6.7, SD=2.2, t(43)=5.13, p<.0001

Fig. 2. Sample displays and similarity ratings for four trials of Experiment 1. The t-scores are tests of significant differences from 5.0, a neutral rating ("1 and 2 are equally similar to A") on the transformed 1-9 scale (rating <5 indicates A:1 are more similar; >5 indicates A:2 are more similar). A neutral rating for equal comparison distances is demonstrated in (a), while the closer document point is rated as more similar in (b). The vertical illusion occurs in (c), while an emergent cluster occurs in (d); (M=arithmetic mean, SD=standard deviation, t=t-score [degrees of freedom], p=probability difference is due to chance)

To interpret these data more systematically, we calculated mean correlations of the ratio of direct distance between A:1 and A:2 *with* rated similarity. A Pearson's correlation coefficient was calculated separately for each participant, with the number of pairs of data points equal to the number of trials (10 in this case). These correlations were positive for 37 of the 44 participants. The correlations were normalized by Fisher's r-to-z transformation [29]. The arithmetic mean of the resulting z-scores was calculated by aggregating across participants. The mean z-score, transformed back into a correlation coefficient, equaled .43 (given how the data were coded, a positive correlation indicates that the closer pair is seen to be more similar). Based on a t-test calculated on the z-scores, this correlation is significantly greater than 0, t(43) = 6.31, p<.0001. This correlation was equally strong for female and male participants, not significantly differing as a function of participant gender, F(1, 42) = 0.24, p>.05. The correlation also did not significantly differ as a function of the order in which participants saw the block of point-display trials, F(2, 41) =0.40, p>.05. This null result is important, because it demonstrates that whether participants viewed network and/or region displays before viewing point displays did not much influence their interpretation of those point displays.

Participants took a mean of 12.2 seconds (SD = 7.2 s) to respond to each point-display trial (the fastest response was 4.9 s, the slowest was 40.6 s). Female and male participants answered nearly equally quickly; response-time not significantly differing as a function of participant gender, $F(1, 42) = 0.28$, $p > .05$. Response-time did vary significantly as a function of the order in which participants saw the block of point-display trials, $F(2, 41) = 5.58$, $p < .01$. Participants who saw point-display trials third (after the network and region trials) were fastest (8.2 s), those who saw point-display trials second were slowest (16.5 s), while those who saw point-display trials first were intermediate in their speed of response (13.2 s). In general across all types of displays, most participants responded slowest near the beginning of the experiment and sped up after that. For some reason, a few participants answered exceptionally slowly to point trials during the second block.

Finally, we examined participants' self-reports of how they interpreted the point displays. In response to the post-test question "How much did you consider the distance between documents in your answers?," participants marked a mean response of 2.4 on a 3-point scale that included "not at all," "somewhat," and "very much" (one participant did not answer). In order to find out if these self reports actually related to how participants interpreted the displays, we correlated participants' self reports with their distance-similarity correlations (in z-score form). This correlation was .32, a modest yet significant relationship ($p < .05$), suggesting that participants were able to judge their own interpretation strategies with at least some accuracy.

3   Experiment 2

In our second experiment, we attempted to replicate and extend the results we obtained with point-display spatializations in our first experiment. In particular, we examined the distance-similarity relationship by systematically varying distances between the two pairs of comparison documents; the range of variation was rather restricted and varied haphazardly in Experiment 1, with the comparison distances actually equal in 6 of the 10 trials. We also wanted to replicate and further explore the emergent-feature effect we found in Experiment 1, wherein two document points within a cluster of points were seen to be part of a common structure and therefore seen to be more similar. In addition to this emergent cluster, however, we unexpectedly found possible evidence for an emergent-line effect in Experiment 1. Two trials had one or two points intervening between either A:1 or A:2. In at least one case, the similarity ratings suggested that participants may have seen a linear sequence of points as an emergent linear feature, which might have influenced similarity ratings as did the emergent cluster. Finally, in this experiment we varied the size of the viewed displays. If participants compared distances between the document points in order to arrive at similarity judgments, we wondered if only relative distance was important, or whether absolute distance might play a role. The possible role of spatial scale in perceptual and cognitive tasks has been discussed in recent literature [e.g., 30]. Does the scale of the display influence the impact of differing relative distances?

3.1   Methods

3.1.1 Participants. Forty-eight students (27 males and 21 females) from an undergraduate introductory human geography class took part in the experiment, with a mean age of 21.5. They received a small amount of course credit in return for their participation. None had participated in Experiment 1.

3.1.2 Materials. As in Experiment 1, participants viewed computer displays composed of different graphical elements. However, only point and region displays were tested in this experiment (and both black-and-white and colored regions were included). All displays again included black points with three points to be compared for similarity labeled as 'A,' '1,' and '2.' Participants performed the same similarity judgments using the same scale as in Experiment 1.

The block of point displays consisted of 16 trials in this experiment. This block was presented in a counterbalanced order with three other blocks (involving region trials), for a total of 94 test trials. The point displays were again systematically varied to allow comparisons of the effects of different visual metaphors on judgments of similarity. In this experiment, the point displays were varied to allow examination of the effects of (1) straight-line distance, (2) emergent clustering of points, and (3) the number of intervening points between documents (i.e. a possible emergent linear feature). Straight-line distances among points were varied systematically so that the ratio of the distance between A:2 to the distance between A:1 varied from 1.0 (equal) to 1.5, 2.0, 2.5, or 3.0 times as long. To avoid possible confounding by the vertical illusion, comparisons between A:1 and A:2 were matched in orientation on the screen (i.e., both vertical, horizontal, or diagonal).

Participants responded to five practice trials in Experiment 2, similar to those from Experiment 1 but including a couple dealing with color. After the main test trials, participants answered 56 post-test questions including the same 11 questions used in Experiment 1 about their personal backgrounds. The additional post-test questions were adapted from Experiment 1 to account for the new display types.

We used the same equipment and setup as in Experiment 1, with one important addition. In Experiment 2, we varied the size of the projected image of the displays. Two sizes were used, *Small* and *Large*. Half of the participants viewed the small image from a distance of 1.6 meters, projected to be .6 meters wide and .4 meters high, at a height of 1.2 meters above the floor; this resulted in a horizontal viewing angle of approximately 20°. The other half of the participants viewed the large image, which was projected the same as in Experiment 1: 1.8 meters wide, 1.4 meters high, at 0.6 meters above the floor. Viewed from 2.7 meters in front of the screen, the horizontal viewing angle was approximately 37°.

3.1.3 Procedure. Participants were tested exactly as in Experiment 1. The main test trials were again organized into blocks of the same display type, so that participants responded to all trials of one type before proceeding to another type. Block order was fully counterbalanced, and trial orders within blocks were randomized for each participant.

3.2   Results and Discussion

As in Experiment 1, similarity ratings were scored so that a mean rating less than 5 indicates that participants saw A:1 as more similar, while a mean rating greater than 5 indicates that participants saw A:2 as more similar. Because several of the trials in this experiment were specifically designed to examine emergent-feature effects, which we discuss below, we begin here by restricting our examination to the five trials designed only to vary direct distance over relatively unfilled space. On these trials, we systematically varied the ratio of the distances between A:2 and A:1 from a ratio of 1/1 (equal distances) to a ratio of 3/1 (A:2 3 times as far apart as A:1). On the trial where this ratio equaled 1.0, participants in fact rated the relative similarity of the two pairs as 4.9, which is not significantly different from 5.0. For the remaining trials, where the ratio got systematically larger (A:1 got relatively closer), the similarity ratings declined accordingly (approached greater similarity for A:1): 4.4, 3.7, 3.3, 3.4. These were all significantly less than 5.0. Interestingly, the pattern supports a linear relationship between direct distance and similarity across the scale of the experiment, with the exception that the most extreme distance difference between A:2 and A:1 (ratio of 3/1) led to A:1 being rated as no more similar than the next most extreme difference (ratio of 2.5/1). This small difference in relative distance was either imperceptible to participants, or exceeded some threshold at which distance differences no longer influence perceived similarity.

As in Experiment 1, we calculated mean correlations for each participant of direct distance ratios A:1/A:2 with rated similarity across trials, restricting the correlations just to the five trials described above. These correlations were positive for 37 of the 48 participants, again indicating that the closer pair is seen to be more similar. (Correlations for two participants could not be calculated, as they rated all five trials as having equal similarity). The mean correlation (based on z-score transformed values) equaled a robust .70, which is statistically significantly greater than 0, $t(45) = 6.06$, $p<.0001$. This correlation was nearly identical for female and male participants, $F(1, 44) = 0.04$, $p>.05$. The correlation also did not significantly differ as a function of the order in which participants saw the block of point-display trials, $F(3, 42) = 0.37$, $p>.05$. As in Experiment 1, whether participants viewed region displays (with or without colors or black borders) before viewing point displays did not influence their interpretation of those point displays.

We next examined the seven trials specifically designed to test for emergent-cluster effects. These trials placed two of the three comparison document points within a cluster of points in the display. Figure 3a shows an example where the direct distance between A:1 is much less than between A:2, but A and 2 are inside a cluster of points. Participants gave this a mean rating of 6.4, indicating significantly greater similarity of A:2 than A:1. On all of the trials where A shared a cluster with either 1 or 2, but not both, the pair of documents sharing the cluster were rated as significantly more similar, even when they were not as close together. As Figure 3b shows, this emergent cluster effect occurred even when the cluster was formed by a fairly sparse concentration of points.

The final four trials were designed to test for emergent-feature effects of a different kind, namely emergent linear features produced by points intervening between the comparison points. In all four trials, the direct distances between A:1 and A:2 were equal. In one trial, a single point intervened between A and 1. That did not produce an

326   Daniel R. Montello et al.

emergent-feature effect, as participants rated the relative similarity of the two pairs as 4.9. Two intervening points in another trial produced no effect either, being rated as 5.1 (Figure 3c). But both three and four intervening points (four shown in Figure 3d) produced a noticeable emergent linear feature: The mean similarities for the two trials were 3.8 and 3.4, both significantly less than 5.0.

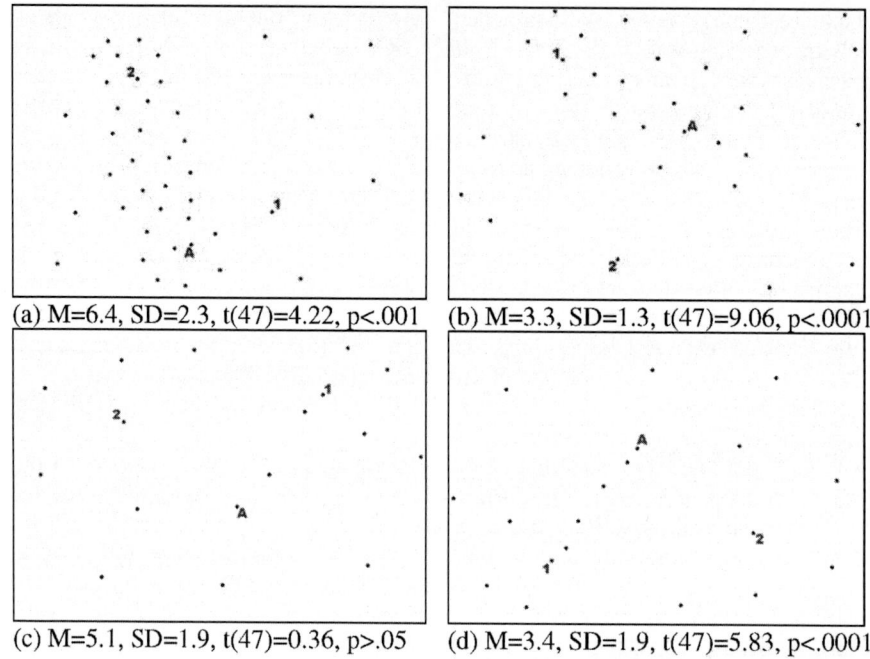

(a) M=6.4, SD=2.3, t(47)=4.22, p<.001   (b) M=3.3, SD=1.3, t(47)=9.06, p<.0001

(c) M=5.1, SD=1.9, t(47)=0.36, p>.05   (d) M=3.4, SD=1.9, t(47)=5.83, p<.0001

Fig. 3. Sample displays and similarity ratings for four trials of Experiment 2. The t-scores are tests of significant differences from 5.0, a neutral rating ("1 and 2 are equally similar to A") on the transformed 1-9 scale (rating <5 indicates A:1 are more similar; >5 indicates A:2 are more similar). Emergent-cluster effects are demonstrated in (a) and (b). An emergent-linear effect does not occur in (c) but does in (d); (M=arithmetic mean, SD=standard deviation, t=t-score [degrees of freedom], p=probability difference is due to chance)

Participants took a mean of 8.5 seconds (SD = 3.8 s) to respond to a point-display trial in this experiment (the fastest response was 3.3 s, the slowest was 21.5 s). This is 3-4 seconds faster than in Experiment 1. Female and male participants answered nearly equally quickly, F(1, 46) = 0.07, p>.05. As in Experiment 1, response times varied significantly as a function of the order in which participants saw the block of point-display trials, F(3, 44) = 6.12. Participants who saw point-display trials first were slowest (11.9 s), followed by those who saw them second (8.1 s), while those who saw them third or fourth were nearly identically fast in their responses (6.9 and 7.0 s, respectively). This fits a general pattern in which participants get faster as the experiment progresses, and they get accustomed to the displays and similarity rating task.

In this experiment, participants were assigned to view the displays at one of two sizes, Small or Large. Mean similarity ratings averaged over all trials did not significantly differ (4.0 for Small, 3.9 for Large), nor did mean response-times (8.3 s for

Small, 8.6 s for Large). However, display size did influence the strength of the distance-similarity relationship on the five trials that varied direct distance systematically. For these trials, participants viewing Small displays had a mean distance-similarity correlation of .52, while those viewing Large displays had a mean distance-similarity correlation of .83. This difference is statistically significant, $F(1, 44)=4.93$, $p<.05$. Display size influenced the operation of the distance-similarity metaphor; larger images strengthened the correspondence between direct distance and perceived similarity. Display size apparently did not influence the emergence of clusters or linear features; only one of the 11 trials that tested these effects revealed a significant difference in similarity ratings for the two viewing conditions (which is not unlikely, even given only chance variation).

Finally, as in Experiment 1, we examined self-reports of how participants interpreted the point displays. In response to the post-test question "How much did you consider the distance between documents in your answers?," participants marked a mean response of 2.9 on an expanded 5-point scale that included "not at all," "slightly," "somewhat," "very much," and "exclusively." We again correlated participants' self reports with their correlation of direct distance and similarity for the five appropriate trials. This correlation was only .16, a weaker relationship than in Experiment 1 and statistically nonsignificant. However, in response to the post-test question "How much did you consider the presence of documents grouped or clustered around the documents A, 1 and A, 2?," participants marked a mean response of 3.3 on the 5-point scale; these responses correlated .57 ($p<.0001$) with the similarity rating given to the trial shown in Figure 3c, a trial that led to a strong emergent cluster. Similarly, in response to the post-test question "How much did you consider the presence of intervening documents between documents A, 1 and A, 2?," participants marked a mean response of 2.7 on the 5-point scale (one participant did not answer). Responses to this question correlated -.35 ($p<.05$) with the similarity rating given to the trial shown in Figure 3d, a trial that led to a strong emergent linear feature. Both correlations of post-test self-report questions with similarity ratings are in the direction indicating that participants were able to judge their own interpretation strategies with some accuracy.

4   General Discussion and Conclusions

The two experiments reported here investigate how users interpret point-display spatializations to infer similarity relationships. These displays represent information-bearing entities, such as library documents, as a collection of points. The spatial and graphical relations among points are intended metaphorically to represent the semantic similarities among the entities. Probably the most common way this has been done is to locate points so that more similar points are closer together. We have dubbed the notion that people will think closer things are more similar the *first law of cognitive geography,* by analogy to the well-known *first law of geography.* In the context of information displays, this law operates as the *distance-similarity metaphor,* wherein viewers of information displays equate direct (straight-line) distance with the dissimilarity between the reference entities supposed to be represented by the graphical elements.

The results of our two experiments provide qualified support for the validity of the first law of cognitive geography and its operation as the distance-similarity metaphor in the interpretation of information spatializations. When people view point representations of documents that appear against a fairly uniform background, either blank or evenly filled with other points, they equate metric direct distance to semantic similarity. This relationship is approximately linear over the ranges we varied in these experiments, though we did uncover evidence of a threshold at which further exaggeration of the contrasts between two comparison distances will not further enhance similarity contrasts. In cases such as these where people apply the distance-similarity metaphor, graphical characteristics that produce phenomena in which perceived distance varies from actual distance will influence judged similarity. An example of such a phenomenon is the vertical illusion, in which vertically-oriented linear features will be perceived as relatively longer than horizontally-oriented features.

However, when the background context against which people view comparison points is not uniform, having uneven groupings of points, visual organization of the points can negate the normal operation of the distance-similarity metaphor. A higher-density collection of points against a sparser background of points is seen as a cluster; if two comparison document points are seen to be inside this cluster, they are seen as more similar, part of a common feature, even if another comparison point is in fact much closer in direct distance. Similarly, a string of points against a background of unorganized points is seen as a linear feature; if two comparison document points are part of this linear feature, they are seen as more similar, again part of a common feature (we showed in Experiment 2 that a set of at least three intervening points, a string of five points, is required to produce the perception of a line). We call these *emergent-feature effects*. They can specifically override the distance-similarity metaphor, leading to a *feature-similarity metaphor* wherein elements that are part of the same feature share greater similarity. In terms of the mechanism of the emergent-feature effect, it is interesting to note that the perceived cluster or linear feature is a phenomenon of visual perception as the Gestalt psychologists described [e.g., 31]. The fact that people viewing the display assign greater similarity to points that are part of the same feature may reflect higher-level *conceptual* organization, however. In particular, such an interpretation by users is consistent with an expression of categorical (particularly regional) cognition [see 32, 33, 34].

Another perceptual phenomenon that may be expected to influence the interpretation of graphical displays is the *filled-interval illusion,* wherein the interval separating comparison points has other points within it and thus appears longer. Either of our emergent features, clusters or lines, might be expected to produce this effect; although we do not yet have data substantiating it, a look at the sample displays in Figures 1 and 2 suggests that this does happen (in 2d, A:2 looks further apart than A:1, even though they are equally distant). It is important to note that the implication of the filled-interval illusion for similarity judgments should be exactly opposite from the implications of the emergent-feature effect. Thus, concentrations of points may produce the appearance of greater distance separating points, but they apparently also lead to the impression of greater similarity, demonstrating a clear dissociation of distance and similarity relationships.

In Experiment 2, we found a previously unreported effect of display size on the relationship of judged similarities to relative distances between the comparison points. For baseline trials where direct distance determined participants' judgments of similarity, the distance-similarity metaphor operated more strongly for participants who viewed the Large display than for those who viewed the Small display; the mean correlation of distance ratios with similarity was a considerable .52 for the Small display but a considerably larger .83 for the Large display. This difference occurred even though the size difference between Small and Large was not as great as it could be—the Large display was just three times wider and filled less than twice the width of the visual field than did the Small display. The result leaves open several interesting questions about its precise mechanism. Is it the difference in sizes between the displayed images that matters (distal size), or the difference between the portions of the visual field filled by the images (proximal size)? Does the larger display exaggerate the perceived distance differences between the points, or does it simply increase the strength of correspondence between relative distances and similarities? These questions call for further research. In any case, this constitutes a novel empirical demonstration that scale matters to the psychology of spatial information [30], and it potentially has important implications for information visualization and communication.

As noted above, the two experiments discussed in this paper included displays based on other spatialization metaphors, specifically networks and regions. In these experiments, along with a third that included only network displays, we have examined the operation of spatial metaphors used to represent similarity relationships. These experiments have also examined the possible role of several graphical variables that are not spatial *per se,* namely various aspects of color (hue, value, and saturation) and line thickness. These results will appear in a forthcoming paper. They are interesting because they demonstrate how nonspatial (nongeometric) aspects can influence the interpretation of information displays. They are also interesting because they allow a comparison of different types of distance other than direct metric distance, such as network distance, that may stand for similarity in information displays.

We are currently finishing a fourth experiment that addresses the distance-similarity metaphor directly by having participants estimate distance instead of just similarity. In this experiment, as in our other experiments, three groups of participants view point, network, and region displays. One group of participants is randomly assigned to rate similarity as in the experiments described in this paper; because they are not told what to base those similarity judgments on, we call this "default similarity." A second group of participants is randomly assigned to estimate distances between points; because they in turn are not told what type of distance to estimate, we call this "default distance." The third group of participants is also randomly assigned to estimate distances between points, but in this condition, they are told to estimate direct or straight-line distance; we call this "direct distance." A comparison of ratings in these three conditions will address some of the issues brought up by the research in this paper, such as the possibility that emergent features can increase perceived similarity among documents at the same time they increase perceived distance •a clear exception to the first law of cognitive geography. These results should help us contribute further to the newly-developing area of scientific research on the use of information spatializations. Research in this area will help address basic questions about the cognitive processing of information displays and help inform the design of human-information interfaces that are more effective, efficient, accessible, and enjoyable to use.

Acknowledgements

We thank Amy Linker and Heather Alexander for help with data collection, and the UCSB students who provided data by serving as research participants in the experiments. We also thank David Mark and Corinne Jörgensen, who have collaborated on this research program since its inception. Three anonymous reviewers provided feedback that improved the paper. The National Imagery and Mapping Agency generously provided funding to support this research.

References

1. International Cartographic Association (ed.): Multilingual Dictionary of Technical Terms in Cartography. ICA, Stuttgart (1973)
2. Card, S.K., Mackinlay, J.D., Shneiderman, B.: Readings in Information Visualization: Using Vision to Think. Morgan Kaufmann, San Francisco (1999)
3. Spence, R.: Information Visualization. Addison Wesley, Boston (2001)
4. Fabrikant, S.I.: Evaluating the Usability of the Scale Metaphor for Querying Semantic Spaces. In: Montello, D.R. (ed.): Spatial Information Theory: Foundations of Geographic Information Science. Lecture Notes in Computer Science, Vol. 2205. Springer, Berlin (2001) 156-172
5. MacEachren, A.M.: How Maps Work: Representation, Visualization, and Design. Guilford Press, New York (1995)
6. Slocum, T.A., Blok, C., Jiang, B., Koussoulakou, A., Montello, D.R., Fuhrmann, S., Hedley, N.R.: Cognitive and Usability Issues in Geovisualization. Cart. & Geog. Infor. Sci. 28 (2001) 61-75
7. Couclelis, H.: Worlds of Information: The Geographic Metaphor in the Visualization of Complex Information. Cart. & Geog. Infor. Sys. 25 (1998) 209-220
8. Kuhn, W., Blumenthal, B.: Spatialization: Spatial Metaphors for User Interfaces. Depart. Geoinfor., Tech. Univ. Vienna (1996)
9. Fabrikant, S.I., Buttenfield, B.P.: Formalizing Semantic Spaces For Information Access. Annals Assoc. Amer. Geog. 91 (2001) 263-280
10. Shepard, R.N.: Representation of Structure in Similarity Data—Problems and Prospects. Psychometrika 39 (1974) 373-422
11. Skupin, A.: From Metaphor to Method: Cartographic Perspectives on Information Visualization. In: IEEE Symposium on Information Visualization, InfoVis 2000, Salt Lake City, UT (2000) 91-97
12. `Tobler, W.R.: A Computer Movie Simulating Urban Growth in the Detroit Region. Econ. Geog. 46 (1970) 234-240
13. Fabrikant, S.I.: Spatial Metaphors for Browsing Large Data Archives. Unpublished Ph.D. Diss., Depart. Geog., Univ. Colorado-Boulder (2000)
14. Chalmers, M.: Using a Landscape Metaphor to Represent a Corpus of Documents. In: Frank, A.U., Campari, I. (eds.): Spatial Information Theory: A Theoretical Basis for GIS. Lecture Notes in Computer Science, Vol. 716. Springer-Verlag, Berlin (1993) 377-390
15. Wise, T.A.: The Ecological Approach to Text Visualization. J. Amer. Soc. Infor. Sci. 53 (1999) 1224-1233
16. Dieberger, A., Frank, A.U.: A City Metaphor for Supporting Navigation in Complex Information Spaces. J. Visual Lang. Comp. 9 (1998) 597-622
17. Berendt, B., Jansen-Osmann, P.: Feature Accumulation and Route Structuring in Distance Estimations - An Interdisciplinary Approach. In: Hirtle, S.C., Frank, A.U. (eds.): Spatial Information Theory: A Theoretical Basis for GIS. Lecture Notes in Computer Science, Vol. 1329. Springer-Verlag, Berlin (1997) 279-296

18. Hartley, A.A.: Mental Measurement in the Magnitude Estimation of Length. J. Exp. Psych.: Human Perc. Perf. 3 (1977) 622-628
19. Montello, D.R.: The Perception and Cognition of Environmental Distance: Direct Sources of Information. In: Hirtle, S.C., Frank, A.U. (eds.): Spatial Information Theory: A Theoretical Basis for GIS. Lecture Notes in Computer Science, Vol. 1329. Springer-Verlag, Berlin (1997) 297-311
20. Goldstone, R.L.: Similarity, Interactive Activation, and Mapping. J. Exp. Psych.: Learn., Mem., Cog. 20 (1994) 3-28
21. Medin, D.L., Goldstone, R.L., Gentner, D.: Respects for Similarity. Psych. Rev. 100 (1993) 254-278
22. Tversky, A.: Features of Similarity. Psych. Rev. 84 (1977) 327-352
23. Buffardi, L.: Factors Affecting the Filled-Duration Illusion in the Auditory, Tactual, and Visual Modalities. Perc. & Psychophys. 10 (1971) 292-294.
24. Thorndyke, P.W.: Distance Estimation from Cognitive Maps. Cog. Psych. 13 (1981) 526-550
25. Amstrong, L., Marks, L.E.: Differential Effects of Stimulus Context on Perceived Length: Implications for the Horizontal-Vertical Illusion. Perc. & Psychophys. 59 (1997) 1200-1213
26. Gregory, R.L.: Eye and Brain: The Psychology of Seeing (3rd edn.). McGraw-Hill, New York (1978)
27. Goldstein, E.B.: Sensation and Perception (3rd edn.). Wadsworth, Belmont, CA (1989)
28. Gregory, R.L. (ed.): The Oxford Companion to the Mind. Oxford University Press (1987)
29. Cohen, J., Cohen, P.: Applied Multiple Regression/Correlation Analysis for the Behavioral Sciences. Lawrence Erlbaum Ass., Hillsdale, NJ (1975)
30. Montello, D.R.: Scale and Multiple Psychologies of Space. In: Frank, A.U., Campari, I. (eds.): Spatial Information Theory: A Theoretical Basis for GIS. Lecture Notes in Computer Science, Vol. 716. Springer-Verlag, Berlin (1993) 312-321
31. Masin, S.C.: Absolute and Relative Effects of Similarity and Distance on Grouping. Perc. 31(2002) 799-811
32. Friedman, A., Brown, N.R.: Reasoning about Geography. J. Exper. Psych.: Gen. 129 (2000) 193-219
33. Hirtle, S.C., Jonides, J.: Evidence of Hierarchies in Cognitive Maps. Mem. & Cog. 13 (1985) 208-217
34. Huttenlocher, J., Hedges, L.V., Duncan, S.: Categories and Particulars: Prototype Effects in Estimating Spatial Location. Psych. Rev. 98 (1991) 352-376

Constructing Semantically Scalable Cognitive Spaces

William Pike and Mark Gahegan

GeoVISTA Center, Department of Geography, Pennsylvania State University
302 Walker Building, University Park, PA 16802 USA
{wpike,mng1}@psu.edu

Abstract. This paper describes a technique for creating generalizable depictions of cognitive spaces from natural language documents, and presents a Web-based system that uses this procedure to visualize structure in geographic discourse. We implement a concept abstraction routine that leverages a lexical ontology to infer the semantics of discussion terms at increasing levels of generalization. A Web discussion medium that uses the Delphi method to guide geographic discourse serves as the framework from which concept structures are elicited. Delphi discussants explore these structures using two Web-enabled visualization schemes: Self-Organizing Maps and concept graphs. These visualization tools rely on a set of concept similarity measures tailored to conceptual information at multiple levels of abstraction. The cognitive spaces produced using this system can reveal key themes in a domain, and can help guide the creation of domain ontologies. We apply these tools to explore concept structures in the field of human-environment interaction.

1   Introduction

Geographic concepts are often constructed and refined through dialogical inquiry within a community of experts. Such inquiry takes the form of a discourse that may occur over time scales of minutes (a brief discussion with a peer) to years (contributions to journals or conferences). Regardless of the temporal scale, discourse is vital to the growth of a domain community, the identification of key research themes within it, and the progress of its collective thinking. An understanding of the structure of a domain – together with the concepts and people that constitute it (and the relationships among these constituents) – can benefit from examination of the discourse that occurs within the community that defines it. This paper describes computational aids to discourse analysis that can reveal the evolution of a domain's structure as reflected by the interactions among its members, with emphasis on the field of human-environment relations. The contributions of the research presented here are twofold: in a practical sense, this work helps participants in online discussions visualize cognitive spaces that reveal key human-environment themes; this work also contributes to the theory supporting the creation of visual summaries of concept relationships and the construction of domain ontologies.

The structure of a domain inferred through analysis of discourse has both temporal and spatial components. Temporal structure refers to the evolution of domain concepts over time (including changes in the sub-concepts that contribute to them). Spatial structure is defined by the relationships between concepts in some n-dimensional

W. Kuhn, M.F. Worboys, and S. Timpf (Eds.): COSIT 2003, LNCS 2825, pp. 332–348, 2003.
© Springer-Verlag Berlin Heidelberg 2003

concept space bounded by the range of concepts involved in discussion. Together, these structures help address one of the primary research needs of geographic information science: the creation of ontological relationships that can underpin a semantics of geospatial information. Ontologies can support interoperability of geographic concepts, but significant problems remain (1) how to generate these ontologies for particular domains, applications, or tasks [1], (2) how to capture ontological evolution over time, and (3) how to reflect the roles of different researchers in the creation and application of concepts. This research presents one approach to these problems, wherein core concepts across several levels of generalization are abstracted from the discourse that already occurs within a community.

This approach to constructing concepts and relationships has several advantages. First, instead of imposing new (and possibly foreign) ontology management tools on a panel of domain experts, we attempt to integrate the collection of conceptual information into existing collaborative tools. Second, since discussions are by nature collaborative; we are able to examine not only concepts as individuals see them, but to aggregate individual ideas together to detect points of group agreement and disagreement. Lastly, ongoing discussions are an effective way to tap into the evolution of ideas and can help experts maintain ontologies. As discussion evolves, it should be possible to detect when certain concepts or relationships have been modified, superceded, or used in different contexts. The "ontologies" that result from analysis of discourse may not be robust enough to be immediately applicable to all problems of data or knowledge interoperability, but they are a valuable first step toward creating more rigorously defined ontologies.

Here, we describe the implementation of a Web-based system that allows teams of collaborators to collectively define and explore cognitive spaces through discussion. Our Web tools afford the depiction of concept relationships across semantic scales – from the most specific instantiation of a concept to its most general representation – using two visualization techniques: Self-Organizing Maps and concept graphs. Together, these tools enable discussants to navigate evolving cognitive spaces, and we illustrate their application to the field of human-environment interaction – a domain marked by diverse, and often conflicting, perspectives. This Web system demonstrates an early attempt to deploy concept visualization tools in a real-world, real-time setting.

2   Background

When attempting to represent the cognitive spaces inherent in a geographic domain, an initial concern is selecting a suitable corpus from which to draw domain information. Such a corpus could include transcripts of face-to-face meetings, email messages, printed articles, quantitative data sets, or any combination of these sources. However, source characteristics make each more or less suitable to the elucidation of robust conceptual relationships. For example, real-time meetings often do not persist long enough to demonstrate well-founded concept structures or evolving concept trajectories, nor may they include sufficient context to aid interpretation. Visualization or classification tools that have been applied to such discussion (e.g., [2]) are often used *post hoc*, showing summary views of a concept space at the end of a meeting. At the other extreme of temporal scale and community size are domain discourses drawn from printed sources such as journal or news articles. In this case, the corpus may contain many hundreds or thousands of documents, each with a rich concept set and contextualized arguments, and can reflect changing themes over the course of years or even centuries. While such documents may include references to other works that help infer concept relationships, the documents themselves are generally not structured with an evolving discourse in mind.

Here, we use the Delphi method [3] as a discourse framework that balances the benefits of both interpersonal communication and longer-term scientific discourse. The Delphi method describes an approach to group collaboration that fosters the exploration and distillation of expert opinion, and simply consists of a set of guidelines for structuring group communication. The problems to which the method is applied are generally complex and lack a deterministic or readily agreed-upon solution. The method allows individuals to express and defend their beliefs about possible solutions with the aim of generating a body of expert opinion that bounds the breadth and depth of a problem space.

This research develops and implements a Web-based Delphi tool (e-Delphi, further information and access available at http://hero.geog.psu.edu/eDelphi) that enables panels of experts to collectively explore a domain. Our e-Delphi system is extended with text processing, concept capture, and visualization routines that assist in the construction and exploration of cognitive spaces. Because the Delphi process is highly structured and organized around exploratory discourse, it can be an effective method for knowledge elicitation and acquisition. Moreover, Delphi activities complement qualitative discussion with quantitative voting to express support for or disagreement with particular concepts. Our e-Delphi system was designed for use in the Intelligent Networking Environment program within the Human-Environment Regional Observatory (HERO; http://hero.geog.psu.edu) project.

In addition to research on computer-mediated communication, this work draws from research streams in natural language processing, information visualization, and geographic ontology creation. Many attempts to extract concept relationships from text commonly rely on techniques such as latent semantic indexing that operate on matrices of all terms that appear in any document. A data reduction procedure, often singular value decomposition, is then applied to prune infrequent concepts and relationships. In an alternative approach, Fabrikant and Buttenfield [4] use document keywords alone to visualize the relationships among geoscience journal articles. In cases where keywords do not exist – such as in online e-Delphi discussions – there must be another means of discovering the concepts contained in text. Miller et al. [5] apply wavelet transforms to full-text news articles to detect and visualize associations between key terms as they appear in the text. Prior work has also addressed the need to reveal changes in concept occurrence and strength over time [6], producing linear depictions of term importance (interpreted through frequency of occurrence) over the period covered by the corpus. Feldman et al. [7] also account for temporal affects in text mining, producing trend graphs that help reveal flows and discontinuities in documents. Language processing theory has also been extended to handle concept structures unique to discourse [8]. The current research contributes to this body of work by integrating the synthesis and depiction of concept relationships into the discourse framework in which they are created, thereby allowing visualized cognitive spaces to play a role in the continuing development of concept structures. In addition, we support navigation of cognitive spaces at levels of abstraction other than those which the text passages themselves consider.

Prior work in creating geospatial ontologies has considered how individuals conceive of geographic relationships [9] and has addressed the role of semantic scale in these relationships [10], [11]. Further, Rodriguez and Egenhofer [12] have dealt with issues of semantic similarity across concept representations at different levels of specificity. We extend these studies by examining how such ontological structures can be inferred from evolving discourse among human-environment geographers, although our results should have applicability to other fields of science.

3   Abstracting Concepts from Geographic Discourse

There are two broad approaches to retrieving concepts from text documents. The first relies on latent statistical associations between terms in a corpus, treating each term as equal in importance (but generally excluding a stoplist of terms such as prepositions and conjunctions). This approach requires the least sophistication, as it does not consider factors like part of speech, interaction with nearby terms, or synonymy to determine a term's role in a text passage, and it does not address the *meaning* of individual terms. As a brute-force approach, it is capable of reflecting the most frequent terms (and usually in their most frequent forms) and is often useful for rapid assessment of overall concept patterns. The alternative approach to text analysis derives from natural language processing, and aims to make use of known linguistic patterns and well-defined term semantics to infer the meaning of each term in a corpus. Unlike latent analysis, which is statistical and thus essentially language-independent, natural language processing requires language-specific rules and dictionaries. Often, a goal of natural language processing is to translate natural language expressions into logical predicates, although this task is well beyond the needs of the current study. Here, it is sufficient to simply compute the most important concepts – not just important words – in a text passage. The process of retrieving these concepts involves two steps: *extraction* and *abstraction*. Figure 1 outlines the procedure for creating cognitive spaces discussed in the following sections. In our e-Delphi implementation, each of these components operates as part of a server-side Web application that manages discussion and processes contributions. A small Java applet that runs through a standard Web browser augments e-Delphi discussions with the visualization capabilities discussed later.

3.1   Extraction

The first stage in creating cognitive spaces from Delphi discussions is to extract terms likely to be the main points in each contribution. Since these passages can range in length from a few words to a few paragraphs, the number of terms extracted for each will vary. A term extraction component first parses each contribution into sentences and words; stopwords and internal punctuation are not immediately excluded, as they are useful for inferring the relationship between nearby words. Using a Java interface to the WordNet lexical ontology [13], each term is tagged with its corresponding parts of speech and a morphological processing routine resolves a standard lemma for each term (e.g., "beautify" is the lemma for "beautified").

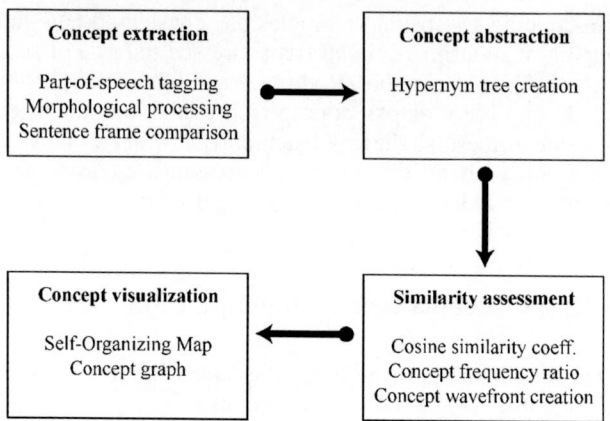

Fig. 1. Components of Web-based cognitive space creation system

After each term has been tagged and lemmatized, sentences are compared against sentence frames for standard English usage. These frames help the extraction module determine which terms are key concepts and which are of lesser importance. Sample frames are of the form "noun-preposition-noun" or "verb-preposition-{definite article | indefinite article}-noun." Comparison against frames is an attempt at sense disambiguation, so that polysemous senses of a word (such as "land") are not conflated; in some instances, "land" may refer to part of the earth's surface, while in others it may refer to the act of bringing an aircraft to the ground. Maintaining a distinction between a term's senses is vital to computing relationships between concepts in cognitive space – "land" as a noun might be closely related to "water" as a noun, but "land" as verb is not.

To illustrate the extraction process, consider the following sentence as input to the extraction module:

```
Farms are particularly vulnerable to drought
```

After lemmatization, the sentence looks like this:

```
farm, be, particularly, vulnerable, [preposition],
drought
```

The parts of speech for each token are then mapped:

```
{noun|verb}, {noun|verb}, adverb, adjective, [preposi-
tion], noun
```

In this example, two terms, "farm" and "be," are ambiguous. "Farm" can refer to a geographic place or to the act of cultivation, while the noun form of "be" is the chemical symbol for Beryllium. By comparing the various possibilities against sentence frames, we determine that both "noun-noun-adverb" and "verb-verb-adverb" are unlikely, as is "verb-noun-adverb". Thus, the sense disambiguation method chooses the following sentence form as the correct one:

```
Noun, verb, adverb, adjective, [preposition], noun
```

The next task is to determine the key concepts in this sentence. With the same sentence frames used to disambiguate the original input, it is possible to map the subject, predicate, and object of a phrase and eliminate those words that are not part of the core meaning of the passage. In our current implementation, the assessment of "important" words is done conservatively, such that more terms are included than are perhaps necessary. This cautiousness arises from a hermeneutical concern over interpreting what each Delphi participant might have intended a word to mean. Consider again the original sentence:

```
Farms are particularly vulnerable to drought
```

This statement could have been made in response to one or more alternative ideas (stated elsewhere in the discourse or not), each of which would place emphasis on a different term in the original sentence. We might contrast the original sentence against

```
Cities are particularly vulnerable to drought
```

in which case the most important term in the original sentence might be "farms" (in contrast to "cities"). Another possible contrast is

```
Farms are particularly immune from drought
```

in which case the emphasis should be placed on "vulnerable" in contrast to "immune." Finally, in the case of

```
Farms are particularly vulnerable to floods
```

the emphasized concept would be "drought" in contrast to "floods." Since it is a difficult task to determine which of these alternatives is most likely, we instead define all of the concept terms ("farm," "particularly," "vulnerable," and "drought") as key themes in the sentence. The hope is that when the system later looks for overall patterns of important terms, the excess words (e.g., "particularly") will be pruned away.

3.2   Abstraction

Many contemporary text spatialization systems represent the important concepts in a corpus solely through the terms that appear in the text. While specific terms are useful, when constructing a cognitive space it is desirable to explain concepts in more abstract terms. Geographers might describe the objects of their study as "cities," "trees," "families," or "governments," but when using these terms to explain the domain of geography, it can be more instructive to speak of "locations" or "natural phenomena" or "social organizations." Ultimately, users of a concept space should be able to browse it at a range of abstractions to suit their preferences and perspectives.

The goal of text abstraction is simply to create generalized representations of concepts. Text abstraction can also be called semantic generalization or semantic scaling (by semantic scale, we mean a continuum from the most specific expression of a concept to the most general). Just as we might zoom out from a conventional digital map to see a more generalized view of terrain free from the minutiae of individual streets, we ought to be able to zoom out from a cognitive space to apprehend the overall concept patterns free from the minutiae of individual low-level terms.

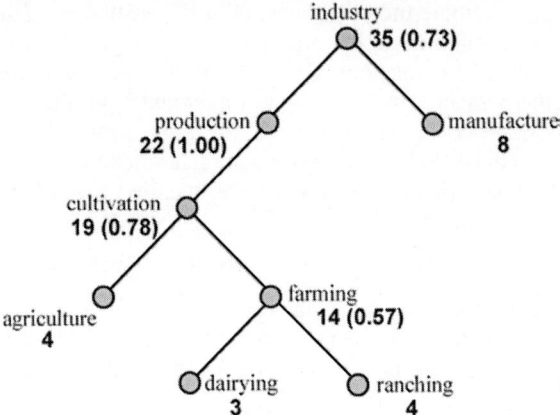

Fig. 2. Hypernym tree for particular senses, derived from WordNet, of terms occurring in a sample e-Delphi activity. "Agriculture" and "farming" are kinds of "cultivation," for example Numbers below each term indicate its weight and, in parentheses, its frequency ratio

To create scalable representations of cognitive spaces, we use WordNet to find hypernyms for the terms extracted from e-Delphi discussions. Hypernym trees describe hierarchies of "kind-of" relationships that link increasingly generalized descriptions of concepts (there are other ways to generalize concepts, for example through meronymy or hyponymy, but these are not treated here). Figure 2 shows the hypernyms for particular senses of the words "farm" and "grassland"; the concepts have a common parent at "location." WordNet is not a geographic ontology, so the classification of a farm as a geographic point and grassland as a geographic area is not necessarily consistent with how e-Delphi contributors may have intended the terms. No ontology will be universally applicable, however, and WordNet's linguistic richness makes it suitable for this application. For each term parsed by the text extraction module, a text abstraction module builds a hypernym structure that traces the term's relationship back to WordNet's top-level categories. Disambiguating a term's part of speech is handled by sentence frames, but even within a part of speech a term may be polysemous. Where a term, such as "production," has multiple senses as a noun, we attempt to find a common hypernym among all the senses. If none can be found, we use the hypernym tree for the most common sense of the word.

4   Concept Similarity Measures

Following text extraction and abstraction, we are left with a set of key concept terms and their hypernyms for each e-Delphi passage. Turning these data structures into representations of cognitive spaces requires a set of concept similarity measures that can map each term's location in concept space and determine its relations to similar terms. This section presents a set of similarity measures that together enable the scalable concept spaces introduced in Section 5.

4.1   Cosine Similarity Coefficient

The familiar cosine similarity measure is based on a vector space model of concept relations that treats each text passage (whether a document or an e-Delphi contribution) as a vector through a multidimensional space defined by all of the terms occurring in the corpus. Each concept constitutes a distinct dimension, with the position of the vector in each dimension given by the weight of that concept in the passage [14]. The cosine of the angle between two passages x and y is defined by:

$$similarity(x, y) = \frac{\sum_{i=1}^{n}(x_i \cdot y_i)}{\sqrt{\sum_{i=1}^{n}(x_i)^2 \cdot \sum_{i=1}^{n}(y_i)^2}} \quad . \tag{1}$$

The weight of each term i in x and y is given by x_i and y_i. Through pairwise comparisons it is possible to determine the relationships among each document in a corpus. In a Delphi activity, it is also possible to group passages together by author/researcher to determine the similarity between participants in concept space. We also extend the definition of cosine similarity to address the semantic scalability afforded by text abstraction; here, each vector represents the concepts in a passage *at a given level of abstraction*. By generalizing concepts to different representations (for instance, from "farm" and "grassland" to "point" and "region," respectively), we create a vector through a more abstract space of slightly lower dimensionality. Thus, there will be a different cosine similarity coefficient between each pair of passages at each level of abstraction.

4.2   Concept Frequency Ratio and Concept Wavefronts

Not every level of generalization in the WordNet hierarchy is of equal expressiveness in a given corpus. In the case of "farm" and "grassland," which can be aggregated under common parents "location" and "physical thing," which abstraction is most useful in a particular context? To find the most informative abstractions, we use concept frequency ratios to create *concept wavefronts* [15]. A concept wavefront is a set of generalized terms at a particular level of abstraction. To compute a concept frequency ratio, we first determine the weight of each concept, at each level of generalization. The weight of a concept C is defined as the sum of a term's frequency in a corpus and the frequency of its children [14] (Equation 2).

$$weight(C) = frequency(C) + \sum_{C_1 \in child(C)} frequency(C_1) \quad . \tag{2}$$

The concept frequency ratio is given by:

$$ratio(C) = \frac{\underset{C_1 \in child(C)}{Max}\ weight(C)}{\sum_{C_1 \in child(C)} weight(C)} \quad . \tag{3}$$

Concept wavefront generation starts from WordNet's top-level categories (e.g., "physical thing") and proceeds toward more specific terms whenever $ratio(C)$ is greater than a predefined threshold. If at any point in the traversal of the hypernym tree $ratio(C)$ falls below this threshold, the wavefront algorithm stops and collects the current term as a *fuser concept* (a fuser concept can be defined as one whose children contribute equally to the concept's weight [15]). The set of fuser concepts for a corpus constitutes a concept wavefront; this wavefront provides a set of informative generalizations, which, in contrast to simply collecting every generalization for a term, contains semantic abstractions appropriate to the domain defined by the corpus.

For selected terms in Figure 2, the corresponding concept frequencies and ratios are shown. "Farming" has a frequency count of 14 – its children occurred seven times in the sample e-Delphi passages, and the term itself appeared seven times. Its frequency ratio of 0.57 indicates that it is a useful generalization of "dairying" and "ranching," terms that contributed approximately equally to its weight. (In general, lower concept frequency ratios indicate useful generalizations.) By contrast, "cultivation" is a less useful generalization, because its children are differentially weighted. Initially, the $ratio(C)$ threshold is set high enough (through experimentation, approximately 0.75) to find a maximally general representation. Once the first wavefront is found (which would include the fuser concept "industry" in Figure 2, which has a frequency ratio of 0.73, and all other concepts in other hypernym trees with similar ratios), the $ratio(C)$ threshold can be lowered and the wavefront generation routine begun again from just those concepts included in the previous wavefront. In this example, we might lower $ratio(C)$ to 0.60, and try to find the next more specific fuser concept, which would be "farming." Through iterative adjustment of the concept frequency ratio, generally moving in increments of 0.15, our e-Delphi tool collects useful generalizations across several levels of abstraction.

5   Visualizing Scalable Cognitive Spaces for Human-Environment Geography

Once extraction, abstraction, and similarity assessment have been performed, it is possible to create visual representations of concept relationships that help reveal spatial and temporal structure in geographic discourse across several levels of semantic generalization. This section describes two visualization schemes, Self-Organizing Maps and concept graphs, which are incorporated into our e-Delphi implementation. We present the results of applying these visualization tools to real e-Delphi discussions by a team of human-environment geographers. Each of the visualization approaches is available to e-Delphi participants as they work with the Web-based tool; users can manipulate the visualizations in their browsers and interact with the e-Delphi activity through them.

5.1   e-Delphi Case Study

The e-Delphi system is used by teams of collaborators to help identify the core elements of a problem, from defining protocols for low-level analysis tasks to outlining the characteristics of higher-order domains. Delphi participants gradually focus their discussion such that patterns of common understanding or disagreement can be revealed over time. The examples in the following sections are drawn from a Delphi exercise undertaken by a team of twelve human-environment geographers. The team aimed to define the core elements of human-environment interaction crucial to understanding the local impacts of environmental change. The original e-Delphi implementation used by these participants was extended with the text processing techniques described above, enabling the construction of cognitive spaces as discussion evolved.

5.2 Self-organizing Maps

The first visualization scheme available through e-Delphi makes use of concept weight values. This approach uses Self-Organizing Maps to reduce a multidimensional concept space, revealing clusters of related concepts as well as overall patterns of conceptual agreement within a discourse. A Self-Organizing Map (SOM) is a type of neural network, consisting of a network of artificial neurons (nodes) each possessing a vector of weights [16],[17]. A two-dimensional (2-D) network is commonly used to find (construct) classifiers and hence provide a continuous topological mapping between the n-dimensional feature space and the 2-D network space. Such a network can perform the task of data reduction or clustering suitable for unsupervised learning.

In training, the weight vectors are initialized randomly, then the numeric input vectors consisting of weights for each concept in the corpus at a given level of abstraction are repeatedly introduced to the network. For each input vector the closest-matching node is located, and its weights are adjusted towards those of the input vector. The amount of adjustment is typically calculated as some function of the n-dimensional Euclidean distance between the two vectors. Adjustments are also made to all nodes within a certain predefined distance.

When performing data reduction or clustering, the behavior and structure of the SOM work together to preserve, where possible, the various proximity relationships inherent within the original data; that is to say, if two clusters are close to each other in the original n-D space, they are likely to be similarly close in the 2-D transformation of that space. Thus, a visualization of the transformed space provides insight into the more complex n-D space. The actual distance between clusters is, however, not a measure of cluster proximity. To see that, we need to visualize the error component associated with each node, calculated by summing the distances between that node and all input data that falls within its neighborhood. In the figures shown below, this value is encoded using a grayscale ranging from dark (most error) to light (least error), so lighter shades between concepts indicate a similarity and darker areas show dissimilarity or isolation. (A Sammon mapping [18] can be used to more accurately preserve the actual distances between nodes).

The SOM tool used in e-Delphi is a component of GeoVISTA *Studio* (http://www.geovistastudio.psu.edu); the SOM component plugs in to the discussion management functionality of e-Delphi, allowing SOMs to be trained on the server and distributed as imagemaps to client browsers. Figure 3 shows a series of snapshots taken of the example e-Delphi activity at three points in its evolution. The terms included within the maps were extracted from the discussion but appear here prior to any abstraction. Figure 3a was produced early in the discussion and shows only a few obvious structures; labels that suggest a likely category for these four clusters have been applied manually. Figure 3b shows the state of the discussion after approximately one week; the *ecology* and *climate trends* clusters have been replaced by a set of concepts related to *environmental perception*, while *demographics* and *governance* remain significant themes. Figure 3c was produced at the conclusion of the Delphi discussion and suggests that the discourse had begun to focus around three key areas: *demographics, governance,* and *land development.* While prior work (e.g., [19]) has applied the SOM to relationships between Web documents, it does not attempt to capture the evolution of concept spaces that result. Moreover, rather than presenting a single static depiction of a text corpus, our tool continually updates these maps – live on the Web – in response to participant contributions.

It is also possible to use the SOM to display relationships between participants in a Delphi activity; Figure 4a-b shows a series of snapshots taken at the beginning and end of the Delphi discussion, with the input vectors consisting of the weights for each concept discussed by participants (whose names have been changed here). Because the SOM does not preserve cluster proximity, these maps are most useful for assessing the overall patterns of agreement between participants. In Figure 4a, the dark bands between each participant's cluster indicate significant conceptual distance, which is expected at the early stages of discourse. After several weeks (Figure 4b), the dark bands have begun to fade, indicating that conceptual distances have decreased. Only one participant ("Hank") maintained stark differences from the others, suggesting that he tended to discuss an idiosyncratic set of concepts. An alternative to the SOM that affords visualization of the proximity between individual participants is shown in Figure 4c. This graph structure foviates initially on the current user (in this case, "Violet" has logged in, so her node is labeled "Me") and shows the conceptual distance to other participants based on cosine similarity values. The display can be refocused on other users, allowing discussants to interpret the discourse from another's perspective.

While the SOM representations are useful for depicting synoptic patterns of low-level terms in concept space, they do not excel when used to simultaneously represent concept relationships at different levels of abstraction. Because the input data to the SOM change with the level of generalization, the map topology differs with semantic scale. Moreover, each map can only display one level of generalization at a time. One immediate effect of this shortcoming is that in our implementation, cluster labels – representing a sort of hypernym – must be inferred by the map user, and cannot be applied automatically by the SOM.

5.3   Concept Graphs

In response to the shortcomings of the SOM to display semantic scalability, an alternative visualization scheme based on concept graphs [20] is available through the e-Delphi system. Whereas the SOM excelled at detecting clusters of low-level terms, concept graphs exploit the full set of e-Delphi's text abstraction techniques and concept similarity measures introduced above, helping to reveal the more general themes in a discourse.

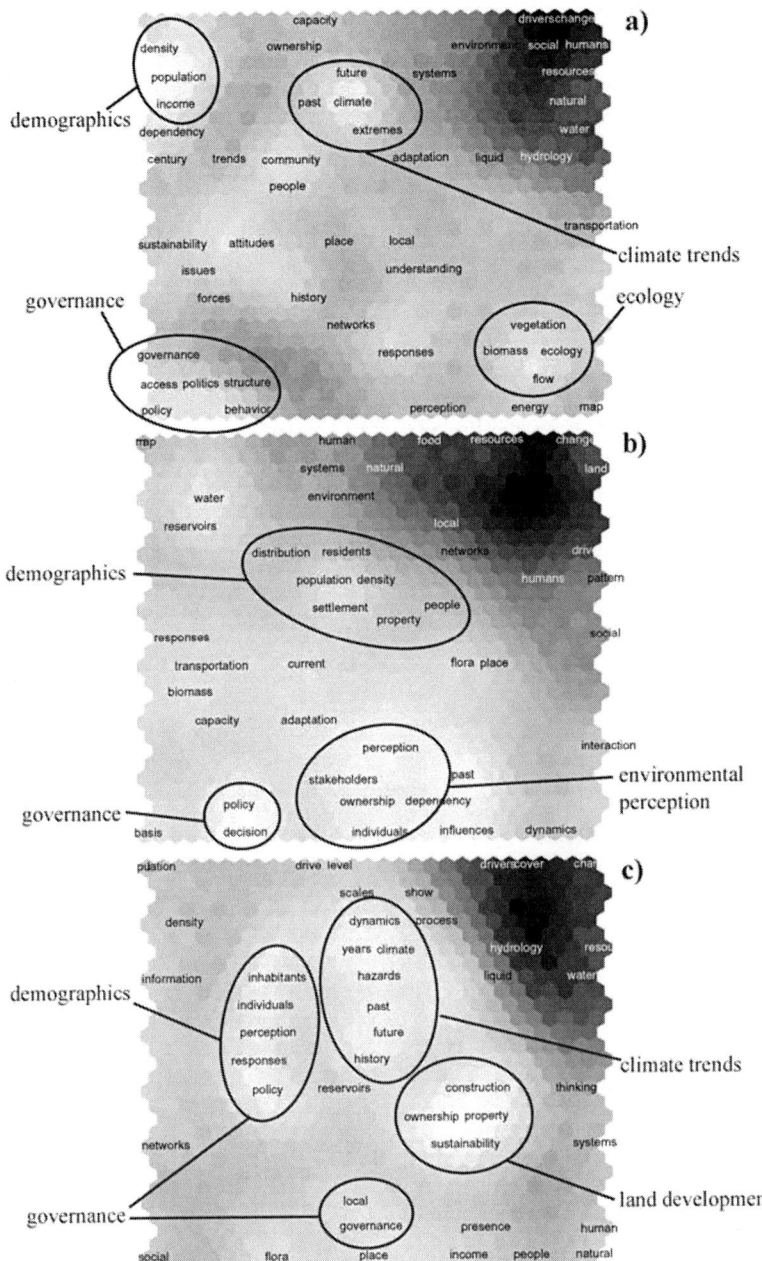

Fig. 3. Clusters of key terms in SOM of Delphi discussion on human-environment interaction. **a)** After 1 day of discussion; **b)** After 1 week of discussion; **c)** After 4 weeks of discussion

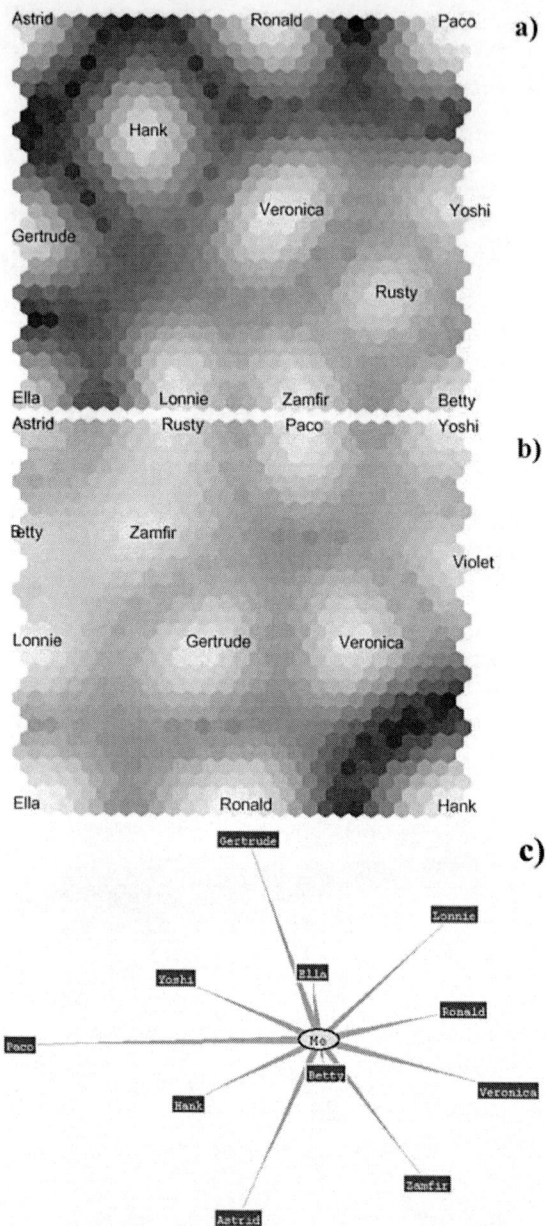

Fig. 4. a) SOM of participant relationships after 1 week of discussion. **b)** After 4 weeks of discussion. **c)** Graph of participants after 1 week of discussion, using cosine similarity to indicate conceptual distance between discussants

Figure 5 shows a concept graph that displays a portion of the WordNet hypernym tree for all terms in the discussion after one week's time. Such graphs provide a top-down view on the concept space that emerges from a discourse, in contrast to the bottom-up, term-level SOM approach. In Figure 5, the hypernym structure for the Delphi corpus contains six top-level concepts that together constitute the "universe" of concepts involved in the discourse at a given point in time. A user can expand each concept branch to drill down to ever more specific representations. Two top-level concepts, "Physical thing" and "Phenomenon," have been expanded to show some of the child concepts from which they were generated. An e-Delphi user can open such a graph in a popup browser window adjacent to the main e-Delphi interface, and use the graph to summarize and navigate the concept space that results from the present discussion. Clicking on a node in the graph queries the e-Delphi system for all text passages that contain that concept; clicking on "land," for instance, will display all passages that might discuss land use or land cover, or any concept for which land is a hypernym. Clicking on "physical object" will display the passages containing "land" as well as those containing concepts under the "living thing" branch. Browsing a concept structure in this manner allows users to focus on the concepts important to them (e.g., the branch for "Human activity" can be hidden) while considering their relations at whatever level of abstraction they feel appropriate (e.g., stopping the expansion at "Body of water" instead of dealing with the individual terms such as "Lake" or "Pond" that make up this concept). Semantic scalability is achieved as the system instantly redraws a concept graph to reflect both parents and children of the concepts on which a user foviates. Arcs can be redrawn to connect nodes at a given level of generaliz tion, with the length of each arc determined by the cosine similarity value for that pair; this technique provides a visual cue to the strength of associations between terms across semantic scales.

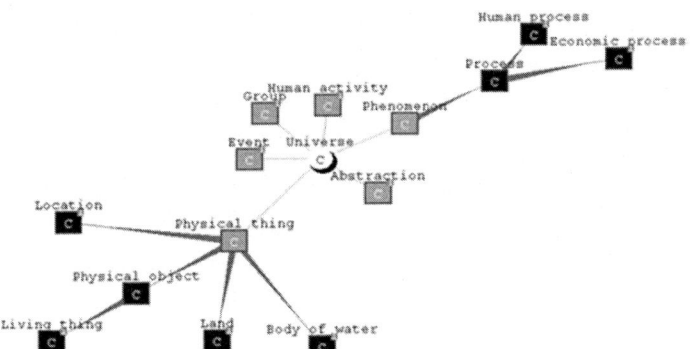

Fig. 5. Concept graph used for online navigation of e-Delphi concept spaces, showing hypernym relationships between concepts. Top level concepts (gray) surround the "Universe" node; more specific expressions (black) extent to the upper left and lower right. In the online implementation, clicking on a node displays the discussion passages that relate to each concept

In Figure 6, the concept wavefront algorithm has been applied, resulting in a graph that displays only the most informative fuser concepts for a discussion, rather than every entry in the hypernym tree. This visualization approach helps Delphi participants quickly navigate only the key themes in a changing discourse. A wavefront generally appears as a ring of nodes around the perimeter of the graph. In this example, the first (most general) concept wavefront the system detected beyond the top level is displayed for the user (dark nodes). (The light nodes represent the top-level concepts and have been grayed out to emphasize the wavefront). The wavefront display also shows the relative ranking of these themes within the discourse by scaling the size of the nodes according to the support inferred for each concept through a combination of its frequency in the text and the votes it received in Delphi balloting.

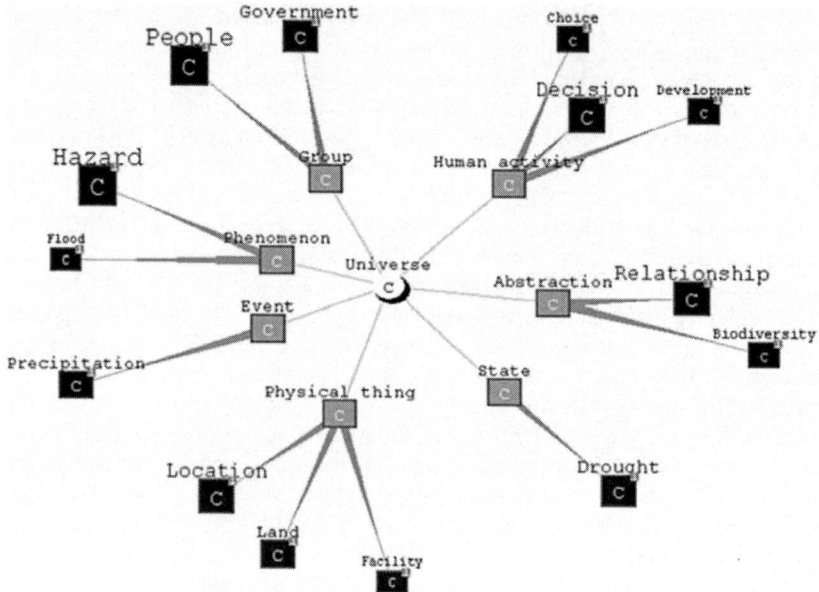

Fig. 6. Concept wavefront (dark nodes) surrounding top-level concepts Each wavefront node can be expanded to display the next more specific wavefront

The wavefront depiction is a method of generating ontologies of core concepts for other applications, skipping over less meaningful generalizations. In this example, one could define a tentative ontology for human-environment relations based on the terms contained in the first concept wavefront, enriching the hierarchy with ever more specific terms derived from subsequent wavefronts.

Currently, analysis of temporal structure in e-Delphi discussions is limited to qualitative comparison of SOM snapshots or concept graphs created at different points in time. Users can ask the e-Delphi system to produce a SOM or graph to represent the discussion at any point in its history, and can use these historical visualizations to assess the evolution of concept spaces. While the degree of temporal change is not quantified, such quantification is difficult and the benefits to users are at present unclear.

6   Conclusions

This paper describes a technique for creating semantically generalizable cognitive spaces from natural language documents and presents a Web-based system that employs this procedure to visualize structure in geographic discourse in real-time. Concept extraction and abstraction techniques are developed to elicit key themes from discussion and relate these themes to more abstract parent categories. Two visualization tools are used to explore the concept spaces that result: Self-Organizing Maps and concept graphs. The SOM does not facilitate the traversal of semantic scales and is best used to find "bottom-up" structure at the level of individual terms as they appear in the text. Concept graphs derived from a pair of similarity measures – cosine similarity and concept wavefronts – allow for top-down navigation of a concept space. Wavefronts enable users to find sets of related concepts at a given level of generalization that serve as the most informative abstractions, where informativeness is measured against an adjustable concept frequency ratio threshold.

A core research direction in geographic information science has been the problem of generating ontological hierarchies that support interoperability of geographic concepts. These ontologies must enumerate the key ideas involved in a task or domain, and the wavefront technique applied here is capable of extracting a possible ontological structure from discourse. The concepts deemed important to a domain will not remain constant over time, however, and construction of cognitive spaces through discourse allows concept structures to evolve with a community's thinking. The concept relationships inferred through the methods discussed here are limited to one type of semantic relationship (hypernymy), but could be extended to address more sophisticated types of relations (e.g., functional) using other thesauri in addition to WordNet. Nonetheless, ontological information cannot always be gathered from existing resources alone; frequently, we must consult domain experts, and our approach provides one way to elicit this information in an unobtrusive, familiar medium.

The natural language approach described here has several shortcomings that will be addressed with further research. Most notably, the WordNet lexical database used for concept extraction is domain independent and does not include specialized terms or definitions unique to geographic research (or any other scientific field). Jargon and compound terms such as "land use" are not handled well by the current implementation. In addition, a close analysis should be made of the benefits of real-time concept visualization on the course of an online discussion. Lastly, it is necessary to develop and test techniques to evaluate the accuracy of the sentence parsing and sense disambiguation tools. One possibility would be to combine a human-aided evaluation system with a machine learning approach to refining our system's ability to determine the correct sense of a term. For example, when processing a user's contributions, the system could request that they indicate which of an ambiguous term's senses the writer intended. Our current implementation, however, suggests that language processing and concept similarity methods promise to be valuable tools in the construction of scalable concept spaces from the sorts of discussion in which scientists are often already engaged. Although the techniques we apply are designed to support geographic knowledge discovery, we anticipate their applicability to a wide range of semantic issues beyond human-environment interaction.

Acknowledgements

This research was supported by NSF grants BCS-9978052 (HERO), ITR (BCS)-0219025, and ITR (EAR)-0225673 (GEON). Thanks to Isaac Brewer for early thoughts on the applications of the Delphi method to human-environment interaction and to Sachin Oswal for work on the concept graphing tool.

References

1. Guarino, N., Understanding, building, and using ontologies. International Journal of Human-Computer Studies, 1997. 46: p. 293-310.
2. Chen, H., et al., Information visualization for collaborative computing. IEEE Computer, 1998: p. 75-82.
3. Turoff, M. and S. Hiltz, Computer based Delphi processes, in Gazing into the Oracle: The Delphi Method and its Application to Social Policy and Public Health, M. Adler and E. Ziglio, Editors. 1996, Kingsley: London.
4. Fabrikant, S. and B. Buttenfield, Formalizing semantic spaces for information access. Annals of the Association of American Geographers, 2001. 91(2): p. 263-280.
5. Miller, N., et al. Topic Islands - A wavelet-based text visualization system. in Proceedings of IEEE Visualization 1998. 1998. p. 189-196.
6. Havre, S., B. Hetzler, and L. Nowell. ThemeRiver: Visualizing theme changes over time. in Proceedings of IEEE Symposium on Information Visualization, InfoVis 2000. 2000. p. 115-123.
7. Feldman, R., et al. Trend graphs: Visualizing the evolution of concept relationships in large document collections. in Second European Symposium on Principles of Data Mining and Knowledge Discovery (PKDD 1998). 1998. Berlin: Springer-Verlag. p. 38-46.
8. Redeker, G., Coherence and structure in text and discourse, in Abduction, Belief and Context in Dialogue, H.C. Bunt and W.J. Black, Editors. 2000, John Benjamins: Philadelphia. p. 233-264.
9. Smith, B. and D.M. Mark, Geographical categories: an ontological investigation. International Journal of Geographical Information Science, 2001. 15(7): p. 591-612.
10. Fonseca, F., et al., Semantic granularity in ontology-driven geographic information systems. Annals of Mathematics and Artificial Intelligence, 2002. 36(1-2): p. 121-151.
11. Frank, A.U., Tiers of ontology and consistency constraints in geographical information systems. International Journal of Geographical Information Science, 2001. 15(7): p. 667-678.
12. Rodriguez, M.A. and M.J. Egenhofer, Determing semantic similarity among entity classes from different ontologies. IEEE Transactions on Knowledge and Data Engineering, 2003. 15(2): p. 442-456.
13. Fellbaum, C., ed. WordNet: An electronic lexical database. 1998, MIT Press: Cambridge, MA. 433 p.
14. Mani, I., Automatic summarization. Natural Language Processing. 2001, Philadelphia: John Benjamins. 285 p.
15. Lin, C.-Y. Topic identification by concept generalization. in Proceedings of the 36th Annual Meeting of the Association for Computational Linguistics (ACL 95). 1995. New Brunswick, NJ: Association for Computational Linguistics. p. 308-310.
16. Kohonen, T., Self-organizing maps. 1997, New York: Berlin. 426.
17. Garson, G., Neural networks: an introductory guide for social scientists. 1998, London: Sage. 194.
18. Sammon, J., A nonlinear mapping for data structure analysis. IEEE Transactions on Computers, 1969. C-18(5): p. 401-408.
19. Kaski, S., et al., WEBSOM - Self-organizing maps of document collections. Neurocomputing, 1998. 21: p. 101-117.
20. Sowa, J., Knowledge Representation: Logical, Philosophical, and Computational Foundations. 2000, Pacific Grove, CA: Brooks/Cole. 594 p.

Route Adaptive Selection of Salient Features

Stephan Winter

Institute for Geoinformation, Technical University Vienna
Gusshausstr. 27-29, 1040 Vienna, Austria
winter@geoinfo.tuwien.ac.at
http://geoinfo.tuwien.ac.at

Abstract. Human communication on wayfinding makes extensive use of landmarks. With a formal model of salience, route planning services can include landmarks as well. Such a model was presented considering visual, semantic, and structural properties of spatial features. This model measures saliency independent from a given route. Our hypothesis is that an additional factor is cognitively relevant for the selection of appropriate salient features: advance visibility for a person approaching a destination point. We will propose a computational measure for advance visibility. The new measure is used to identify *suited* salient features at route decision points: a feature is suited for a wayfinding instruction if it is (a) salient, and (b) in advance visible. The relevance of advance visibility is tested by a comparison of wayfinding success with instructions made with and without this additional measure. Computational effort is observed to check feasibility.

1 Introduction

The use of landmarks in wayfinding instructions improves the wayfinding success rate [1,2,3]. Thus wayfinding instructions provided by route planning services should make use of landmarks. A service needs a formal model characterizing the *saliency* of spatial features to simulate the otherwise informal concept of landmarks. According to a recent proposal, we can automatically select salient features from databases [4]. The authors had in mind to enrich wayfinding instructions at decision points or along route segments by salient features. Salience was characterized by measures observing the visual, semantic, and structural properties of features in urban environment, according to a classification of landmark properties by Sorrows and Hirtle [5]. They implemented their model for façades, scoring their visual and semantic salience, and produced automatically a set of landmarks highly correlated with human judgements [6]. This paper continues to consider façades, being interested in urban features that are visible from a distance. Other urban features are neglected. Dealing with visibility aspects, the paper concentrates on users who are not visually impaired (for wayfinding support of visually impaired see, e.g., [7]).

The proposed measures turn out to characterize only the global properties of features. No attempt was made to relate the salience to a given route. For example, we can assume that people refer to a salient feature if the feature is clearly visible ahead along a route, but they might skip the same feature in their directions of another route if it is unfavourably oriented for this route. Mainly two reasons support the hypothesis that a global measure of salience oversimplifies the problem of wayfinding by landmarks:

W. Kuhn, M.F. Worboys, and S. Timpf (Eds.): COSIT 2003, LNCS 2825, pp. 349–361, 2003.
© Springer-Verlag Berlin Heidelberg 2003

1. People feel comfortable if they recognize reference features early, before arrival at a decision point. They feel confirmed that they are on track, and they do not need to break movement at the decision point, but can interpret the next wayfinding instruction in advance. Furthermore, they are more sensitive to visual information in their fronts than in other directions. But if a traveler is required to look all around at a decision point because the most salient feature might be located in any direction, these psychological and biometrical considerations are neglected.
2. At decision points in most cases the required re-orientation are one of *turn left*, *turn right*, or *go ahead*, that means the turns take place in a range of $[-90°, 90°]$. That means reference features ahead of the entering route segment are more often spatially connected to route continuation than features that lie behind, and are thus easier to combine with an instruction.

These arguments give motivation to complement the salience of a feature at a decision point by its *advancing visibility*, with respect to a specific advancing direction. The hypothesis of this paper is that a measure characterizing how early a feature becomes visible along a route completes the salience measure for a cognitive relevant aspect, and therefore supports the success of mobile navigation services in reducing the stress and cognitive workload of its users. Hence, a feature suited for enriching a wayfinding instruction should have both properties, salience *and* advancing visibility. The hypothesis needs strong support from human subject testing.

I will propose a computational measure for advancing visibility. At each decision point the salient features are considered for their advancing visibility with respect to a given route. Then the feature with the best product of visibility and salience will be selected for a wayfinding instruction, instead of the most salient feature. First, this measure will be defined formally (Section 3), and then checked whether its calculation is feasible (Section 4) and cognitively relevant (Section 5). Feasibility requires a formal specification of advancing visibility that is provable correct, efficient, and robust in all kinds of urban environments. The computational load of the added measure should be low compared to the computation of salience. The measure will be tested on a real world data set. Cognitive relevance means that observing the advancing visibility should significantly improve user satisfaction and wayfinding success. This will be tested by comparing the wayfinding success with instructions referring to the most salient features, and instructions referring to the most suited, i.e., salient and visible features. We expect that the new selection method is better with regard to user satisfaction or wayfinding success.

2   Previous and Related Work

In this section the concepts are collected that are needed for the following sections, or are related to the concepts used. Especially salience and visibility measures are reviewed.

Raubal and Winter have proposed a model for the automatic selection of salient features from a spatial database [4]. This model was implemented and tested [8,6]. The model realizes the three characteristics of landmarks according to Sorrows and Hirtle [5]: their visible, semantic, and structural properties. The authors propose a computational model, based on individual measures and a scoring method, to determine a numerical value of salience for any façade of a building in a database. None of their measures is related to a route. The formal model of salience was tested whether it matches with cognitive salience [8]. The subjects were shown panoramic views at decision points: they had to decide for the most salient feature in this complete 360°-view. That means the test is also independent from a route or a physical orientation of the observer. The test was successful in so far as the highest ranks in the computed list of salient features matched with the highest ranks of the survey.

This model of salience is static. It allows offline calculation of salience measures for on-the-fly generation of wayfinding instructions. Among the considered visual properties is the visibility of the feature, which will be discussed next.

An *isovist* is the field that can be seen from a vantage point, and is a property of this point [9,10]. The calculation of the isovist of each point in space is only feasible in discrete spaces. Batty and Biang propose for this purpose multi-agent models which operate in cellular space [11]. The algorithm is a region-growing approach, limited by line-of-sight constraints.

Raubal and Winter's model of salience [4] considers a visibility parameter of spatially extended features which is an extension of the concept of an isovist. Their visibility parameter describes the area from which the façade is visible and recognizable, at least to a certain extent. The calculation is done by a raster line-of-sight algorithm taking some thresholds into account: a largest distance that is considered for recognizability, a certain angle under which the façade has to be seen, and a minimal part of the façade that has to be visible [8]. This extended isovist will be re-used for advancing visibility.

To my knowledge, partial coverage of landmarks was never investigated in the cognitive literature. Thresholds are somehow arbitrary as long as parts of a landmark can stand for the whole, e.g., the tower of St. Stephen's Cathedral in Vienna is visible from many places in downtown and beyond, but it is refered to in human communication as the "St. Stephen's Cathedral", and not as the tower of the cathedral.

The measure of advancing visibility is related to *visibility graphs* as well [12]. Visibility graphs are constructed in a planar space with polygonal obstacles. Edges of a visibility graph represent a visibility relationship between pairs of nodes. Visibility graphs are complex, being computed in $\mathcal{O}(n^2 \log n)$, i.e., they assume problems with decent numbers of vertices. Typically visibility graphs are used to search for shortest paths (the shortest path in this space is always a path in the visibility graph); hence, they do not yield any areal interpretation. However, for a node at least a conservative estimation of its isovist could be provided, calculating the convex hull of the end nodes of all its outgoing edges. Given a façade ground line with two corners, the intersection of the two corners' isovists would be a conservative estimation of the visibility area of the (whole) façade. By this way one would have a vector solution.

In contrast to visibility graphs and isovists we are interested in a measure that considers visibility with respect to a given route, i.e., visibility from a specific direction. In the next section such a measure is proposed. The measure will be based on the visibility area as reviewed here.

All the considered measures deal well with a planar space, which is a simplification of architectural or urban space. They do not consider height of human-made or topographical barriers, like terrain elevation, and thus, they deal not properly with visibility across barriers. In the literature of landscape planning and classification we find methods for viewshed analysis, which provide visibility areas in modulated terrain [13,14]. These algorithms are also line-of-sight algorithms, only the complexity is higher considering the third dimension. Without limiting generality we exclude this aspect in this paper, for the reason that our example is a flat area with five- to six-storied buildings. In other environments it might be necessary to do the more complex calculations. A visibility area from a three-dimensional analysis may fall apart into disconnected components. The following model handles even disconnected visibility areas without change.

3   Advance Visibility

In this section a formal definition for advance visibility is introduced. The measure will be derived from two parameters: the properties of a feature considered to be located at a decision point, and the route segment that enters the considered decision point. The measure will be implemented and tested in the following sections.

In the following let us limit ourselves to nodes in a street network that have the potential to become decision points in routing contexts. Decisions will be neccessary at nodes that offer more than one possibility to continue traveling, i.e., at nodes with a degree higher than 2. The degree of a node is the number of its incident edges. Furthermore, only those features that are located *at* decision points shall be considered as candidates for referencing in route instructions. The spatial proposition *at* is formalized in the following way. Generally, all faces incident to a node are considered. Now let us embed the node in a cadastral map the center of the represented intersection. A minimal circle covering at least one parcel in every incident face identifies the parcels *at* the intersection. The buildings on these parcels are considered as buildings *at* the decision point. All their façades that face towards an edge incident to the decision point are considered as façades *at* the decision point. A refinement collects all façades from this set for which their visibility area contain the potential decision point (point-in-polygon test, see, e.g., [15]).

Having an unambiguous relationship between a potential[1] or real decision point and the façades at this decision point, the second parameter can be considered: the route segment that enters the considered decision point. Routes are considered here as sequences of nodes and segments. Route nodes are the start node, the end node, and decision points. Decision points are a subset of the street network nodes. Route segments are directed links between two consecutive decision points. A route segment forms a logical unit, reflected by the fact that route instructions consist of one direction per route segment. Note that the street network is of finer resolution, i.e., route segments have generally a $1 : n$-relationship with street segments. In other words a route segment corresponds geometrically to an open polygon formed by street segments. For each route segment, the first polygon edge is an outgoing street segment of the previous decision point, and the last polygon edge is an incoming street segment of the decision point ahead. The concept of advance visibility is related to the decision point ahead. The following geometric model considers the street polygon when speaking of the *entering route segment*.

---

[1] In the following we will speak simply of decision points, with the general meaning of potential ones, as long as there is no specific route.

From the two parameters at a decision point, the properties of related features and the entering route segment, we will derive now two measures for each feature: the route *coverage* of the feature's visibility area, and the *orientation* of the feature with respect to the heading of the current route segment. A combination of both will provide a measure of advance visibility.

Route coverage is the total coverage of the entering route segment by a feature's visibility area. It is proportional to the advance time the feature comes into the field of view of a moving person. The earlier the traveler can perceive the feature, the more comfortable the traveler will feel. Consider Fig. 1: in the left case, the feature's visibility area covers some of the entering route segment. However, the coverage is small and requires from the moving person to reach the decision point first before she has a chance to perceive the feature. This case shall be low ranked by a measure. In the right case, the moving person can perceive the feature early along the entering route segment. The visibility area covers much more of the entering route segment. This case shall be higher ranked.

The measure can be constructed by the proportion of the covered part and the total length of the entering route segment. Let p_s, p_e be the start and end point of the entering route segment, and p_i its intersection with the boundary of the visibility area. Then the coverage c is:

$$c = \frac{|p_i p_e|}{|p_s p_e|} \ .$$

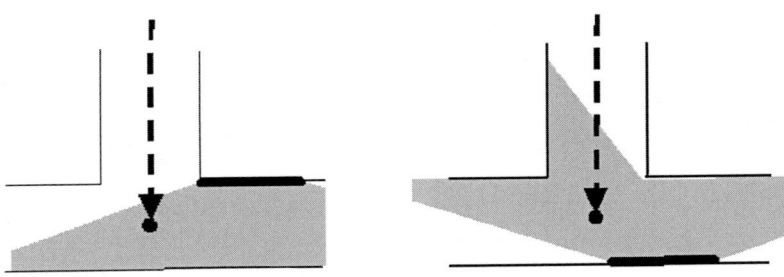

Fig. 1. A feature at a decision point with low advance visibility (left), and high (right).

It is possible that the feature comes into sight, and the sight gets lost again. In these cases we would find a set of $p_{i1}, ..., p_{in}$ (where n is odd). In these cases it makes sense to always select the first point, p_{i1}. Once a moving person got sight and thus, the global direction, the wayfinding is anchored. The calculation of p_i (or p_{i1}) can be done by a sweep-line algorithm [12] that traces the polygon of the entering route segment and the boundary polygon of the visibility area.

The coverage implies still a field of view of the moving person that covers 360°. Consider for example again Fig.1, left: if a person reaches the intersection point of visibility area and entering route segment, the perception of the feature would require substantial head turn, which is uncomfortable and hence unlikely happening. There are other spatial configurations with a similar low coverage value c, which can be much easier detected by the person. Consider Fig. 2: on the left, the situation is quite similar to the discussed one. On the right, the coverage is not significantly higher. But the feature is oriented towards the route, and therefore a moving person needs less effort to detect it. Therefore, an additional measure is proposed: the orientation of a feature. This measure amplifies the selectivity of c in these cases.

Calculate the cardinal direction of the normal vector on the façade's ground line, d_f. The direction is defined by the endpoints of its frontline, oriented towards the visibility area. The cardinal direction of the entering route segment, d_r, is more difficult to define since the segment is an open polygon. We choose here p_s, p_e, taking the global approaching direction into account, but another reasonable choice would be the orientation of the incoming edge only. Now the normed orientation o of the feature with respect to the entering route segment is:

$$o = \frac{|d_f - d_r|}{180} .$$

In the optimal case they are oriented towards each other, $o = 1$. In the worst case, they show in the same direction, $o = 0$.

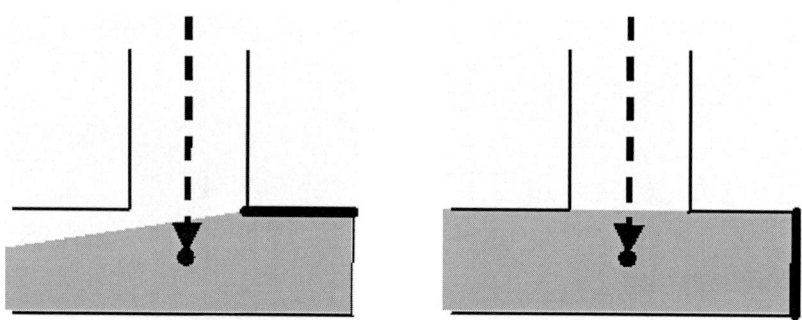

Fig. 2. A feature at a decision point facing ahead the route (left), and to the route (right).

Now the measure v of advance visibility can be constructed using coverage and orientation:

$$v = c \cdot o .$$

Advance visibility is normed. Features with a value close to $v = 1$ are oriented towards the approaching moving person, and are early visible, maximally from the previous decision point. Features with low values lack at least one of these properties. Combined they model the observation preferences of moving persons in urban space: looking ahead towards their expected destination.

From a computational point of view advance visibility is cheap. The only complex step is the determination of the visibility area of each feature. But this step was made for the salience measure already, such that the result can be reused here. Having the shape of the visibility areas, the sweep-line algorithm is applied for the coverage; it takes $\mathcal{O}(n \log n + p \log n)$ [12], with n the number of boundary and route segments (relatively small), and p the number of intersections (most often $p = 1$). The orientation is calculated by cross-products. An interesting aspect is that, although the measure is route dependent, it can be calculated off-line and stored in a database together with salience measures, as long as the network is static.

4   Application and Test

The presented advance visibility measure will be used now on top of the framework by Raubal and Winter to identify salient features. For each decision point along a route a feature will be provided that is both, in advance visible and salient. This feature will be used as a local reference in wayfinding instructions.

Besides scoring for salience [6], for each feature its advance visibility measure can be calculated, separate for each possible entering route segment in a decision point. This would be a vector v_r of $3...n$ measures, n being the degree of the decision point, if every entering street segment belongs to only one route segment. This is generally not the case, but the number of attached route segments will be small. Typically, n itself is limited to a small integer by the grid character of street networks [16]. In the street network of Vienna, for example, $\deg_{max} = 6$. Then, given any specific entering route segment r, for each feature at a decision point the product of salience s and advance visibility v_r can be calculated, providing a ranking of the salient features that are in advance visible.

The strategy can be explained with the following application example. Let us consider again the route of Raubal and Winter, from the café *Diglas* to the restaurant *Novelli* in downtown Vienna. For a wayfinding instruction at the decision point *Stock-im-Eisen-Platz* we need an optimal reference feature. The decision point is an intersection in the pedestrian zone with four adjacent street segments. The local features at the decision point are shown in the panoramic view of Fig. 3. The salience values of the local features are: 1.7, 1.1, 1.0, 0.7, 0.7, 0.6, 0.6, 0.6, 0.5 [8], i.e., there is a really flashy building (the modern *Haas* building, 1.7), but there are also some other distinctive buildings (1.1, 1.0).

Fig. 3. A panoramic view (360°) of the street intersection *Stock-im-Eisen-Platz*, Vienna.

Now consider Fig. 4 which shows a map of the *Stock-im-Eisen-Platz*, and a part of the specific route. The route enters from the north into the intersection, and continues to the west. The intersection itself is a decision point ($\deg = 4$), that means for all local features the salience measures as well as the advance visibility for any of the four entering route segments are stored. It turns out that the most salient feature, the *Haas* building, has neither the best coverage of the entering route segment, nor a good orientation. Instead, the *Bank Austria* building is the one the entering route segments approaches straight, which makes coverage $c = 1$ (the building is visible even from the previous decision point), and orientation o close to 1. It turns out that the salience measure for this building is the second highest. A product of salience and advance visibility is higher than all other products. The wayfinding instruction will be constructed referring to this building, instead of the more salient *Haas* building.

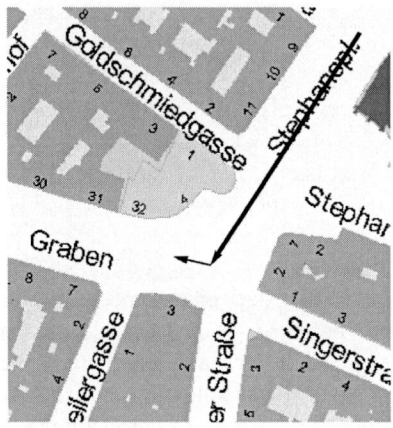

Fig. 4. The same intersection (unnamed in the map), with some references (Map: ©Magistrat Wien).

The same procedure can be applied to all decision points of the route. The result is a set of reference features for wayfinding instructions alternative to the set based on maximal salience, as proposed by Raubal and Winter. It is still to show that this set of features leads to better wayfinding success.

5   Human Subject Tests

In this section we give evidence that wayfinding instructions enriched with salient *and* in advance visible features lead to more successful or easier human navigation than wayfinding instructions enriched with the most salient features. For that reason several experiments were accomplished. Besides the necessity to identify the rank of local landmarks, the main question was for experiences with a route description with late visible or not appropriatly oriented landmarks.

In the first set of experiments the test candidates were 15 graduate students of the University of Redlands. The surveys were conducted on January, 27th-30th on campus; the test candidates were familiar with the environment being in their sixth month of their master program. Participation was part of the course work.

The first survey should identify the landmarks on campus. The question *List the landmarks on campus* was answered by lists between 5 and 21 features. The interpretation yields a ranked list of more prominent features, or commonly accepted landmarks (Fig. 5)

In fact, the top ranked landmarks form the axis of the campus, and are on every postcard from here. – The total set of nominated features consists of 37 elements, which means that the individual lists are quite heterogeneous. The number of nominations can be used to replace automatically derived salience measures for the features on campus, which are not available. The numbers are normed for this reason.

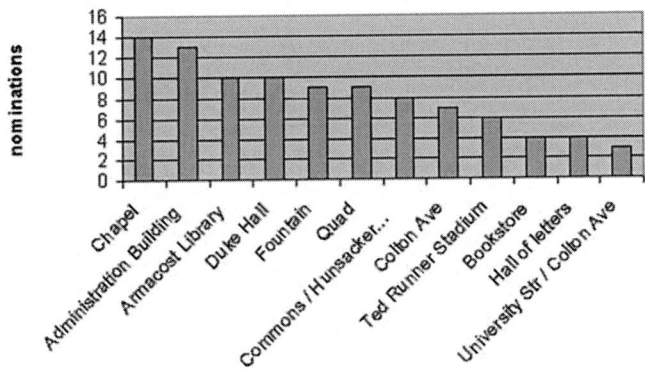

Fig. 5. Nominations of landmarks on the campus of University of Redlands.

In a second survey the participants are asked to describe a route on campus to a person who is unfamiliar with the environment: *Describe me a route from my apartment, 1111 Central Avenue, to the fitness center.* The participants were animated to walk this route while writing, i.e., the descriptions were developed on location. The primary interest in this survey was to check whether the global landmarks were used in wayfinding instructions at all. The start and destination of the route were selected in a way that several of the most prominent features should be touched.

The handed-in wayfinding instructions use between 7 and 29 references to salient features. To interpret the numbers of nominations is difficult, because the group choose between four different routes. Surprisingly, a subgroup of four participants had chosen a route around the campus, which is the route one would drive by car. Without these exceptions, all of the most prominent landmarks are located along the routes. However, it turns out that the most frequently cited salient features in the wayfinding instructions are different ones (Fig. 6). Only the Duke Hall appears in both sets on prominent places. The other prominent landmarks have comparably low rankings.

In wayfinding instructions landmarks are selected for their specific role, either as a global landmark, used for overall orientation, a route mark, used for confirmation, or as a local landmark at decision points, used for re-orientation [3]. Habel has shown that in wayfinding instructions landmarks appear most frequently at decision points [17]. Then we cannot expect to find the most prominent features in our wayfinding instructions in any case. But if a prominent feature is located at a decision point, one can expect that this one is selected.

This assumption cannot be confirmed from the survey. The Chapel is a route mark on these routes, but the Quad and the Administration building are located at decision points.

Furthermore, the latter two features are spatially extended, and smaller features were preferred improving the precision of an instruction. It seems that there is a difference between the salience of a feature, and the degree it is perceived to be useful in a route description. Salience alone does not decide if it is actually used in wayfinding instructions or not.

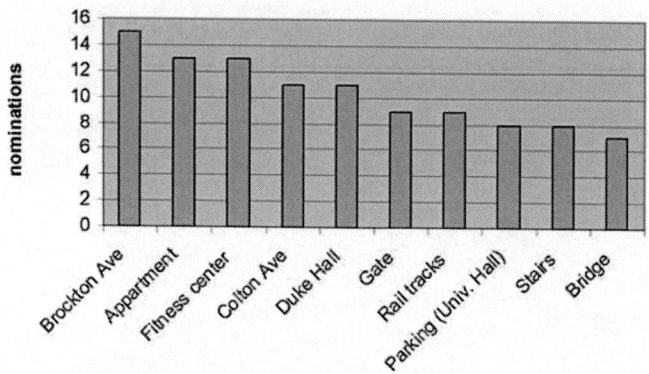

Fig. 6. Nominations of landmarks in descriptions of a route on campus.

In a third survey the participants got one of three different sets of wayfinding instructions, all guiding the same route. They should imagine that they are unfamiliar with the place, and rely on the instructions instead of their survey knowledge [18] of the campus. Afterwards they should report whether they felt always comfortable with the instructions or would like to propose improvements. Between the three sets of wayfinding instructions especially the selected salient features were varied in a way that if one group had a reference to a salient feature with high advance visibility, a control group got at the same decision point a feature with unfavorable orientation or coverage.

The wayfinding instructions were designed in the following way. Coverage and orientation were estimated on location, and no salience was computed. 86% of the subjects of the second survey have used building names in their wayfinding instructions, instead of visual or semantic properties. Therefore, the current survey considers any building as salient feature, and refers to them in some wayfinding instructions by name, and in others by their visual or semantic properties. The instructions provided no metric information.

The reports have shown that the selection of landmarks with unfavorable orientation or coverage was penalized nearly always by proposals for improvements, or even by loss of track. However, also the selection of landmarks with good advance visibility were criticized, but to a significant lesser extent, and partly due to unclear language. There was no loss of track from a landmark with good advance visibility.

Where the critical descriptions are identified correctly (in any case at least once), the proposed instruction improvements do seldom consider better in advance visible buildings (one response). They refer more often to (a) shorter routes (the given one was not a direct route; five responses), (b) the general preference of building labels instead of visual or semantic properties (three responses), and (c) some route marks if the distance to the next destination point was not cognizable (three responses). Only the latter can be taken as evidence supporting our hypothesis; the other proposals are indifferent in this respect although they might be caused by the same conflicts.

A second test was performed at the University of Münster on July 2nd. Participants were nine students; the test was again part of coursework.

In this case, the subjects were shown panoramic views of fourteen street intersections in Münster. The intersections were shown in no specific order. The subjects should mark their favorite landmarks, and comment for the reasons of their choices. The comments support the model of Raubal and Winter by naming their properties as relevant for a decision. That means as long as no route is considered, global salience is an appropriate measure to select distinctive features. Surprisingly, the subjects did choose not only façades, but also a gate, an open place, stairs, or a street sign.

Then the subjects were given a route instruction guiding from their institute to the cathedral of Münster. This route consists of nine segments, or nine directions. The text was strictly formal: "Turn [left/right] and walk until you reach this façade.", and accompanied by a photo from a façade local to a decision point. Photos communicate the visual properties of features without ambiguity from language. The photos were made in orthogonal perspective. In most cases, this perspective was different to the perspective from the decision point. The subjects were asked to report their experiences with the chosen landmarks.

The salience of the façades were estimated by best guess of the author. All the nine intersections were also part of the first questionnaire such that the salience of the selected façades could be evaluated afterwards. Only in three cases the selection and the judgement of the subjects coincided. In the six other cases the selected façades were not the most salient ones. However, the subjects reported in all cases that the selected façades were easy to identify and useful as reference features.

Some of the façades showed unfavorable orientation or coverage with respect to the route. These façades could be unambiguously identified from the comments of the subjects. All of the nine subjectss report difficulties at exactly these decision points, and in any case at least 50% of the subjects identified the coverage or the orientation (perspective) of the chosen façade as the reason. Again, some subjects reported loss of track which means that the selection of unfavourable landmarks was unexpected. To a larger portion the rest of the comments propose alternative landmarks; among the reasons given is again the better (advance) visibility.

An interesting comment refers to the landmark distribution: two decision points were in a distance of only $15m$, which lead to confusion in the absence of any geometric information. Another interesting response indicated that among otherwise equally suited landmarks the ones were preferred that are spatially related to route continuation. For example, if the subject shall move to the right, then the landmark to the right should be preferred to the landmark to the left. At a higher level of detail, when a subject moves along the right side of the street, she prefers façades on the left street side, due to the more convenient observation distance and viewing horizon. All these aspects give reason for future research.

The first experiment, although confused by too many confounding variables, showed that salience is not the only criteria for the selection of landmarks. The second test was more focused, and gives direct evidence for the hypothesis that advance visibility – route coverage and orientation – are additional critical factors in the selection process. This result can be supported by a consideration of the cognitive workload of potential users. If the next reference feature in a wayfinding instruction is not visible, the user has to cross a gap of information in space [18]. But if it is visible in advance, the so-called *environmental space* shrinks to a *vista space* [19], and wayfinding works more on affordance [20] than on cognizant strategies.

6   Discussion

This paper proposes a measure of advance visibility for salient features at decision points. The measure can be combined with the measure of salience to identify features that are salient and visible in advance as well. Features will be preferred that are relatively good in salience and advance visibility; features with high salience but bad advance visibility will be lower ranked, and features that are not salient either are still not considered. The paper applies the measure to real world cases and shows the changes in wayfinding instructions that were based on maximal saliency. It also shows evidence that wayfinding instructions using advance visibility improve wayfinding success and client satisfaction.

Having a method to select suited salient features – salient features that are early recognizable along a specific route – advanced questions can be posed. For example, it is another question how to make the best use of the selected features in wayfinding instructions [21,22]. It might be even useful to change the selected salient feature when advancing to the decision point in the moment a more salient one comes into view. Another open question is how to decide between salient, but not visible, and visible, but not salient features. Within the framework of Raubal and Winter in principle all features can be addressed by their individual visual or semantic properties, even if they are not considered salient. Another interesting question occurs on routes that visit a decision point more often than once: there are good reasons to remember the previously used feature instead of the actually most suited. All these questions need closer investigation.

Acknowledgement

I gratefully acknowledge critical comments from the reviewers that helped to improve the paper.

References

1. Deakin, A.K.: Landmarks as navigational aids on street maps. Cartography and Geographic Information Systems **23** (1996) 21–36
2. Denis, M., Pazzaglia, F., Cornoldi, C., Bertolo, L.: Spatial discourse and navigation: An analysis of route directions in the city of Venice. Applied Cognitive Psychology **13** (1999) 145–174

3. Michon, P.E., Denis, M.: When and why are visual landmarks used in giving directions? In Montello, D.R., ed.: Spatial Information Theory. Volume 2205 of Lecture Notes in Computer Science. Springer, Berlin (2001) 292–305

4. Raubal, M., Winter, S.: Enriching wayfinding instructions with local landmarks. In Egenhofer, M.J., Mark, D.M., eds.: Geographic Information Science. Volume 2478 of Lecture Notes in Computer Science. Springer, Berlin (2002) 243–259

5. Sorrows, M.E., Hirtle, S.C.: The nature of landmarks for real and electronic spaces. In Freksa, C., Mark, D.M., eds.: Spatial Information Theory. Volume 1661 of Lecture Notes in Computer Science. Springer, Berlin (1999) 37–50

6. Nothegger, C., Winter, S., Raubal, M.: Computation of the salience of features. Technical report, Institute for Geoinformation, Technical University Vienna

7. Golledge, R.G., Klatzky, R.L., Loomis, J.M., Speigle, J., Tietz, J.: A geographical information system for a gps based personal guidance system. International Journal of Geographical Information Science 12 (1998) 727–749

8. Nothegger, C.: Automatic Selection of Landmarks for Pedestrian Guidance. Diploma thesis, Vienna University of Technology (2003)

9. Batty, M.: Exploring isovist fields: space and shape in architectural and urban morphology. Environment and Planning B 28 (2001) 123–150

10. O'Sullivan, D., Turner, A.: Visibility graphs and landscape visibility analysis. International Journal of Geographical Information Science 15 (2001) 221–237

11. Batty, M., Jiang, B.: Multi-agent simulation: new approaches to exploring space-time. In: Annual Meeting of GISRUK, Southampton (1999)

12. Berg, M.d., Kreveld, M.v., Overmars, M., Schwarzkopf, O.: Computational Geometry. 2nd edition edn. Springer, Berlin (2000)

13. Fisher, P.F.: Extending the applicability of viewsheds in landscape planning. Photogrammetric Engineering and Remote Sensing 62 (1996) 1297–1302

14. Fisher, P.F.: An exploration of probable viewsheds in landscape planning. Environment and Planning B: Planning and Design 22 (1995) 527–546

15. Pavlidis, T.: Algorithms for Graphics and Image Processing. Springer, Berlin (1982)

16. Taaffe, E.J., Gauthier, H.L., O'Kelly, M.E.: Geography of transportation. 2nd edn. Prentice Hall, Upper Saddle River, NJ (1996)

17. Habel, C.: Prozedurale Aspekte der Wegplanung und Wegbeschreibung. In Schnelle, H., Rickheit, G., eds.: Sprache in Mensch und Computer. Westdeutscher Verlag, Opladen (1988) 107–133

18. Siegel, A.W., White, S.H.: The development of spatial representations of large-scale environments. In Reese, H.W., ed.: Advances in child development and behavior. Volume 10. Academic Press, New York (1975) 9–55

19. Montello, D.R.: Scale and multiple psychologies of space. In Frank, A.U., Campari, I., eds.: Spatial Information Theory. Volume 716 of Lecture Notes in Computer Science. Springer, Berlin (1993) 312–321

20. Raubal, M., Egenhofer, M.: Comparing the complexity of wayfinding tasks in built environments. Environment and Planning B 25 (1998) 895–913

21. Lovelace, K.L., Hegarty, M., Montello, D.R.: Elements of good route directions in familiar and unfamiliar environments. In Freksa, C., Mark, D.M., eds.: Spatial Information Theory. Volume 1661 of Lecture Notes in Computer Science. Springer, Berlin (1999) 65–82

22. Gillner, S., Mallot, H.A.: Navigation and acquisition of spatial knowledge in a virtual maze. Journal of Cognitive Neuroscience 10 (1998) 445–463

Referring to Landmark or Street Information in Route Directions: What Difference Does It Make?*

Ariane Tom and Michel Denis

Groupe Cognition Humaine, LIMSI-CNRS
BP 133, 91403 Orsay Cedex, France
{Ariane.Tom,Michel.Denis}@limsi.fr

Abstract. When describing routes in urban environments, speakers usually refer to both street names and visual landmarks. However, a navigational system can be designed which only refers to streets or, alternatively, only to landmarks. Does it make any difference which type of information users are provided with? The answer to this question is crucial for the design of navigational aids. We report two experiments. The first one showed that in a wayfinding task, route directions referring to streets were less effective than those referring to landmarks for guidance purposes. The second experiment showed that when people generate route directions, they tend to produce less street than landmark information. These studies provide a further illustration of the critical role of landmarks in route directions.

Keywords: Spatial cognition, route directions, streets, landmarks, urban navigation.

1   Introduction

Spatial discourse is generally considered to be a particularly good way of externalizing people's knowledge about their environment. Specifically, the interactions between space and language can profitably be investigated by studying route directions. The intent underlying route directions is twofold. First, route directions are instances of *procedural discourse*, since they are intended to guide a person – a pedestrian or a car driver – from a starting point to a distant target (Golledge, 1993; Taylor & Tversky, 1992). This procedure requires first of all that the actions to be taken and the places where they are to be executed should be identified. Landmarks in the environment are typically used to refer to these places. By this very fact, route directions also belong to the class of *descriptive discourse*, since people often describe the visual features of the environment to be traversed in addition to prescribing actions. These prescriptive and descriptive components of route directions are tightly intertwined, which makes them especially difficult to analyze, not only for psychologists, but also for the computer scientists who face the task of designing user-friendly artificial navigational aid systems (Allen, 2000; Daniel, Tom, Manghi, & Denis, in press; Denis, Pazzaglia, Cornoldi, & Bertolo, 1999; Streeter, Vitello, & Wonsiewicz, 1985).

* This research was supported by a doctoral research grant from the Délégation Générale pour l'Armement (DGA).

W. Kuhn, M.F. Worboys, and S. Timpf (Eds.): COSIT 2003, LNCS 2825, pp. 362–374, 2003.
© Springer-Verlag Berlin Heidelberg 2003

Studies of route directions have shown that *landmarks* are of major importance for guiding people. The frequency with which they are mentioned is subject to individual differences. For instance, it has been clearly established that women refer to visual landmarks more frequently than men do (Bell, Fischer, & Baum, 1996; Denis, 1997; Galea & Kimura, 1993). Despite these differences, landmarks are nearly always referred to in descriptions of routes (Allen, 2000; Denis, 1997; Fontaine & Denis, 1999; Michon & Denis, 2001). Their role as navigational aids has been amply documented. Michon and Denis (2001) found that their use in urban environments is unequally distributed along the route described. They tend to be concentrated at locations where reorientations are called for, and also at any other places where reorientations *could* occur (such as crossroads). These results are congruent with the assumption that landmarks allow moving people to anticipate the critical nodes where they need to pay special care.

Another means of conveying spatial information is to focus on the *paths* along which people are moving. Surprisingly, this topic has received very little consideration. One reason is probably that the role of landmarks has mostly been documented in environments such as campuses (Cornell, Heth, & Skoczylas, 1999; Rossano & Reardon, 1999), buildings (Kirasic & Mathes, 1990; Taylor, Naylor, & Chechile, 1999), and underground environments (Fontaine & Denis, 1999). These environments are biased towards references to landmarks, and in such environments references to paths, which are unlabelled and indistinctive, are unlikely.

Nevertheless, even studies carried out in urban environments reflect some rather limited spontaneous reference to streets (Michon & Denis, 2001). Streets commonly have proper names. It is well established that proper names are more difficult to retrieve from memory than common names (e.g., Cohen & Burke, 1993; Izaute, 1999). Street names have been shown to be hard to memorize, even for expert taxi drivers, when the corresponding streets cannot be retrieved by the use of a chunking strategy (Kalakoski & Saariluoma, 2001). Even when they have been well memorized, their long-term retention is poor compared to that of landmarks (Bahrick, 1983).

The reference to street names has some limitations. Street name plates are sometimes not visible from a distance. Furthermore, as mentioned above, street names are often proper names. These names are arbitrary and do not convey any spatial or descriptive features of the streets that have to be taken. But, on the other hand, street names are a concise form of information. They also have the advantage of resolving any problem of reference since, for instance, there may be several parks in an environment, but there is definitely only one Général Leclerc Avenue. More generally, increasing the level of determinacy of the discourse is known to favor the processing of spatial discourse (Ioerger, 1994; Mani & Johnson-Laird, 1982; Schneider & Taylor, 1999).

When studying route instructions, one has also to consider a related factor, namely, the *direction* of walking. Directionality is encoded in spatial mental representations (e.g., Thorndyke & Hayes-Roth, 1982). While navigating, the visibility and hence the relevance of landmarks may differ considerably depending on whether the route is traversed from A to B or from B to A (Cornell, Heth, Kneubuhler, & Sehgal, 1996; Heth, Cornell, & Alberts, 1997). We also considered this factor in our research.

Two studies are reported in this article. The first was intended to test the guiding value of streets and landmarks in a spatial discourse referring to an urban environment. For this purpose, we recorded behavioral indices reflecting the navigational performance of participants, depending on whether they were provided with instructions referring to the streets or to the visual landmarks of the environment. We used a procedure which allowed us to record the occurrence of cognitive anticipation of information. We also used a map-drawing task as a convenient mode of externalizing the mental representation of the route traversed. In the second study, we asked participants to describe the same route, while navigating either from A to B or from B to A. These verbal productions were analyzed to determine any differences in the reference to street or landmark information. Thus, the two experiments considered the same street/landmark contrast during navigation episodes, but in one case, route instructions were being used by the participants, whereas in the second experiment, the participants were generating these instructions.

2   Experiment 1: Using Route Directions in Navigation

This study was designed to test the contrast between the use of landmarks and street names as components of route directions in a wayfinding task. We focused mainly on the participants' behavior during locomotion, but we also subsequently collected information about how participants would represent the newly learned route on a map.

2.1   Method

Route. The route was located in an inner suburb of Paris. It was 480 meters long. As depicted in Figure 1, it was entirely located outdoors and included six turns. Hence, six reorientations were required.

Participants. The participants were 40 undergraduates (20 females and 20 males; mean age: 24.40, $SD = 3.11$). They were all unfamiliar with the route to be traversed. They were randomly assigned to one of two experimental conditions, namely the Landmark-based or the Street-based condition.

Route Directions. Route directions were constructed in two stages. First, we collected route directions from a sample of 20 people living in the study area. This corpus allowed us to identify the most frequently cited streets and landmarks. Then, two sets of instructions were constructed in parallel, so that the same portions of the route were referred to either by a street name or by a landmark. Each version of the description comprised 25 instructions, of which 15 were common to both sets and 10 were different in the two sets. These differences resulted exclusively from the use of a street name or a landmark. For instance, participants were invited either to "turn right into Jean Bouveri Street" or to "turn right toward the phone booths". Each instructional item was printed on a separate page of a booklet. The two versions of the instructions are shown in Table 1.

Procedure. The experimental sessions were run individually for each participant. After they had been guided to the start point of the route, the participants were given a booklet in which the instructions were printed. The participants were told that they would have to find their way with only the printed instructions to help them, and that the experimenter escorting them would be walking a few meters behind. They were instructed to stop walking and turn a page each time further information was needed to continue. These stops were used by the experimenter to note on a map of the route the exact place where the consultations took place. While they were navigating, three indices were recorded. *Directional errors* included all deviations from the intended route. *Checkings* referred to a verbally expressed intention to check some feature of a piece of information (for instance, a street name). Finally, *stops* were defined as pauses lasting more than 2 seconds. The frequency and duration of the stops and checkings were recorded. Only the occurrences of directional errors were noted, as in this case the participants were stopped and taken back to the place where the error occurred, to avoid off-route consultations. After the navigation episode, participants were asked to draw a map of the route. This map was intended to be a helpful guide to a pedestrian unfamiliar with these surroundings.

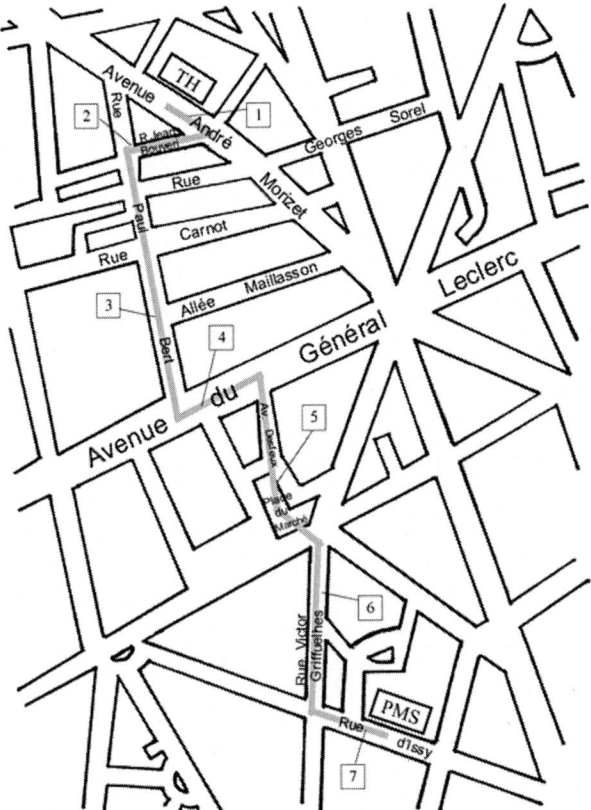

Fig. 1. The route used in Experiments 1 and 2 (TH = Town Hall; PMS = Park Maintenance Service; segments are numbered #1 to #7)

2.2   Results

The data were submitted to a one-way analysis of variance (ANOVA) with Instruction type as the between-participant factor.

Table 1. Route directions used in Experiment 1

Street-based instructions	Landmark-based instructions
1 - Stand with your back to the Town Hall	1 - Stand with your back to the Town Hall
2 - Turn left into **Andre Morizet Avenue**	2 - Turn left toward **the police station**
3 - Cross the avenue at the first pedestrian walkway	3 - Cross the avenue at the first pedestrian walkway
4 - Turn right into **Jean Bouveri Street**	4 - Turn right toward **the phone booths**
5 - Keep going to the end of the street	5 - Keep going to the end of the street
6 - You are now in **Paul Bert Street**	6 - You are now in front of **a bakery**
7 - Turn left	7 - Turn left
8 - Go straight on	8 - Go straight on
9 - Cross **Georges Sorel Street**	9 - Go past **the occupational medicine building**
10 - Continue straight on	10 - Continue straight on
11 - Pass **Carnot Street**	11 - Pass **the coffee shop Au Bon Accueil**
12 - Go past **Maillasson Avenue**	12 - Go past **Glycines Square**
13 - Keep going to the end of the street	13 - Keep going to the end of the street
14 - This runs into **General Leclerc Avenue**	14 - Pass in front of **a grocery** store on the corner
15 - Turn left	15 - Turn left
16 - Cross at the first pedestrian walkway	16 - Cross at the first pedestrian walkway
17 - Turn left	17 - Turn left
18 - Take **Desfeux Avenue** on the right	18 - Take the street on the right after **Pizza Hut** ®
19 - You are now in the market square	19 - You are now in the market square
20 - Cross the square	20 - Cross the square
21 - Head for **Victor Griffuelhes Street** which is almost in front of you	21 - Pass in front of **Bijou Hotel** which is almost in front of you
22 - Turn right into the street	22 - Turn right into the street
23 - Go straight on	23 - Go straight on
24 - Turn left into **Issy Street**	24 - Turn left after the entrance to **the skating rink**
25 - The Park Maintenance Service is a little bit farther on, on the left	25 - The Park Maintenance Service is a little bit farther on, on the left

Table 2. Average number and duration of directional errors, checkings and stops for each condition (*SDs* are in parentheses)

	Number		Duration (seconds)	
	Street-based condition	Landmark-based condition	Street-based condition	Landmark-based condition
Directional errors	0.70 (0.80)	0.45 (0.69)	–	–
Checkings	1.25 (0.97)	0.20 (0.41)	29.73 (19.70)	5.90 (13.78)
Stops	1.70 (1.13)	0.30 (0.58)	15.01 (8.86)	3.18 (5.81)

Behavioral Indices. The numbers and durations of the three behavioral indices considered in this study are shown in Table 2. While directional errors were not found to differ in the two conditions [$F(1,38) = 1.12$, $p > 0.05$], checkings and stops were significantly less frequent in the Landmark-based than in the Street-based condition [$F(1,38) = 19.99$, $p < .0001$, and $F(1,38) = 24.50$, $p < .0001$, respectively]. Furthermore, the durations of these last two behavioral indices were significantly shorter in the Landmark-based than in the Street-based condition [$F(1,38) = 19.64$, $p < .0001$, and $F(1,38) = 24.93$, $p < .0001$, respectively].

Consultations. The locations where each new piece of information was consulted were compared in the two conditions. For each instruction, we subtracted the distance from the start point to the point of consultation with the Street-based instructions from the same measurement for the Landmark-based instructions. Figure 2 shows an example of how such differences were calculated. This computation was made for each instruction, which resulted in a total of 25 differences. A negative value indicated that the corresponding instruction was consulted earlier in the Landmark-based than in the Street-based condition. A positive value indicated the reverse.

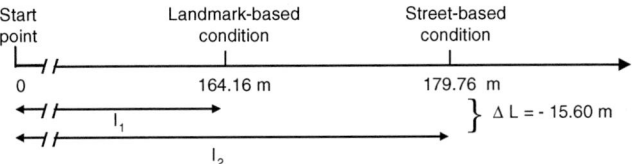

Fig. 2. Example of a calculated difference (Instruction #13)

The results are shown in Figure 3. The first two differences were null, indicating that the corresponding instructions were consulted at exactly the same locations (in this case, the starting point). Twenty differences were negative, and 8 of these were significant differences. Three values were positive, only one being significantly different. Because the consultation of a new instruction means that the preceding one is no longer relevant, we had to consider the content of every instruction that preceded those which induced significant differences. We found that of the 9 significant differences, 3 referred to instructions common to both descriptions, and 6 to instructions that differed in the two descriptions. This last finding indicates that the corresponding landmarks cited in the description made it possible to consult the next instruction sooner. As landmarks are bigger than street name plates, they are identified earlier, and consequently the next instruction is consulted earlier in this condition. This, however, does not suffice to account for all the differences. For instance, the skating rink was not identifiable from a distance as its entrance and name plate faced away from the direction of navigation. Nevertheless, the participants still consulted the next instruction earlier in the Landmark-based condition.

Map Drawing. The maps that participants drew were analyzed according to several criteria. As shown in Table 3, we focused here on the number of items recalled. Further, we considered only the 10 streets or 10 landmarks that had been mentioned in the descriptions. Actually, the participants reported very few other pieces of information in their drawings. These were only landmarks, and they were given only by the participants of the Landmark-based condition. In the Street-based condition, participants drew more streets than landmarks [$F(1,38) = 31.53$, $p < .0001$], and the reverse pattern emerged in the Landmark-based condition [$F(1,38) = 212.42$, $p < .0001$]. Interestingly, in the Street-based condition, the participants drew fewer streets than their counterparts drew landmarks in the Landmark-based condition [$F(1,38) = 60.48$, $p < .0001$].

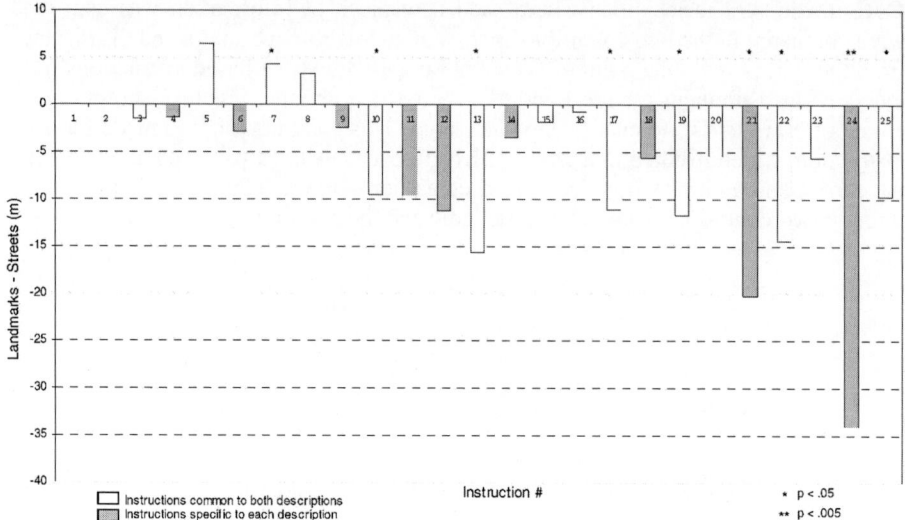

Fig. 3. Differences in consultation locations for both conditions

Table 3. Average number of streets and landmarks drawn on the map for both conditions

	Street-based condition	Landmark-based condition
Streets	2.24 (1.78)	0.00 (0.00)
Landmarks	0.30 (0.80)	6.60 (1.76)

2.3   Discussion

Experiment 1 clearly showed that street-based instructions were much less efficient than landmark-based ones for guiding someone unfamiliar through an urban environment. This was first demonstrated by the analysis of behavioral indices collected during navigation. The frequency of errors was not affected by the use of landmark or street information. Since the participants were allowed to consult the instructions while they were walking, the probability that they would stray from the intended route was almost zero. However, we found that stops and checkings were both more frequent and longer when participants were being guided by street-based information. Further, the Street-based description tended to be followed step by step, since no anticipation in consultation was seen. Finally, map drawings showed that the memory of the route was also impaired when participants had been guided by street-based information. Overall, our findings support the contrast between streets and landmarks in the use and memory of spatial information, and the critical role of landmarks in processing route directions.

3   Experiment 2: Generating Route Directions during Navigation

Results from Experiment 1 showed that landmarks are more efficient than street names for guiding people along a novel route in an urban environment and helping them memorize information about it. However, one can wonder whether this differential effect in processing the two sets of information is confirmed when people are *describing* urban routes. In particular, would their descriptions refer less frequently to streets than to landmarks?

3.1   Method

Route. The route was the same as the one used in Experiment 1.

Participants. The participants were 24 undergraduates (12 females, 12 males; mean age: 23.33, $SD = 5.06$). None of them had participated in the previous experiment.

Procedure. The experimental sessions were run individually for each participant. The participants were first accompanied by the experimenter to the starting point (the Town Hall, for one half of the participants, and the Park Maintenance Service for the other half). Those who started from the Town Hall were informed that the target was the Park Maintenance Service (Direction A), and those who started from the Park Maintenance Service were informed that the target was the Town Hall (Direction B). The participants were then invited to follow the experimenter along the route, and to describe it aloud in such a way that these instructions would allow other people unfamiliar with the surroundings to find their way. The participants were equipped with a Dictaphone and a microphone. When they reached the end point, the participants were brought back to the starting point (by a different route). Once they were back at the starting point, the participants were asked to navigate along the same route a second time, and to describe it again. This time, the participants were followed by the experimenter.

3.2   Results

Coding the Verbal Productions. First, all 48 descriptions were transcribed. They were analyzed according to the procedure designed by Denis (1997). All statements were formatted into a set of minimal units of information. These standardized propositions were then divided into 12 classes:

- Class 1: Action ("Go straight on")
- Classes 2 / 3: Action with reference to a landmark / a street ("Pass a laundry", "Take Victor Griffuelhes Street on the left")
- Classes 4 / 5: Introduction of a landmark / a street ("Face a bar", "There's another street")
- Classes 6 / 7: Description of a landmark / a street ("The bar is named Black and White", "It's a narrow street")
- Classes 8 / 9: Location of a landmark relative to another landmark / a street ("The pedestrian walkway is on the left of the Town Hall", "Pizza Hut is at the corner of the street")
- Classes 10 / 11: Location of a street relative to a landmark / another street ("Victor Griffuelhes Street is just after the Bijou Hotel", "The street is perpendicular to Victor Griffuelhes Street")
- Class 12: Commentaries ("You can't miss it")

Quantitative Analyses. Data were submitted to an ANOVA with Directions (A or B) as the between-participant factor and Descriptions (first or second) as the within-participant factor.

We first considered the number of propositions that the participants generated during navigation (cf. Table 4). The number of propositions was similar in both descriptive episodes: $F(1,22) = 1.67$, $p > .05$. However, *post hoc* analyses showed that in the second description, fewer propositions were used for Direction B than for Direction A ($p = .01$).

Table 4. Average number of propositions for both descriptions and both directions

	Description 1	Description 2
Direction A	66.50 (31.40)	67.00 (27.89)
Direction B	59.17 (24.75)	48.42 (15.80)

We wondered whether this difference was true for the whole route, or whether it only emerged for specific portions of it. We therefore divided the route into 7 segments delimited by the 6 reorientation points. However, because the first segment of the route was very short, data from segments 1 and 2 were merged. We therefore analyzed 6 segments, considered as repeated measures. As shown in Figure 4, the differences between descriptions were mainly located in specific parts of the route, and even appeared as early as the first description for a specific portion of the route.

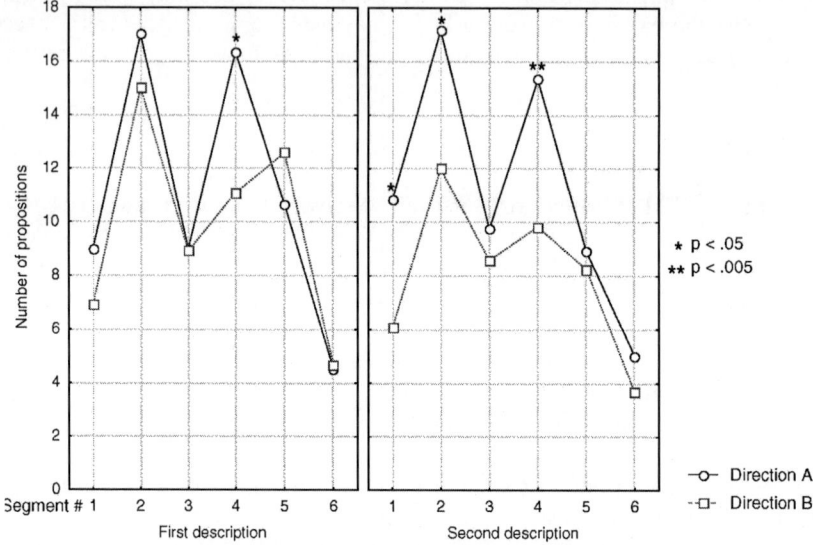

Fig. 4. Average number of propositions per segment for both directions and both descriptions

The correlation between the number of propositions for Descriptions 1 and 2 was significant: $r(22) = 0.72$, $p < .005$. Table 5 shows that this correlation was confirmed for almost all parts of the route. Thus, although the protocols were quite different in length, the differences between participants appeared to reflect reliable individual characteristics.

Table 5. Correlations between numbers of propositions for each segment of the route

	Segment #	1	2	3	4	5	6	1	2	3	4	5	6
		Description 1						Description 2					
Description 1	1	1											
	2	.66**	1										
	3	.66**	.87**	1									
	4	.70**	.70**	.63**	1								
	5	.48*	.85**	.78**	.62**	1							
	6	.52**	.60**	.63**	.43*	.60**	1						
Description 2	1	.67**	.46*	.43*	.72**	.24	.28	1					
	2	.38	.55**	.31	.65**	.43*	.20	.48*	1				
	3	.39	.59**	.49*	.63**	.47*	.24	.42*	.77**	1			
	4	.60**	.58**	.46*	.80**	.43*	.22	.64**	.72**	.59**	1		
	5	.27	.53**	.47*	.52**	.49*	.37	.34	.50**	.64**	.32	1	
	6	.48*	.63**	.64**	.61**	.54**	.36	.59**	.52**	.51**	.52**	.49*	1

$* p < .05$ $** p < .005$

Content Analyses. The further point of interest was the distribution of propositions across the 12 classes defined earlier. Table 6 shows this distribution. It was found to be consistent for both descriptions, for both directions of the route, and for all 6 segments of the route, since the main effects of these factors were not found to be significant (all $ps > .05$). *Post hoc* analyses were performed on these data. In particular, we considered the presence of clusters. A cluster was defined as a group of classes of which the frequency of occurrence did not significantly differ from each other, but did differ significantly with respect to the other classes. Three clusters emerged (see Table 6). The first one comprised actions alone or actions associated with a landmark or a street, together with introductions of landmarks. Interestingly, streets were used as frequently as landmarks when they were associated with an action. The second cluster included descriptions of landmarks and introductions of streets. Streets were therefore introduced less frequently than landmarks ($p < .0001$). The third cluster included all the remaining classes. This cluster notably showed that the participants again felt it less necessary to describe a street than a landmark ($p < .005$). Finally, when considering the five classes that included references to landmarks, their summed frequencies amounted to 53.33%, whereas the corresponding number was only 27.21% for the classes that focused on streets.

Analysis of Directional Errors. Though the experiment was not designed to examine the behavior of participants during navigation, we nonetheless recorded the number of directional errors committed by the participants during the second walk. If they strayed from the intended route, the participants were immediately stopped and taken back to the place were the error occurred. As shown in Table 7, there were few errors, but an interesting result emerged, in that all errors were committed by the participants navigating in Direction B. The difference between Directions A and B was thus significant: $F(1,22) = 7.32, p < .01$. Furthermore, the errors always occurred at the same place (in segment 2).

Table 6. Ranked frequency of occurrence of each class

	Class	Frequency (%)
Action with reference to a landmark	2	20.57
Action	1	17.54
Introduction of a landmark	4	16.62
Action with reference to a street	3	15.73
Description of a landmark	6	9.65
Introduction of a street	5	6.30
Description of a street	7	3.62
Location of a landmark relative to a street	9	3.44
Location of a landmark relative to another landmark	8	3.25
Commentary	12	1.73
Location of a street relative to a landmark	10	1.11
Location of a street relative to another street	11	0.45

Note: Limits of clusters are shown by double lines

Table 7. Number of directional errors

	Direction of walking	
	A	B
Directional errors	0.00 (0.00)	0.47 (0.60)

3.3   Discussion

The analysis of descriptions first showed that the direction of walking along the same route resulted in differences in the length of the descriptions. These differences appeared from the first walk for a portion of the route, and may be explained by an environmental asymmetry in the frequency of salient landmarks. This asymmetry was also responsible for the more frequent directional errors for Direction B. As a matter of fact, it is much easier to walk straight on in the direction of a big avenue (Direction A) than to come from this avenue and then take the fourth perpendicular street on the left (Direction B). Interestingly, in the second walk, differences in the length of the descriptions also emerged for two other parts of the route. This time, these differences could be attributed to a change in the content of the instructions. For instance, instead of enumerating a series of landmarks, participants could refer to a single avenue and so could be more concise. Furthermore, the variability observed in the length of descriptions was consistent for all parts of the route. This result probably reflects a cognitive style.

The second point of interest was the content of the directions, and particularly the reference to either landmarks or streets. Streets were less often introduced and described than landmarks, and they were associated with actions to be executed less frequently (although non-significantly) than landmarks were. The articulation between actions and places is known to be the core prescriptive component of route directions. Further, the frequency of occurrence of this class of items accounts for the efficiency of instructions (e.g., Daniel et al., in press; Denis, 1997). The most striking result is that directions contained twice as many references to landmarks as to streets.

4   Conclusions

Overall, our studies have confirmed the existence of a marked contrast between street and landmark information in route directions. Although street names offer an ideal solution to many referential problems, they appear to be poor guides, at least for a person totally unfamiliar with an environment. Furthermore, the spatial knowledge they provide is not as helpful as that provided by landmarks (Allen, 2000; Denis, 1997; Fontaine & Denis, 1999; Michon & Denis, 2001). Street names are nonetheless used in route directions, although less often than landmarks. In addition, our data show that the direction of walking does not affect this phenomenon.

Any effort to enhance the quality of urban route directions must consider these findings. Behavioral data of the kind reported here can be used as a basis for developing a formal model of landmark saliency. For instance, Raubal and Winter (2002) have considered cognitive and perceptual concepts to define attractive landmarks that allow newcomers to a city to successfully reach their destination. Recent work has also shown that the selection of salient features can be computed automatically (Nothegger, 2003). Our findings can therefore contribute to the improvement of route planning services by including appropriate features for the cognitive expectations of the human users.

References

1. Allen, G. L. (2000). Principles and practices for communicating route knowledge. *Applied Cognitive Psychology, 14*, 333-359.
2. Bahrick, H. P. (1983). The cognitive map of a city: Fifty years of learning and memory. In G. H. Bower (Ed.), *The psychology of learning and motivation: Advances in research and theory* (pp. 125-163). New York: Academic Press.
3. Bell, A. B., Greene, T. C., Fischer, J. D., & Baum, A. (1996). *Environmental psychology*. Fort Worth: Harcourt Brace College Publishers.
4. Cohen, G., & Burke, D. M. (1993). Memory for proper names: A review. *Memory, 1*, 249-263.
5. Cornell, E. H., Heth, C. D., Kneubuhler, Y., & Sehgal, S. (1996). Serial position effects in children's route reversal errors: Implications for police search operations. *Applied Cognitive Psychology, 10*, 301-326.
6. Cornell, E. H., Heth, C. D., & Skoczylas, M. J. (1999). The nature and use of route expectancies following incidental learning. *Journal of Environmental Psychology, 19*, 209-229.
7. Daniel, M.-P., Tom, A., Manghi, E., & Denis, M. (in press). Testing the value of route directions through navigational performance. *Spatial Cognition and Computation*.
8. Denis, M. (1997). The description of routes: A cognitive approach to the production of spatial discourse. *Current Psychology of Cognition, 16*, 409-458.
9. Denis, M., Pazzaglia, F., Cornoldi, C., & Bertolo, L. (1999). Spatial discourse and navigation: An analysis of route directions in the city of Venice. *Applied Cognitive Psychology, 13*, 145-174.

10. Fontaine, S., & Denis, M. (1999). The production of route instructions in underground and urban environments. In C. Freksa & D. M. Mark (Eds.), *Spatial information theory: Cognitive and computational foundations of geographic information science* (pp. 83-94). Berlin: Springer.
11. Galea, L. A. M., & Kimura, D. (1993). Sex differences in route learning. *Personality and Individual Differences, 14*, 53-65.
12. Golledge, R. G. (1993). Geographical perspectives on spatial cognition. In T. Gärling & R. G. Golledge (Eds.), *Behavior and environment: Psychological and geographical approaches* (pp. 16-46). Amsterdam: North Holland.
13. Heth, C. D., Cornell, E. H., & Alberts, D. M. (1997). Differential use of landmarks by 8- and 12-year-old children during route reversal navigation. *Journal of Environmental Psychology, 17*, 199-213.
14. Ioerger, T. R. (1994). The manipulation of images to handle indeterminacy in spatial reasoning. *Cognitive Science, 18*, 551-593.
15. Izaute, M. (1999). De la dénomination: la spécificité des noms propres. *L'Année Psychologique, 99*, 731-751.
16. Kalakoski, V., & Saariluoma, P. (2001). Taxi drivers' exceptional memory of streets names. *Memory and Cognition, 29*, 634-638.
17. Kirasic, K. C., & Mathes, E. A. (1990). Effects of different means for conveying environmental information on elderly adults' spatial cognition and behavior. *Environment and Behavior, 22*, 591-607.
18. Mani, K., & Johnson-Laird, P. N. (1982). The mental representation of spatial descriptions. *Memory & Cognition, 10*, 181-187.
19. Michon, P.-E., & Denis, M. (2001). When and why referring to visual landmarks in direction giving? In C. Freksa & D. M. Mark (Eds.), *Spatial information theory: Cognitive and computational foundations of geographic information science* (pp. 292-305). Berlin: Springer.
20. Nothegger, C. (2003). Automatic selection of landmarks. Unpublished master's thesis (Diplomarbeit), Vienna Technical University, Vienna, Austria.
21. Raubal, M., & Winter, S. (2002). Enriching wayfinding instructions with local landmarks. In M. J. Egenhofer & D. M. Mark (Eds.), *Geographic information science* (pp. 243-259). Berlin: Springer.
22. Rossano, M. J., & Reardon, W. P. (1999). Goal specificity and the acquisition of survey knowledge. *Environment and Behavior, 31*, 395-412.
23. Schneider, L. F., & Taylor, H. A. (1999). How do you get there from here? Mental representations of route descriptions. *Applied Cognitive Psychology, 13*, 415-441.
24. Streeter, L. A., Vitello, D., & Wonsiewicz, S. A. (1985). How to tell people where to go: Comparing navigational aids. *International Journal of Man-Machine Studies, 22*, 549-562.
25. Taylor, H. A., Naylor, S. J., & Chechile, N. A. (1999). Goal-specific influences on the representation of spatial perspective. *Memory and Cognition, 27*, 309-319.
26. Taylor, H. A., & Tversky, B. (1992). Spatial mental models derived from survey and route descriptions. *Journal of Memory and Language, 31*, 261-292.
27. Thorndyke, P. W., & Hayes-Roth, B. (1982). Differences in spatial knowledge acquired from maps and navigation. *Cognitive Psychology, 14*, 560-589.

Extracting Landmarks with Data Mining Methods

Birgit Elias

Institute of Cartography and Geoinformatics
University of Hannover
Appelstr. 9a
30161 Hannover, Germany
`Birgit.Elias@ikg.uni-hannover.de`

Abstract. The navigation task is a very demanding application for mobile users. The algorithms of present software solutions are based on the established methods of car navigation systems and thus exhibit some inherent disadvantages: findings in spatial cognition research have shown that human users need landmarks for an easy and successful wayfinding. Typically, however, an object is not a landmark per se, but can be one relative to its environment. Unfortunately, these objects are not part of route guidance information systems at the moment.

Therefore, it is an aim of research to make landmarks for routing instructions available. In this paper we focus on a method to automatically derive landmarks from existing spatial databases. Here a new approach is presented to investigate existing spatial databases and try to extract landmarks automatically by use of a knowledge discovery process and data mining methods. In this paper two different algorithms, the classification method ID3 and the clustering procedure Cobweb, are investigated, whether they are suitable for discovering landmarks.

1   Introduction

The wayfinding process in unknown environments is a very difficult and challenging task. Consequently, the interest to solve this problem by automated computed routing is enormous. Route guidance systems for vehicles are an established technology that are available even in middle class cars today.

Automatically produced routing directions are based on the data and concepts of current car navigation systems: the user gets turning instructions and metric distance measures. This concept works quite well, because of the restriction of automobiles to the road network. Currently, to improve usability navigation databases are enriched with additional information. Locations of petrol stations, pharmacies, public buildings and restaurants - so called points of interest - are part of the database. Another recent development is to transform the presentation of the route in the car into 3D-portrayals with models of historic buildings merged into the database to satisfy the touristic interest of the user.

W. Kuhn, M.F. Worboys, and S. Timpf (Eds.): COSIT 2003, LNCS 2825, pp. 375–389, 2003.

The increasing amount of small and mobile technologies leads to a new group of user: the pedestrian. This evokes new problems for the automatic, computer driven navigation: pedestrians are not tied to the road network.

It makes no sense to simply adopt the current concepts of car navigation to a mobile application for pedestrians. The discipline of spatial cognition investigates the human wayfinding concepts. Findings reveal, that humans need salient objects in the environment - so called *landmarks* - to orient and navigate themselves through space. But these objects are not per se a landmark, because being a landmark is a relative property depending on the local environment. Besides all that the capability of humans to estimate metric distances correctly are very poor.

Many research approaches in the field of spatial cognition have been undertaken about the wayfinding process and theory of landmarks, but only a few deal with the practical procedure to extract landmarks automatically from existing databases. In this approach it is investigated, whether the salient objects in the environment can be extracted from spatial databases with methods of spatial data mining.

This paper is organized as follows: the basic theory and related work of navigation with landmarks and the principles of data mining are presented. The approach using knowledge discovery with data mining methods for discovering landmarks in the data is outlined. After that the existing database which is used for the extraction process is introduced and examined. The extraction approach is presented including the data preparation and preprocessing steps. Finally the applicability of the approach to discover landmarks by use of automatic procedures is evaluated.

2   Related Work

2.1   Basic Theory of Navigation with Landmarks

There are two different kinds of route directions to convey the navigational information to the user: either in terms of a description (verbal instructions) or by means of a depiction (route map). According to (Tversky & Lee 1999) the structure and semantic content of both is equal, they consist of landmarks, orientation and actions. Using landmarks is important, because they serve multiple purposes in wayfinding: they help to organize space, because they are reference points in the environment and they support the navigation by identifying choice points, where a navigational decision has to be made (Golledge 1999). Accordingly, the term landmark stands for a salient object in the environment that aids the user in navigating and understanding the space (Sorrows & Hirtle 1999). In general, an indicator of landmarks can be particular visual characteristic, unique purpose or meaning, or central or prominent location.

Furthermore landmarks can be divided into three categories: visual, cognitive and structural landmarks. The more of these categories apply for the particular object, the more it qualifies as a landmark (Sorrows & Hirtle 1999).

This concept is used by (Raubal & Winter 2002) to provide measures to specify formally the landmark saliency of buildings: the strength or attractiveness of landmarks is determined by the components visual attraction (e.g. consisting of façade area, shape, colour, visibility), semantic attraction (cultural and historical importance, explicit marks (e.g. shop signs)) and structural attraction (nodes, boundaries, regions). The combination of the property values leads to a numerical estimation of the landmarks' saliency.

A study of (Lovelace, Hegarty & Montello 1999) includes an exploration of the kinds and locations of landmarks used in directions. It can be distinguished between four groups: choice point landmarks (at decision points), potential choice point landmarks (at traversing intersections), on-route landmarks (along a path with no choice) and off-route landmarks (distant but visible from the route). A major outcome of the study is that choice point and on-route landmarks are the most used ones in route directions of unfamiliar environments.

2.2   Principles of Data Mining

With the increasing amount of digital data stored, the demand raises to interpret and derive useful knowledge from these 'data mountains'. Therefore, a new generation of tools and techniques for automated and intelligent database analysis has been developed, called knowledge discovery in databases (KDD). This term can be defined as the discovery of interesting, implicit, and previously unknown knowledge from large databases (Frawley, Piatetsky-Shapiro & Matheus 1991). This research field combines topics from machine learning, pattern recognition, databases, statistics, artificial intelligence, knowledge acquisition for expert systems and data visualization. An overview about the topic is given in (Fayyad, Piatetsky-Shapiro & Smyth 1996 a).

The term 'knowledge discovery in databases' comprises the overall process of finding and interpreting patterns from data, while 'data mining' only refers to the stage of data analysis without the additional steps. So the KDD process includes preprocessing of data, data mining itself and postprocessing, as well as the interpretation of potentially discovered patterns. The KDD process is interactive, iterative and consists of the following steps (according to (Fayyad et al. 1996 a)):

Selection: developing prior knowledge, goals of end-user, creating a data set

Preprocessing: data cleaning (removal of noise), deciding on strategies for handling missing data, data reduction (find the representing features), data transformation (harmonize different scales and attribute types)

Data mining: choose the data mining method, choose the data mining algorithm, data mining process

Interpretation: interpretation of mined patterns, verifying discovered knowledge

Data mining methods are algorithms designed to analyse data or to extract from data patterns into specific categories (Fayyad, Piatetsky-Shapiro, Smyth & Uthurusamy 1996 b). Basic models of data mining are clustering, regression models, classification, summarization, link and sequence analysis. The algorithms used are taken from the fields of machine learning, neural networks, rough set and fuzzy set theory or statistics (Cios, Pedrycz & Swiniarsky 1998).

Theses procedures can be applied to data sets consisting of collected attribute values and relations for objects. Attribute types can be nominal (values are categories with no ranking), ordinal (values have meaningful ordering) and numeric (values can be continuous, discrete, interval). Potential problems to be solved in the preprocessing stage are either incomplete or missing data in the data set or different attribute types to be handled.

In the last years the scope of data mining from relational databases has been extended to spatial databases. The knowledge discovered through spatial data mining can be of various forms: providing characteristic and discriminant rules, extraction and description of prominent structures or clusters, spatial associations, and others (Koperski, Adhikary & Han 1996), (Anders & Sester 2000).

The algorithms can be devided in two basic techniques. According to the terminology of the machine learning community, there are methods for:

- learning from examples (*supervised learning*)
- learning from observation (*unsupervised learning*)

One very prominent example for supervised learning techniques is the ID3 algorithm by Quinlan (Quinlan 1986). It needs a set of classified examples and counter-examples and generates a decision tree, which can be used for classification of unknown examples. The method only deals with complete and nominal data sets and subdivides the training set of examples into homogeneous groups using discriminating attributes, whose selection is based on information gain calculation (Koperski et al. 1996). There are also extensions that enable the treatment of quantitative values as well. An application of knowledge acquisition for automatic interpretation of spatial data using ID3 is given in (Sester 2000).

The incremental conceptual clustering algorithm Cobweb is an example for unsupervised learning: clustering procedures divide examples in natural groups with similar features depending on a specific neighborhood concept. The chosen approach used here divides the examples into hierarchical groups. It is able to process data records in the form of symbolic attribute-value lists and builds up a decision tree (Witten & Eibe 1999).

3   An Approach of Extracting Landmarks

The above mentioned approach to enrich wayfinding instructions with landmarks (Raubal & Winter 2002) follows the categories visual, semantic or structural attraction, introduced by (Sorrows & Hirtle 1999) (see Sect. 2.1). Additionally, for grouping the structural elements the acknowledged structure elements of a city (nodes, boundaries, regions) named by (Lynch 1960) are used. The result is a catalogue of properties for each attraction element, which has to be filled with real data to make an automatic identification of landmarks possible. A problem could lie of course in data retrieval. Either there exists no database with the needed information or it is not available for greater areas. Maybe it is distributed over different databases and the collecting of all is too time-consuming and expensive like in the case of points of interest.

One way is to investigate existing topographic data sets: we investigate the information of available digital databases and extract the salient objects by analysing the real data. A first study to use existing databases for landmark extraction is given in (Elias 2002). In that study a set of specific feature types have been characterized as landmarks and have subsequently been extracted from the data sets. The problem is, however, that the quality as a landmark is relative: a multi-stored building may be an excellant landmark in a small city with only one such building - in New York however, it would be useless as a navigation aid. Therefore, in this paper the relative uniqueness of an object in its environment is taken into account. Thus, we start to investigate data mining methods for discovering landmarks.

The idea is that objects, which have a unique attribute in a certain environment, qualify as landmarks. Therefore, the underlying model is to compare the attribute values of all data records: These objects with distinct or even unique values differing from the global mean have to be something special. The procedure will also lead to an attribute ranking according to their importance for the model. If the chosen attributes are suitable for developing a global object schema and outliers from this schema present something particular, it is possible to determine landmarks through statistics and data mining methods.

Consequently, our approach passes through the stages of a knowledge discovery process introduced by Sect. 2.2. As the goal is to extract the landmarks from a database, we need a data set. In this case we concentrate on the building objects. We enrich our explicit given information about buildings by deriving attributes and relations with the help of spatial analysis. The combination of different attribute-values leads to derived attributes, e.g. building length to width ratio. The data have to be preprocessed to unify the different attribute types, missing values have to be handled.

There are several open-source data mining and statistics tools available in the internet, for our investigation we use the WEKA implementation (Witten & Eibe 1999). This tool provides different data mining modules including preprocessing options.

4 Landmarks Discovery in Building Data

4.1 Data Source and Selection

In this paper we focus on the digital cadastral map of lower saxony to use the benefits of an object oriented vector database of area-wide availability. This digital map includes buildings, parcels and their land use. Besides the geometry of the objects, the following attributes are available:

- building use: residential, public, underground, outbuildings

- land use types: public purposes, residential, commerce and service, industrial, mixed land uses, traffic, park, garden, sports, etc.
- building labels: name or function of building (e.g. town hall, kindergarten, church)
- special building parts with a roof: winter garden, car port etc.

To keep the first approach on a simple level, we simplify the conditions. We assume that for this analysis process the objects present a homogeneous group with predominantly identical attributes and that it is not possible to compare buildings with roads for example, because they have both important attributes not existing in the other group (building use: residential, official, etc.; road classification: highway, community road, etc.). In this first step, we decide to investigate the group of buildings and compose a complete attribute-value list for each building object.

The derived attribute values are calculated by combining all explicit given information of the used data sets. In most cases the provided attributes describe geometric or topologic properties of the buildings, additionally some semantic information is given.

The next step is to determine the situation, in which a landmark is needed for the wayfinding process and to convey the spatial cognition findings to our study. According to (Michon & Denis 2001) pedestrians simply progress along a route by directing themselves towards a landmark, thus landmarks are especially needed, when a turn on the route is required. Therefore, we try to provide a landmark at every (potential) choice point (or close to it) on the route. At every turning point of the route we start an examination, whether there is an applicable object in our data sets.

The amount of data makes it necessary to select the affected data sets with respect to the users perspective and visibility within a predefined neighborhood. An overview is given in Fig. 1 showing a city situation with two different neighborhoods. The inner circle represents a distance of 50 m from the point of view and contains in this case nearly 20 buildings, the outer circle stands for a 100 m radius and more than 30 buildings, which can be used for searching for a salient one. Depending on the built-up area situation different kinds of visibility zones or neighborhoods can be established. If there is a dense, cropped situation a narrow neighborhood (for example the 50 m distance) is sufficient for the analysis. In this scenario we disregard the fact that a user usually move towards a certain direction and therefore wants a landmark in front of him, not behind him. We only investigate the situation, that the user is standing in the middle of the road and searches for landmarks to support his current decision and guide his next movements.

4.2 Preprocessing: Providing Attribute-Value Lists

With respect to the formerly mentioned cadastral data set, the attribute-value table has to be composed for the selected buildings. All existing information

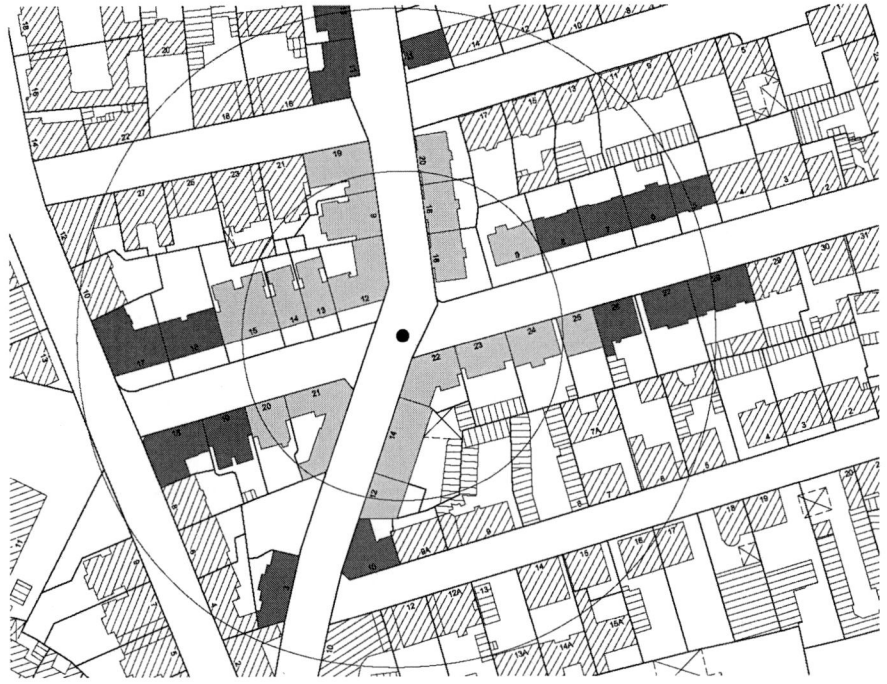

Fig. 1. Selected buildings: neighborhood 50 m and 100 m

about the objects has to be extracted from the database: information about semantics and geometry of the object itself, but also information about the topology, in this case neighborhood relations to other buildings and other object groups (for examples the distance to roads, to the parcel boundary etc.) are collected.

The derived attributes available from the digital cadastral map are shown in Tab. 1. One profound disadvantage of the database content is the lack of height values for the buildings.

Table 1: Attributes and relations of buildings

no	attribute	description	graphic
1	building use	public, residential, outbil-dung ...	
2	building label	name or function of build-ing	
3	size of building	length b * width a in $[m^2]$	

4	elongation	derivation to typical building form; ratio length/width	
5	number of corners	counting quoins (normal: 4 or 6)	
6	single building	all alone, single in a row, one neighbour, two neighbours	
7	building moved away from road	closest distance [m]	
8	ratio of building area to parcel area	$\frac{building\ ground\ area}{parcel\ area}$	
9	density of buildings (direct neighbourhood)	$\frac{number\ of\ buildings}{area\ (100*100m)}\left[\frac{1}{m^2}\right]$	
10	density of buildings (district)	$\frac{number\ of\ buildings}{area\ (500*500m)}\left[\frac{1}{m^2}\right]$	
11	orientation to road	along (length towards road), across (width), angluar, building at corner [grad]	
12	orientation to north	angle building length to north t [rad]	
13	orientation to neighbor	difference angle to neighbour (t1 - t2) [rad]	
14	perpendicular angle in building	derivation of angles to normal [rad]	

15	parcel land use	public, residential, commerce and service, industrial, traffic, ...	
16	number of buildings on parcel	counting buildings	
17	special building objects on parcel	number of car ports, winter garden etc.	
18	neighbor parcel land use	difference yes, no	
19	form of parcel area	number of corners, number of neighbors	

4.3   Data Preprocessing

Table 1 provides values in different units: there is information in nominal categories (building use: public, residential, ...) or ordinal values (number of neighbors: 0, 1, 2) or even different metric values (distance to road [m], orientation to north [rad]). Depending on the chosen data mining algorithm different attribute types and units are needed, so the data has to be transformed. One procedure is for example the changeover from metric values to categories: the size of the buildings can be given in squaremetres or size classes can be set up (small, large). We have to take care with the transformation of different attribute types, because the type of transformation may influences the results of the whole analysis process. The specifications of needed attribute types change from algorithm to algorithm, but in data mining software automatic transformation operations are available (Witten & Eibe 1999). To avoid the above mentioned problems and make the interpretation of the results easier, we simplify our approach in this first study by creation and use of a synthetic data set, which relies mostly on the existing data model of cadastral data. We build up a data set with limited attributes of the fictitious situation given in Fig. 2. As described in Sect. 4.1, we assume that the user stands at a decision point and searches for a landmark in the immediate environment. In this case the next neighborhood in a specified radius consists of 10 buildings used for the extraction procedure. The single objects are characterized by the following features with nominal values:

- building use: residential, public, outbuilding
- size of building: small, large

- number of immediate neighbors: 0, 1, 2
- orientation towards road: parallel, across, angular, corner
- distance from road: 0 or 3 metres
- building height: 12, 15 or 17 metres

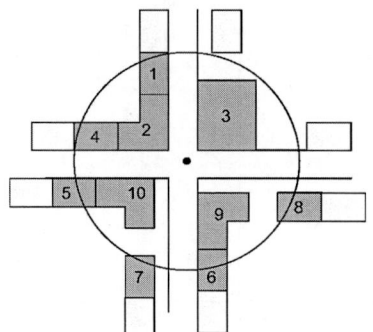

no	building use	size	neigh.	orientat.	dist.	height
1	residential	small	2	parallel	0	15
2	residential	small	2	corner	0	15
3	residential	large	0	corner	0	15
4	residential	small	2	parallel	0	15
5	residential	small	2	parallel	0	15
6	residential	small	2	parallel	0	15
7	residential	small	1	parallel	3	15
8	residential	small	1	parallel	3	15
9	residential	small	1	corner	3	15
10	residential	small	1	corner	3	15

Fig. 2. Synthetic test data

Obviously, building number 3 is somehow singular and different to all other buildings: it has no direct contact to neighbored buildings and is the only large sized object. Thus, in the local environment this building is something particular and we expect the data mining methods to reveal this knowledge.

4.4   Data Mining

We apply our test data set to the introduced data mining algorithms in Sect. 2.2: a classification method using ID3 and a clustering approach using Cobweb. The used software WEKA is described in (Witten & Eibe 1999). Both procedures process the attribute values as nominal categories, thus no transformation is needed.

Classification with ID3. This algorithm is a method for supervised learning and therefore needs classified examples. The result is a decision tree providing the shortest optimal description possible for a classification into the given classes. As in our study there are no classified examples, we use this procedure iteratively and enhance our data set with the class landmark (values: yes, no). Therefore, we iteratively hypothesize each building to be a landmark, whereas all the other buildings are no landmarks. 'Real' / true landmarks then are identified by yielding the most simple / shortest description.

In Fig. 3 the results for three different buildings (1, 8, 3) are shown. In the boxes the discriminating attribute (e.g. neighbors) is shown, at the branches and leafs the attribute values (0, 1, 2) and the decision (landmark yes, no or null) are given. 'Null' indicates that no existing examples with such values are existing in the training data set.

The decision tree for identifying building number 8 as landmark (see middle of Fig. 3) reads as follows: If the number of neighbors is 1 and the orientation is parallel, then the object is a landmark.

On the left hand side of Fig. 3 the first building is determined being a landmark, but the decision tree provides no decision rule to determine a landmark positively at all. Consequently, this decision tree makes no sense and building number one is no landmark.

In contrast the result for building 3 (given at the right hand side of Fig. 3) shows the discriminating attribute 'size', which immediately distinguishes between landmark and no landmark. It needs only a very short description corresponding to less branching in the tree to detect that a landmark discriminates from all other objects by only one attribute. And in fact, in our synthetic data set this record holds the only attribute-value being singular.

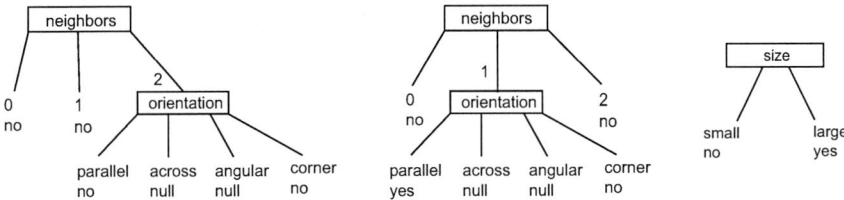

Fig. 3. ID3 - results of records 1, 8 and 3

The remaining data records provide decision trees comparable to the left and the middle parts of Fig. 3, each providing either no positive landmark decision rule at all or only a longer desription with more branchings in the tree than the decision tree of record number 3. The software provides standard deviations for the calculation, admitting that in most cases no complete classification of all examples was possible.

The conclusion can be drawn that a landmark is characterized by a short decision tree with only a few levels that leads to positive landmark decision. From this result, the guiding instruction for that junction would be e.g. 'pass the large building' (or 'turn left at large building').

Clustering with Cobweb. This is a hierarchical clustering algorithm and thus an unsupervised learning method. Using this technique needs no explicit examples. Unclassified examples are parted in a hierarchy of natural groups by this procedure.

The results of our data set computed with Cobweb are given in the decision tree in Fig. 4. The depicted tree shows the discrimination of the 10 instances in different branches. On the first level the records are diverted in three clusters consisting of 1, 4 and 5 instances. The latter two were further separated in two clusters containing only instances containing completely identical examples, where all attributes have the same values. The boxes enumerate the values of all attributes in the given cluster, below the boxes the corresponding instances are referenced to make a comparison with Fig. 2 possible.

As it is clearly visible, the procedure aggregates similar instances into clusters. The more deviations in attribute values exist, the greater the distance in the hierarchical tree gets. Considering the cluster of the first level (see Fig. 5) the grey circles reveals the similiarity of the instances within the clusters. The groups built with 4 and 5 instances on the first level have 5 of 6 attribute values in common. Comparing the clusters of level one to each other, decreasing similiarity becomes clear. There are 2 or 3 attribute values identical, respectively this difference leads to the subdivision into 3 branches. The attributes of building number 3 differ very clearly from the other instances and therefore this record is isolated from the other clusters in one of the first levels.

Since the algorithm subdivides the records into similar groups, an instance with strongly different attribute values is separated from the others at a very high level in the decision tree. Because of its singularity, it is all alone in its group.

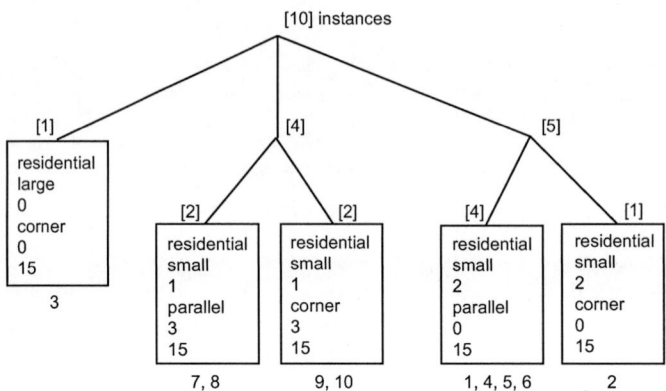

Fig. 4. Results of Cobweb

5   Conclusion and Future Work

This paper presented an approach to automatically discover landmarks in spatial databases. In this study we concentrate on a building database. Based on two different data mining methods - classification and clustering - the algorithms ID3 and Cobweb are used to analyse synthetic test data of a spatial situation. The results reveal the potential of the methods for discovering landmarks: they are

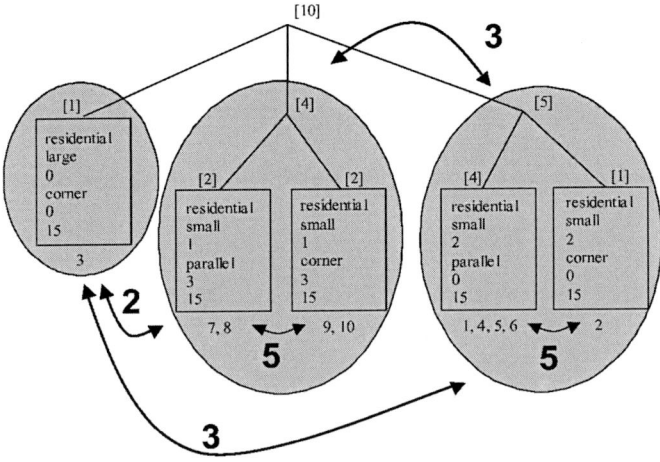

Fig. 5. Coweb decision tree: attribute values in common

characterized by being salient in their environment and the procedures reveal such singularity of objects either by the shortest possible decision tree (ID3) or by a cluster separated at a very high level containing only one element (Cobweb). Both methods seem to be suitable for the discovering process. To check the findings, whether they can be generalized for more applications, they have to be verified by applying it to real data basing on the proposed attributes in Tab. 1.

The results with these data mining methods look very promising. Both lead to the desired result of identifying locally salient objects, that are likely to be distinguishable by (a set of) simple attributes. The advantage of the ID3 algorithm is the fact, that the discriminating attributes are directly identified in the process and can directly be used as guiding instruction. Another option is to merely present a depiction, where the landmark objects are highlighted (Sester 2002). With Cobweb, in contrast, only the landmark object is identified. Its (visually) characterizing attributes have still to be extracted.

These methods provide a mechanism to fully automatically extract objects with a relative uniqueness in a given environment. It is very flexible as it allows the introduction of arbitrary objects with arbitrary attributes - and still leads to the identification of the most unique object adapted to a given situation. It could even be used to accommodate to certain external conditions like seasons (limited visibility because of dense vegetation in the summer) or day-light, which leads to a pre-selection of certain attributes and thus to different landmarks.

Further investigations on the procedure to select the users appropriate neighborhood have to be done. The moving direction of the user will be taken into account. In addition, the analysis process has to be extended to different object types (traffic constructions, parks, sporting facilities ...). Methods for data processing of different object types and categories has to be developed. The reliability of the extracted landmarks has to be determined by a quality measure to avoid ambiguous landmarks misleading the user.

The long-term objective is to provide an approach to discover the needed landmarks within spatial databases in one single step.

References

Anders, K.-H. & Sester, M. [2000], Parameter-Free Cluster Detection in Spatial Databases and its Application to Typification, *in:* 'International Archives of Photogrammetry and Remote Sensing', Vol. 33(Part B4/1), ISPRS Congress, Amsterdam, pp. 75–82. Comm. IV.

Cios, K., Pedrycz, W. & Swiniarsky, R. [1998], *Data Mining Methods for Knowledge Discovery*, Kluwer Academic Publishers.

Elias, B. [2002], Automatic Derivation of Location Maps, *in:* 'GeoSpatial Theory, Processing and Applications', Vol. 34/4, Ottawa, Canada.

Fayyad, U. M., Piatetsky-Shapiro, G. & Smyth, P. [1996a], *Advances in Knowledge Discovery and Data Mining*, AAAI Press/The MIT Press, Menlo Park, Californien, chapter From Data Mining to Knowledge Discovery: An Overview, pp. 1–34.

Fayyad, U. M., Piatetsky-Shapiro, G., Smyth, P. & Uthurusamy, R., eds [1996b], *Advances in Knowledge Discovery and Data Mining*, AAAI Press/The MIT Press, Menlo Park, Californien.

Frawley, W. J., Piatetsky-Shapiro, G. & Matheus, C. [1991], *Knowledge Discovery in Databases*, AAAI Press/The MIT Press, Menlo Park, Californien, chapter Knowledge Discovery in Databases: An Overview, pp. 1–27.

Golledge, R. D. [1999], *Wayfinding Behavior*, John Hopkins Press, chapter Human Wayfinding and Cognitive Maps, pp. 5–45.

Koperski, K., Adhikary, J. & Han, J. [1996], Spatial Data Mining: Progress and Challenges, *in:* 'Research Issues on Data Mining and Knowledge Discovery', Montreal, pp. 55–70.

Lovelace, K., Hegarty, M. & Montello, D. [1999], Elements of Good Route Directions in Familiar and Unfamiliar Environments, *in:* C. Freksa & D. Mark, eds, 'Spatial Information Theory: Cognitive and Computational Foundations of Geographic Information Science', Springer Verlag, pp. 65–82.

Lynch, K. [1960], *The Image of the City*, The MIT Press, Cambridge.

Michon, P.-E. & Denis, M. [2001], When and Why Are Visual Landmarks Used in Giving Directions?, *in:* D. Montello, ed., 'COSIT 2001', Springer Verlag, pp. 292–305.

Quinlan, J. R. [1986], 'Induction of Decision Trees', *Machine Learning*.

Raubal, M. & Winter, S. [2002], Enriching Wayfinding Instructions with Local Landmarks, *in:* M. Egenhofer & D. Mark, eds, 'Geographic Information Science', Vol. 2478 of *Lecture Notes in Computer Science*, Springer Verlag, pp. 243–259.

Sester, M. [2000], 'Knowledge acquisition for the automatic interpretation of spatial data', *International Journal of Geographical Information Science* 14(1), 1–24.

Sester, M. [2002], Application Dependent Generalization - The Case of Pedestrian Navigation,, *in:* 'GeoSpatial Theory, Processing and Applications', Vol. 34/4, Ottawa, Canada.

Sorrows, M. & Hirtle, S. [1999], The Nature of Landmarks for Real and Electronic Spaces, *in:* C. Freksa & D. Mark, eds, 'Spatial Information Theory: Cognitive and Computational Foundations of Geographic Information Science', Springer Verlag, pp. 37–50.

Tversky, B. & Lee, P. [1999], Pictorial and Verbal Tools for Conveying Routes, *in:* C. Freksa & D. Mark, eds, 'Spatial Information Theory: Cognitive and Computational Foundations of Geographic Information Science', Springer Verlag, pp. 51–64.

Witten, I. H. & Eibe, F. [1999], *Data Mining: Practical Machine Learning Tools and Techniques with Java Implementations*, Morgan Kaufmann, San Francisco.

Visual Attention during Route Learning:
A Look at Selection and Engagement

Gary L. Allen and K. C. Kirasic

Department of Psychology
University of South Carolina
Columbia, SC 29208 USA
{allen-gary,kck}@sc.edu

Abstract. Two aspects of visual attention, the selection of environmental features and the engagement of attention on those features, were examined in an
experimental study using a slide-presentation simulation of route experience.
Results showed that (a) after learning, viewers' knowledge of spatial relations
among high-information regions was more accurate than their knowledge of
spatial relations among low-information regions; (b) during learning, viewers
were more selective when looking at high-information regions than when looking at low-information regions; (c) during learning, viewers were slower to disengage attention when looking at high-information regions than when looking
at low-information regions; and (d) during learning, the most common type of
visual activity when viewers saw high-information regions were saccades between landmarks and the path's vanishing point in the scene. These findings indicate that although route learning is a relatively simple and wellpracticed task,
it involves attention in terms of the selection of highly informative regions for
in-depth coding of spatial relations.

Keywords: route learning, attention, distance knowledge.

1   Attention and Route Learning

Route learning is a common means of acquiring information relevant to wayfinding in
large-scale environments (Allen, 1999). Traversing a route involves simply making
one's way from a starting point to a destination while responding to the locomotor
affordances of the path, a feat that basically involves avoiding obstacles. Accordingly,
route learning involves a change in potential wayfinding effectiveness as a result of
previous experience with the environment. This change can be assessed in behavioral
tests focusing on errors and time to destination or in cognitive tests of spatial or procedural knowledge.

The results of route learning have been characterized in the spatial cognition literature in a number of different ways. Often, route knowledge is characterized as a series
of stimulus-response (or condition-action) associations. According to a literal interpretation of this view, route knowledge is spatial in its consequence, not in its content.
However, route knowledge can also be characterized as a type of spatial representation. Memory for a traveler's movement through the environment, both translations of
position and rotations of heading, is spatial in content, and this, too, is an aspect of
route knowledge. Another spatial aspect of route knowledge is memory for fixed

W. Kuhn, M.F. Worboys, and S. Timpf (Eds.): COSIT 2003, LNCS 2825, pp. 390–398, 2003.

spatial relations among environmental features as perceived by the traveler along a path. Thus, route learning can involve a variety of psychological processes with quite different scientific traditions, including associative learning, procedural learning, and perceptual learning.

The issue addressed in this study concerned the role of attention in acquiring route knowledge by means of perceptual learning. Previous studies have demonstrated that viewers are quite active cognitively as they learn spatial relations among features along a route by means of visual experience alone. They parse the route into segments (Allen, 1981, 1985; Berendt & Jansen-Osmann, 1997), organize information about the geometry of the route (Lindberg & Gärling, 1983), and differentiate high- and low-information regions along the path of observation (Allen, Kirasic, Siegel, & Herman, 1979; Allen, Siegel, & Rosinski, 1978). This last activity is particularly interesting in terms of its implications for the allocation of attention. If travelers differentiate between high- and low-information regions along a route, then logically they should allocate more attention to the high-information regions in order to facilitate route learning.

1.1   Attention: Two Connotations

Attention has a variety of widely accepted connotations, two of which were central to this investigation. First, attention involves selectivity. Attention connotes the selection of particular aspects of the current task situation for the purpose of either responding to them immediately (a perception-action context) or referring to them later (a learning context). That which is selected may be in the realm of things that can be immediately perceived, which is the case in most psychological experiments, or in the realm of internally represented objects and events. The obvious consequence of selection is that only a subset of things perceived or recalled can be readily available to the observer at any point in time. The implication for route learning is straightforward: While learning a route, an observer inevitably attends to a subset of perceivable environmental features.

The second connotation of attention central to this study is engagement, which refers to the degree to which effort is allocated towards the immediate task situation. Engagement as allocation of effort makes it less likely that critical aspects of the task will go unnoticed and greatly increases readiness to respond. Thus, whether the context is perception-action or learning, engagement of attention has implications for task performance. The more demanding the task, the greater need for engagement. The degree to which attention has been engaged is reflected directly in the time necessary to disengage. Functionally speaking, the greater the investment of effort in one task, the more time it will take to direct effort to a different task, given no practice in coordinating the two tasks. The implications for route learning are again relatively clear. During acquisition of route knowledge, an observer may vary engagement of attention in response to the importance of the spatial information at various viewpoints along the path.

1.2   Attention during Route Learning

Previous research supports the idea that attentional selection and engagement are important influences on the acquisition of route knowledge. Studies that focused specifically on what travelers are looking at during route experience (Carr & Schissler, 1969) and what travelers recall after route experience (Allen et al., 1979; Beck & Wood, 1976; Lovelace, Hegarty, & Montello, 1999; Denis, Pazzaglia, Cornoldi, & Bertolo, 1999) yield similar conclusions. During route learning, observers selectively attend to paths, prominent landmarks, and choice points. These features correspond exactly to those viewpoints along a path of observation that have reliably been identified as most useful in providing information about where the observer stands in relation to other locations along the route and where he or she should go in order to complete the route successfully (Allen et al., 1978, 1979). Accordingly, these are referred to as high-information regions of a route. It has also been demonstrated that observers can identify viewpoints that provide little useful information about the route (Allen et al., 1979). The areas visible from these viewpoints are referred to as low-information scenes.

In addition, the need for substantial engagement of attention during route learning has been shown in a variety of experimental studies. Although it has been demonstrated that implicit learning (that is, acquisition of information about the route without specific learning instructions and thus, without intentional engagement of attention) does indeed occur (Anooshian & Seibert, 1996), ample evidence from studies using concurrent-task procedures indicates that attentional engagement can significantly impact the accuracy of what is learned. When routes are learned by means of moving through an environment, memory for inter-route locations—but not necessarily for the geometry of the route per se—is significantly impaired if travelers perform a concurrent task that requires engagement of attention (Lindberg & Gärling, 1981, 1982, 1983; Sholl, 1990). During perceptual learning of a route, memory for viewpoints of the environments as well as for inter-viewpoint relations are impaired if viewers perform an attention-engaging concurrent task (Allen & Willenborg, 1998). Overall, it is clear that engagement of attention is an important factor in route learning. However, these studies do not emphasize a link between process and content. Additional inquiry would be useful to determine whether attentional engagement is uniformly beneficial or whether its benefits are directly tied to the utility of what has been selected for attentional allocation.

1.3   Hypotheses for Evaluation

The purpose of this experiment was to determine if observers' selective attention and attentional engagement differed as a function of the informativeness of environmental region during visually based route learning. Selective attention was examined by recording viewers' eyemovements while they viewed a slide presentation simulating a realworld route. Area of a scene scanned by the viewer served as an operational measure of selectivity. The greater the area scanned, the less selective the allocation of attention to that scene. It was hypothesized that viewers would be less selective in allocating their attention when seeing viewpoints that showed low-information regions than when seeing viewpoints that showed high-information regions.

Engagement of attention was examined by recording viewers' response times in a concurrent task that required them to respond when they heard a signal. Time to respond served as an operational measure of engagement. The faster the response, the less attentional engagement in route learning. It was hypothesized that viewers would be faster in responding to the signal for the concurrent task when they were seeing viewpoints that showed low-information regions than when seeing viewpoints that showed high-information regions.

2   Method

In this experiment, analyses were conducted on four measures of performance comparing high-information and low-information regions along a pictorialized route: (a) correlations of estimated-to-actual walking distance among viewpoints; (b) number of scene points visually inspected as shown by eye-monitoring records; (c) response time to an auditory signal; and (d) type of eye-movement pattern.

2.1   Participants

Data were collected from 10 university students (five men and five women) between the ages of 18 and 25 years (mean = 20.5 years), who participated voluntarily to receive research participation credit in Psychology courses. None of the participants was familiar with the environment depicted in the slide presentation showing the route.

2.2   Route Learning Task

A videotaped presentation was prepared that showed 61 successive color scenes along a one km route through a heterogeneous urban neighborhood. Scenes subtended approximately 52° visual angle in the environment. They depicted viewpoints that were 20 m from each other, except when turns were encountered. During turns, there was 50% visual overlap between scenes until rotation of viewing angle was complete. Each scene was shown for approximately 5 sec; the entire presentation required approximately 5 minutes.

The pictured route began in a wooded park along a serpentine path and then veered left, exiting the park into a small parking lot. Travel through the parking lot led to a large green on a university campus. The route made a large semicircle on that green and then continued on a straight sidewalk that ended at an intersection. Crossing the intersection resulted in travel through two blocks of apartment houses, where the street was under repair. This area ended at an intersection with a wide urban thoroughfare. The route continued after this street was crossed and passed a large church on the left. A block later, the route involved a right-angle turn into a residential neighborhood and continued several blocks before ending in sight of another large church.

2.3   Low and High-Information Regions

On the basis of a preliminary study involving 50 raters, low- and high-information regions within the route were identified. These raters were informed that they were to rate scenes from the walk in terms of how useful they were in helping a traveler know where he or she was along the walk and get to the end of the walk successfully. Each rater saw the presentation twice before rating each scene. During rating, scenes were presented in random order, and raters used a five-point scale, ranging from 1—not at all helpful/lowest information value to 5— most helpful/highest information value. The criterion for selection as a low-information scene was a mean rating that was one standard deviation or more below the group mean; seven low-information regions were identified using this criterion. The criterion for selection as a highinformation scene was a mean rating that was one standard deviation or more above the group mean; 12 high-information regions were identified using this criterion.

2.4   Eye-Movement Data

Participants' eye-movements were monitored and recorded during the presentation of the videotape portraying the route. The equipment was an ISCAN corneal reflection unit with autocalibration software. This unit featured a dual-camera set-up with external infrared source. The system provided a record of time-coded eye-movement and fixation patterns superimposed on the scenes comprising the pictorialized route. For scoring the number of areas within a scene scanned during the route learning task, a nine-section grid was imposed on the computer monitor, and an experimenter simply counted the number of sections in which eye movements or fixations occurred.

2.5   Concurrent Task

The concurrent task involved a simple response-to-signal procedure. Participants were instructed to press a computer key as rapidly as possible whenever they heard an auditory signal. Five practice trials were provided. During the route learning task, twenty-five signals were timed to occur 5 sec after the appearance of the 12 scenes showing highinformation regions, the seven scenes showing low-information regions, and six scenes selected at random. Mean response times to signals occurring during the viewing of low- and high-information regions were computed for each participant.

2.6   Distance Estimation Task

The distance estimation task was designed to assess participants' knowledge of inter-location distances between viewpoints along the pictorialized route. Participants estimated distance using a magnitude estimation procedure from the beginning of the route and from the end of the route, respectively, to each scene depicting a low-information region and each scene depicting a high-information region. They were told to estimate distance using a range from 1 to 1000 units. No standard metric (feet, yards, meters) was mentioned. Accuracy of estimates was assessed by means of estimated-to-actual distance correlations for the two sets of scenes.

2.7   Procedure

Participants were tested individually. Initially, the general purpose of the study and the functioning of the eye-movement equipment was explained. The concurrent task was then introduced and practice was provided in responding to the auditory signal.

The need for calibrating the eye-movement monitoring equipment was then explained, and the autocalibration procedure was followed. This procedure required participants to look at geometric forms in specific locations onscreen while coordinates were calibrated automatically. Once the equipment was calibrated, the route learning task was administered twice. After the second administration, participants engaged in the distance estimation task. This task concluded the procedure.

3   Results

The comparisons important to the test of the hypotheses are shown in Table 1. Each comparison is explicated as follows.

3.1   Knowledge of Inter-location Distance

Analyses comparing low- and high-information scenes would be meaningless if the validity of this distinction were not established initially. One way of validating the distinction is to compare interlocation distance knowledge for groups of low- and high-information viewpoints. The estimated-to-actual distance correlations obtained from each participant when estimating distances to low-information scenes and high-information scenes were compared using a related-measures one-tailed t-test. The results showed that the mean correlation was greater for estimates to high-information scenes than for low-information scenes, $t(9) = 2.82$, $p < .05$, thus confirming the expectation that observers would learn more accurate spatial relations among highinformation regions than among low-information regions. Highinformation scenes are those that contain significantly more information pertinent to spatial learning than do low-information scenes.

3.2   Number of Areas Scanned

The mean number of areas (out of nine) scanned in low- and highinformation scenes was compared using a related-measure one-tailed t test. Results showed that the mean number of areas scanned was greater for the low-information scenes than for the high-information scenes, $t(9) = 2.54$, $p < .05$, thus confirming the hypothesis that observers would scan more areas in low-information scenes. During route learning, observers were more selective in attending to specific features within highinformation regions than they were within low-information regions.

3.3   Response Time to Concurrent Task

The mean response times to the signal in the concurrent task while the observers viewed low-information scenes and high-information scenes were compared using a related-measures one-tailed t-test. Results showed that the mean response time was faster for the low-information scenes than for the high-information scenes, $t(9) = 3.48, p < .01$.

3.4   Scanning Patterns

Eye-movement recordings for each low- and high-information scene were examined for the purpose of identifying scanning patterns in relation to the environmental features visible in each scene. The most frequent scanning pattern in high-information scenes involved saccades from landmarks to the vanishing point of the path in the scene, following the path. In scenes involving turns, the vanishing point was the area where the path disappeared behind occluding features such as buildings. This pattern accounted for 54% of the scorable eye-movements in highinformation scenes and 36% of the scorable eye-movements in lowinformation scenes. The most frequent scanning pattern in lowinformation scenes involved saccades moving from one peripheral portion of the scene to another peripheral portion. This pattern accounted for 40% of the scorable eye-movements in low-information scenes and 26% of the scorable eye-movements in high-information scenes.

Table 1. A comparison distance estimation performance (route knowledge), number of areas scanned (selective attention), response time to concurrent task signal (attention engagement), and landmark-to-path vanishing point scanning for high- and low-information scenes.

Performance Measure	High-Information Scenes (Mean)	Low-Information Scenes (Mean)
Distance Estimation (correlation coefficient)	.96	.82
Number of Areas Scanned (maximum = 9)	4.3	5.9
Response Time to Signal (ms)	495	389
Landmark-to-Path Scanning (%)	54	36

4   Discussion

The results supported the idea that not all parts of a route are equally important in the process of acquiring knowledge of intra-route spatial relations. Distance estimates among viewpoints designated by raters as high-information scenes were more accurate than were estimates among viewpoints designated by raters as low-information scenes. This finding suggests that accurate route knowledge may depend on how highinformation scenes are encoded by observers. Data reflecting the number of areas scanned and the pattern of landmark-to-path scanning indicate differences in how attention is allocated in high-information and low-information scenes. Results from the concurrent response time task further indicate differences in attentional engagement.

4.1   The Hypotheses Revisited

Consistent with the idea that high-information regions along routes require more selectivity and greater engagement of attention, it had been hypothesized that observers would focus attention on selective features within the route. Consequently, it was anticipated that during the route learning task, their eye movements would be constrained to fewer areas in high-information regions than in low-information regions. This hypothesis was confirmed.

It had also been hypothesized that high- and low-information regions would result in differential engagement of attention, with more effort being dedicated to the high-information regions. Consistent with this hypothesis, the data showed that it took longer to disengage attention when observers were scanning high-information scenes than when they were scanning low-information scenes.

4.2   General Implications

In providing new empirically based insight into the role of attention in route learning, this study makes a small step toward the objective of achieving a process analysis of spatial learning in large-scale environments. First, the results replicate and confirm previous findings that not all perceivable features in large-scale environments are equally important in the process of distance knowledge acquisition. Prominent landmarks and explicit boundaries clearly attract attention and influence the cognitive organization of route knowledge (Allen & Kirasic, 1985; Berendt & Jansen-Osmann, 1997).

In addition, although route learning is frequently referred to in the spatial cognition literature as a unitary phenomenon, it is clear that there are multiple processes involved, including associative learning, motor learning, and perceptual learning. The evidence further indicates that the attentional requirements of these processes may not be identical. For example, motor learning may require less attentional engagement than does perceptual learning. Furthermore, some aspects of route knowledge, such as inter-location spatial relations, may require more attention than do others, such as recognition of features and reproduction of movement patterns.

These findings also help to establish that attention is not a unitary process in the context of spatial learning. Selectivity results in the allocation of attention to high-information regions of the environment and thus facilitates the acquisition of spatial information concerning these regions. Engagement of attention results in the application of mental effort towards the task of encoding information. Not surprisingly, such attention dedicated to high-information regions provides an efficient means of encoding critical information about inter-location relations. In the future, investigations of how these two aspects of attention are related to each other may prove useful. Furthermore, it would be informative to examine how the allocation and engagement of attention are top-down processes, that is, driven by knowledge. Further examination of these issues will provide new understanding of the factors contributing to success and failure in wayfinding and insight into the basis of individual and age-related differences in spatial learning.

Acknowledgement

This research was supported in part by a Biomedical Research Support Grant from the U.S. National Institutes of Health through the University of South Carolina.

References

Allen, G. L. (1981). A developmental perspective on the effects of 'subdividing' macrospatial experience. *Journal of Experimental Psychology: Human Learning and Memory*, 7, 120-132.

Allen, G. L. (1999). Spatial abilities, cognitive maps, and wayfinding: Bases for individual differences in spatial cognition and behavior. In R. Golledge (Ed.), *Wayfinding Behavior: Cognitive Maps and Other Spatial Processes* (pp. 46-80). Baltimore: Johns Hopkins University Press.

Allen, G. L., & Kirasic, K. C. (1985). Effects of the cognitive organization of route knowledge of judgments of macrospatial distance. *Memory & Cognition*, 13, 218-227.

Allen, G. L., Kirasic, K. C., Siegel, A. W., & Herman, J. F. (1979). Developmental issues in cognitive mapping: The selection and utilization of environmental landmarks. *Child Development*, 50, 1062-1070.

Allen, G. L., Siegel, A. W., & Rosinski, R. R. (1978). The role of perceptual context in structuring spatial knowledge. *Journal of Experimental Psychology: Human Learning and Memory*, 4, 617- 630.

Allen, G. L., & Willenborg, L. J. (1998). The need for controlled information processing in the visual acquisition of route knowledge. *Environmental Psychology*, 18, 419-427.

Anooshian, L. J., & Seibert, P. S. (1996). Diversity within spatial cognition: Memory processes underlying place recognition. *Applied Cognitive Psychology*, 10, 281-300.

Beck, R. J., & Wood, D. Cognitive transformation from urban geographic fields to mental maps. *Environment and Behavior*, 8, 199-238.

Berendt, B., & Jansen-Osmann, J. (1997). Feature accumulation and route structuring in distance estimations: An interdisciplinary approach. In S. C. Hirtle & A. U. Frank (eds.), *Spatial Information Theory: A Theoretical Basis for GIS* (pp. 279-296). *Lecture Notes in Computer Science*; Vol. 1329. Berlin: Springer-Verlag

Carr, S. & Schissler, D. (1969). The city as a trip: Perceptual selection and memory in the view from the road. *Environment and Behavior*, 1, 7-36.

Denis, M., Pazzaglia, F., Cornoldi, C., & Bertolo, L. (1999). Spatial discourse and navigation: An analysis of route directions in the city of Venice. *Applied Cognitive Psychology*, 13, 145-174.

Lindberg, E., & Gärling, T. (1981). Acquisition of locational information about reference points during locomotion with and without a concurrent task: Effects of number of reference points. *Scandinavian Journal of Psychology*, 22, 109-115.

Lindberg, E., & Gärling, T. (1982). Acquisition of locational information about reference points during locomotion with and without a concurrent task: The role of central information processing. *Scandinavian Journal of Psychology*, 22, 207-218.

Lindberg, E., & Gärling, T. (1983). Acquisition of different types of locational information in cognitive maps: Automatic or effortful processing. *Psychological Research*, 45, 19-38.

Lovelace, K. L., Hegarty, M., & Montello, D. R. (1999). Elements of good route directions in familiar and unfamiliar environments. In C. Freksa & D. M. Mark (Eds.), Spatial information theory: *Cognitive and computational foundations of geographic information science* (pp. 65-82). *Lecture Notes in Computer Science*; Vol. 1661. Berlin: Springer-Verlag,

Sholl, M. J. (1990). *The role of attention in the acquisition of spatial knowledge*. Presented at meetings of the Psychonomic Society, New Orleans.

Author Index

Lecture Notes in Computer Science

For information about Vols. 1–2730
please contact your bookseller or Springer-Verlag

Vol. 2776: V. Gorodetsky, L. Popyack, V. Skormin (Eds.), Computer Network Security. Proceedings, 2003. XIV, 470 pages. 2003.

Vol. 2777: B. Schölkopf, M.K. Warmuth (Eds.), Learning Theory and Kernel Machines. Proceedings, 2003. XIV, 746 pages. 2003. (Subseries LNAI).

Vol. 2778: P.Y.K. Cheung, G.A. Constantinides, J.T. de Sousa (Eds.), Field-Programmable Logic and Applications. Proceedings, 2003. XXVI, 1179 pages. 2003.

Vol. 2779: C.D. Walter, Ç.K. Koç, C. Paar (Eds.), Cryptographic Hardware and Embedded Systems – CHES 2003. Proceedings, 2003. XIII, 441 pages. 2003.

Vol. 2781: B. Michaelis, G. Krell (Eds.), Pattern Recognition. Proceedings, 2003. XVII, 621 pages. 2003.

Vol. 2782: M. Klusch, A. Omicini, S. Ossowski, H. Laamanen (Eds.), Cooperative Information Agents VII. Proceedings, 2003. XI, 345 pages. 2003. (Subseries LNAI).

Vol. 2783: W. Zhou, P. Nicholson, B. Corbitt, J. Fong (Eds.), Advances in Web-Based Learning – ICWL 2003. Proceedings, 2003. XV, 552 pages. 2003.

Vol. 2786: F. Oquendo (Ed.), Software Process Technology. Proceedings, 2003. X, 173 pages. 2003.

Vol. 2787: J. Timmis, P. Bentley, E. Hart (Eds.), Artificial Immune Systems. Proceedings, 2003. XI, 299 pages. 2003.

Vol. 2789: L. Böszörményi, P. Schojer (Eds.), Modular Programming Languages. Proceedings, 2003. XIII, 271 pages. 2003.

Vol. 2790: H. Kosch, L. Böszörményi, H. Hellwagner (Eds.), Euro-Par 2003 Parallel Processing. Proceedings, 2003. XXXV, 1320 pages. 2003.

Vol. 2792: T. Rist, R. Aylett, D. Ballin, J. Rickel (Eds.), Intelligent Virtual Agents. Proceedings, 2003. XV, 364 pages. 2003. (Subseries LNAI).

Vol. 2794: P. Kemper, W. H. Sanders (Eds.), Computer Performance Evaluation. Proceedings, 2003. X, 309 pages. 2003.

Vol. 2795: L. Chittaro (Ed.), Human-Computer Interaction with Mobile Devices and Services. Proceedings, 2003. XV, 494 pages. 2003.

Vol. 2796: M. Cialdea Mayer, F. Pirri (Eds.), Automated Reasoning with Analytic Tableaux and Related Methods. Proceedings, 2003. X, 271 pages. 2003. (Subseries LNAI).

Vol. 2798: L. Kalinichenko, R. Manthey, B. Thalheim, U. Wloka (Eds.), Advances in Databases and Information Systems. Proceedings, 2003. XIII, 431 pages. 2003.

Vol. 2799: J.J. Chico, E. Macii (Eds.), Integrated Circuit and System Design. Proceedings, 2003. XVII, 631 pages. 2003.

Vol. 2801: W. Banzhaf, T. Christaller, P. Dittrich, J.T. Kim, J. Ziegler (Eds.), Advances in Artificial Life. Proceedings, 2003. XVI, 905 pages. 2003. (Subseries LNAI).

Vol. 2803: M. Baaz, J.A. Makowsky (Eds.), Computer Science Logic. Proceedings, 2003. XII, 589 pages. 2003.

Vol. 2804: M. Bernardo, P. Inverardi (Eds.), Formal Methods for Software Architectures. Proceedings, 2003. VII, 287 pages. 2003.

Vol. 2805: K. Araki, S. Gnesi, D. Mandrioli (Eds.), FME 2003: Formal Methods. Proceedings, 2003. XVII, 942 pages. 2003.

Vol. 2807: V. Matoušek, P. Mautner (Eds.), Text, Speech and Dialogue. Proceedings, 2003. XIII, 426 pages. 2003. (Subseries LNAI).

Vol. 2810: M.R. Berthold, H.-J. Lenz, E. Bradley, R. Kruse, C. Borgelt (Eds.), Advances in Intelligent Data Analysis V. Proceedings, 2003. XV, 624 pages. 2003.

Vol. 2812: G. Benson, R. Page (Eds.), Algorithms in Bioinformatics. Proceedings, 2003. X, 528 pages. 2003. (Subseries LNBI).

Vol. 2815: Y. Lindell, Composition of Secure Multi-Party Protocols. XVI, 192 pages. 2003.

Vol. 2816: B. Stiller, G. Carle, M. Karsten, P. Reichl (Eds.), Group Communications and Charges. Proceedings, 2003. XIII, 354 pages. 2003.

Vol. 2817: D. Konstantas, M. Leonard, Y. Pigneur, S. Patel (Eds.), Object-Oriented Information Systems. Proceedings, 2003. XII, 426 pages. 2003.

Vol. 2818: H. Blanken, T. Grabs, H.-J. Schek, R. Schenkel, G. Weikum (Eds.), Intelligent Search on XML Data. XVII, 319 pages. 2003.

Vol. 2819: B. Benatallah, M.-C. Shan (Eds.), Technologies for E-Services. Proceedings, 2003. X, 203 pages. 2003.

Vol. 2820: G. Vigna, E. Jonsson, C. Kruegel (Eds.), Recent Advances in Intrusion Detection. Proceedings, 2003. X, 239 pages. 2003.

Vol. 2821: A. Günter, R. Kruse, B. Neumann (Eds.), KI 2003: Advances in Artificial Intelligence. Proceedings, 2003. XII, 662 pages. 2003. (Subseries LNAI).

Vol. 2822: N. Bianchi-Berthouze (Ed.), Databases in Networked Information Systems. Proceedings, 2003. X, 271 pages. 2003.

Vol. 2823: A. Omondi, S. Sedukhin (Eds.), Advances in Computer Systems Architecture. Proceedings, 2003. XIII, 409 pages. 2003.

Vol. 2824: Z. Bellahsène, A.B. Chaudhri, E. Rahm, M. Rys, R. Unland (Eds.), Database and XML Technologies. Proceedings, 2003. X, 283 pages. 2003.

Vol. 2825: W. Kuhn, M. Worboys, S. Timpf (Eds.), Spatial Information Theory. Proceedings, 2003. XI, 399 pages. 2003.

Vol. 2827: A. Albrecht, K. Steinhöfel (Eds.), Stochastic Algorithms: Foundations and Applications. Proceedings, 2003. VIII, 167 pages. 2003.

Vol. 2830: F. Pfenning, Y. Smaragdakis (Eds.), Generative Programming and Component Engineering. Proceedings, 2003. IX, 397 pages. 2003.

Vol. 2832: G. Di Battista, U. Zwick (Eds.), Algorithms – ESA 2003. Proceedings, 2003. XIV, 790 pages. 2003.

Vol. 2834: X. Zhou, S. Jähnichen, M. Xu, J. Cao (Eds.), Advanced Parallel Processing Technologies. Proceedings, 2003. XIV, 679 pages. 2003.

Vol. 2836: S. Qing, D. Gollmann, J. Zhou (Eds.), Information and Communications Security. Proceedings, 2003. XI, 416 pages. 2003.

Vol. 2837: N. Lavrač, D. Gamberger, H. Blockeel, L. Todorovski (Eds.), Machine Learning: ECML 2003. Proceedings, 2003. XVI, 504 pages. 2003. (Subseries LNAI).

Vol. 2838: N. Lavrač, D. Gamberger, L. Todorovski, H. Blockeel (Eds.), Knowledge Discovery in Databases: PKDD 2003. Proceedings, 2003. XVI, 508 pages. 2003. (Subseries LNAI).

Vol. 2839: A. Marshall, N. Agoulmine (Eds.), Management of Multimedia Networks and Services. Proceedings, 2003. XIV, 532 pages. 2003.